Algorithms for Intelligent Systems

This book series publishes research on the analysis and development of algorithms for intelligent systems with their applications to various real world problems. It covers research related to autonomous agents, multi-agent systems, behavioral modeling, reinforcement learning, game theory, mechanism design, machine learning, meta-heuristic search, optimization, planning and scheduling, artificial neural networks, evolutionary computation, swarm intelligence and other algorithms for intelligent systems.

The book series includes recent advancements, modification and applications of the artificial neural networks, evolutionary computation, swarm intelligence, artificial immune systems, fuzzy system, autonomous and multi agent systems, machine learning and other intelligent systems related areas. The material will be beneficial for the graduate students, post-graduate students as well as the researchers who want a broader view of advances in algorithms for intelligent systems. The contents will also be useful to the researchers from other fields who have no knowledge of the power of intelligent systems, e.g. the researchers in the field of bioinformatics, biochemists, mechanical and chemical engineers, economists, musicians and medical practitioners.

The series publishes monographs, edited volumes, advanced textbooks and selected proceedings.

Indexed by zbMATH.

All books published in the series are submitted for consideration in Web of Science.

Lopa Mandal · Joao Manuel R. S. Tavares ·
Valentina E. Balas
Editors

Proceedings of International Conference on Computational Intelligence, Data Science and Cloud Computing

IEM-ICDC 2021

Springer

Editors
Lopa Mandal
Department of Information Technology
Institute of Engineering and Management
Kolkata, West Bengal, India

Valentina E. Balas
Department of Automation and Applied Informatics, Department of International Relations, Programs and Projects, Intelligent Systems Research Centre
Aurel Vlaicu University of Arad
Arad, Romania

Joao Manuel R. S. Tavares
Department of Mechanical Engineering, Faculty of Engineering
University of Porto
Porto, Portugal

ISSN 2524-7565 ISSN 2524-7573 (electronic)
Algorithms for Intelligent Systems
ISBN 978-981-19-1656-4 ISBN 978-981-19-1657-1 (eBook)
https://doi.org/10.1007/978-981-19-1657-1

This Springer imprint is published by the registered company Springer Nature Singapore Pte Ltd.
The registered company address is: 152 Beach Road, #21-01/04 Gateway East, Singapore 189721, Singapore

Committees

Program Committee

Patron

Dr. Satyajit Chakrabarti, President, Institute of Engineering and Management, Kolkata, India

General Chair

Dr. Satyajit Chakrabarti, Director, Institute of Engineering and Management, Kolkata, India

Convener

Dr. Lopa Mandal, Institute of Engineering and Management, Kolkata, India

Co-conveners

Dr. Baisakhi Das, Institute of Engineering and Management, Kolkata, India
Dr. Maumita Chakraborty, University of Engineering and Management, Kolkata, India

Editorial Board

Research/Corporate Keynote Speakers

Dr. Mike Hinchey, University of Limerick, Ireland
Dr. Miodrag Mihaljevic, Serbian Academy of Sciences and Arts, Belgrade, Serbia
Dr. Kouichi Sakurai, Kyushu University, Japan
Mr. Rajiv Banerjee, CMCI, USA

Advisory Committee

Dr. Valentina E. Balas, Aurel Vlaicu University of Arad, Romania
Dr. Joao Manuel R. S. Tavares, University of Porto, Portugal
Dr. Mike Hinchey, University of Limerick, Ireland
Dr. Arnab Bose, University of Chicago, USA
Dr. Miodrag Mihaljevic, Serbian Academy of Sciences and Arts, Belgrade, Serbia
Dr. Kouichi Sakurai, Kyushu University, Japan
Dr. Sudipta Roy, Washington University Saint Louis, Missouri, USA
Mr. Rajiv Banerjee, CMCI, USA
Dr. Biplab Sikdar, National University of Singapore
Dr. Kaushik Deb, CUET, Bangladesh
Dr. Parth Gupta, Applied Scientist, Amazon, USA
Dr. Siddhartha Bhattacharyya, VSB Technical University of Ostrava, Czech Republic
Dr. Pramatha Nath Basu, Jadavpur University, India
Dr. Nabarun Bhattacharya, CDAC, India
Dr. Subir Kumar Sarkar, JU, Kolkata, India
Dr. Samar Bhattacharya, JU, India
Dr. Paramartha Dutta, Visva Bharati University, India
Dr. Karan Singh JNU, India
Dr. J. P. Singh, NIT Patna, India
Dr. Sujata Dash, North Orissa University, India
Dr. Samiran Chattopadhyay, Jadavpur University, India
Dr. Gaurav Trivedi, IIT Gwahati, India
Dr. Ashok Kumar Das, IIIT Hyderabad, India
Dr. Matangini Chattopadhyay, Jadavpur University, India
Dr. Diptendu Dutta, Aneesha Knowledge Technology, India

Dr. Aniruddha Nag, CSI Kolkata, India
Mr. Subimal Kundu, CSI Kolkata, India
Dr. Abhishek Das, Aliah University, India
Dr. Indraneel Mukhopadhyay, IEM, Kolkata, India
Ms. Banani Chakrabarti, IEM, Kolkata, India
Ms. Gopa Goswami, IEM, Kolkata, India

Technical Committee

Dr. Kouichi Sakurai, Kyushu University, Japan
Dr. Sudipta Roy, Washington University Saint Louis, Missouri, USA
Dr. Biplab Sikdar, National University of Singapore
Dr. Siddhartha Bhattacharyya, VSB Technical University of Ostrava, Czech Republic
Dr. Kaushik Deb, CUET, Bangladesh
Dr. Nabendu Chaki, University of Calcutta, India
Dr. Sankhayan Choudhury, University of Calcutta, India
Dr. Debashis De, MAKAUT, India
Dr. Sourav Saha, IBM, India
Dr. Koushik Ghosh, Burdwan University, India
Dr. Sourav De, Coochbehar Government Engineering College, India
Dr. Matangini Chattopadhyay, Jadavpur University, India
Dr. Debasish Jana, TEOCO Software Pvt. Ltd., India
Dr. Paramartha Dutta, Visva Bharati University, India
Dr. Bijan Sarkar, Jadavpur University, India
Dr. Karan Singh, JNU, India
Dr. Kaushik Majumder, IBM, India
Dr. Amitava Nag, CIT, Kokrajhar, India
Dr. Saprarshi Goswami, Bangabasi Morning College, Kolkata, India
Dr. J. P. Singh, NIT Patna, India
Dr. Parth Gupta, Applied Scientist, Amazon, USA
Dr. Sourav Saha, IBM, India
Dr. Lopa Mandal, IEM, Kolkata, India
Dr. Swarnendu Chakraborty, NIT AP, India
Dr. Saiyed Umer, Aliah University, India
Dr. Souvik Sengupta, Aliah University, India
Dr. Rik Das, Xavier Institute of Social Service, Jharkhand, India
Dr. Tapan Mandal, Government College of Engineering and Textile Technology, Serampore, India
Dr. Amit Das, Calcutta University, India
Dr. Gaurav Trivedi, IIT Guwahati, India
Dr. Parama Bhowmik, JU, India
Dr. Baisakhi Das, IEM, Kolkata, India
Dr. Maumita Chakraborty, UEM, Kolkata, India

Dr. Saswati Mukherjee, Jadavpur University, India
Dr. Punyasha Chatterjee, JU, India

Organizing Chairs

Dr. Lopa Mandal, Institute of Engineering and Management, Kolkata, India
Dr. Baisakhi Das, Institute of Engineering and Management, Kolkata, India
Dr. Maumita Chakraborty, University of Engineering and Management, Kolkata, India
Mr. Partha Sarathi Paul, Institute of Engineering and Management, Kolkata, India
Mr. Susovan Jana, Institute of Engineering and Management, Kolkata, India

Organizing Committee

Dr. Avipsita Chatterjee, Institute of Engineering and Management, Kolkata, India
Dr. Amit Kumar Mandal, Institute of Engineering and Management, Kolkata, India
Mr. Subhabrata Sengupta, Institute of Engineering and Management, Kolkata, India
Ms. Shreyashi Dutta, Institute of Engineering and Management, Kolkata, India
Mr. Partha Bhattacharyya, Institute of Engineering and Management, Kolkata, India
Mr. Rabi Narayan Behera, Institute of Engineering and Management, Kolkata, India
Ms. Kajari Sur, Institute of Engineering and Management, Kolkata, India
Mr. Pulak Baral, Institute of Engineering and Management, Kolkata, India
Mr. Swagatam Basu, Institute of Engineering and Management, Kolkata, India
Mr. Subindu Saha, Institute of Engineering and Management, Kolkata, India

Preface

In the present scenario, computational intelligence, data science and cloud computing are becoming more and more relevant in our society in all possible ways. To address this need of the hour, International Conference on Computational Intelligence, Data Science and Cloud Computing (IEM-ICDC) 2021 was organized by the Department of Information Technology of Institute of Engineering and Management, Kolkata, India, held online using virtual platform during December 22–23, 2021. The two-day conference provided a platform where researchers from all over the world, from academia and industry, took part in discussion and exchanged their views to make an intelligent, smart and secured society. Participation was from twenty different countries around the world. IEM-ICDC 2021 has been made possible with the generous support of the sponsors: Springer, Computer Society of India (CSI) and Centre for Development of Advanced Computing (CDAC). Most significant new findings about AI and robotics, image processing and NLP, cloud computing and big data analytics as well as cybersecurity, blockchain and IoT and various allied fields were highlighted through insightful invited talks and innovative technical sessions. Every detail of the conference has been reported in the proceedings with the hope that this will be helpful and be a motivating force for future research in the practical designs of cutting-edge systems in related fields. The book is organized into four parts. Part I discusses research findings on "Computational Intelligence." Part II and Part III present "Image and Natural Language Processing" and "Cyber Security and Blockchain," respectively. Part IV highlights research works on "Cloud and ICT." We thank all the sponsors, authors, supporters and organizing team of the conference for the grand success of this event. In the current dire situation of the world, we hope the conference has come out with important research contributions to the technological aspects to combat the crisis and help the society at large.

Kolkata, India — Lopa Mandal
Porto, Portugal — Joao Manuel R. S. Tavares
Arad, Romania — Valentina E. Balas

Contents

About the Editors

Prof. Dr. Lopa Mandal is currently Professor of the Department of Information Technology, Institute of Engineering and Management. She has completed her M.Tech./IT and Ph.D. (Engineering) from Jadavpur University, Kolkata, India. Her research interest includes artificial intelligence, data analytics, distributed and cloud computing, mobile computing, learning technologies. She has got many national and international journal and conference publications and book chapters to her credit. She has been a technical program committee member of different international conferences and editorial/review board member of International Journal. She has been Volume Editor of *Proceedings of IEM-ICDC 2020* published in Springer Nature Book Series "Lecture Notes in Data Engineering and Communication Technologies" and Guest Editor of *American Journal of Advanced Computing* IEM-ICDC 2020 Special issue on "Computational Intelligence and Smart Information and Communication Technology" published by Smart Society, USA, in January 2021. She has been the convener and organizing chair and also organizing committee member of different international conferences. She is also a member of Computer Society of India. She has got around 20 years of work experience in industry and academia.

Dr. João Manuel R. S. Tavares is graduated in Mechanical Engineering from University of Porto, Portugal (1992). In 1995, he obtained a M.Sc. in Electrical and Computer Engineering in the field of Industrial Informatics, in 2001, a Ph.D. degree in Electrical and Computer Engineering, and in 2015, the Habilitation in Mechanical Engineering from the same university. From 1995 to 2000, he was a researcher at the Institute of Biomedical Engineering. Since 2001, he has been a senior researcher and project coordinator at the Institute of Mechanical Engineering and Industrial Management. He was Assistant Professor in the Department of Mechanical Engineering of the Faculty of Engineering of the University of Porto between 2001–2011, Associate Professor in the same department between 2011–2021 and has been Full Professor in the same department since 2021/8. Since 2001, he has been supervisor and co-supervisor of several M.Sc. and Ph.D. thesis. He is co-author of more than 650 articles in national and international journals and conferences and co-editor of more than 55 international books and guest-editor of several special issues of international journals. Also, he has been evolved in several research projects, both as researcher and as scientific coordinator. He has been co-chairman of various international conferences and of numerous mini-symposia, workshops, and thematic sessions, and a member of organizing and scientific committees of several national and international conferences. He has been a member of several national and international journal editorial boards and editor-in-chief of the T&F journal *Computer Methods in Biomechanics and Biomedical Engineering: Imaging and Visualization*, Editor-in-Chief of the journal *Computer Methods in Biomechanics and Biomedical Engineering: Imaging and Visualization* published by Taylor & Francis and Co-editor of the Springer book series "Lecture Notes in Computational Vision and Biomechanics." His main research areas include computational vision, medical imaging, biomechanics, scientific visualization, human–computer interaction, and new product development.

Dr. Valentina E. Balas is currently full Professor in the Department of Automatics and Applied Software at the Faculty of Engineering, "Aurel Vlaicu" University of Arad, Romania. She holds a Ph.D. in Applied Electronics and Telecommunications from Polytechnic University of Timisoara. Dr. Balas is author of more than 250 research papers in refereed journals and international conferences. Her research interests are in intelligent systems, fuzzy control, soft computing, smart sensors, information fusion, modeling and simulation. She is the editor-in-chief to *International Journal of Advanced Intelligence Paradigms* (IJAIP) and to *International Journal of Computational Systems Engineering* (IJCSysE), member in editorial board member of several national and international journals and is the director of Intelligent Systems Research Center in Aurel Vlaicu University of Arad. She is a member of EUSFLAT, SIAM and a senior member IEEE, Member in TC—Fuzzy Systems (IEEE CIS), member in TC—Emergent Technologies (IEEE CIS), member In TC—Soft Computing (IEEE SMCS).

Part I
Computational Intelligence

Chapter 1
AutoDrone: Shortest Optimized Obstacle-Free Path Planning for Autonomous Drones

Prithwish Jana and Debasish Jana

1 Introduction

Drone [1] is the common name of an aerial vehicle; that is, unmanned has no pilot to make them fly, no cabin crew to serve on air and no passenger to fly along. This was originally thought to serve aerospace applications or military use. Nowadays, the usage has been in personal use as well as for reaching out to places where human may find it difficulty to reach. Drone needs to be operated remotely by a human on ground through a screen-enabled remote-controlled device within a short range. Drones promise to be the mode-of-delivery in India's future smart city project. They can be used to deliver things to the desired destination. In case of autonomous drone (AutoDrone), an intelligent drone may have to fly without being remotely controlled by a human to follow order. Rather, a drone may be used to fly on their own for traversing a trajectory toward a desired destination. It may rely on artificial intelligence backbone to imitate or act like an efficient human pilot. This may be especially useful in search or relief operations across rough terrains or inaccessible corners of monumental ruins in an area devastated by disasters like earthquake or tsunami. In such places, time is crucial and radio signals may be impenetrable through impassable frontiers or unassailable mountains—as such, autonomous system is need of the hour.

To take control of in-flight situations, the decision engine within a drone may be embedded. This decision engine must be intelligent enough to interpret raw sensor data meaningfully. With camera-guided real-time vision and interpretation of the path and obstacle, drone can fly to destination using the shortest hindrance-free path

P. Jana (✉)
Indian Institute of Technology Kharagpur, Kharagpur, India
e-mail: pjana@ieee.org

D. Jana
TEOCO Software Pvt. Ltd., Kolkata, India

L. Mandal et al. (eds.), *Proceedings of International Conference on Computational Intelligence, Data Science and Cloud Computing*, Algorithms for Intelligent Systems,
https://doi.org/10.1007/978-981-19-1657-1_1

to travel. Thus, the intelligent engine within the drone should be able to choose the shortest travel trajectory when there is no preset list of defined paths (i.e., drone has no prior information about the terrain it is going to enter), alongside avoiding the obstacles. If we map the whole terrain to a rectangular grid of certain precision, where some grid cells represent obstacles and the drone should avoid collision with those, then we can reduce the set-up of the problem to a two-dimensional Euclidean planar graph [2]. Ideally, there could be an unobstructed direct route, connecting source and destination. Using GPS-guided coordinate system, an autonomous intelligent drone can easily find out the shortest path and angle along which it should drive. In reality, there could be obstacles as buildings, monuments, elevated footpaths, electricity towers, non-flyable areas, and on-air hindrances that may creep in dynamically.

Our work focuses on finding out the shortest path without obstacle from a source to a chosen destination. The implementation testbed for simulation is written in C++ [3] on Ubuntu platform. The rest of the paper is organized as follows: Sect. 2 jots down the related work in this area. Section 3 elaborates the proposed method. This follows simulation experimental results in Sect. 4. Finally, we conclude in Sect. 5.

2 Related Work

From a graph-theoretic [4] perspective, a directed or undirected connected graph consists of several nodes (or vertices) and a set of edges that link the nodes. The shortest path from a source node to a target node is the route joining them via edges such that, the number of edges (in unweighted graph) or the sum of the respective weights of the connecting edges (in weighted graph) is minimal. Similarly, a map of connecting roads may be represented as a Manhattan graph of intersections (vertices) and the road connecting two intersections (edges). Finding the shortest path in such a simplified road map can be essentially done by reducing it to graph-theoretic shortest path probing. As such, each edge is weighed in accordance with the length of the road segment that it represents.

Gallo et al. [5] have listed eight popular algorithms that show the way to solve the shortest path problem on a tree or a directed graph. Dijkstra's shortest path algorithm has been explained by Johnson [6]. There are many algorithms for finding the shortest path in a graph, e.g., the popular ones include Bellman–Ford's algorithm [7], Floyd–Warshall algorithm [8] and Dijkstra's algorithm [9]. Bellman–Ford algorithm is used to find the shortest paths in a graph, from a particular source vertex to all other vertices. Floyd–Warshall algorithm finds the shortest paths between all the existing pairs of vertices in a graph, where each connecting edge present in the graph has some nonzero weight, positive, or negative. Dijkstra's algorithm deals with single-source shortest path and, as such, is more appropriate in our problem domain. But for our problem setup, this cannot be applied directly as-is in the original form directly because paths are not well-defined in an Euclidean plane and there may be obstacles blocking the line-of-sight paths.

Chai et al. [10] predicted spatial context through computation of distance map between a designated pixel to its spatially nearest boundary of an object. We may envisage a road map of connecting aerial routes and erected buildings as a graph of vertices (identified objects) and edges (connecting aerial connections) to signify *connectivity*. Thus, the presence of an edge between two nodes means a drone can fly among these two nodes aerially in Euclidean space. Mutherjee et al. [11] worked on semantic segmentation on images to do an effective annotation at pixel-level. Thus, a road map with erected buildings and monuments of a city-scape or village-scape can be semantically segmented into a map, and eventually to a graph in Euclidean space. And the task for the AutoDrone boils down to the task to find the single-source shortest path in a directed graph in Euclidean space. We may assume that in an Euclidean graph, the estimated weight of each connecting edge can be provided by the Euclidean distance between its two points.

Chakrabarty et al. [12] demonstrated the interpretation of autonomous traversing of a drone from a designated source A to a designated target point B, when there is no obstacle in-between. As such in reality, it may not be completely aerially clear between two points but it may be among few other intermediate nodes as multiple hops where each hop is going through some aerially clear edge. Within the unmanned aircraft system traffic management (UTM) ecosystem, a drone is able to fly across an aerially clear path through multiple hops. Jana [13] has argued that the shortest path traversal needs to be through 'visually clear' path without any obstacles, be it for autonomous driver-less car or pilot-less aircraft like drone. According to Koch et al. [14], while mapping mostly a landscape scene, be it spacious or flat scenes, flight planning for drone can be thought of simple grid-like structure where drones can follow two-dimensional patterns of nodes and edges within 2D space. Sakurai et al. [15] proposed a spiking neural network (SNN) algorithm to compute plan for appropriate path having several moving obstacles around target agents that itself is attempting to minimize its own traversal path to the destination. In practical scenarios, there could be many other moving obstacles. As for example, this may include other cars for a driver-less autonomous car and human or building obstacles for a moving AutoDrone or robot. In such situations, getting an effective shortest path is not a trivial task, because the dynamic behavior of the obstacles may not be known or there could be a large number of other candidate paths to choose from. But from a drone's perspective that flies at a certain height, obstacles are more or less static except for birds and other drones. So, in this work we tackle such obstacles and propose a path-planning algorithm in 3D space that can maneuver the drone from its current position to an intended destination.

3 Proposed Methodology

In this section, we first list out perception techniques to map surrounding neighborhood to a 2D layout. We then provide the problem description and describe our proposed method of path planning. Finally, we transform this into 3D Euclidean space to accomodate source–destination pairs at different altitudes.

3.1 Perception and Real-Time Formation of Maps

At the starting location, the drone can first move vertically upwards to a high altitude from where almost all obstacles would be at a lower altitude. From that position, the drone's camera can capture high-resolution RGB images that represent a bird's-eye view of the extended region surrounding the drone. Fast semantic segmentation on RGB images through fully convolutional network [16] or superpixel-based classification [11] can label each pixel to their corresponding class as building, tree, road, tower, etc. Further, panoptic segmentation [17] separates different instances of the same object (e.g., two different buildings). In instances of fog, smog, or any other technical glitch, when captured RGB image is blurry with lot of noises, binarization [18, 19] techniques can be applied on fuzzy image segmentation results to get a cleaner per-pixel binary classification of obstacle/not obstacle. All these, individually or in combination, give an accurate 2D aerial map delineating the positions of obstacles. In addition, airborne 3D LiDAR [20] fitted on the drone can proffer depth readings that can inversely signify the height of the various obstacles. Fusing such multi-sensor information, by methods like bimodal learning [21] from RGB-D information or unified co-attention networks [22] can augment the spatial maps with depth information. Global Positioning System (GPS) helps in identifying the intended destination coordinate in the captured image. As such, the drones can determine at which height it should move depending on at which height it would encounter more percentage of open-area without hindrances.

3.2 Problem Design

We consider that the drone locomotes across an area defined by a two-dimensional region of size R meters $\times$ C meters, as exemplified in Fig. 1. There are a set of destinations $D = \{d_1, d_2, \ldots\}$ where the drone needs to deliver items. Additionally, there are obstacles spread throughout the area that hinders the straight line-of-sight path from the drone base station to each of the destinations. First, with regard to the precision (p meters) of motion desired, we discretize the 2D space into a $\lceil \frac{R}{p} \rceil \times \lceil \frac{C}{p} \rceil$ grid of (p meters $\times$ p meters) unit squares. Identifying the lower-leftmost point of the grid as the origin (0, 0), individual destination locations are defined by 2D Cartesian coordinates $d_i = (x_i, y_i)$. Further, a drone's obstacles can be buildings or trees that occupy a polygonal area (convex or concave), when seen through bird's-eye view. At the height a drone is set to fly, each obstacle may look arbitrary-shaped. So, the drone should avoid the region defined by the convex hull of that obstacle, to avoid collisions. Thus, if some grid cell overlaps with the convex hull to such polygons, it is considered as an *obstacle-grid-cell*. In our design, there are no partially obstacled grid cells; i.e., each cell is either wholly occupied by an obstacle or it is not. Neighboring obstacle-grid-cells together can make up a larger obstacle of arbitrary rectilinear polygonal shape.

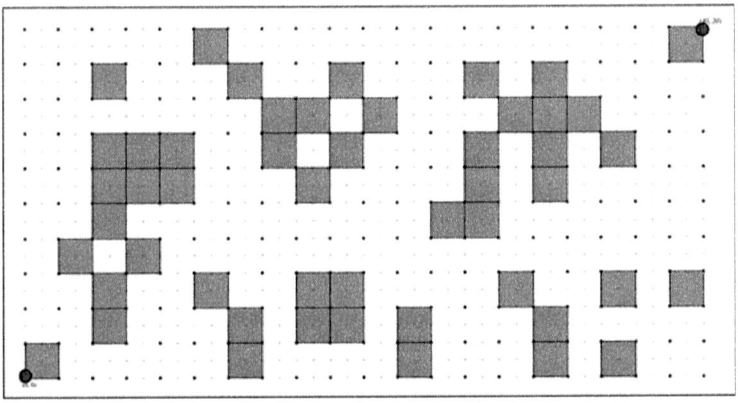

Fig. 1 Example of 2D grid formed from a 10 m × 20 m area. Each cell (denoted within black dot corners) is 1 m × 1 m. Shaded cells denote obstacles. As such, the straight-line joining source at the bottom-left to destination at the top-right (shown as blue circles) is not obstacle-free and drone has to find a path that avoids such obstacles

3.3 Destination to Base Station Shortest Path

We intend to find the shortest source–destination path from the drone's base station to a destination d_i. Now, when there are no obstacles blocking the line-of-sight between source and the destination, the shortest Euclidean path will be a straight-line joining them. But, a drone may have to deviate from this line-of-sight path when obstacles adorn the drone's flying area. Instead of going straight, the drone would have to change its direction at specific coordinates to avoid collision and reach safe. In this paper, we provide a fast and efficient approach to compute an optimized source–destination shortest path that a drone would take, in order to avoid collision with obstacles. To solve this efficiently, we take a computational geometry-based approach, as described in the following paragraphs.

Obstacle Graph Formation. As we said before, the drone has to shift the direction at specific coordinates (*points of deflection*). And for the drone to follow the shortest path, it should change directions not just anywhere but only at obstacle corners. This is because the shortest source–destination path can be imagined as a tight string between source and destination that when in its fully contracted position will only touch some of the obstacle vertices that do not allow it to contract further. It may be noted here that the shortest path may not always be convex in nature. So, we firstly form a graph known as the *obstacle graph* G_O whose vertices are the set of all corner-points corresponding to obstacle-grid-cells. If the shortest source–destination path contains any position from which the drone has to deflect its direction, it will definitely be one among the vertices in G_O. Further, any vertex in G_O may be shared among one, two, three, or four obstacle-grid-cells. The vertices that are shared by four obstacle-grid-cells are completely interior to a bigger obstacle formed by the neighboring obstacle-grid-cells. And therefore, they cannot be points of deflection. We mark these special vertices and may safely exclude them in the visibility graph computation (explained in the subsequent paragraph) because these vertices cannot be 'seen' from any other vertex. This way, we reduce the time to probe whether line-of-sight between two points is obstructed or not. In G_O, we add an edge between obstacle vertices $u, v \in G_O$ if edge (u, v) is an obstacle edge, i.e., a bounding edge of an obstacle-grid-cell.

Visibility Graph Formation. Subsequently, a *visibility graph* is formed from here. It is an Euclidean graph where vertices connected by edges are mutually visible from one another; i.e., their straight-line path is not blocked by any obstacle. In this case, the drone can move unobstructed among these two points and the path will also be the shortest Euclidean path. On the contrary, the drone may need to change directions in-between when there does not exist an edge between two vertices. As we mentioned before, a vertex that lies on the shortest path from source S to destination D includes S, D and one or more unmarked obstacle vertices $v \in G_O$, and nothing else. As such, we improve upon the overall computation time by not probing 'visibility' status among all $\binom{\lceil \frac{R}{P} \rceil \times \lceil \frac{C}{P} \rceil}{2}$ pairs, but only for each pair of vertices in G_O, which in general is much smaller than the former.

Checking Visibility between Vertex-Pairs. Now that we fixed the vertices for our visibility graph, we need to add undirected edges between the pairs that can mutually 'see' each other. Such edges will be *intermediate subpaths* of the overall S-D shortest path. Taking each vertex on the visibility graph as the intermediate source (S_{in}), we systematically check if the other vertices placed to its right (intermediate destination, D_{in}) are visible from the former or not. So for each S_{in}, we consider a shifted coordinate system with S_{in} at origin $(0, 0)$ and check only those D_{in} that lie in the first and fourth quadrants. This will complete visibility computation for potential intermediate subpaths having slope between $-90°$ to $90°$. D_{in} in the second and third quadrants; i.e., potential intermediate subpaths having slope between $90°$ to $270°$ will be automatically completed by symmetricity of undirected edges. In particular, four distinct mutually exclusive cases can arise when checking visibility between S_{in} and D_{in} (all direction specifications are with respect to a 2D plane):

1. *D_{in} is vertically above or below S_{in}*. A drone can fly along edges of an obstacle except those that are contained completely within an obstacle. These latter edges can be computed in linear time from the obstacle graph when we find those edges that are shared by two obstacles. We call these *blocking-edge*s because when it is present between two points, these edges block their mutual line-of-sight and there is no visibility of one from another. So, if some vertical blocking-edge overlaps (fully or partially) the vertical-line joining S_{in} and D_{in}, then they are not mutually intervisible. If no such blocking edge is there, S_{in} and D_{in} are intervisible.
2. *D_{in} is horizontally rightwards to S_{in}*. Similar to the previous case, we check from G_O whether there exists some horizontal blocking-edge that overlaps (fully or

partially) the horizontal-line joining S_{in} and D_{in}. If not, they are are mutually intervisible.

3. D_{in} *is diagonally rightwards to* S_{in} *i.e., makes a* 45° or −45° *angle*. From obstacle graph G_O, we check if the line joining S_{in} and D_{in} contains any obstacle vertices which is also the left-bottom or left-top corner of an obstacle-grid-cell. If not, they are are mutually intervisible.
4. D_{in} *is in a generic position w.r.t* S_{in} *other than the previous cases*. To check whether two vertices S_{in} and D_{in} are intervisible, we need to find out the presence of an obstacle edge that blocks the straight line-of-sight joining S_{in} and D_{in}. The brute-force approach is to check every obstacle edge in G_O, for each S_{in} and D_{in} pair. But this induces a huge computational complexity, especially when the number of obstacles is huge, e.g., in a metropolitan city or a disaster-affected landscape. Here, we propose a rotating sweep line-based approach where only a minimal subset of obstacle-edges need to be checked. Considering S_{in} as the origin, we sort the vertices in V = (set of unmarked obstacle vertices in G_O lying in the first and fourth quadrants for this shifted coordinate ∪ original intended destination D) in a decreasing order of slope. In situations where the slope is equal for two intermediate destinations, to break the tie we consider an increasing order of distance from source. Then, starting at the point with the highest slope, we go on probing points in decreasing order of slope, with a clockwise-rotating sweep line pivoted at S_{in}. Further, we maintain a dynamic list L_{cr} of *critical edges*. At each point probed, we add to L_{cr} the set of clockwise edges attached to the point and exclude from L_{cr} the set of anti-clockwise edges attached to the point. Thereby, the list of critical edges is never allowed to be large enough. For every new edge getting added in L_{cr}, we check whether it intersects the line joining S_{in} and D_{in}. This is continued until the slope of the rotating sweep line becomes less than the line joining S_{in} and D_{in}. This is because, after that point there cannot be any edge that obstructs the line-of-sight between them. As such, S_{in} and D_{in} are intervisible if there are no critical edge that intersects their line-of-sight till this cut-off point.

The visibility graph G_V is created by adding an edge between each pair of vertices whom we found to be intervisible by the previous steps. The corresponding edge weight is the Euclidean distance of the line-of-sight between the vertices.

Computing Shortest Path in Visibility Graph. The visibility graph G_V is effectively an Euclidean graph where intervisible vertices are joined by edges. For the pair of vertices that are mutually intervisible, the shortest path between them is the straight line-of-sight whose distance is the Euclidean distance. If two pairs of vertices are not connected by an edge, we need a path of edges whose cumulative weight is minimum such that the traversed distance is shortest. Further, the source vertex S is fixed because it is the position where the drone is currently located. The destination vertex D is where the drone intends to go. So, we execute Dijkstra's single-source shortest path algorithm on the visibility graph G_V to compute the shortest distance from S to D. The drone can go along edges on the shortest path thus computed, because the edges indicate safe obstacle-free paths. Needless to say, the drone would have to change its heading angle of motion only when it transits from one edge to another. As the shortest path indicates maximum stretches of straight-line motions (as the ideal path, i.e., the direct line-of-sight path, is itself a single stretch of straight-line motion), it indicates lesser positions where the drone needs to maneuver mid-air steering. This effectively indicates that the time lost due to such maneuver is less, and there are lesser chances of mid-air malfunctions due to sharp turns.

3.4 Long Journeys and Transformation to 3D Space

A drone is usually not intended for covering large distances as it has to make a round-trip journey to its base station. Further, the battery consumption goes to the higher side when there are items that the drone has to carry. But, our approach can be extended to large distances by incorporating a trade-off between optimized and greedy. This can be done by keeping few *stops* across the whole journey. When the drone reaches its first intermediate stop, it can repeat its maneuver of going to a high altitude and capturing RGB-D information. As such, it creates a new map layout with the first intermediate stop as its new source and continues its journey. In this case, the journey between intermediate stops is optimal and choice of stops is overall greedy. Further, in long journeys of the drone, even slow-moving obstacles can show substantial change in position between the start and end of flight of the drone. Such motions can be efficiently detected by unsupervised object tracking [23] on videos and the map layout can be dynamically modified on the run.

It is known that finding the shortest path in the 3D space and 3D path planning is a NP-Hard problem [24]. But 3D path-planning may be required when source and destination are at different altitudes. Our approach can be extended to 3D aerial space for drones through some heuristics. One approach is that the drone changes height (moves in z-axis) only at the stops and chooses a height that has least number of obstacle-grid-cells, through its perception subsystem. And at each height, it will use the proposed 2D optimized path planning. Another alternative would be to consider multiple discretely rotated 2D planes (at small angles) pivoted at the source. The drone can then run the proposed path-planning technique on each plane and choose the plane which gives the least distance.

4 Experimental Results

The implementation included a simulation testbed for finding out the shortest obstacle-free path in a $m \times n$ rectangular grid with obstacles shown as shaded regions. Obstacles are square-shaped blocks denoted by shaded grid cells. The shortest path is shown as a red line. It may be noted that drones are of insignificant size w.r.t. obstacles, so drones may travel through boundaries between two separate objects

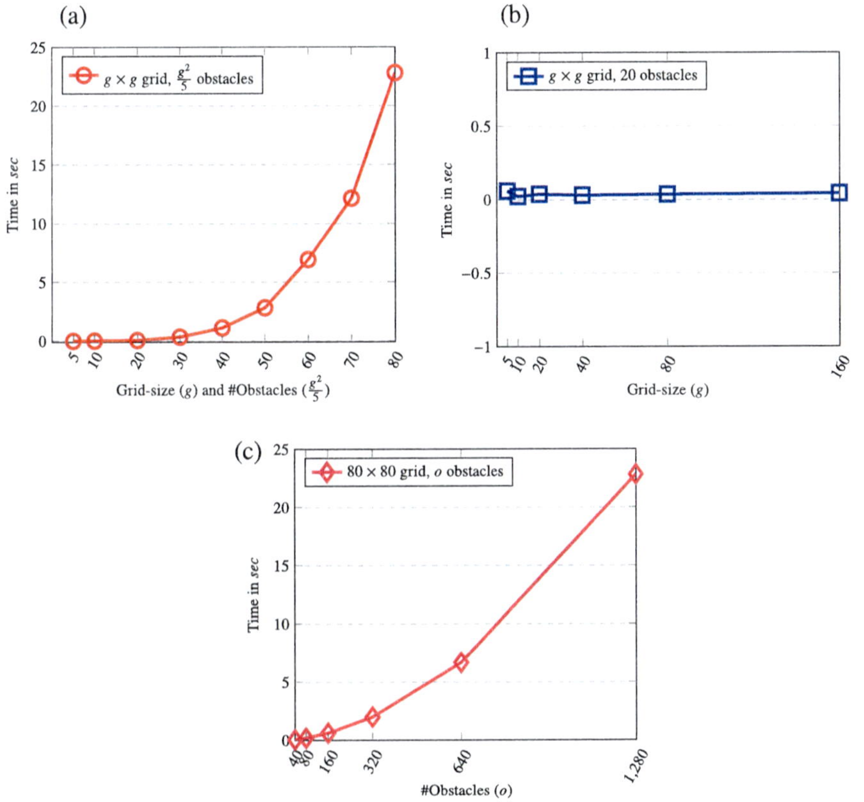

Fig. 2 Variation in execution time with **a** simultaneous change in grid-size and number of obstacles, where obstacle occupancy is 20%, **b** change in grid-size when number of obstacles is fixed, **c** change in number of obstacles, with fixed grid-size

that contain sufficient space in-between. The source code is written in C++ [3] on Ubuntu platform.

While running the simulation experiments for the shortest obstacle-free path, we measure the elapsed execution time by varying grid-size, number of obstacles and observe the effect in Fig. 2. With simultaneous change of grid-size and number of obstacles in Fig. 2a, we observe that as expected, when grid-size and number of obstacles increase, the execution time increases exponentially. Keeping the number of obstacles fixed in Fig. 2b, as we increase grid-size, we see time does not vary much. Thus, we may infer that execution time depends on number of obstacles and not much on grid-size. This can be well illustrated in Fig. 2c where execution time varies almost linearly with varying number of obstacles keeping the grid-size same.

In Figs. 3, 4, and 5, we present some illustrations of the simulations done on different grid-sizes for proportionally increasing number of obstacles. In Fig. 3, there are 5 obstacles spread over a 6 × 4 rectangular grid, each cell being of size 3 units ×

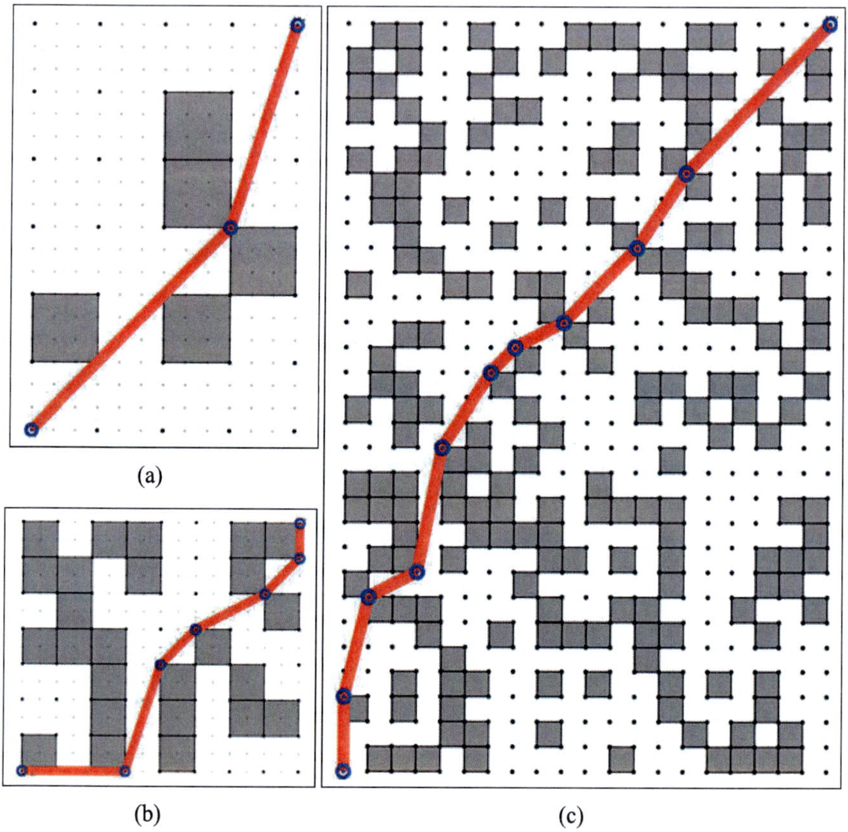

Fig. 3 Simulations on **a–b** small- and **c** medium-scale maps. Obstacles are shown as shaded squares, shortest safe path as red line and deflection points as blue circles

3 units. The shortest path is of length $\sqrt{162} + \sqrt{90} = 22.21$ units. On a similar note, Fig. 3 shows a 7×8 rectangular grid with each cell of size 2units $\times$ 2units, and 25 obstacles. Shortest path is of length $\sqrt{36} + \sqrt{40} + \sqrt{8} + \sqrt{20} + \sqrt{8} + \sqrt{4} = 24.45$ units. Figure 3 is a medium-scale grid of 30×20 with 200 obstacles, and each cell is of unit size. Here, the shortest path is of length $\sqrt{9} + \sqrt{17} + \sqrt{5} + \sqrt{26} + \sqrt{13} + \sqrt{2} + \sqrt{5} + \sqrt{18} + \sqrt{13} + \sqrt{72} = 38.05$ units.

Figures 4 and 5 are examples of simulation on even larger grid-sizes with more number of obstacles. While Fig. 4 shows application of the proposed method on a grid of size 70×90 with 2500 obstacles ($\sim$40% obstacle occupancy), Fig. 5 shows it on a 100×150 grid with 7000 obstacles ($\sim$47% obstacle occupancy). The shortest path lengths are 118.53 units and 195.39 units, respectively.

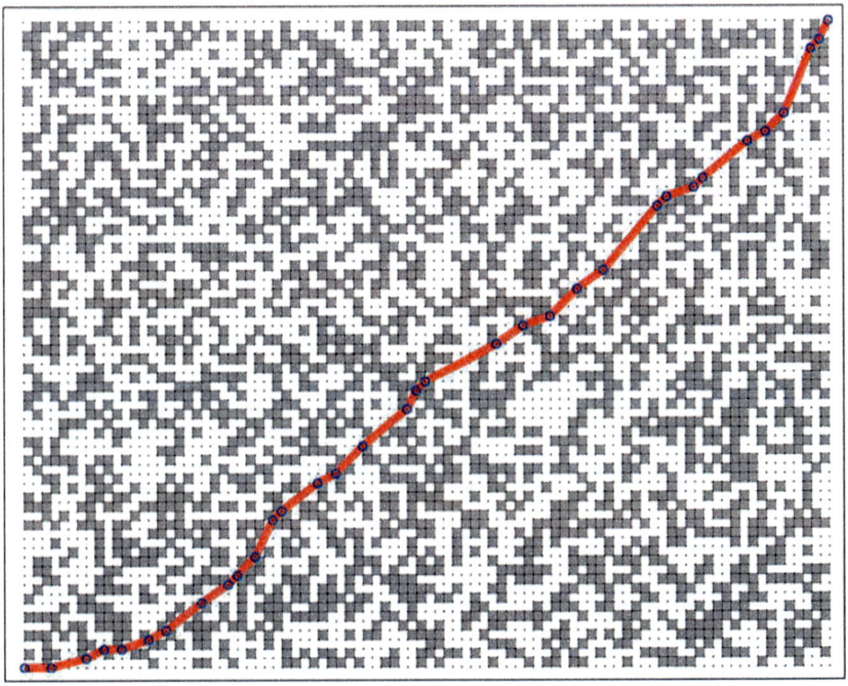

Fig. 4 A 70 × 90 rectangular grid with each cell of size 1 × 1, and 2500 obstacles

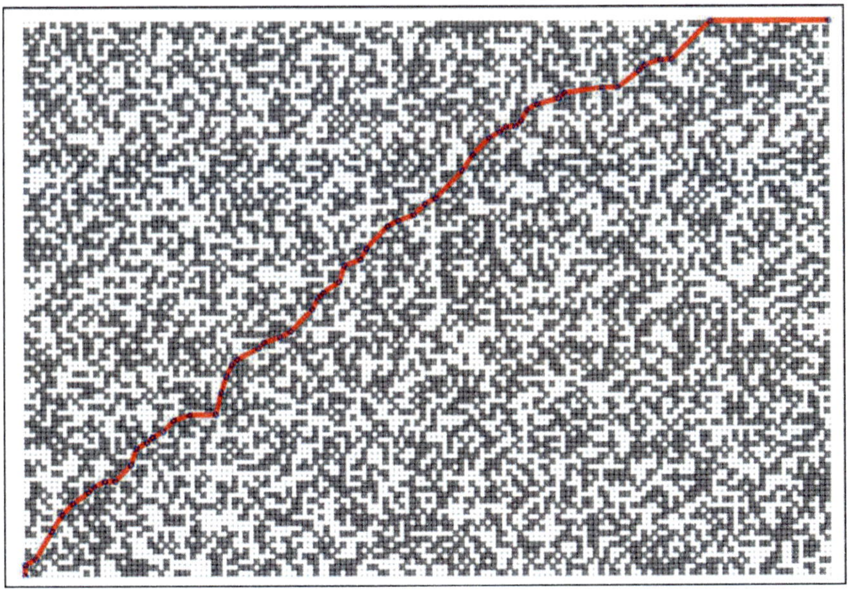

Fig. 5 A 100 × 150 rectangular grid with each cell of size 1 × 1, and 7000 obstacles

5 Conclusion and Future Scope

In this work, we have achieved finding the shortest obstacle-free path from a given source node to a designated target node via multiple node hops, through an unobstructed flight-route with clear visibility. AutoDrones following this technique will encounter less number of angular direction changes while attempting to fly in this route. This approach has many societal benefits of utilizing AutoDrones for reaching target in the shortest aviation path, thus becoming energy-efficient too. Through perception subsystem, the proposed approach refrains from using any preset maps and instead, constructs dynamic maps during travel. We have presumed that all the obstacles are fully-static like buildings and monuments or at least static till the drone again computes dynamic route at its intermediate stop. However, in reality, there could be moving obstacles in the flyway when the drone is flying at a low altitude, like walking humans, another drone, or flying bird. The computation of the next move could be more dynamic in nature to tackle such cases, which we aim to address in our future works.

References

1. Cauchard JR, Jane LE, Zhai KY, Landay JA (2015) Drone & me: an exploration into natural human-drone interaction. In: Proceedings of the 2015 ACM international joint conference on pervasive and ubiquitous computing, pp 361–365
2. Bounceur A, Bezoui M, Euler R (2018) Boundaries and hulls of euclidean graphs: from theory to practice. CRC Press
3. Jana D (2014) C++ and object-oriented programming paradigm. PHI Learning Pvt Ltd
4. Deo N (2017) Graph theory with applications to engineering and computer science. Courier Dover Publications
5. Gallo G, Pallottino S (1988) Shortest path algorithms. Ann Oper Res 13(1):1–79
6. Johnson DB (1973) A note on Dijkstra's shortest path algorithm. J ACM (JACM) 20(3):385–388
7. Sulaiman OK, Siregar AM, Nasution K, Haramaini T (2018) Bellman Ford algorithm—in routing information protocol (RIP). J Phys Conf Ser 1007:012009
8. Burfield C (2013) Floyd-Warshall algorithm. Massachusetts Institute of Technology
9. Chen JC (2003) Dijkstra's shortest path algorithm. J Formalized Math 15(9):237–247
10. Chai D, Newsam S, Huang J (2020) Aerial image semantic segmentation using DCNN predicted distance maps. ISPRS J Photogramm Remote Sens 161:309–322
11. Mukherjee A, Jana P, Chakraborty S, Saha SK (2020) Two stage semantic segmentation by SEEDS and fork net. In: 2020 IEEE Calcutta conference (CALCON). IEEE, pp 283–287
12. Chakrabarty A, Stepanyan V, Krishnakumar KS, Ippolito CA (2019) Real-time path planning for multi-copters flying in UTM-TCL4. In: AIAA Scitech 2019 forum, p 0958
13. Jana P (2020) Shortest path traversal by autonomous vehicles and drones in euclidean space. In: Young scientists' conference (YSC) at 6th India International Science Festival (IISF), Govt. of India. IISF-GoI
14. Koch T, Körner M, Fraundorfer F (2019) Automatic and semantically-aware 3D UAV flight planning for image-based 3D reconstruction. Remote Sens 11(13):1550
15. Sakurai M, Ueno Y, Kondo M (2021) Path planning and moving obstacle avoidance with neuromorphic computing. In: 2021 IEEE international conference on intelligence and safety for robotics (ISR). IEEE, pp 209–215

16. Wu G, Shao X, Guo Z, Chen Q, Yuan W, Shi X, Xu Y, Shibasaki R (2018) Automatic building segmentation of aerial imagery using multi-constraint fully convolutional networks. Remote Sens 10(3):407
17. Kirillov A, He K, Girshick R, Rother C, Dollár P (2019) Panoptic segmentation. In: Proceedings of the IEEE/CVF conference on computer vision and pattern recognition, pp 9404–9413
18. Jana P, Ghosh S, Sarkar R, Nasipuri M (2017) A fuzzy C-means based approach towards efficient document image binarization. In: Proceedings of the 9th international conference on advances in pattern recognition (ICAPR). IEEE, pp 332–337
19. Jana P, Ghosh S, Bera SK, Sarkar R (2017) Handwritten document image binarization: an adaptive K-means based approach. In: 2017 IEEE Calcutta conference (CALCON). IEEE, pp 226–230
20. Rajagopal A, Stier N, Nelson W, Chandrasekaran S, Brown AP (2020) DeepOSM-3D: recognition in aerial LiDAR RGBD imagery. In: Geospatial informatics X, vol 11398. International Society for Optics and Photonics
21. Parajuli B, Kumar P, Mukherjee T, Pasiliao E, Jambawalikar S (2018) Fusion of aerial LiDAR and images for road segmentation with deep CNN. In: Proceedings of the 26th ACM SIGSPATIAL international conference on advances in geographic information systems, pp 548–551
22. Zhou H, Qi L, Wan Z, Huang H, Yang X (2020) RGB-D co-attention network for semantic segmentation. In: Proceedings of the Asian conference on computer vision (2020)
23. Jana P, Bhaumik S, Mohanta PP (2021) Unsupervised action localization crop in video retargeting for 3D ConvNets. In: 2021 IEEE region 10 conference (TENCON). IEEE
24. Yang L, Qi J, Xiao J, Yong X (2014) A literature review of UAV 3D path planning. In: Proceeding of the 11th world congress on intelligent control and automation. IEEE, pp 2376–2381

Chapter 2
Transfer Learning and Transformer Architecture for Financial Sentiment Analysis

Tohida Rehman, Raghubir Bose, Samiran Chattopadhyay, and Debarshi Kumar Sanyal

1 Introduction

Financial sentiment is a very important business indicator. Positive financial sentiments boost business while negative financial sentiment are catastrophic to business. This is because financial sentiment influences enterprises to stay ahead in business vision. It is understood that the definition of "**new and relevant information**" might change as updated information retrieval technologies become available and early adoption of such technologies does provide an advantage. The sentiment analysis models trained on general aggregated dataset are not fully suitable in a financial domain. This is because financial texts have a specialized language coupled with unique vocabulary. They also have a tendency to use vague expressions instead of easily identified negative/positive words. Impact of global pandemics like COVID on business sentiments and its cascading impact has become relevant. Approaches from Loughran and McDonald [1] may seem a solution because they use "word counting" methods. The modern approach of efficient and effective use of transfer learning and transformation architecture methods appears to be more plausible solution to the challenges mentioned above and is the focus of this paper.

T. Rehman (✉) · S. Chattopadhyay
Department of IT, Jadavpur University, Kolkata, India
e-mail: tohida.rehman@gmail.com

R. Bose (✉)
BFSI BTAG, Tata Consultancy Services, Kolkata, India
e-mail: raghubir.bose@tcs.com

S. Chattopadhyay
Institute for Advancing Intelligence (IAI), TCG Crest, Kolkata, India

Jadavpur University, Kolkata, India

D. K. Sanyal
Indian Association for the Cultivation of Science, Kolkata, India

L. Mandal et al. (eds.), *Proceedings of International Conference on Computational Intelligence, Data Science and Cloud Computing*, Algorithms for Intelligent Systems,
https://doi.org/10.1007/978-981-19-1657-1_2

2 Objective

The objective of this paper is to use an efficient mechanism to train language models with financial dataset. Once trained, the downstream models are activated with the weights derived from the language modelling activity. This way it is possible to formulate a much better model for financial sentiment prediction. This approach ranging from a single word embedding layer to the whole model provides a solution to the scarcity of labelled data. Language models do not require any labels, as they are designed to predict the next word. Semantic representation is possible using language models. Hence, we need to fine-tune the labelled data so that the language model can use this information to predict the labels. Using transfer learning the language model can work on domain-specific unlabelled corpus. Due to the contextual nature of the language used in financial domain, fine-tuned context-sensitive models are very effective.

The goal of this paper is to use fine-tune pre-trained language models for financial domain for sentiment analysis including the effect of COVID pandemic. To achieve the goal, the sentiment of a sentence from a financial news article will be tried to be predicted, using the Financial PhraseBank created by Malo et al. [2] and FiQA sentiment scoring dataset [3] and the World Bank COVID financial sector response [4].

The main objectives of this paper are the following. We evaluated the use multi-directional transfer and transformation model [BERT] on multiple financial sentiment analysis datasets and achieve an outperforming modern sentiment scoring structure. This we intend to execute using a pipeline based on BERT model. We also conduct several experiments to investigate catastrophic forgetting and hence fine-tuning without significant loss of performance.

3 Related Work and Challenges

The major deterrent for sentiment analysis in the financial domain is the lack of a substantial aggregated content of labelled dataset. This is the major reason for which it is not easy to apply proper neural network algorithms on the financial information available. The full potentiality of neural network is yet to be fully leveraged in financial domain as far as sentiment analysis is concerned. The problem of language models is that they mostly use a left context or a right context. Historically language model used the concept of end-to-end one-way transfer to find the probability distribution and once trained words cannot reach themselves as in a bidirectional encoder. But now the natural language understanding is bidirectional that means it can capture the left context as well as the right context. The next effort is to have a masked language model or MaskedLM [5]. We use a k% masked data and try to predict the same by some means as seen below. The problem with the model is too much training is required and not enough context may be available as shown in Fig. 1.

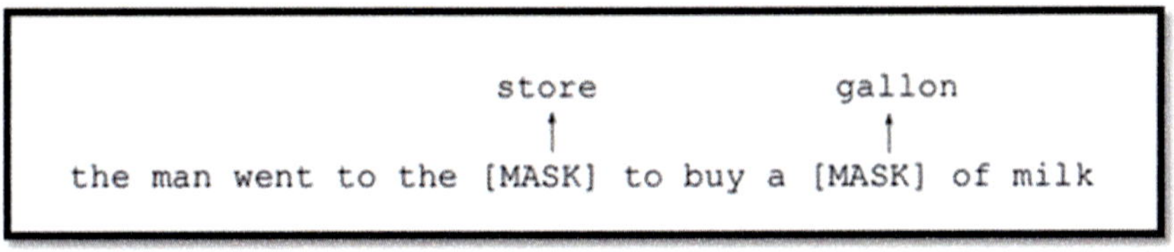

Fig. 1 Masked LM scenario

ELMo [6] or (Embeddings from Language Models) is one of the first successful researches on modern approaches where a pre-trained model was used extensively. ELMo converts every input into its contextualized representation. Using the pre-trained weights of ELMo, contextualized word embeddings can be calculated for any word present as the input sentence. Initializing embeddings for downstream tasks with any arbitrary input were shown to improve performance on most tasks when compared to other approaches such as GloVe [7] or word2vec [8]. In ELMo learning, the first layer is not fully transferred to the next layer. ULMFiT uses novel approaches such as progressive unfreezing, slanted triangular learning rates, and discriminative fine-tuning. ULMFit can be fine-tuned to a whole pre-trained language model and can also be used for further pre-training a language model on a domain-specific aggregated data. This is based on the assumption that the target task information comes from a different source with respect to the general aggregated content used for training the model.

4 BERT

Bidirectional encoder representations from transformer (BERT) [5] uses bidirectional transformers for language models. The key algorithmic segments of BERT is as follows:

1. Pre-training from unlabelled texts.
2. Bidirectional contextual models and transformation architecture
3. Text entailment or next sentence prediction.

The key proposition of BERT is as follows:

BERT defines the activities of language modelling such as predicting randomly masked tokens in a sequence rather than the next token in addition to classifying sentences as following each other or not. BERT uses an efficient token-based masked language Model. BERT uses a [CLS] token as a sequence approximate. The user can get the first token and uses it for sequence prediction rather than a token prediction. BERT is a large neural network trained on an large aggregated data. To explain simple, let us refer to Fig. 2.

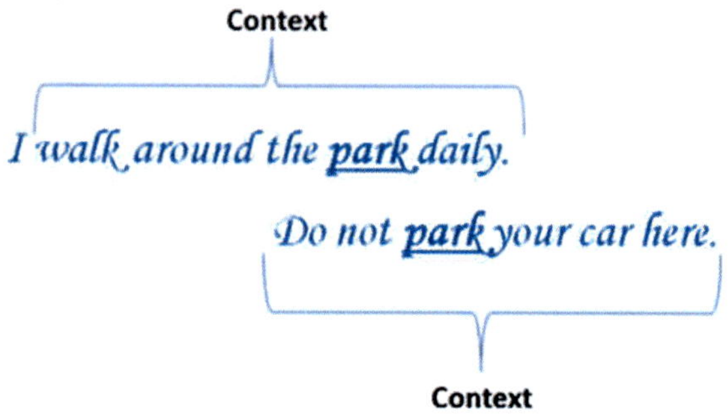

Fig. 2 Analysing context of a sentence

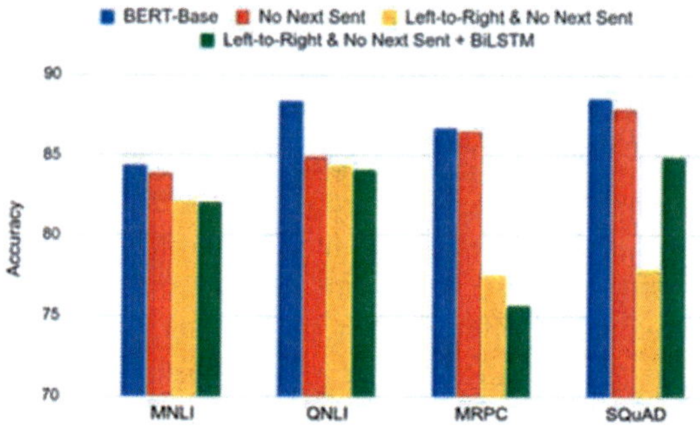

Fig. 3 Effect of BERT pre-training

If we try to predict the nature of the word "park" in the example above by only taking either the left or the right context, then we will have different views. One way to explain this better is to consider both the left and the right context before making a prediction. BERT can also be fine-tuned to execute NLP tasks including natural language inference or question answering. How pre-training influences BERT is shown in Fig. 3.

BERT uses the principle of transfer learning [9] as well.

5 Implementation Methodology

The overall methodology can be generically looked at the following steps.

1. Setup the environment for the experiments. We will set up a virtual environment for the same. This includes setup of BERT models.
2. Bidirectional contextual models and transformation architecture preparation of the dataset from various sources.
3. Training model is created. Classifier is run.
4. Prediction from an input data.
5. Conclusion from the prediction.

There are various controllable input parameters which are passed as configurations to the overall framework. The input parameters are then processed through the train of bidirectional encoded transformers. The outcome of the work is a sentiment analysis score.

6 Research Question and Experimental Setup

This paper focusses on the following research questions

- Comparison of various pre-trained NLP models.
- Sentiment analysis with discrete and continuous targets.
- How different training impacts model performance.
- Selection of the right encoder.
- Sentence, word and character embedding [10].
- How many layers are introduced for model tuning.
- The event of catastrophic forgetting.

After the [CLS] token, a dense layer is added. This is then applied on a labelled dataset for pre-training. The dataset for the paper is:

- TRC2-Financial. It is a collection of 1.8 million news articles published by Reuters [11].
- Financial PhraseBank: Made available by [2].
- COVID-19 Finance Sector-Related Policy Responses: World Bank 2020 [4].

The overall model can be explained as in Fig. 4: The stack of bidirectional encoders with masked LM is applied on the train of data till the sentiment position is reached. It has been seen [12] that model performance for classification improves with fine-tuning and additional pre-training. This is the reason why this paper has considered four sets of pre-training data which is essentially derived from a large corpus of financial knowledge. The effect of training transfers across layers. One of the known risks that this paper intends to explore is the area of catastrophic forgetting and training the model to go past the same.

The metrics that are considered in this paper are: accuracy, macro-F1 average, loss due to cross-entropy, mean squared error loss.

In this paper, we are considering the following:

- Warm-up proportion of 0.21
- Dropout per cent probability of 0.12
- Maximum sequence length of 64 tokens
- Learning rate of $2e-5$
- Mini batch size of 64
- Number of epoches is taken as 10
- Discrimination rate is 0.87.

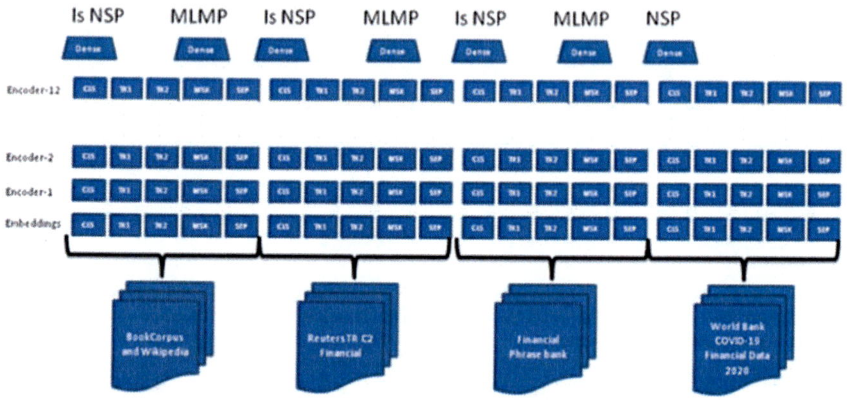

Fig. 4 BERT process

The environment channels, dependencies are

pytorch, anaconda, conda-forge - jupyter 1.0.0, pandas 0.23.4, python 3.7.3, numpy 1.16.3, nltk, tqdm, ipykernel 5.1.3, pip: textblob, joblib 0.13.2, pytorch-pre-trained-bert 0.6.2, scikit-learn 0.21.2, spacy 2.1.4, torch 1.1.0.

Google Cloud platform and Google Colab research project have been used for this scenario. The above models have been implemented in Pytorch and have been tested on several data sets. The performances match with the associated TensorFlow implementations. This paper has used the following configurations for the model as shown in Fig. 5.

The pre-trained model which is used in BERT is used to store the configuration of the models. The pre-trained model also handles methods for model lifecycle management. In addition to the same, also a few methods common to all models for the following purpose:

1. Resizing of the input embeddings.
2. Prune heads in the self-attention heads.

There is also a pre-trained configuration module available in BERT. It is the base configuration classes of BERT. The BERT configuration module is common to all models' configurations as well as lifecycle methods for managing the models. When the model is loaded with a configuration file, the weights are not loaded. This process only affects the configuration of the model. The typical configuration of a BERT model is that with 12-layer along with 12-heads, 768-hidden and 110M parameters. Lower case English is used to train the model. We take the following text and try to visualize the BERT models. The input sentence being "COVID-19 has hit the world economy very badly", when subjected to a 12-layer configuration the corresponding visualizations of the BERT models can be seen from Figs. 6 and 7.

S.N	Configuration Variable	Description
1.	data_directory: str	Path for the training and evaluation datasets.
2.	bert_model: BertModel	The BERT model to be used. For example: BertForSequenceClassification.from_pretrained(...)
3.	model_directory: str	The path where the resulting model will be saved.
4.	max_seq_length: int	The maximum length of the sequence to be used. Default value is 64.
5.	train_batch_size: int	The batch size for the training. Default value is 32.
6.	eval_batch_size: int	The batch size for the evaluation. Default value is 32.
7.	learning_rate: float	The learning rate. Default value is 5e5.
8.	num_train_epochs: int	Number of epochs to train. Default value is 4.
9.	warm_up_proportion: float	During the training, the learning rate is linearly increased. This value determines when the learning rate reaches the intended learning rate. Default value is 0.1.
10.	no_cuda: bool	Determines whether to use gpu. Default is False.
11.	do_lower_case: bool	Determines whether to make all training and evaluation examples lower case. Default is True.
12.	seed: int	Random seed. Defaults to 42.
13.	local_rank: int	Used for number of gpu's that will be utilized. If set -1, no distributed training will be done. Default value is -1.
14.	gradient_accumulation_steps: int	Number of gradient accumulations steps. Defaults to 1.
15.	fp16: bool	Determines whether to use 16 bits for floats, instead of 32.
16.	output_mode: 'classification' or 'regression'	Determines whether the task is classification or regression.
17.	discriminate: bool	Determines whether to apply discriminative fine-tuning.
18.	gradual_unfreeze: bool	Determines whether to gradually unfreeze lower and lower layers as the training goes on.
19.	encoder_no: int	Starting from which layer the model is going to be finetuned. If set 12, whole model is going to be fine-tuned. If set, for example, 6, only the last 6 layers will be fine-tuned.

Fig. 5 Configuration data

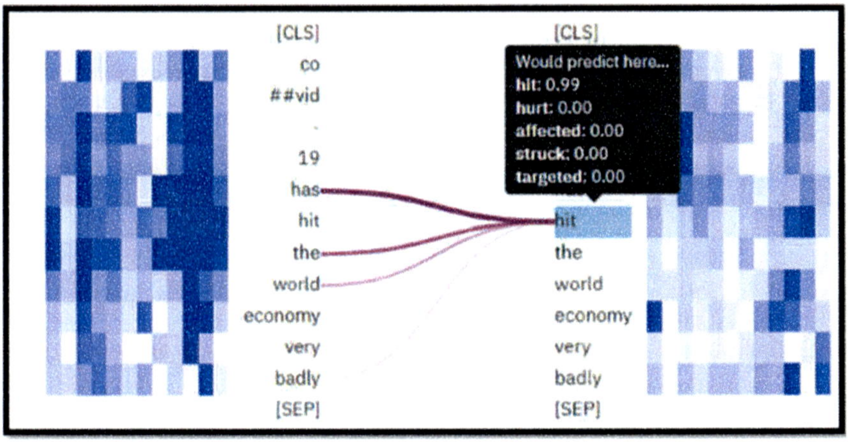

Fig. 6 Visual analysis of BERT

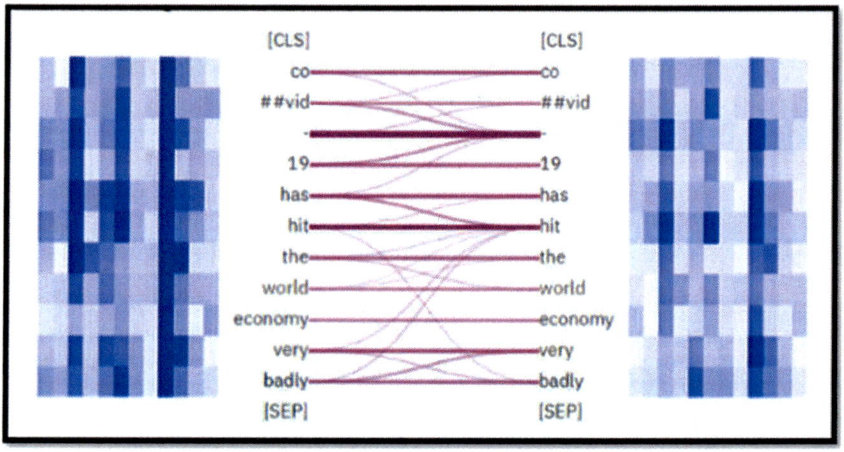

Fig. 7 Visual analysis of BERT (12 layers)

7 Detailed Flow

1. The BERT object has a method called prepare_model which sets up the structure in terms of processor of the dataset, cardinality of the labels, usage of GPUs, distributed training, steps of gradient accumulation and activity tokenizing.
2. The list of values of the labels in the dataset, i.e. "positive", "negative", "neutral".
3. The utility.py hosts all the utility classes used in the paper.
4. FinSentProcessor is responsible for processing all the data processing needs of the BERT object.
5. FinSentProcessor has the processing methods for the model.
6. The Usage of BertTokenizer from_pre-trained function needs a discussion is to instantiate a PreTrainedBertModel from a pre-trained model file and download and cache the pre-trained model file if needed.
7. The BertTokenizer includes tokenization of punctuation, splitting and wordpiece.
8. The get_data function gets the data for training or evaluation. It returns the data in the format that pytorch can use.
9. The create_the_model functioncreates the structure sets the model to be trained and the optimizer.
10. Discriminative fine-tuning is used.
11. BERTAdam is identified as an optimizer.
12. The get_Loader function creates a data loader object for a dataset.
13. The convert_examples_to_features function is used to convert a data file into input feature list.
14. The get_metrics function is used to provide an endpoint for accuracy and precision–recall for different sentiments.

8 Results

The initial performance results obtained by training different layers BERT have 12 training transformer encoder layers. At different, depth loss and accuracy are different. This loss-versus-accuracy for initial training can be seen in Fig. 8.

When this performance data is correlated between loss and accuracy, we get the following correlation for BERT the results as shown in Fig. 9.

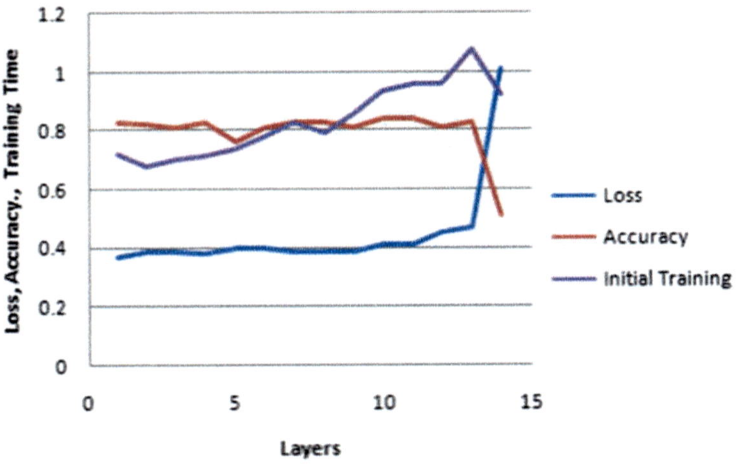

Fig. 8 Loss versus accuracy

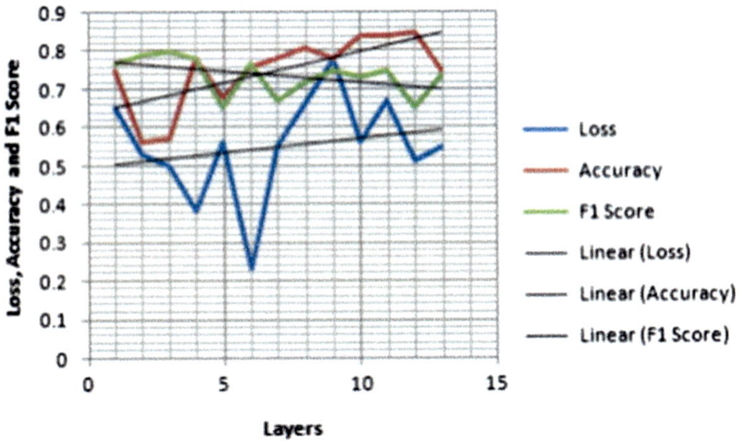

Fig. 9 Performance data

9 Challenges

1. The problem with this approach using BERT is the size. The model used in BERT-base contains 110M parameters. There are 340M parameters in BERT-large. It is difficult to deploy a model of such size into many environments with limited resources, such as mobile or embedded systems.
2. At times BERT tends to overfit when subject to fine-tuning.
3. Significant training and inference times. Training times do not impact the main process.
4. The inference times required by BERT vary depending on the model and needed hardware available; but in many cases, this significantly limits us on quantity, cost and speed.
5. Currently, BERT language parameters are extremely large. Not all attention heads are always required.
6. There are known model limitations to BERT specifically when it comes to semantics and long-term contexts.
7. There is a high chance of catastrophic forgetting.
8. The main challenge is interpretability of the transfer learning and transformation architecture—model-driven interpretation is still a challenge.
9. The combination of interpretability and transferability to downstream tasks is the major challenge that needs to be solved.
10. An adaptive and layered batch optimization technique [LAMB] [13] reduces BERT training time from 3 days to just 76 min.

10 Future Scope

BERT-based fine-tuning has huge prospects in many areas of sentiment analysis in finance. In the area of finance, there is a strong need to have a continuous sentiment analysis for any business data. The field of continuous fine-grained financial sentiment analysis will lead to a far more empathetic finance system which should be able to maximize human value. Further, improvements can be done on a real-time stock market data and apply BERT-based sentiment analysis to the same and predict the target price of the stock in a far accurate way. This way financial investment companies will be able to provide differentiated value to their end customers by minimizing the risk using right sentiment analysis technique using transfer learning and transformer architecture. The more the size and coverage of the pre-trained dataset more in the accuracy and efficiency of the continuous sentiment analysis. There is a research work done in the area of combining the three different kinds of pre-training LM like bidirectional, left to right and seq-to-seq.

References

1. Loughran T, McDonald B (2020) Textual analysis in finance. Ann Rev Finan Econ 2:357–375
2. Malo P, Sinha A, Takala P, Korhonen P, Wallenius J (2014) Good debt or bad debt: detecting semantic orientations in economic texts. J Assoc Inf Sci Technol 65(4):782–796
3. Kiros R, Zhu Y, Salakhutdinov R, Zemel RS, Torralba A, Urtasun R, Fidler S (2015) Skip-thought vectors. In: NIPS'15: Proceedings of the 28th international conference on neural information processing systems, vol 2, pp 3294–3302
4. Erik F, Alonso Gispert T, Kliatskova T, Mare DS (2020) Taking stock of the financial sector policy response to COVID-19 around the world. Policy Research Working Paper; No. 9497
5. Devlin J, Chang M-W, Lee K, Toutanova K (2019) BERT: pretraining of deep bidirectional transformers for language understanding. Proc NAACL-HLT 2019:4171–4186
6. Peters ME, Neumann M, Iyyer M, Gardner M, Clark C, Lee K, Zettlemoyer L (2018) Deep contextualized word representations
7. Brochier R, Guille A, Velcin J (2019) Global vectors for node representations. In: The world wide web conference—WWW'19
8. Mikolov T, Chen K, Corrado G, Dean J (2013) Efficient estimation of word representations in vector space
9. Tan C, Sun F, Kong T, Zhang W, Yang C, Liu C (2018) A survey on deep transfer learning
10. Reimers N, Gurevych I (2019) Sentence-BERT: sentence embeddings using Siamese BERT-networks
11. Lewis DD, Yang Y, Rose TG, Li F (2004) Rcv1: a new benchmark collection for text categorization research. J Mach Learn Res 5:361–397
12. Howard J, Ruder S (2018) Universal language model fine-tuning for text classification
13. You Y, Li J, Reddi S, Hseu J, Kumar S, Bhojanapalli S, Song X, Demmel J, Keutzer K, Hsieh C-J (2020) Large batch optimization for deep learning: training BERT in 76 minutes

Chapter 3
Classification of Fashion Items Based on Shape Features Using Machine Learning

Nikhil S. Tengli and Suvarna Nandyal

1 Introduction

Fashion and clothing-related issues are becoming increasingly essential in people's daily life. The fashion business has become an essential economy sector, with extensive industrial chains that include garment design, manufacture, and distribution. Indeed, there has been an increase in global demand for clothes in recent years [1]. Primarily on the internet, where a significant amount of fashion-related business is undertaken. With the development of technological breakthroughs every day, new fashion portals are emerging. Customers are also interested in the online shopping system. They do not have to go to shopping malls and spend hours looking for the clothes they want. Instead, consumers can use online shopping platforms to find and order their preferred clothes from anywhere, using their mobile device or any other platform. As a result, internet businesses should make it simple for shoppers to find the clothes they want.

Second, people get bored or need help to find appropriate clothes to wear in their daily lives. Wearing clothes that complement each other in terms of color, texture, style, skin tone, and so on is a crucial part of identity and style [2]. Choosing an outfit is a typical challenge people usually find complex daily. This issue is extensive, and it includes a large number of implicit and abstract visual and social aspects. Thus, there is a need for cloth or different outfits classification that can be future used for applications like similar outfit retrieval, matching with same colors, style, categorization of formal and casual dress codes.

N. S. Tengli (✉) · S. Nandyal
PDA College of Engineering, Kalaburagi, India
e-mail: nikhil.tengli039@gmail.com

L. Mandal et al. (eds.), *Proceedings of International Conference on Computational Intelligence, Data Science and Cloud Computing*, Algorithms for Intelligent Systems,
https://doi.org/10.1007/978-981-19-1657-1_3

Intuitively, any image that includes people will very certainly feature clothing as well. Thereby, applying image classification algorithms in the fashion field is valuable because it can be helpful commercially, such as organizing a design firm's collections or recommending specific types of clothing to people. The general issues of image classification, including variations and the gap between computer and human understanding, are acquired and magnified in this field since the characterization of the visual properties of the clothes in terms of color, texture, and shape is challenging [3]. Apart from that, the cloths can be folded, occluded, or the shape of the body-worn are considered the geometric transformations.

The most generalized approach of the studies above takes the same approach: They retrieve feature descriptors and use them for further classification and prediction. It is also critical to carefully choose the feature extraction schematic because it significantly impacts effectiveness. In this paper, we focus on classifying various types of clothes into seven categories: jacket, shirt, suit, sweater, t-shirt, polo shirt. We have a segmented region of interest (ROI) based on pixel information [4]. Based on the shape features of cloth, we have combined three different types of shape descriptors such as Zernike moments, Fourier shape descriptors, and chain code. These feature vectors are trained using an SVM classifier. The database is created by acquiring images from an online platform, social media, and Google image. The result shows that the proposed system can correctly classify the sample on average 73.57%.

The remaining portion of this paper is organized as follows. The second section discusses the literature survey. The third section explains the proposed method for the recognition and classification of clothes. The experiment and outcome of this work are explained in section four. Section five is where we can find the conclusion.

2 Literature Survey

Many scholars have worked on developing automatic clothes classification systems. To recognize the woven fabric, Pan et al. [6] presented a BP neural network. Ben et al. [7] identified woven cloth depending on textural features and an SVM classifier. Yamaguchi et al. [8] labeled superpixels derived from image segmentation using a conditional random field model to describe clothes. Liu et al. [9] proposed classifying clothes into 23 categories depending on postures estimates and variables such as color, SIFT, and HOG. Bourdev et al. [10] presented a method that uses nine binary features such as male/female and t-shirt and long length hair to define people's appearance. For clothes segmentation, a method for segmenting clothing in fashion store databases was presented by Manfred et al. [11]. Hu et al. [12] explained a new segmentation of cloths method based on constrained Delaunay triangulation (CDT) that does not require a pre-defined clothing model. Occlusions must be dealt with in a specific way and diverse positions that humans have. Weber et al. [13] explained a novel way to get the clothing mask beginning with a group of trained pose detectors. Color, pattern, neck style, sleeve length, and other characteristics can also classify clothing.

Chen et al. [14] created a technique to generate a list of nameable properties for clothes in photographs that are not confined. Lorenzo-Navarro et al. [15] presented an experimental investigation on the LBP, HOG descriptors, and color's capacity to classify clothing attributes.

3 Methodology

Image feature descriptors for clothing classification are introduced in this paper. These technologies can be used to develop autonomous devices that can recognize everyday clothing. The proposed classification approach consists of four steps: (i) Segmentation identifies and locates the cloth in an image by pixel annotation; (ii) preprocessing by adopting gray and binary conversion with resizing of the image (iii) the feature extraction, Zernike moments, Fourier shape descriptors, and chain codes are applied to each segmented image to create a feature vector; (iv) the classification is performed by support vector machine (SVM). Figure 1 shows the architecture of the cloth classification method.

3.1 Image Segmentation

All fashion products are the object of interest in our work; an image may contain jeans, jackets, and other fashion items. However, there always be additional information such as background, surrounding objects, and less significant noise. Segmentation, also known as object detection, is used to find the object of interest in an image. Segmentation separates an image into many regions depending on similar pixels gathered together in one region. However, it does not achieve the categorization of pixels in general.

On the other hand, semantic segmentation accomplishes pixel-by-pixel semantic grouping based on certain label classes [16]. In this work, a pixel-wise annotation is enabled for semantic segmentation. Semantic segmentation aims to produce semantically meaningful results that can be used in various applications [17].

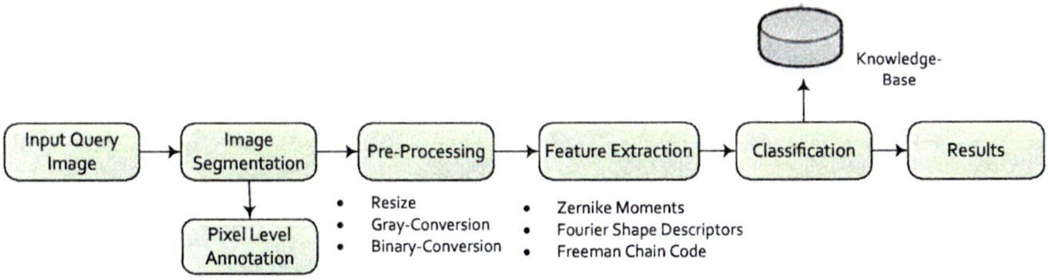

Fig. 1 Architecture of cloth classification model

3.1.1 Ground Truth Generation Using Pixel-Level Annotation

A label is assigned to each pixel in an image, such as shirt, and semantic pixel-wise segmentation labels each pixel. The appropriate labels linked with relevant pixels are kept after the processing is completed. Some generated ground truth masks are shown in Fig. 2. For visualization, we use different colors to represent different categories, as depicted in Fig. 3. For the training model, the ground truth masks are transformed to the format of a 2-D array where each element is the class number that the original image pixel belongs to ground truth. Figure 4 shows the sample results of the segmentation applied to the database.

3.2 *Pre-processing*

We are dealing with real-time images as they appear on the internet throughout this study. This presents many image quality difficulties, such as varying lighting conditions, different image scales. We address these variations in the first pre-processing phase by converting images into grayscale, binary format and resizing each image to 128 × 128 pixels size of the image using the below equations.

$$I = \text{imresize}(\text{Img}, [128\ 128]) \tag{3.1}$$

Resize the segmented image to 128 × 128.

$$I = \text{im2gray}(\ \text{RGB}\ ) \tag{3.2}$$

Transforms the specified RGB input image to a grayscale intensity image I.

$$\text{BW} = \text{im2bw}(\text{RGB}, \text{level}) \tag{3.3}$$

Fig. 2 Instances of ground truth with pixel-level annotations

Fig. 3 Colors representing each category

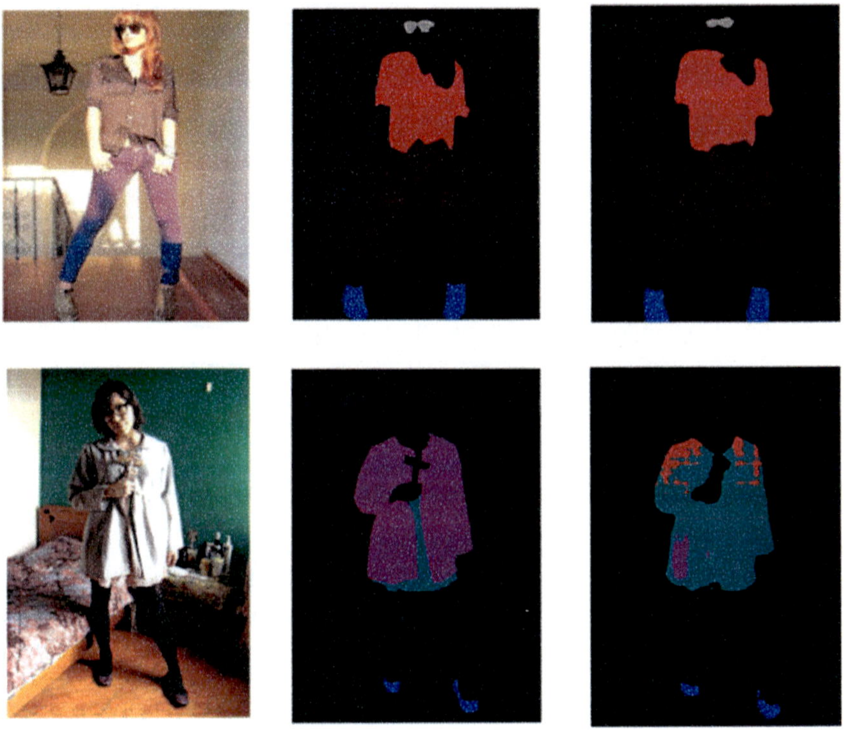
(a) (b) (c)

Fig. 4 Segmentation results, **a** Input image, **b** Ground truth, **c** Predicted segmentation

Transforms the segmented RGB input image to a binary image.

3.3 Feature Extraction

Based on the shape features of cloth, we have combined three different types of shape descriptors such as Zernike moments, Fourier shape descriptors, and chain code.

3.3.1 Freeman Chain Code

The contours which are obtained define the shape or pattern of various dress codes. This means that characteristics of a pattern can be utilized to differentiate one dress code from another. These pattern features can be extracted via search or inbound contour tracing methods using chain codes [18, 19]. Chain code is built on the concept of following the object's outer edge and storing the direction of movement.

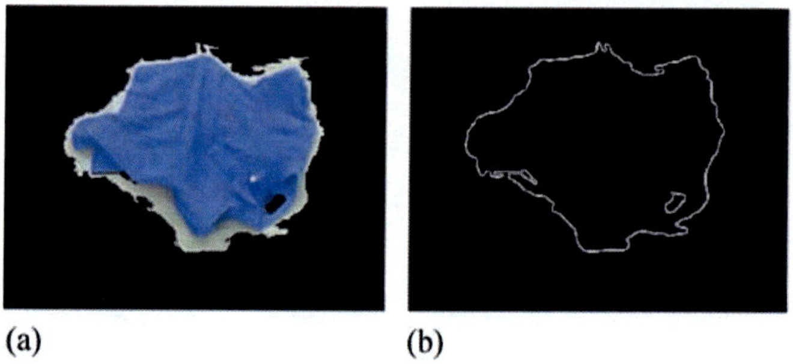

Fig. 5 Contour extraction using chain code, **a** Segmented input image, **b** Contour image

The direction is determined by using a four- or eight-direction chain code. The histogram of pixels is used to determine the chain code. In the feature extraction stage, shape approximation techniques, particularly chain coding, have been widely employed to encode the boundary line because of its simplicity and low storage requirement.

The boundary of dress is represented by chain codes (cc), where the codes describe the direction of where the next pixel from the current position is located. A grid is placed on a digital image's boundaries. The boundary points are calculated to be as close as possible to the nearest grid point. After that, they pick a beginning point that yields a sampled image. A four-directional chain code and an eight-directional chain code can be used to create a chain code. To generate the chain code, following steps are followed: To begin, locate the pre-processed numerical image's boundary. Second, search the image from top left to bottom right until find the first pixel belonging to the region. Thirdly, search for anything near the current pixel of the boundary in a clockwise direction. Finally, the detected inner border is represented by direction code, process how chain code is calculated. This work extracts 64 chain code features for each segmented input image to discriminate between different dress codes. Figure 5 shows the sample result of the contour extraction using chain code.

3.3.2 Fourier Shape Descriptors

Shape analysis frequently employs the Fourier transformation, and these descriptors characterize the object's shape in the frequency domain. The lower frequency descriptors provide information about the shape's general characteristics. In comparison, the higher frequency descriptors get data about the shape's more delicate intricacies. Even though the number of coefficients transform generates is typically considerable, a subset of them is sufficient to obtain the shape's overall characteristics. It can be ignored because very high-frequency information provides little shape characteristics and is not helpful in shape discrimination. As a result, the Fourier descriptor's dimensions of the Fourier descriptors used to identify shapes have been significantly reduced.

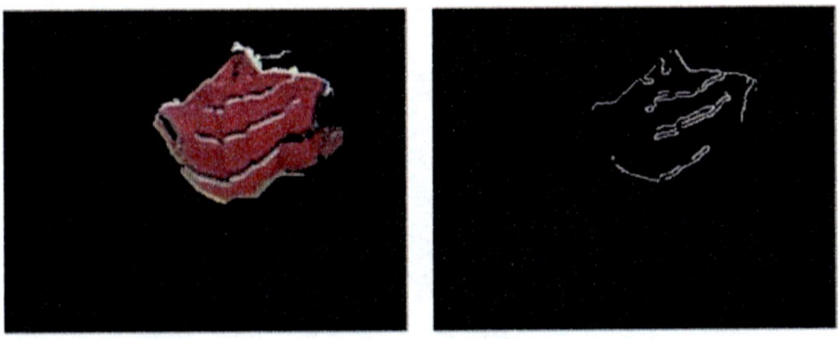

(a) (b)

Fig. 6 Contour extraction using Fourier transformation, **a** Segmented input image, **b** Contour image

The Fourier transform (FT) on a shape signature function produced from shape boundary coordinates $\{(x(t), y(t)), t = 0, 1, \ldots, N-1\}$ yields one-dimensional FD. The centroid distance function, which is defined as the distance between the border points and the form's centroid (x_c, y_c), is a standard shape signature function.

$$r(t) = \left([x(t) - x_c]^2 + [y(t) - y_c]^2\right)^{1/2}, \quad t = 0, 1, \ldots, N-1 \tag{3.4}$$

where $x_c = \frac{1}{N}\sum_{t=0}^{N-1} x(t)\ y_c = \frac{1}{N}\sum_{t=0}^{N-1} y(t)$.

The Fourier converted coefficients are then obtained by applying one-dimensional FT to $r(t)$.

$$a_n = \frac{1}{N}\sum_{t=0}^{N-1} r(t)\exp\left(\frac{-j2\pi nt}{N}\right), n = 0, 1, \ldots\ldots\ldots, N-1 \tag{3.5}$$

Shape descriptors, or Fourier descriptors, are the magnitudes of the coefficients $a_n(n = 0, 1, \ldots, N-1)$ normalized by the value of the first coefficient a_0 and are used as shape descriptors, called Fourier descriptors. For the cloth classification system, we extract 128 Fourier coefficients, and Fig. 6 shows the sample results of obtaining contour images by applying Fourier transformation. The acquired FDs are translation, rotation, and scale-invariant.

3.3.3 Zernike Moments

The quantitative amounts at a specific distance between a reference point or an axis are defined by moments. They are the integer quantities that are utilized to understand and derive the characteristics of a function. It is a term that is commonly used to describe the shape. As a result, moments can be used to extract characteristics [20, 21]. Zernike first proposed this hypothesis in 1934. Zernike moments are representations of image attributes that include little duplication and overlap information. They can

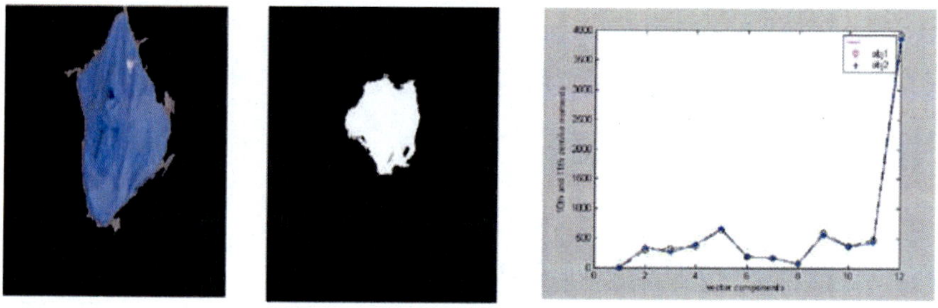

(a) (b) (c)

Fig. 7 Zernike moments values applied on a binary image, **a** Segmented image, **b** Binary image, **c** Plot of Zernike moments

thus be used to extract features from the image. Zernike proposed a set of complex polynomials that forms an entirely orthogonal set over the unit circle's interior.

The origin is taken as the center of the given image for moment computation, and pixel coordinates are mapped in the range of a circle with unit distance. At the moment of computation, pixels outside the unit distance circle are ignored. The function $f(x, y)$ equal to 1 or 0 is used in Zernike moments (binary image). Our study extracts 11 particular Zernike moments for each segmented image, and Fig. 7 depicts the 12 components of the 10th and 11th Zernike moments applied to binary images of segmented objects.

3.4 Classification Using Support Vector Machines

SVM are supervised machine learning techniques that are used for classification and regression. They are classified as generalized linear classifiers. SVM is a pattern recognition and object identification approach that is widely utilized. The goal of the support vector machine is to use an optimization strategy to build a hyperplane as the decision surface to maximize the margin of separation between positive and negative samples. In the feature space, linear functions are commonly utilized as a separating hyperplane. Many kernel functions like polynomial and radial basis functions are employed to improve performance. A polynomial function is used in this research.

The system in the proposed work starts with a training sample $\{(x_i, y_i)\}_{i=1}^{N}$, where x_i is the training vector and y_i is its class label. The proposed method aims to obtain the separating hyperplane with ideal weight vector w and the bias b such that

$$\begin{aligned} & y_i\left(w^T\varphi(x_i)+b\right)\geq 1-\xi_i, \quad \forall_i \\ & \xi_i \geq 0, \quad \forall_i \end{aligned} \tag{3.6}$$

with w and the slack variables ξ_i minimizing the cost function given below

$$\varphi(w, \xi_i) = \frac{1}{2} w^T w + C \sum_{i=1}^{N} \xi_i \tag{3.7}$$

where the slack variables ξ_i indicates data error measures, C represents the value assigned to the errors, and $\phi()$ is a kernel mapping that transfers the data into a higher-dimensional feature space.

4 Experimental Results and Analysis

In this segment, we explain the clothing datasets utilized for validating the cloth classification system. After that, we discussed the experimental settings and analyzed the results, which are represented using a confusion matrix.

4.1 Dataset

We have narrowed down the focus of this paper to be fashion data classification. We have built a clothing dataset of 1000 images where 500 images belong to male dress codes and 500 female dress codes with 25 classes: coat, poncho, blouse, dress, shirt, vest, lingerie, t-shirt, uniform, suit, sweater, jacket, sports sweater. There are around 400 images in the training dataset, and Fig. 8 shows the sample images from our database.

Fig. 8 Sample images from the dataset

4.2 *Experimental Setup and Performance Analysis*

Radial basis function (RBF) kernel is considered the kernel for the SVM classifier utilized in this work. RBF kernel has been utilized as a linear and sigmoid kernel by adding specific elements. As explained in the methodology, three various feature extraction methods are proposed in this work. The workflow is explained in Fig. 9. For each input image, according to the information of pixels, we are segmenting ROI, i.e., dress region in the image. Then images are pre-processed by resizing the image into 128 × 128 for extracting Fourier descriptor. Zernike moments are extracted by first converting the segmented image into a grayscale image. The Freeman chain code is obtained from the binary converted image. We are extracting 203 features from Zernike moments (11), Fourier shape descriptors (128), and Freeman chain code (64). The feature vector is trained and stored in the database. For the input query, the image SVM classifier classifies into a specified dress code.

Table 1 displays our sample dataset with seven classes, out of which training images are 700 and testing images are 245.

4.2.1 Performance Analysis

Training of support vector machine (SVM) combines vectors of all three feature descriptors with linear kernel. We have obtained an accuracy value of 82.36%, as written in Table 3. The graphical representation of the confusion matrix is seen in Table 2, where columns indicate prediction class and rows indicate actual class.

The proposed method is compared with KNN and Naïve Bayes method. And Table 4 shows the performance results.

5 Conclusion

We have suggested a new cloth classification method in this paper. The critical contribution of the proposed method is the usage of shape information for the classification of different outfits. We have employed Zernike moments, Fourier shape descriptors, Freeman chain coding, and support vector machines as recognition methods to reach this goal. The performance analysis of the proposed feature descriptors using shape information considerably increased the recognition rate of clothes with various outfits. Comparing with KNN and Naïve Bayes methods, the proposed method gives the promising results. But still, we want to expand our database in the future and categorize clothing into male and female outfits.

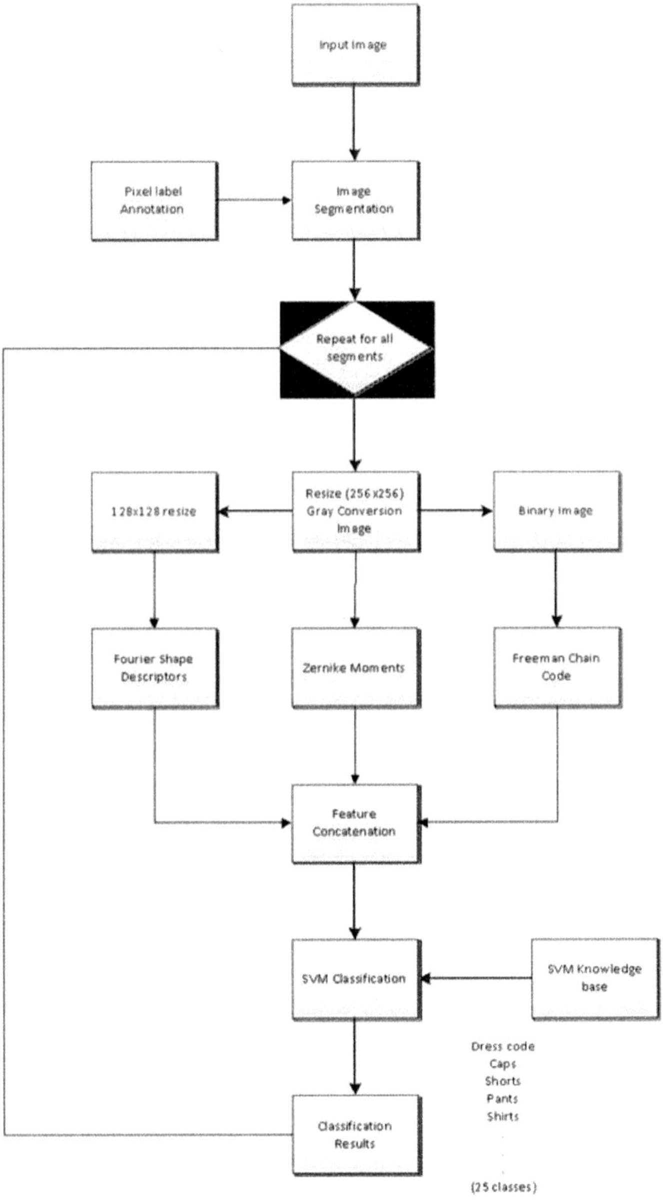

Fig. 9 Workflow of the cloth classification model

Table 1 Seven classes from dataset

	Training set	Test
Jacket	100	35
Shirt	100	35
Suit	100	35
Sweater	100	35
T-shirt	100	35
Polo shirt	100	35
Tank top	100	35

Table 2 Confusion matrix of our proposed

	Jacket	Shirt	Suit	Sweater	T-shirt	Polo shirt	Tank top
Jacket	31	1	0	0	0	1	1
Shirt	0	25	1	2	1	0	0
Suit	2	2	31	1	2	1	0
Sweater	0	3	1	29	0	1	0
T-shirt	1	1	1	2	30	2	0
Polo shirt	1	2	0	1	1	28	0
Tank top	0	1	1	0	1	2	34

Table 3 F-score report of our proposed

	Precision (%)	Recall (%)	F-score (%)
Jacket	82.81	88.72	88.02
Shirt	78.72	73.04	80.20
Suit	87.52	87.72	88.23
Sweater	74.66	77.14	78.04
T-shirt	84.26	81.72	80.05
Polo shirt	85.95	77.71	74.12
Tank top	82.57	82.92	84.15
Overall	82.36	81.28	81.83

Table 4 Precision score comparison of our proposed method with KNN and Naïve Bayes

	Proposed	KNN	Naïve Bayes
Jacket	82.81	76.22	81.32
Shirt	78.72	71.90	67.84
Suit	87.52	75.73	84.72
Sweater	74.66	69.24	76.94
T-shirt	84.26	71.55	77.82
Polo Shirt	85.95	67.32	79.51
Tank top	82.57	76.25	76.32
Overall	82.36	72.60	77.78

References

1. Zhang Y, Liu X, Shi Y, Guo Y, Xu C, Zhang E, Tang J, Fang Z (2017) Fashion evaluation method for clothing recommendation based on weak appearance feature. Sci Program 2017:1–12. https://doi.org/10.1155/2017/8093057
2. Pandit A, Goel K, Jain M, Katre N (2020) A review on clothes matching and recommendation systems based on user attributes. Int J Eng Res Technol (IJERT) (2020)
3. Willimon B, Walker I, Birchfield S (2013) Classification of clothing using midlevel layers. Department of Electrical and Computer Engineering, Clemson University, Clemson, SC 29634, USA
4. Yamaguchi K, Kapoor MH, Ortiz LE, Berg TL (2012) Parsing clothing in fashion photographs. In: Proceedings of IEEE conference on computer vision and pattern recognition. IEEE, pp 3570–3577
5. Liu S, Feng J, Domokos C, Xu H, Huang J, Hu Z, Yan S (2014) Fashion parsing with weak color-category labels. IEEE Trans Multimedia 16(1):253–265
6. Pan R, Gao W, Liu J, Wang H (2011) Automatic recognition of woven fabric pattern based on image processing and bp neural network. J Textile Inst 102(1):19–30
7. Salem YB, Nasri S (2010) Automatic recognition of woven fabrics based on texture and using SVM. SIViP 4(4):429–434
8. Bourdev L, Maji S, Malik J (2011) Describing people: a pose let-based approach to attribute classification. In: Proceedings of the IEEE international conference on computer vision. IEEE, pp 1543–1550
9. Manfredi M, Grana C, Calderara S, Cucchiara R (2014) A complete system for garment segmentation and color classification. Mach Vis Appl 25(4):955–969
10. Hu Z, Yan H, Lin X (2008) Clothing segmentation using foreground and background estimation based on the constrained Delaunay triangulation. Pattern Recogn 41(5):1581–1592
11. Weber M, Bauml M, Stiefelhagen R (2011) Part-based clothing segmentation for person retrieval. In: Advanced video and signal-based surveillance (AVSS). IEEE, pp 361–366
12. Chen H, Gallagher A, Girod B (2012) Describing clothing by semantic attributes. In: Fitzgibbon A, Lazebnik S, Perona P, Sato Y, Schmid C (eds) ECCV 2012, Part III, vol 7574. LNCS. Springer, Heidelberg, pp 609–623
13. Lorenzo-Navarro J, Castillo M, Ramon E, Freire D (2014) Evaluation of LBP and HOG descriptors for clothing attribute description. In: Distante C, Battiato S, Cavallaro A (eds) VRAM 2014, vol 8811. LNCS. Springer, Heidelberg, pp 53–65
14. Semantic Image segmentation [website]. Date of access: 24 May 2018. https://innovation.ox.ac.uk/licence-details/semantic-image-segmentation/
15. Zhou B, Khosla A, Lapedriza A, Oliva A, Torralba A (2016) I am learning in-depth features for discriminative localization. In: 2016 IEEE conference on computer vision and pattern recognition (CVPR). IEEE, pp 2921–2929
16. Senthilkumaran N, Thimmiaraja J (2014) An illustrative analysis of mathematical morphology operations for MRI brain images. Int J Comput Sci Inf Technol 5:2684–2688
17. Costa LF, Cesar RM (2001) Shape analysis and classification theory and practice. CRC Press LLC, Florida
18. Bane S, Pawar DR (2014) Survey on feature extraction methods in object recognition. Int J Comput Sci Inf Technol
19. Huang Z, Edith JL (2010) Analysis of Hu's moment invariants on image scaling and rotation. Edith Cowan University

Chapter 4
Cardamom Quality Evaluation Employing Electronic Nose

Madhurima Ghosh, Devdulal Ghosh, and Nabarun Bhattacharyya

1 Introduction

Cardamom is ranked third in terms of price, followed by saffron and vanilla. Small cardamom (*Elettaria cardamomum* Maton) is grown mainly in India, Guatemala, Sri Lanka, Nepal, and also found in Tanzania, Indonesia, Vietnam, Thailand, Papua New Guinea, and El Salvador, and is exported worldwide specially to Middle Eastern countries. Cardamom extracts and cardamom essential oil are used as a flavoring agent and have numerous potential therapeutic uses.

Hence, rapid and effective method for quality evaluation of cardamom is the need of the hour. The quality varies with the percentage of essential oil extracted and its constituent chemicals. The major constituents of essential oil are 1,8-cineole and alpha-terpinyl acetate. The most popular practice of aroma estimation of cardamom, and thereby price determination, involves quality assessment by human experts based on physical characteristics, such as freshness, shape, size, color, and aroma. However, this method is subject to biasness of the expert, error prone, and non-repetitive. In order to provide accurate estimation of quality, an alternative approach involves various chemical tests employing analytical instruments, such as gas chromatography (GC) and gas chromatography coupled with a mass spectrometer (GC/MS) [1]. The results of this method confirm that the proportion of various compounds of

M. Ghosh (✉)
St. Thomas' College of Engineering & Technology (Under Maulana Abul Kalam Azad University of Technology), Kolkata, India
e-mail: mghosh.madhurima@gmail.com

D. Ghosh · N. Bhattacharyya
Centre for Development of Advanced Computing, Kolkata, India
e-mail: devdulal.ghosh@cdac.in

N. Bhattacharyya
e-mail: nabarun.bhattacharya@cdac.in

L. Mandal et al. (eds.), *Proceedings of International Conference on Computational Intelligence, Data Science and Cloud Computing*, Algorithms for Intelligent Systems,
https://doi.org/10.1007/978-981-19-1657-1_4

the oil extracted from cardamom differs from each other among different varieties of cardamom [2, 3]. However, this process is time-consuming and requires expensive laboratory equipment and skilled personnel to operate these instruments.

Electronic nose (e-Nose) [4] is one of the prospective solutions for rapid quality estimation of various food products and beverages, such as meat [5], tea [6], coffee, dairy products, alcoholic beverages, oil [7], vinegar, cocoa beans, and many more. This method provides considerable level of accuracy, removes human subjectivities and errors, and is repetitive. In an e-Nose, an array of sensors is embedded in the device to sense and generate an "aroma fingerprint" for each training sample. Since most of these sensors are specific to a class of chemicals and have cross sensitivities to other classes too, it is very difficult to quantify the amount of the constituent chemicals present in the data samples. Instead, the approach employed in e-Noses is to create an "aroma fingerprint" based on the sensor values for each type of food sample. Once the training is complete, the device tries to match the aroma fingerprints of unknown samples under test to that of the pre-trained samples resulting in evaluation of the quality of the unknown samples. In some cases, e-Nose is also used for several disease detection, such as asthma [8], cancer, tuberculosis, cystic fibrosis [9], or even monitoring the health of astronauts in space. Although quality estimation of spices using e-Nose is a relatively new domain of research, some preliminary work on sensor selection and aroma fingerprint generation of cardamom are done by Leela et al. [10] and Ghosh et al. [11].

We have adopted a novel approach by using a handheld electronic nose (HEN) [12] developed by Centre for Development of Advanced Computing (C-DAC), Kolkata, for the quality evaluation of cardamom by estimating the oil yield percentage and measure of 1,8-cineole and alpha-terpinyl acetate in the samples through Partial Least Square (PLS) regression algorithm. In this study, we have dealt with thirteen different samples separately for cardamom with husk and without husk.

2 Materials and Method

2.1 Cardamom Extraction Process

ICAR-Indian Institute of Spices Research (IISR), Kozhikode, contributed in this study through extraction of essential oil [10] and measurement of the percentage of major constituent chemicals, namely 1,8-cineole and alpha-terpinyl acetate of thirteen different samples of cardamom. 20 g of each sample including the husk were grinded and then hydrodistilled for three hours in Clevenger apparatus to extract the essential oil. The extracted essential oil was collected, dried over anhydrous sodium sulfate, and stored in a refrigerator until the analysis was carried out.

2.2 *Chemical Analysis of Samples*

Gas chromatography (GC) [11] examination was performed on a Shimadzu (GC-2010) Gas chromatograph fitted to FID indicator and RTX—5 column (30 m × 0.25 mm, film thickness 0.25 μm). Nitrogen was utilized as the transporter gas at a flow rate of 1.0 ml/min, and the oven was programmed as at 60 °C up to 200 °C at the rate of 3.0 °C/min, again up to 220 °C at the rate of 5.0 °C/min, at which the column was kept up for 5 min. 0.1 μl sample was infused. The injection port temperature was kept up at 250 °C, and the detector temperature was 250 °C. 1:40 was the split ratio. The percentage composition component of the oil was dictated by area normalization.

Afterward, the analysis of essential oil was done using a Shimadzu GC-2010 Gas chromatograph furnished with QP 2010 mass spectrometer (GC/MS). RTX—5 column (30 m × 0.25 mm, film thickness 0.25 μm) covered with polyethylene glycol was utilized. Helium was utilized as the transporter gas at a flow rate of 1.0 ml/min. The stove was modified at 60 °C for 5 min and afterward expanded to 110 °C at 5.0 °C/min, then, at that point, up to 200 °C at the pace of 3.0 °C/min, again up to 220 °C at 5.0 °C/min, at which the column was kept up for 5 min. 0.1 μl sample was infused. The injection port temperature was kept up at 260 °C, and the detector temperature was 250 °C. 1:40 was the split ratio, and ionization voltage was kept up at 70 eV. The retention indices were determined comparative with C8—C20 n-alkanes. Table 1 shows the chemical analysis report for thirteen different samples carried out by IISR.

Table 1 Oil yield and major constituent chemical percentages of different cardamom samples obtained through chemical analysis

Sample	Name	Oil yield (%)	1,8-cineole (%)	Alpha-terpinyl acetate (%)
1	ELA	7	21	44
2	THIRU	5.5	22	44
3	GG1	7	18	48
4	GG2	6	22	42
5	WONDE	6.5	26	46
6	CON1	4	25	48
7	CON2	4.5	30	42
8	CON3	4	27	47
9	CON4	4.5	28	44
10	CON5	4	30	42
11	CON6	5	28	46
12	CON7	4	30	43
13	CON8	4.5	27	45

Chemical analysis is done by IISR, Kozhikode

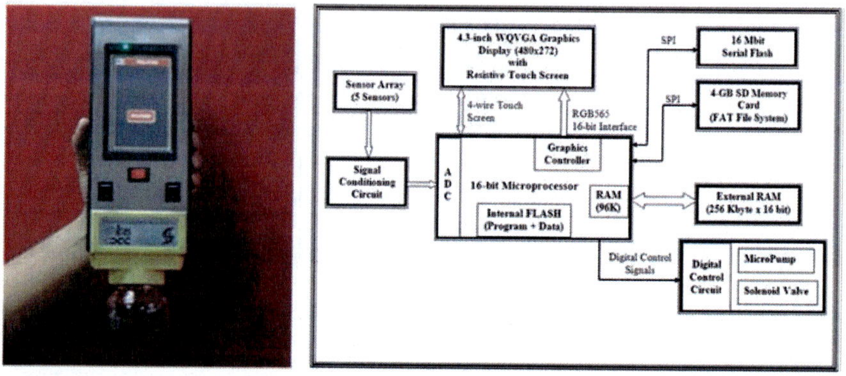

Fig. 1 Block diagram of Handheld Electronic Nose (HEN) architecture

2.3 *Architecture of Handheld Electronic Nose (HEN)*

This study has been conducted employing a Handheld e-Nose (HEN) [13] developed and patented by C-DAC, Kolkata. HEN is a true handheld e-Nose, built on embedded electronics, and is suitable for field operation. The device is battery operated with integrated odor-handling unit. The sample holder can easily be fitted onto the device during testing of samples. A touchscreen graphics display is provided as user interface and displaying of results. Combination of a micro-pump and micro valve serves the sample odor VOC excitation and thereafter delivery to the sensor array. The sensor array is built on five metal oxide semiconductor (MOS) sensors [14], namely TGS 2602, TGS 2620, TGS 816, TGS 823, and TGS 832. Captured sensor data and results are stored in a SD Card in FAT32 file system.

Figure 1 shows the hardware block diagram of HEN. A 16-bit PIC microprocessor is the heart and brain of HEN. After signal conditioning, the captured sensor data is fed to the in-built analog to digital converter of the processor for processing. The 4.3-inch touch screen and graphics display are interfaced to the processor through 4-wire and 16-bit RGB565 lines, respectively. SD memory card and serial Flash memory are interfaced through SPI lines. Control of micro-pump and micro valve is done through digital I/O lines of the processor. Software for HEN is developed on embedded C and graphics portion using Microchip graphics library.

2.4 *Data Collection*

For three consecutive days, the samples were tested on HEN and the corresponding values of the five different sensors were stored. 10 g of each sample with husk was first placed in the sample holder of HEN, and the sensor data was recorded. Then, the husk from these samples were removed, their corresponding weights were measured, and the earlier stated process was repeated. The temperature and humidity recorded on 6th and 7th May were around 30 °C and 57%, respectively. However, on 8th May, the temperature dropped down to 22 °C with heavy rainfall and the humidity increased to about 97%. The responses of MOS sensors are prone to variations in ambient temperature and humidity, and thus, the data taken on 8th May had shown abrupt changes as compared to that of 6th and 7th May. Hence, we discarded the

Table 2 Details about the recorded dataset

Date	Sample	Name	Type	Weight (g)
6th May	1	ELA	With husk	10
			Without husk	7.3
	2	THIRU	With husk	10
			Without husk	7.8
	3	GG1	With husk	10
			With husk	10
			Without husk	7.4
	4	GG2	With husk	10
			Without husk	7.4
			Without husk	7.4
	5	WONDE	With husk	10
			Without husk	7.5
			Without husk	7.5
			Without husk	7.5
	6	CON1	With husk	10
			Without husk	6.3
	7	CON2	With husk	10
			Without husk	7
	8	CON3	With husk	10
			Without husk	6.7
	9	CON4	With husk	10
			Without husk	6.3
7th May	10	CON5	With husk	10
			Without husk	6.5
	11	CON6	With husk	10
			Without husk	7
	12	CON7	With husk	10
			Without husk	6.9
	13	CON8	With husk	10
			Without husk	6.3

records taken on 8th May for training the device and analysis of results. Table 2 displays the details of the data taken for analysis.

2.5 Analysis of Sensor Data

Figure 3 shows the graph of 600 values recorded from the time of opening the valves and until the saturation point of all the five different sensors are reached, for the sample, CON7 (Sample 12) with husk recorded on 6th May. The graphs of all the other samples are similar to Fig. 2, but with variable slopes and different saturation points. However, we have considered the last 20 values of each sample for analysis where the readings are almost stable. The sensor readings are then correlated to group the samples into distinct clusters and for prediction of the oil yield and major constituent chemical percentages responsible for the flavor of cardamom.

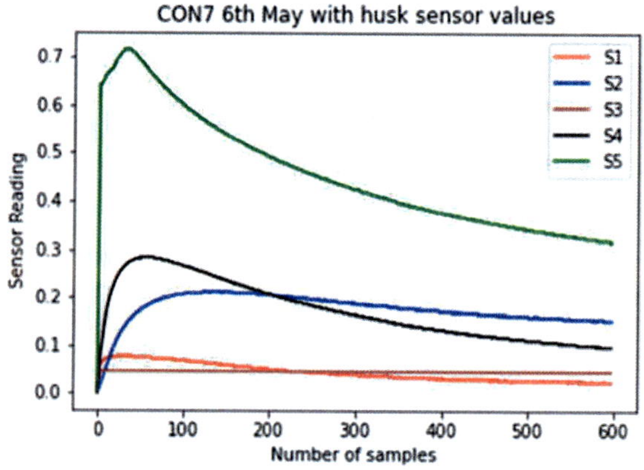

Fig. 2 Five sensor readings of HEN for CON7 (Sample 12)

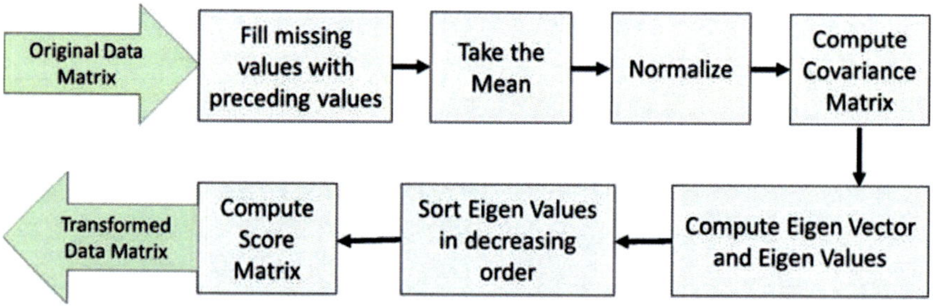

Fig. 3 Algorithm for principal component analysis

2.6 Clustering Algorithms

Clustering refers to division of data points into various groups of similar data or values. Initially, the data for all the samples taken on both the days were fed to two popular clustering algorithms, namely Principal Component Analysis (PCA) and Density-Based Spatial Clustering of Applications with Noise (DB-SCAN). The system is able to segregate the samples into different clusters with considerable accuracy. This indicates that the oil yield percentage and composition of constituent chemicals of essential oil varies significantly and the data is fit for classification of samples into different categories.

2.6.1 Principal Component Analysis (PCA)

The dataset generated from HEN is difficult to interpret and visualize as it has dimensionality of five (five parameters). Hence, the data is visualized in 2D using Principal Component Analysis (PCA) [15] with minimum data loss to check whether the samples could be clustered into distinct groups. The algorithm of PCA is given in Fig. 3.

2.6.2 Density-Based Spatial Clustering of Applications with Noise (DB-SCAN)

DB-SCAN (Density-Based Spatial Clustering of Applications with Noise) [16] is able to find arbitrary shaped clusters and also discards noise as outliers. Moreover, it does not need to provide the total number of clusters the dataset needs to be grouped to and hence, even in the absence of domain knowledge, it would provide good clustering results. However, the distance that specifies neighborhoods and the minimum number of data points in each cluster much be specified for the algorithm to execute. The algorithm for DB-SCAN is given in Fig. 4.

Fig. 4 Algorithm for DB-SCAN

- Label points as core, border and noise
- Eliminate noise points
- For every core point p that has not been assigned to a clusters
  - –Create a new cluster with the point p and all the points that are density-connected to p.
- Assign border points to the cluster of the closest core point.

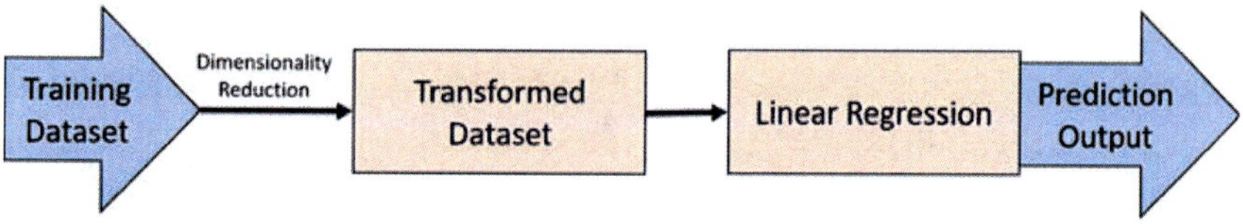

Fig. 5 Algorithm for partial least square regression

2.7 *Regression Algorithm*

Regression models are used to establish relationship between the independent variables, i.e., the sensor values, and the dependent variables, i.e., the oil yield, and measures of 1,8-cineole and alpha-terpinyl acetate in this use case of the training dataset. The established relationship is then utilized to predict the target(s), i.e., the oil yield, 1,8-cineole, and alpha-terpinyl acetate percentages for unknown samples. We have used Partial Least Square (PLS) Regression algorithm for analyzing the data.

2.7.1 Partial Least Squares Regression (PLS)

Partial Least Square (PLS) [17] is an optimal regression algorithm based on covariance. PLS applies dimension reduction of the samples (similar to PCA) before applying a linear regression algorithm on the transformed data. In initial study, it was observed that the data is highly correlated, and hence, PLS is chosen for prediction of the target(s). It also gives accurate results in spite of noisy data. Figure 5 depicts the algorithm for PLS Regression.

3 Result and Analysis

Table 3 depicts the amount of data explained by principal components 1 and 2 for different datasets in order to reduce the dimension using PCA. It is observed that there is significant amount of data loss (approximately greater than 10%) when the

Table 3 Percentage of variance explained by principal components

Date	Type	PC1 (%)	PC2 (%)	Data Loss (%)
6th May	With husk	60.5	29.1	10.4
	Without husk	61.2	20.9	17.9
7th May	With husk	79.4	18.9	1.7
	Without husk	66.6	32.5	0.9

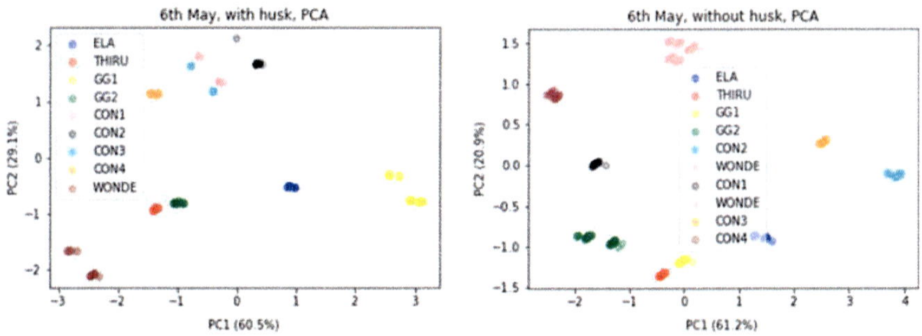

Fig. 6 Clustering using PCA for 6th May with and without husk

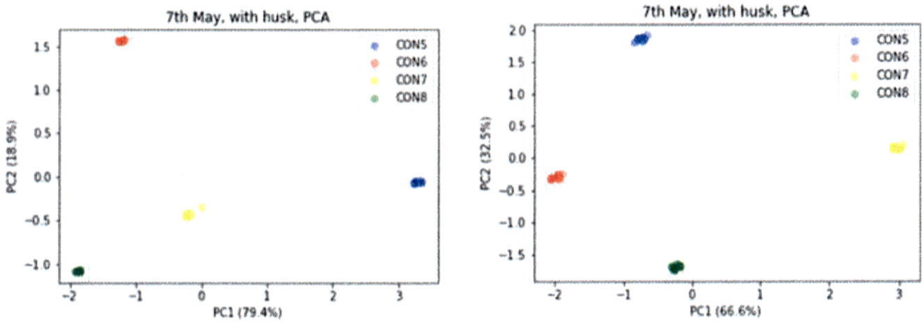

Fig. 7 Clustering using PCA for 7th May with and without husk

number of samples taken under consideration is more than 5. However, this amount of data loss did not affect in proper clustering of the data.

Figure 6 and 7 show the results of clustering using PCA on the samples on 6th and 7th May separately. The algorithm is able to properly cluster the data with considerable accuracy except certain imprecision in the samples CON3 (Sample 8) and CON4 (Sample 9) with husk on 6th May.

Figures 8 and 9 compare the sample number and cluster number (result of DB-SCAN clustering) for the data recorded on 6th May and 7th May. Sample 3 (GG1), 4 (GG2), and 5 (WONDE) were recorded multiple times for experimental purpose. DB-SCAN algorithm could accurately cluster the datasets for with and without husk cardamom samples, except the dataset of with husk on 6th May i.e., the algorithm misplaced Sample 8 (CON3) into Sample 6 (CON1) cluster.

The device was trained using PLS Regression model, and Figs. 10, 11, 12, and 13 show the predictions of oil yield and 1,8-cineole percentages, respectively, by the model for unknown samples.

Table 4 shows the results of PLS regression for estimation (average of 6th and 7th May datasets) of oil yield, 1,8-cineole, and alpha-terpinyl acetate percentages in cardamom samples.

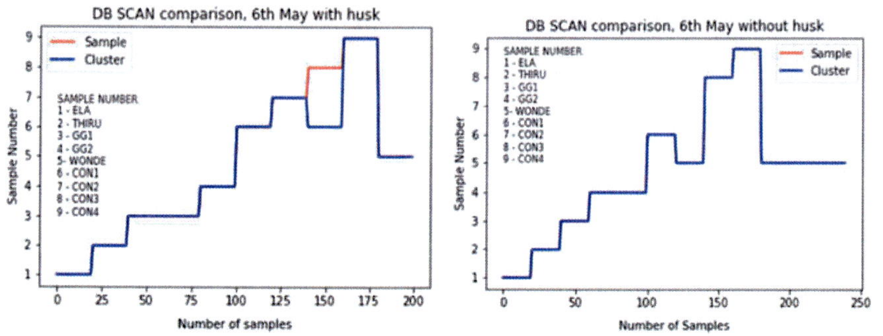

Fig. 8 Comparison of cluster number (result of DB-SCAN) with sample number on 6th May with and without husk[1]

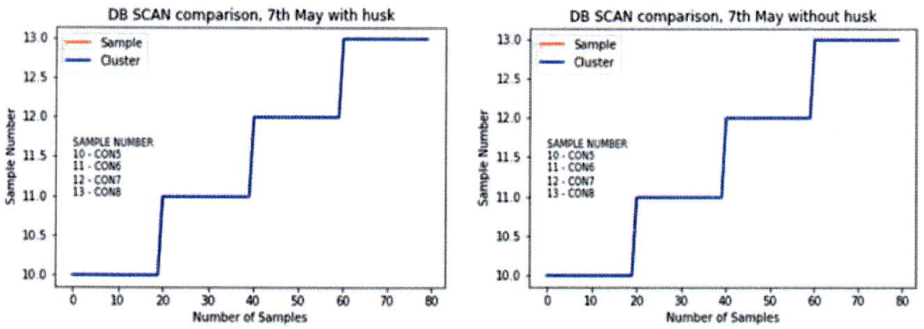

Fig. 9 Comparison of cluster number (result of DB-SCAN) with sample number on 7th May

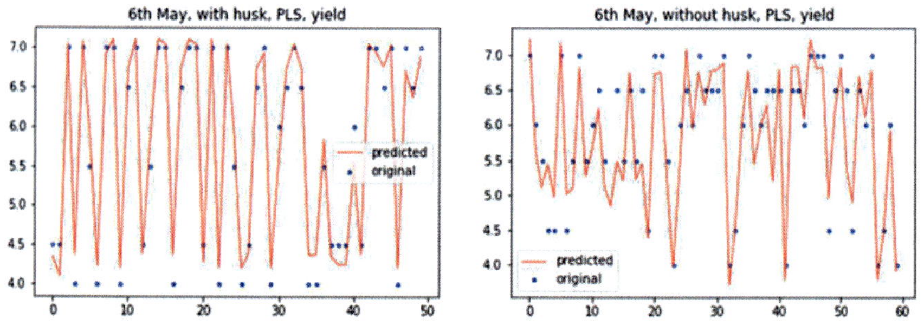

Fig. 10 Prediction of oil yield on 6th May with husk and without husk

Overall, the samples with husk gave better accuracy and precision as compared to the samples without husk. This confirms the fact that the husk of cardamom contains certain amounts of aroma constituents too.

---

[1] Sample 3 (GG1), 4 (GG2), and 5 (WONDE) were recorded multiple times for experimental purpose.

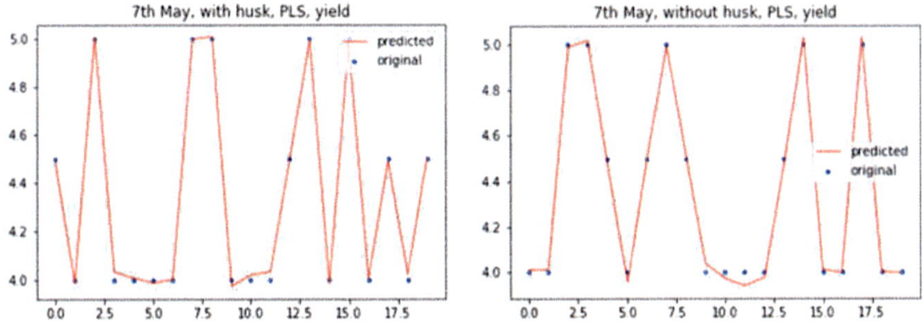

Fig. 11 Prediction of oil yield acetate on 7th May with husk and without husk

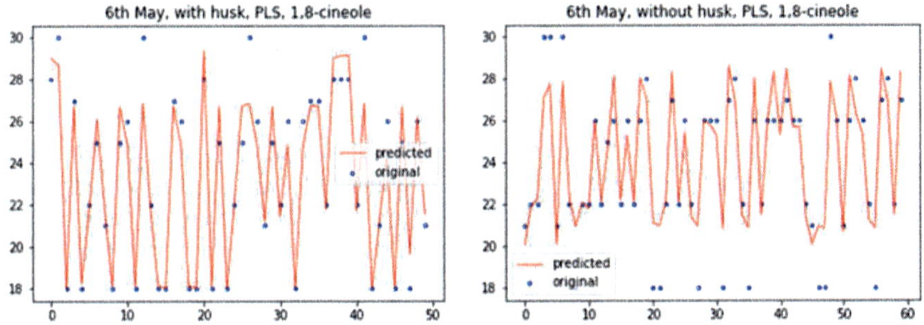

Fig. 12 Prediction of 1,8-cineole on 6th May with husk and without husk

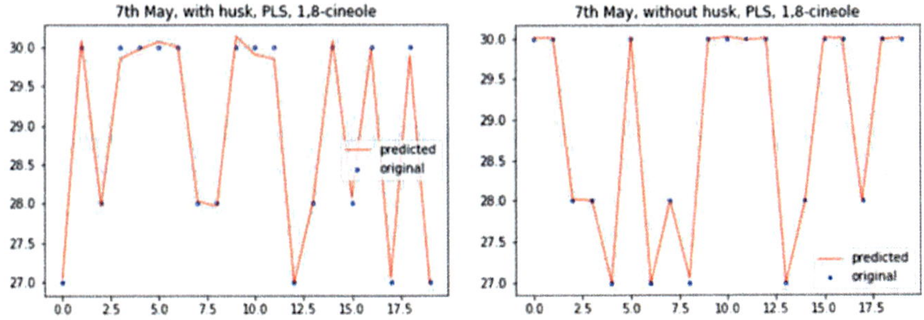

Fig. 13 Prediction of 1,8-cineole on 7th May with husk and without husk

The oil yield percentage prediction gave around 98% accuracy while 1,8-cineole predictions are almost 95% accurate for samples with husk. However, the result for estimation of alpha-terpinyl acetate is not at all satisfactory.

Table 4 Estimation of oil yield, 1,8-cineole, and alpha-terpinyl acetate in cardamom samples by PLS Regression

Prediction	Type	Mean squared error (MSE)	Accuracy (%)
Oil Yield (%)	With husk	0.026	98.2
	Without husk	0.104	89.9
1,8-cineole (%)	With husk	0.688	95.7
	Without husk	1.228	90.2
Alpha-terpinyl acetate (%)	With husk	1.426	69.2
	Without husk	2.494	44.2

All results are average of datasets recorded on 6th and 7th May

4 Conclusion

Our novel approach of cardamom quality estimation using HEN gave satisfactory results for oil yield and one of the major constituent chemicals of essential oil, 1,8-cineole, and is fit for use in rapid evaluation of cardamom quality at affordable cost. However, the estimate for another constituent chemical of essential oil, alpha-terpinyl acetate, is not reasonable.

The MOS sensors used for HEN in this study were specifically selected for quality analysis of tea and were not tuned for cardamom. Sensor array tuned for cardamom aroma constituent chemicals is expected to give more accurate results. A hybrid array of sensors including MOS, Quartz Crystal Microbalance (QCM), Conducting polymer, Electrochemical, etc., which may give optimum results for cardamom is a topic of our further research.

Moreover, the increased size of training dataset might also yield better precision and accuracy. Several other discriminant analysis algorithms and neural network-based regression models can be employed on the datasets to choose the optimum algorithm for our use case. Since the MOS sensors are prone to temperature and humidity change, the parameters of ambient temperature, humidity, and other environmental factors can be added in the feature list of the datasets to generalize the model and avoid training the system multiple times for various environmental conditions. The potential of e-Nose in evaluation of quality of cardamom extract and essential oil, as found in this study, can be explored further to make it suitable for commercial use.

Acknowledgements The authors are thankful to the scientists of ICAR—Indian Institute of Spices Research (IISR), Kozhikode, for providing labeled samples and chemical analysis of the samples.

References

1. Tambe E, Gotmare S (2019) Chemical characterization of three cardamom oils (*Elettaria cardamomum*) By GCMS. World J Pharm Pharmaceut Sci 08:876–885. https://doi.org/10.20959/wjpps201911-14879
2. Ashokkumar K, Murugan M, Dhanya M, Raj S, Kamaraj D (2019) Phytochemical variations among four distinct varieties of Indian cardamom *Elettaria cardamomum* (L.) Maton. Nat Prod Res 34:1919–1922. https://doi.org/10.1080/14786419.2018.1561687
3. Chempakam B, Sindhu S (2008) Cardamom. In: Parthasarathy VA, Chempakam B, John Zachariah T (eds) Chemistry of spices. CABI.com, pp 41–58
4. Karakaya D, Ulucan O, Turkan M (2020) Electronic nose and its applications: a survey. Int J Autom Comput 17:179–209. https://doi.org/10.1007/s11633-019-1212-9
5. Vestergaard JS, Martens M, Turkki P (2007) Application of an electronic nose system for prediction of sensory quality changes of a meat product (pizza topping) during storage. LWT-Food Sci Technol 40(6):1095–1101. https://doi.org/10.1016/j.lwt.2006.06.008
6. Das A, Ghosh A, Ghosh TK, Ray H (2011) Portable electronic nose for tea aroma detection. Int J Comput Appl 6(1):39–43
7. Majchrzak T, Wojnowski W, Dymerski T, Gebicki J, Namiesnik J (2018) Electronic noses in classification and quality control of edible oils: a review. Food Chem 246:192–201. https://doi.org/10.1016/j.foodchem.2017.11.013
8. Dragonieri S, Schot R, Mertens BJ, Le Cessie S, Gauw SA, Spanevello A, Resta O, Willard NP, Vink TJ, Rabe KF, Bel EH, Sterk PJ (2007) An electronic nose in the discrimination of patients with asthma and controls. J Allergy Clin Immunol 120(4):856–862. https://doi.org/10.1016/j.jaci.2007.05.043
9. Robroeks CMHHT, van Berkel JJBN, Dallinga JW, Jobsis Q, Zimmermann LJI, Hendriks HJE, Wouters MFM, van der Grinten CPM, van de Kant KDG, van Schooten FJ, Dompeling E (2010) Metabolomics of volatile organic compounds in cystic fibrosis patients and controls. Pediatr Res 68(1):75–80. https://doi.org/10.1203/00006450-201011001-00143
10. Leela NK, Muneeb AM, Mukherjee S, Ghosh D, Bhattacharya N (2017) Essential oil content of cardamom (*Elettaria cardamomum* Maton) by hand-held electronic nose. J Spices Aromatic Crops 131. https://doi.org/10.25081/josac.2017.v26.i2.882
11. Ghosh D et al (2012) Exploratory study on aroma profile of cardamom by GC-MS and electronic nose. In: 2012 sixth international conference on sensing technology (ICST), pp 399–403. https://doi.org/10.1109/ICSensT.2012.6461708
12. Bhattacharyya N, Bandyopadhyay R, Bhuyan M, Tudu B, Ghosh D, Jana A (2008) Electronic nose for Black Tea Classification and correlation of measurement with "Tea Taster" Marks. IEEE Trans Instrum Measur 57(7)
13. Ghosh A, Ray H, Ghosh T, Das A, Bhattacharyya N (2014) Generic handheld e-nose platform for quality assessment of agricultural produces and biomedical applications. Chem Eng Trans 40:259–264. https://doi.org/10.3303/CET1440044
14. Nabarun B, Bipan T, Arun J, Devdulal G, Rajib B, Manabendra B (2008) Preemptive identification of optimum fermentation time for black tea using electronic nose. Sens Actuators B 131(1):110–116
15. Sehgal S, Singh H, Agarwal M, Bhasker V, Shantanu (2014) Data analysis using principal component analysis. In: 2014 international conference on medical imaging, m-health and emerging communication systems (MedCom), pp 45–48
16. Ester M, Kriegel H-P, Sander J, Xu X (1996) A density-based algorithm for discovering clusters in large spatial databases with noise. In: Proceedings of the second international conference on knowledge discovery and data mining (KDD'96). AAAI Press, pp 226–231
17. Larbi M, Trinchero R, Canavero FG, Besnier P, Swaminathan M (2020) Analysis of parameter variability in integrated devices by partial least squares regression. In: 2020 IEEE 24th workshop on signal and power integrity (SPI), pp 1–4. https://doi.org/10.1109/SPI48784.2020.9218175

Chapter 5
COVID-DeepNet: Deep Convolutional Neural Network Architecture Designed for Early Prognosis of COVID-19 Using Post-anterior View of Chest X-Rays

Debanjan Ghosh, Farhan Hai Khan, Anish Banerjee, Debojyoti Saha, Soumyadip Sarkar, Shivam Kumar, and Malay Gangopadhyaya

1 Introduction

The infamous COVID-19 pandemic emerged from the city of Wuhan in Hubei province, China [1], with over 37.3 million confirmed cases all around the globe as of 12 October 2020 [2]. The disease adversely affects the lungs and causes potentially deadly respiratory syndromes [3]. The virus directly affects and disrupts the ACE2 receptors of the type II alveolar cells in the lungs, thereby creating bilateral nodular and multiple peripheral ground-glass opacities/lesions and consolidations [4]. Expert Radiologists claim that close inspection of CXR Images can help in the early prognosis of the disease and isolation of potential infected patients for contamination containment [5]. Prachi et al. [5] explained that in radiologic terminology, "ground glass" meant the presence of hazy lung opacities that show up on the CXR imaging which are not dense enough to obscure any underlying pulmonary vessels or bronchial walls, while consolidation, on the contrary, refers to dense opacities obscuring vessels and bronchial walls [5].

The RT-PCR is the diagnostic test for COVID-19 [6]. This test can often take hours or days to get processed due to limited availability of resources and testing kits [7, 8]. Since ground-glass opacifications (GGOs) are frequent in COVID-19, chest CT scans are preferred over chest radiographs by radiologists, because chest

D. Ghosh · D. Saha · S. Kumar · M. Gangopadhyaya
Department of Electronics and Communications Engineering, Institute of Engineering & Management, Kolkata, India

F. Hai Khan (✉) · S. Sarkar
Department of Electrical Engineering, Institute of Engineering & Management, Kolkata, India
e-mail: khanfarhanpro@gmail.com

A. Banerjee
Department of Information Technology, Institute of Engineering & Management, Kolkata, India

L. Mandal et al. (eds.), *Proceedings of International Conference on Computational Intelligence, Data Science and Cloud Computing*, Algorithms for Intelligent Systems,
https://doi.org/10.1007/978-981-19-1657-1_5

radiographs may possess limited sensitivity in detecting early changes within the lungs [4, 5]. However, an X-ray being inexpensive and readily available [9] at small and rural hospitals [10] has prevalent usage as a primary diagnostic test everywhere and so it has been considered as the input criterion for the prediction [11] of the novel coronavirus in this study. Lin Lou et al. [21] also proposed that a high number of such opacities increased the probability of a patient suffering from COVID-19.

2 Previous Works

Some diagnostic tests performed several days after decontamination of the COVID-19 patients reveal that the virus is still present and had passed undetected prior to their release [20]. A well-trained deep learning model can detect and contradict this analogy on parameters which the human eye cannot comprehend.

Many recent studies were proposed for the automated multiclass detection using unprocessed Chest X-ray radiology images. Tulin et al. [12] obtained an accuracy of 87.02% using Dark COVID-Net, the COVID-Net model proposed by Wang and Wong [13], obtained 92.4% accuracy in identifying among COVID-19, normal, and pneumonia (non-COVID). The deep learning model developed by Loannis et al. [14], using Transfer Learning with VGG-19, achieved an accuracy of 93.48%. Also, several recent studies emerged on multiclass COVID detection that employs various deep learning models, but with CT images [15–19].

However, most of the already present research and studies neglect the consideration of medical noise present in the CXR images, leading to biased models and overfitting [22, 23], hence defeating the idea of essential generalization in the field of medical imaging detection. Hence, the proposed methodology involves the use of various CXR image enhancement techniques [24] before actually feeding the input to the classifier. Proper Data Augmentation [24] techniques have been applied for balancing the dataset ratio. Lung Masking prior to classification has been considered on account of non-generalized models [22, 23]. The latest advancement in machine learning and artificial intelligence computation shows positive and reliable results of lung segmentation and bone shadow exclusion methods, and deep learning is implemented for analysis of 2D chest X-rays to aid radiologists in noticing suspicious nodules and lesions in patients suffering from lung cancer [28].

3 Detailed Methodology

For this research, creation of the model was primarily focused on keeping the architecture lightweight and at the same time making it accurate enough for immediate deployment in rural areas with minimal resources. Keeping this configuration in mind, the architecture borrows learning/evaluation data from open access data sources and trains a customized U-Net model which requires fewer computational

Table 1 Dataset displaying the sample image distribution for post-anterior chest X-ray images

S. No.	Dataset name	No. of images	Source	Detailed description
M	**Lung Masking Dataset**	**12,704**		
M1	Chest X-ray Masks and Labels [25–27]	704	Kaggle	800 CXRs 704 Masks
M2	SIIM-ACR Pneumothorax Segmentation data 512	12,000	Kaggle	15,205 CXRs 12,000 Masks
C	**Chest X-Ray Classification Data**	**2905**		
C1	COVID-19 RADIOGRAPHY DATABASE	2686	Kaggle	1341 Normal and 1345 Viral Pneumonia
C2	COVID-19 Image Data Collection	219	GitHub	219 COVID-19 (+)

operations and therefore is faster than standard models for Lung Mask Predictions. The superimposed images are then screened for appropriate ratio of the Masking Area and then sent to the proposed COVID-DeepNet Model, which performs Depthwise Seperable Convolution Operations over the images to classify/predict an infected patient as COVID-19, Viral Pneumonia, or Normal.

3.1 Dataset Creation and Preparation

There are two amalgamated datasets used for the two different models, viz., the customized U-Net (**M**) and COVID-DeepNet (**C**) Models, which are summarized in Table 1. Dataset **M**, The Lung Masking Data, is used for Training/Evaluating the lightweight U-Net Architecture and performs Lung Mask Detection, whereas Dataset **C**, The Chest X-Ray Classification Dataset, is used Training/Evaluating the proposed COVID-DeepNet Model, for prediction of infected patients (Fig. 1).

3.2 Solving Data Quality Issues: Augmentation, Transfer Learning, and Data Cleaning

However, as mentioned by Maguolo et al. [22] there exist a lot of models that inaccurately classify the inputs across the Internet. In a study that they conducted, the chest was almost completely blacked out in the CXR images and the test models were heavily dependent upon the dataset. This study attempts to address all of these biased problems effectively. The datasets are limited as compared to other diseases and the CXR images are too few to fit a complex model leading to a shortage of data for proper and unbiased training. The solution to this issue was transfer learning:

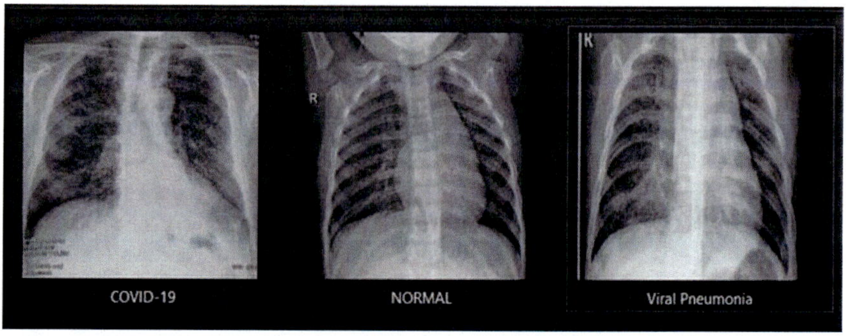

Fig. 1 Sample image overview of the assembled dataset: Dataset Classes—COVID-19, Normal, and Viral Pneumonia

using pre-trained models, training network heads, and then fine-tuning the model for accurate results and faster convergence. Proper Data Augmentation Methodologies were employed to avoid any such potential biases due to limited datasets. The techniques we used were random rotation, horizontal flipping, and CLAHE. These not only balance the data but also are a part of enhanced image pre-processing [24]. A detailed probability distribution with properties of each method is defined in Fig. 3 (Fig. 2).

As hospitals are in complete turmoil during this crisis situation, there are lots of CXRs which are hastily obtained and therefore contain a lot of medical noise (Fig. 5). For this reason, since the novel coronavirus primarily affects the lungs and does not

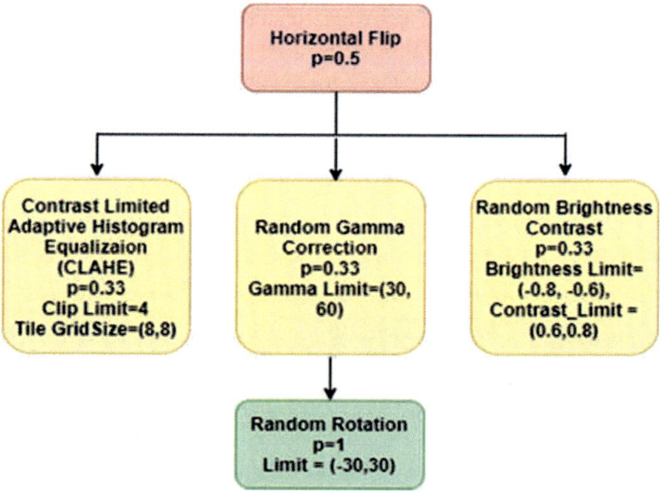

Fig. 2 Data augmentation techniques with descriptive properties: applying horizontal flipping to half the data, CLAHE, GC, and brightness contrast adjustments are optional and any one of them is applied, followed by a random rotation

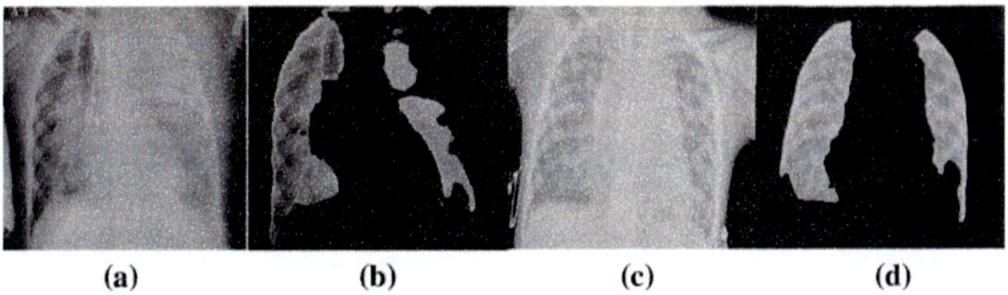

Fig. 3 Noise in medical imaging—lack of clarity, distortions: In rare situations like the above images, there is often a lot of noise in the acquired Data. Figures highlight **a** where right lung is almost not visible in the input image, **b** the corresponding masked output by our custom U-Net is also unclear, **c** the input image has bad lighting conditions, **d** still our custom U-Net architecture has generated the correct masking of the lungs (most visible part)

affect other parts caught up in the CXRs, we used customized lightweight U-Net Lung Segmentation to first identify and separate lungs in the preprocessed CXRs [29] (Fig. 4).

The critical evaluation in Maguolo et al. [22] also served as a source of inspiration for this addition as they experimented on methods used by other individuals by removing lungs from CXRs. Taking inspiration from there, we discovered a novel mechanism to first clean the data itself and remove all such CXR Images which were inappropriate for training. As per our manual observations, after several hyperparameter tunings, for the best results, we define the portion of the masked segment to

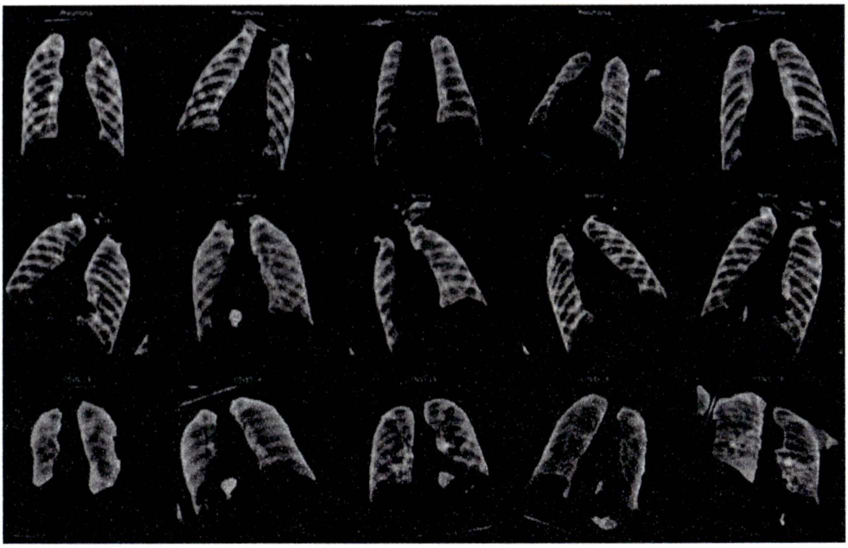

Fig. 4 Sample CXR image overview of the processed dataset: the masked and augmented data is displayed. The infected patients include COVID-19, Normal, and Viral Pneumonia

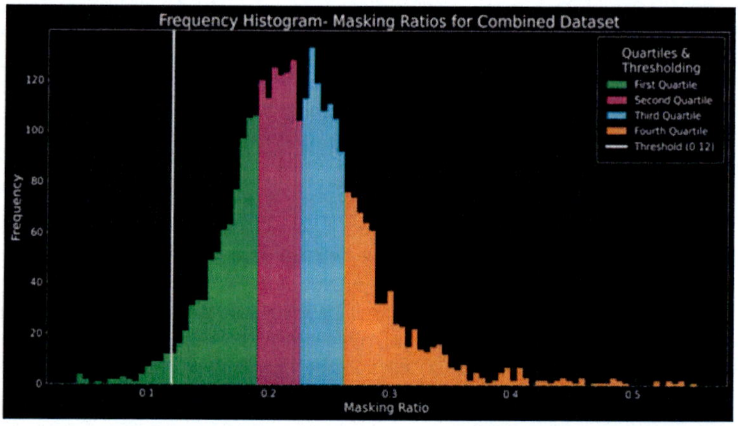

Fig. 5 Masking ratio (μ) frequency distribution plot: ratio of the masking area to the original image area, depicting the clarity of lungs in the image

the total image as the Masking Ratio (μ). Mathematically,

$$\mu_X = A_X / T_X \tag{1}$$

μ_X = Masking Ratio of Image X, A_X = Area of Masked Lung Portion in Image X, T_X = Total Area of Image X.

All input images that had an μ_X **≥ 0.12** (threshold) were accepted as this was a suitable value obtained by viewing the frequency distribution for μ. The Threshold is chosen such that (i) the low-quality images that cannot be processed by the U-net are discarded for training and (ii) most of the dataset is retained at the same time. Keeping both properties in mind a low threshold (0.12) is taken which lies in the initial part of the First Quartile itself and hence causes low data loss. This was used as one of the essential filtration criteria for low resolution and distorted images from the datasets, as most of the CXR images also contain medical noise and hence proper lung mask detection is not obtained (Fig. 6).

This masked image is then sent to the proposed COVID-DeepNet network which ultimately performs the predictive classification. The proposed COVID-DeepNet model based on a depthwise separable convolutional outlook. This structure of increasing the neurons layer by layer helps to increase the non-linearity of the model hypothesis and the (3, 3) strides help reduce model parameters while sustaining a broad receptive field of vision. The customizations of these large complex established models help us reduce the processing time, make the model weights small in size, and overall help in deployable models for mobile devices and CPUs at a cheap cost. The proposed model pipeline is quite simple and derived from pre-trained models and is easily deployable. It has been made for doctors, researchers, and future pioneers. The cloud processor takes less than 5 min to set up and can process images in less than 5 s.

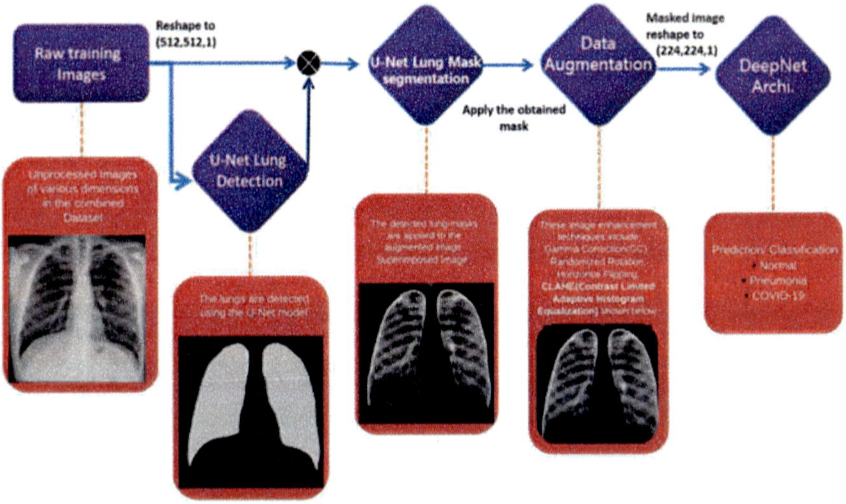

Fig. 6 Schematic diagram: proposed model pipeline and training methodology: the detailed methodology used for training/inference using the models U-Net and COVID-DeepNet

3.3 Lightweight-Customized U-Net Architecture

The main motive of our work is to filter out the point of interest from the X-ray images, which is the section of lungs, in a computationally efficient manner. This is done by designing a lightweight customized U-Net architecture and feeding our images into it which creates a segmentation mapping of the lungs and by superimposing the raw input image with this segmented output we get rid of the unwanted noise and details and highlights the lungs. This output passes through a series of data augmentation processes and is fed to the custom convolutional deep net architecture which generates the final results (Fig. 7).

The input image is convoluted first ($512 \times 512 \times 32$) which is acted upon by max-pooling for dimension reduction (2×2) and is fed to the following convolution network. In this due course, the spatial dimension of the filters reduces with an increase in the depth to analyze deeper details of the image and identify patterns of the desired object (lungs). This repetitive process continues till the dimension becomes $32 \times 32 \times 512$. From this step, upward convolution takes place with the previous step being copied and cropped into the layer to generate the segmentation map of the lungs. This is achieved by Up-Conv2D (2×2). The final result, as shown in Fig. 4 (output segmentation map), is a segmentation map of the lungs. Because of the lesser number of dense layers used, our architecture is computationally faster and at the same time extremely efficient in the said task.

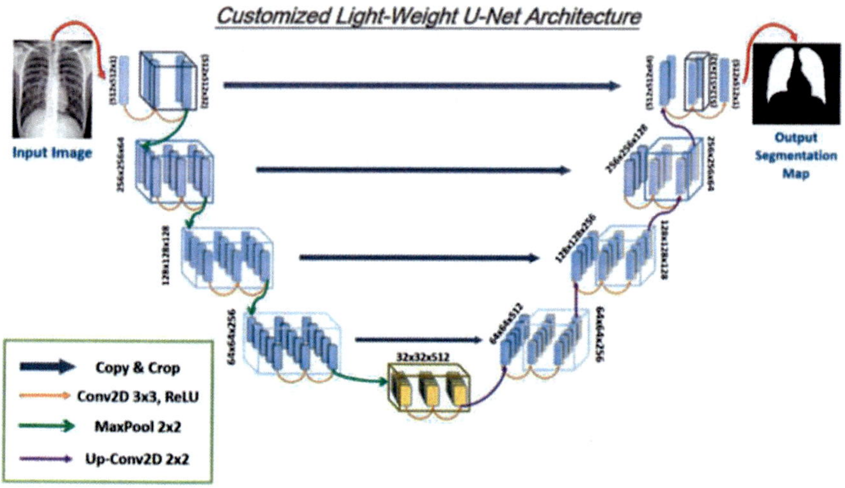

Fig. 7 Detailed U-Net architecture: the customized U-Net structure which is a lightweight version of the Original U-Net

3.4 Detailed COVID-DeepNet Architecture

The segmentation map received out of the U-Net architecture is superimposed with the input image that helps to highlight the segments of the lungs. But still, the problems of quality and noise in the prevailing image are solved using image enhancement techniques like gamma correction, randomized rotation, horizontal flipping, CLAHE, and Random Brightness Contrast. These techniques are not only to enhance the quality of the image but also to increase the diversity of the dataset and input types.

The image produced after superimposition of the segment mapping output of the U-Net architecture and the original image undergo multiple image enhancement techniques. The resultant image is then passed to our DeepNet architecture as input. The architecture is a convolution-based classification architecture as shown in Fig. 8 and depicts an extensive use of depth wise separable convolution layers which saves

Fig. 8 Detailed COVID-DeepNet architecture

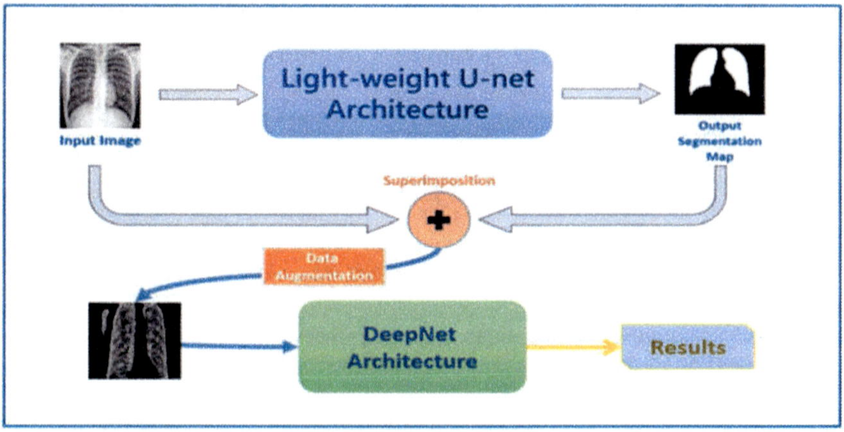

Fig. 9 Network pipeline flow: COVID-DeepNet

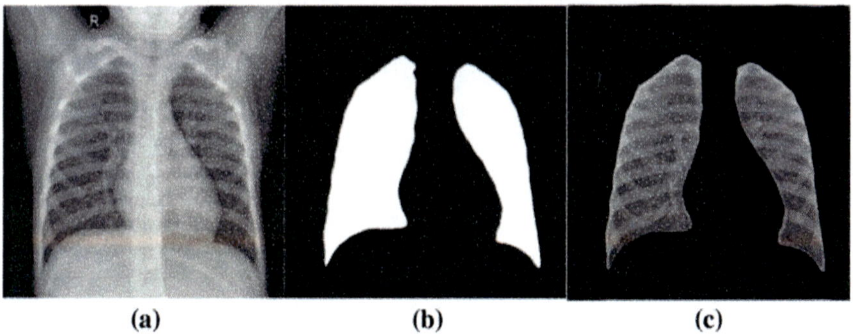

Fig. 10 Various stages of the architectural output: **a** The original input image. **b** The mask segmentation of the lungs. **c** Superimposition of the mask image over the original input image

a lot of computational time and makes the system a lot faster and efficient without compromising on accuracy (Figs. 9 and 10).

4 Results and Analysis

The limited dataset which was very biased was treated with various image enhancement techniques resulting in a balanced dataset consisting of almost equal number of images belonging to all the classes. Each class consisting of approximately 2905 images. The complete pipeline consists of the U-Net and DeepNet successively. Images reshaped to (224, 224, 3) were fed to the DeepNet. The U-Net architecture was pre-trained on a Pneumonia dataset for lung detection. The DeepNet architecture, pre-trained on ImageNet, was trained over a dataset of 7240 images with epochs = 30, batch_size = 16, learning_rate = 0.0001, number of training and validation steps: 447 and 1792, respectively, along with the Adam optimizer. The model was built on TensorFlow library with Keras back-end. Upon training, the final results depict a training accuracy of 93% and validation accuracy of 91% with a loss of 0.2. The test accuracy achieved on this model was 90.6%. The lightweight architecture makes the analysis faster. When the proposed architecture (COVID-DeepNet) was compared to models like ResNet152, SqueezeNet, and VGG-16, the number of true-negative outputs and the number of false-positive outputs came out to be much lesser which indicates minimal wrong diagnosis and an increase in efficient kit utilization, and also the detection accuracy increased manifold (Figs. 11, 12, 13, 14, 15 and 16; Table 2).

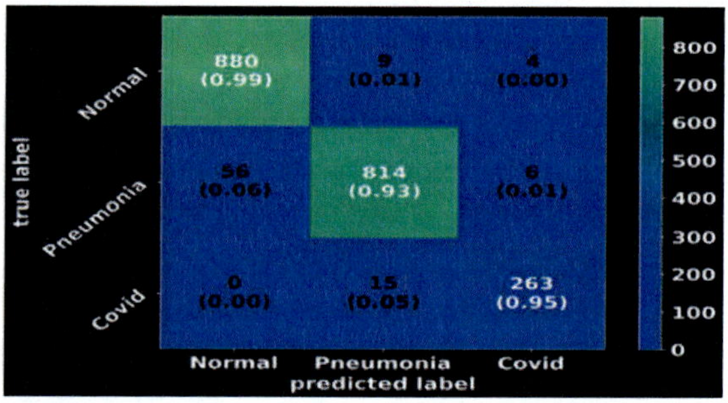

Fig. 11 Confusion Matrix Obtained: The number of images are labeled along with their total percentage in braces. Note that the number of True COVID Cases who are miss classified are nil

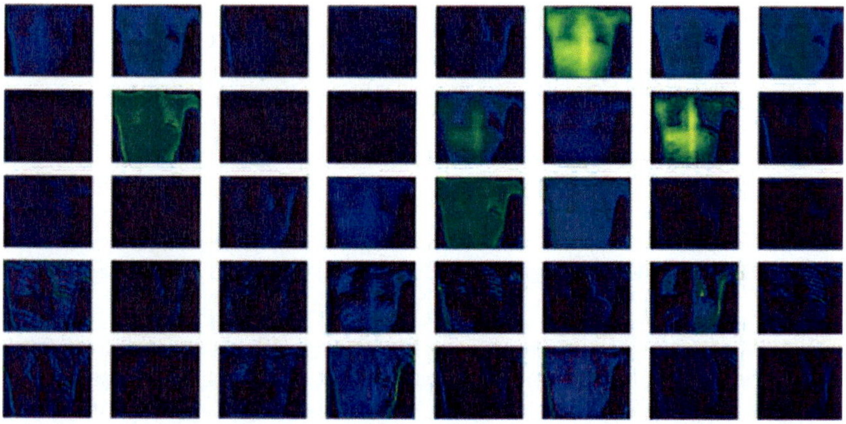

Fig. 12 Layerwise learning of the DeepNet: The COVID-DeepNet architecture learns layerwise different parameters and features. With deeper layers, more intricate and specific parameters and features are learnt and the top layers help in initial structure detection

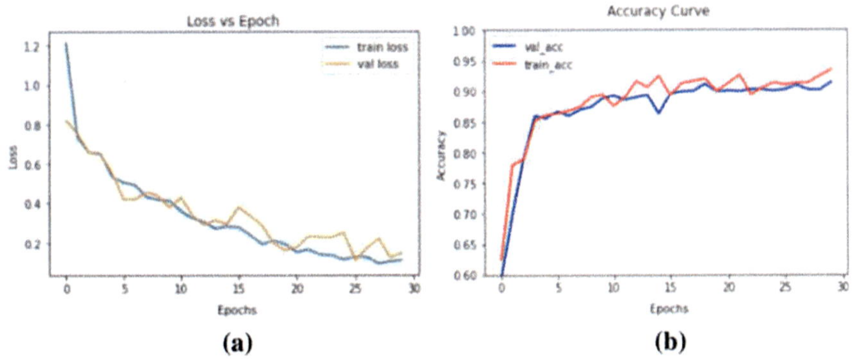

Fig. 13 Loss curve (a) and accuracy curve (b) while training of the extensive dataset

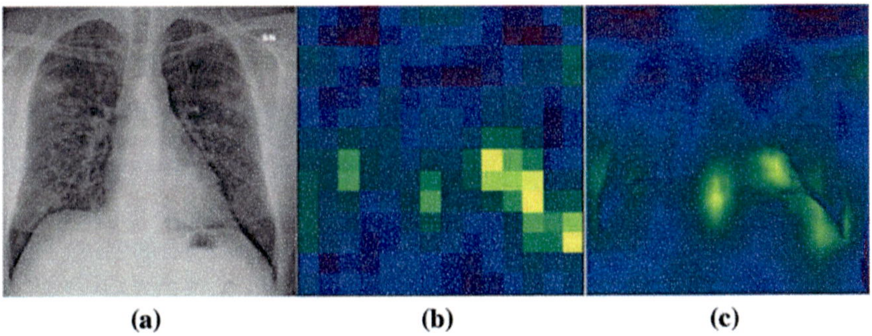

Fig. 14 GradCam Images at different layers of the COVID-DeepNet Model: **a** Input image, **b** Initial layers—highlighting the infected areas, **c** Marking of the infected area at the final fully connected layer

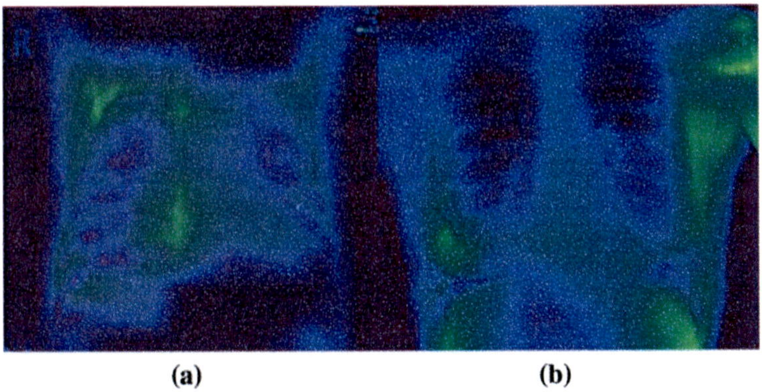

Fig. 15 GradCam visualization by categories: **a** Viral pneumonia, **b** normal

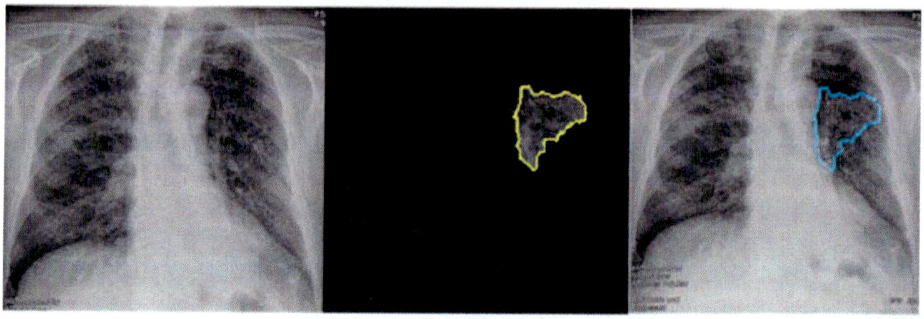

Fig. 16 Lime visualization: Infected areas that were studied while predicting the COVID infection by our algorithm: **a** Input COVID-infected image, **b** Lime visualization to highlight the target area, **c** Superimposed visualization on the original image

Table 2 Comparative analysis of crucial parameters: trainable parameters, training and validation accuracy, and training loss

Architecture	Accuracy (Train)	Accuracy (Validation)	Params (Millions)	Loss
VGG-16	88.1	85.6	14.42	0.181
ResNet152v2	87.4	83.4	58.19	0.232
SqueezeNet	92.3	90.6	0.722	0.174
COVID-DeepNet	**93.2**	**91.4**	**27.15**	**0.159**

For the particular class of interest, COVID-19 CXR Predictions,

• True Positive = 263 • True Negative = (880 + 814) = 1694 • False Positive = (4 + 6) = 10 • False Negative = (15 + 0) = 15	• Sensitivity (Recall) = TP / (TP + FN) = 0.946 • Specificity = TN / (TN + FP) = 0.994 • Precision = TP / (TP + FP) = 0.964

5 Conclusion

Our aim for the work was to develop a way to test the potential of COVID-19 from Chest X-Ray Images of lungs. This is achieved by developing a custom U-Net architecture that detects the lungs area and creates a corresponding segmentation mask image which is superimposed on the original image and then passed through various image enhancement techniques that solves the quality and quantity issues with the dataset. After that when fed to the DeepNet convolutional architecture, it yields the generalized classification results. We use depthwise separable convolutional layers which are a good replacement of normal convolution layers as it introduces a lesser number of trainable parameters and thus reducing the computational time making our approach a lot faster without compromising on the accuracy.

References

1. Novel Coronavirus Events, WHO. https://www.who.int/emergencies/diseases/novel-coronavirus-2019/events-as-they-happen
2. COVID-19 Pandemic Data, Wikipedia. https://en.wikipedia.org/wiki/Template:COVID-19_pandemic_data
3. Huang C et al (2020) Clinical features of patients infected with 2019 novel coronavirus in Wuhan, China. Lancet 395:497–506. https://doi.org/10.1016/s0140-6736(20)30183-5
4. ITN (Imaging Technology News), How Does COVID-19 Appear in the Lungs? and CT Imaging of the 2019 Novel Coronavirus (2019-nCoV) Pneumonia
5. Agarwal P, Chest imaging appearance of COVID-19 infection. https://doi.org/10.1148/ryct.2020200028
6. Corman VM et al (2020) Detection of 2019 novel coronavirus (2019-nCoV) by real-time RT-PCR. Eurosurveillance 25:2000045. https://doi.org/10.2807/1560-7917.es.2020.25.3.2000045
7. The Atlantic, A Dire Warning from COVID-19 Test Providers U.S. coronavirus testing could fail again, as surging demand creates new backlogs and delays
8. The Hindu, COVID-19|Delayed testing led to sudden spike in cases
9. Bercovich E, Javitt MC (2018) Medical imaging: from roentgen to the digital revolution, and beyond. Rambam Maimonides Med J
10. North Central Surgical, The difference between an X ray, CT Scan and MRI
11. Chest X-rays in the ER Can Help Predict Severity of COVID-19 in Younger Patients Automated detection of COVID-19 cases using deep neural networks with X-ray images
12. Ozturk T et al (2020) Comput Biol Med 121:103792. https://doi.org/10.1016/j.compbiomed.2020.103792
13. Wang L, Wong A (2020) COVID-Net: a tailored deep convolutional neural network design for detection of COVID-19 cases from chest radiography images. arXiv preprint
14. Apostolopoulos ID, Bessiana T, COVID-19: automatic detection from X-ray images utilizing transfer learning with convolutional neural networks
15. Song Y, Zheng S, Li L, Zhang X, Zhang X, Huang Z, Chong Y (2020) Deep learning enables accurate diagnosis of novel coronavirus (COVID-19) with CT images. medRxiv
16. Wang S, Kang B et al (2020) A deep learning algorithm using CT images to screen for Corona Virus Disease (COVID-19). medRxiv (2020)
17. Zheng C et al (2020) Deep learning-based detection for COVID-19 from chest CT using weak label. medRxiv
18. Xu X, Jiang X et al (2020) Deep learning system to screen coronavirus disease 2019 pneumonia
19. Barstugan M, Ozkaya U, Ozturk S (2020) Coronavirus (COVID-19) classification using CT images by machine learning methods
20. Science Daily, Some COVID-19 patients still have coronavirus after symptoms disappear
21. Luo L et al (2020) CT differential diagnosis of COVID-19 and non-COVID-19 in symptomatic suspects: a practical scoring method. BMC Pulmonary Med 20, Article number: 129
22. Maguolo G, Nanni L (2020) A critic evaluation of methods for COVID-19 automatic detection from X-Ray images
23. Tartaglione et al (2020) Unveiling COVID-19 from Chest X-ray with deep learning: a hurdles race with small data
24. Ikhsan IAM et al (2014) An analysis of x-ray image enhancement methods for vertebral bone segmentation. In: 2014 IEEE 10th international colloquium on signal processing and its applications. IEEE, pp 208–211

25. Pandey N (2020) Chest X-ray masks and labels, pulmonary chest X-Ray defect detection, National Library of Medicine, National Institutes of Health, Bethesda, MD, USA and Shenzhen No.3 People's Hospital, Guangdong Medical College, Shenzhen, China, Kaggle Datasets – Accessed : 16 July 2020
26. Jaeger S et al (2014) Automatic tuberculosis screening using chest radiographs. IEEE Trans Med Imaging 33(2):233–245
27. Candemir S et al (2014) Lung segmentation in chest radiographs using anatomical atlases with nonrigid registration. IEEE Trans Med Imaging 33(2):577–590
28. Gordienko et al (2018) Deep learning with lung segmentation and bone shadow exclusion techniques for chest X-ray analysis of lung cancer. Adv Comput Sci Eng Educ
29. Yu G et al (2018) Deep learning with lung segmentation and bone shadow exclusion techniques for chest X-ray analysis of lung cancer. Adv Comput Sci Eng Educ, 638–647

Chapter 6
Analytical Study on Feature Selection for Big Data Using Optimization Algorithms

Rachna Kulhare, S. Veenadhari, and Neha Sharma

1 Introduction

Machine learning (ML) is extensively used in a number of useful areas, including information retrieval, text categorization, predictive modeling, and diagnostic picture processing, all of which depend on big databases. Nevertheless, for ML techniques, certain characteristics could be unimportant and deceptive, that increases computing cost by reducing accuracy rate, particularly for highest dimensions' datasets. As a result, features options are required [1].

The primary premise of selecting features is to locate an optimum set of attributes that discriminates from the entire database, as well as the chosen set must maintain or perhaps even improve the initial dataset's classifier efficiency. Features choices are valuable strategies since they may remove unnecessary junk within information, allowing ML algorithms to run relatively quickly as well as efficiently. In other respects, ML approaches that use feature selection may function best despite cutting costs [2].

Feature selection (FS) techniques are categorized into 3 groups, as per the fundamentals: filtering, wrappers, and embedding methods. The filter techniques give a significance value to every other characteristic using a measurable statistic and order the characteristics based on the calculated ratings. The wrapper approaches use classifications to assess chosen subgroups acquired by different classifiers and utilize classification feedback to enhance FS. As a result, the wrapping techniques are more

R. Kulhare (✉) · S. Veenadhari · N. Sharma
Department of CSE, RNTU, Bhopal, MP, India
e-mail: rachna12kulhare@gmail.com

S. Veenadhari
e-mail: veenadhari1@gmail.com

N. Sharma
e-mail: sharmaneha94@gmail.com

L. Mandal et al. (eds.), *Proceedings of International Conference on Computational Intelligence, Data Science and Cloud Computing*, Algorithms for Intelligent Systems,
https://doi.org/10.1007/978-981-19-1657-1_6

accurate than the filtering methods. It can, nevertheless, need to have more computational power. Furthermore, because feature choices are considered a component of the training procedure in ML, embedding techniques are basically special examples of wrapping approaches [3–5].

FS may be thought of as optimization issues where a portion of the entire database is given by an optimization method, and these issues could be addressed using comprehensive and heuristic searching methodologies. However, the computational complexity of applying comprehensive search techniques is typically prohibitive, particularly for datasets with complex geometries of data. As a result, heuristic approaches to tackling feature selection issues may be more plausible. Numerous nature-based optimization techniques could have discovered throughout the research; it is estimated that there have been over 160 distinct methods, as well as modified methods, for getting the optimum solutions on performance prediction. Nevertheless, examining all of these strategies is not our goal. Rather, we will concentrate on bio-inspired methods [6]. As a result, we have limited our assessment to only just few methods including Bat algorithm, Whale optimization, Butterfly optimization, GWO-PSO, salp optimization, Mayfly, GA, BPSO, and Grasshopper algorithm which are some advanced optimization techniques.

2 Related Work

Bidgoli et al. [3] presented a multi-objective feature selection technique for multi-label classification data which selects the optimal set of attributes with the fewest data, maximum accuracy rate, and lowest computational complexity. In this paper, author suggested binary operators worked on majority voting and opposition-based learning. In addition, the suggested technique was tested on eight real-world multi-label databases and contrasted to the NSGA-II FS method on multi-label information.

Larabi-Marie [4] presented an outliers-based bio-inspired optimization technique for selecting features. Nearest neighbor, SVM classifier, and RF classifier are used to evaluate FS utilizing almost 19 datasets. On comparing with classifier without performing FS, the proposed strategies improved accuracy of classifier by a mean of 6.640%. On majority of the databases, this also beats other approaches, and the suggested approaches' findings were statically important, according to data analysis.

Akinyelu [5] proposed a boundaries recognition plus instances selecting approach, in order to improve the processing of ML-based large data classifier methods. Machine learning (ML) methods are one of the most effective techniques for BDA; however, their computational complexity grows as data size grows. ACOISA ML is tested on 5 machine learning models and 10 big datasets, and the findings demonstrate that this has the ability to cut the trained-time of machine learning methods by 94.0% without compromising predicting accuracy. Furthermore, the findings demonstrate that it decreased the storage capacity of massive data through 55% in many situations, resulting in faster big data analysis.

Shahmoradi et al. [6] proposed feature selection for multilayered community using optimization algorithm. The significance of overlaying communities in real networks is next discussed, followed by a full description for overlaying multilayered communities. Additionally, a two-phase strategy has been used to tackle the complexity of multilayered community recognition. To produce a set of answers, a multi-objective quantitative framework is designed and optimized (Pareto front) using a genetic algorithm relying on method (NSGA-II). This suggested approach is capable of detecting the ideal multilayer overlaying communities while overcoming the challenges of genetic representation. Investigations on artificial mLFR networks with key variables as well as six actual data systems demonstrate the algorithm's quality in terms of NMI and multilayered flexibility.

Amarjeet and Chhabra [7] presented a multi-objective optimization strategy for improving the modularization performance of an entity-based system with the least amount of class movement conceivable among existing technology modularization package. NSGA-II, a well-known multi-objective evolutionary approach, is used to do the assessment. To describe the conclusions of the suggested framework, a number of experiments are conducted across four real worlds and two random software applications. The testing findings show that the method is effective in improving the modularization quality of current packaging structures by quickly transferring categories across packages of original software modularization.

Elhariri et al. [8] suggested a method which involves two parts: (1) hand-crafted feature design and CNN-based fully convolutional learning and (2) feature selection using a hybrid filter-wrapper with improved multi-objective salp swarm optimization. The proposed approach is trained and tested using ten example UCI datasets and four crack picture databases. According to the findings, the proposed technique enhances crack severity detection by 37% and 31%, respectively, in terms of identification accuracy rate and F-measure.

Mutual information-based feature selection is translated into a global optimization issue by Han et al. [9], resulting in a novel approach to feature selection. Initially, a singular feature selection technique is proposed, which combines relevancy and redundancies and has good global searching ability and high computing efficiency. A hybrid feature selection architecture is used to arrive at a perfect outcome based on this foundation. On both artificial and actual databases, researchers evaluate the results of our technique with that of existing technique. The suggested technique's efficiency and applicability are demonstrated by simulated results.

By employing a multi-objective evolutionary algorithm, Gu et al. [10] provided a mechanism for constructing ensemble that expressly optimizes classification accuracy and ensemble diversification. Researchers examine the suggested algorithm's Pareto optimum solutions and compare these to the correctness of classifiers. The findings indicate that researchers can discover many classifier ensembles that outperformed single classifiers by deliberately optimizing variety and reliability. The findings also suggest that constructing ensemble that dependably outperformed single classifiers may require a combo of appropriate strategies for achieving and evaluating variety.

Chen et al. [11] employed a novel data-driven healthcare monitoring system based on multi-objective assessment and stacked autoencoders for health indicators. The proposed method comprises conducting multi-objective assessment on a large number of representative features derived from measurement data using an updated non-dominated sort genetic algorithm-II (NSGA-II). Using a layered autoencoder method, the gathered attributes are utilized to construct health indicators. The smallest difference of health indicators across various states and the amount of characteristics are stated as feature selection assessment goals in the modified NSGA-II approach.

Chen et al. [12] proposed the MOFES technique, and it turns feature selection for software defects prediction into a MO optimization issue. The researchers frame the problem as a multi-objective (MO) optimization issue and then propose MOFES, a novel technique based on the concept of searching dependent computing technology. As per the results, this strategy has the advantage of selecting fewer attributes and achieving greater prediction accuracy in most applications, while its processing cost is fair.

3 Feature Selection Using Machine Learning Algorithms

In order to build models for data description, machine learning or statistics employs variable selection or feature selection, variable subsets, or attributes. Characteristics selection has two primary aspects, i.e., minimum redundancy and maximum relevance. Apart from this, feature selection also has other benefits. The users generally use it for making more accurate forecasts, dimensionality reduction, and learning data minimization. The difference between feature selection and extraction lies in their basic principle. In case of feature selection, a group of appropriate features are selected from the feature set. While in the extraction method, new features are made considering feature set transformations or combinations as the basis. When dimensionality related to a recognition system of pattern is considered, function selection is widely used. The general procedure for selecting functions with machine learning methods is shown in Fig. 1. The goal is to minimize the dimensionality of data to only relevant features, regardless of the feature selection types and methods presented. High-dimensional datasets, on the other hand, preserve useless, noisy, and redundant features. The goal of dimensionality reduction is to quickly extract optimal features, because the higher the data dimensionality, the slower the feature optimization becomes. As a result, the major goal of feature selection is to establish just minor features of a specific problem area that needs more efficient classification.

The recognition of genuine entity properties is the fundamental goal of feature selection (where m is less than d). A vast variety of classifier problems use feature selection to enable classifiers better precise and reduce effort. There may be a loud noise in a full set of features. The material is simplified or given with less detail. This is helpful for examining dimensional data or reducing data before using it in a supervised process of learning.

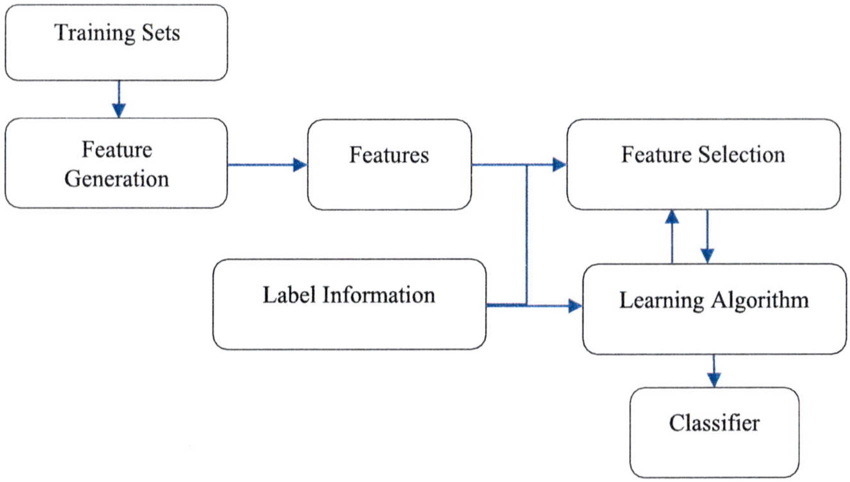

Fig. 1 General procedure for feature selection

Several of these reduction strategies may be used to classify and predict data. Principal component analysis (PCA), linear discriminant analysis (LDA), autoencoders, and other techniques are by far the most used. The next sections go through a few of the approaches for extracting features.

3.1 Principle Component Analysis (PCA)

PCA is a prominent dimensional reduction approach that may allow us get closer to the old dataset although yet preserving some variation. It is a categorization approach for data patterns. PCA is a complex data analysis technique. The number of parameters may be decreased without significant information loss after the data profiles and compact information have also been discovered. The stages of PCA are as follows:

- Subtract the average from the data.
- Covariance matrix is computed.
- Eigenvalues of the covariance matrix are calculated.
- Generate a feature vector by selecting the elements.
- Update the record.

3.2 LDA (Linear Discriminant Analysis)

The LDA method is used in the preprocessing level to reduce dimensionality for ML techniques and model classifiers. To reduce calculation costs and avoid over-regulation, a database is sent in a very simple region and good class separating method is tried to apply. LDA is most commonly used when there is unequal class frequency range as well as the procedures is being investigated using test data produced at irregular intervals. The LDA method maximizes the relationship between class variation for a given records and the evaluated class variances. As a result, there will be a lot of separation. The following are the steps involved in the LDA method:

The 2-D average vectors are calculated for various classes related to a dataset.

The matrices of dispersion are computed. Different values linked with eigenvectors and eigenvalues are evaluated. Where eigenvalues are represented as λ_d and eigenvectors are represented as e_d. The eigenvector classification is completed after the eigenvalues and eigenvectors have been reduced. B is chosen for which the e_d is very big, and a 2-D array $a * b$ is created with the goal of forming w (eigenvectors in w are represented by its columns). The eigenvector $a * b$ array is used to transform samples into subspace.

$$Q = P * R$$

where,

$P =$ matrix $(m * a)$,
$m =$ the number of samples,
$y =$ sample $m * b$ converted into subspace) is a simple representation of this.

3.3 Neighborhood Component Analysis (NCA)

This approach is designed feature reduction of irrelevant dataset, that was designed especially for image datasets. As the NCA is a non-parametric method for selecting features for either classification or regression problems with the goal to maximize the learning efficiency. The main advantage of NCA is that there is no loss of data during feature reduction process. Steps of NCA are as follows:

Select set of labeled data points.
Calculate the Mahalanobis distance among all data points selected randomly based on some probability.
Maximize the objective function for the transformed feature vector.
Select the features by using stochastic rule.

3.4 CA (Correlational-Analysis)

Pearson's correlation analysis: The most often used statistical approach for evaluating the connection between characteristics or values is Pearson's correlation analysis. It is written like this:

$$\text{Pearson}_{\text{CA}} = \frac{n\sum x_i y_i - \sum x_i \sum y_i}{\sqrt{\sum x_i^2 - \left(\sum x_i\right)^2}\sqrt{n\sum y_i^2 - \left(\sum y_i\right)^2}} \tag{1}$$

where x_i and y_i denote feature sets, and n is the number of observations.

Spearman-CA: It is also another statistical method that finds the degree of association among features. It is measured as:

$$\text{Spearman}_{\text{CA}} = 1 - \frac{6\sum d_i^2}{n\left(n^2 - 1\right)} \tag{2}$$

where d_i is the difference in rank between variables P and Q, and n is the sample size.

Kendall-CA: It is also another statistical method (non-parametric) that finds the strength of association among features. It is represented as:

$$\text{Kendall}_{\text{CA}} = \frac{(n_c - n_d)}{\frac{1}{2}n(n-1)} \tag{3}$$

where,

n_c = Number of concordant
n_d = Number of discordant.

3.5 Autoencoders (AE)

AEs are neural networks with a backpropagation process of deep learning that have not been monitored. This is used for representing high-size data in the low-size vector. It creates a low-dimensional interim feature vector and reconstructs an output vector that is comparable to the input vector. This has properties similar to principal component analysis, but principal component analysis is really not applicable in nonlinear predictions, whereas autoencoders can deal with both, which is the advantage of the autoencoders. The three layers of an AE are the output, hidden, and input layer with activating function. AE works in two steps: The first is to represent internal characteristics as in Eq. 4, and the second is to assess output value as in Eq. 5.

$$h(x) = f(W_e x_e + b_e) \tag{4}$$

$$X' = f(W_o h(x) + b_o) \tag{5}$$

The optimization function, as stated in Eq. 6, is being used to minimize errors between both the input feature vector and the output feature vector.

$$\arg\min[K] = \arg\min \sum_{i=1}^{n} \| x_i - x_i' \| \tag{6}$$

3.6 Optimization Algorithms

Optimization techniques have been shown to be effective in solving NP-hard issues. As a consequence of utilizing optimization methods, a feature set that can fulfill the fitness function is produced, that may then be employed in ml algorithms. Depending on a selection method including a regional or global search method, optimization techniques or feature reduction methods look for functionalities in the search space.

4 Feature Selection Using Bio-inspired Algorithms

There is a requirement of FS techniques with increased volume of network traffic, and correspondingly, data grows at a rapid rate. Algorithms that are influenced by nature are known as bio-inspired algorithms. They use basic approaches found in nature to address complicated issues. There are three types of bio-inspired algorithms: evolutionary, swarm-based, and ecology-based [13].

Darwinian principles of nature's ability to develop well-adapted living species in response to their circumstances underpin evolutionary algorithms (EA). Evolutionary algorithms include GA (genetic algorithms), GP (genetic programming), ES (evolutionary strategies), and evolutionary programming. These evolutionary algorithms includes some general steps such as generation, mutation, recombination, and selection. The fitness function provides an evaluation of a population of results that play the role of population members. The population develops as a consequence of the recurrent application of mutation and crossover operators.

Swarm-based algorithms (SBA) are a kind of algorithm that uses a swarm of insects to solve. The behavior of an swarm-based algorithm (SBA) is composed of several distinct swarms engaging with surroundings and with one another individually [14]. Swarms employ self-organization and decentralized control to achieve their objectives. Complex issues are solved using swarm-based systems. Swarms are simple animals with tightly concentrated cognitive skills and restricted communication options. Swarms exhibit cognitive behavior and offer important answers to complicated issues such as finding the shortest path, solving the "knapsack problem," and avoiding predators.

Ecology-based algorithms (EBA): Natural ecosystem mechanisms may be used to tackle difficult engineering and computer science issues. Ecology-based algorithms (EBAs) are made up of communities of people that evolve according to a set of optimization rules. As a result, individuals in each population are changed based on deepening and variety processes, with baseline values unique to each optimization approach. There are two types of ecology-based systems: homogeneous and heterogeneous. All populations in the homogenous model are developed using the same optimization technique with much the same characteristics. Optimization techniques and variables can be modified in the heterogeneous model. The utilization of ecological ideas such as environments, natural connections, and biological successions provides the ecological inspiration. Populations of people recognized in the same region constitute an ecological habitat after they are separated in the search space. A hypersurface might have a variety of ecosystems. Populations can migrate about in their surroundings. At any one period, however, each individual might still belong with one habitat [15].

Some of the works contributing in the field of feature selection are illustrated in Table 1. From Fig. 2, it can observed that most of the feature selection or feature reduction techniques are designed with single objective and mostly for classification problems. Among all bio-inspired algorithms, BPSO [21] algorithm outperforms better that is swarm optimization in nature. But, Mayfly [21] algorithm, an evolutionary algorithm, gives highest accuracy among all. But the most common thing that can be observed among all optimization algorithms for feature selection is that

Table 1 Major bio-inspired algorithms for feature selection

Technique	Year	Bio-inspired	Objective	Type	Accuracy (%)
Bat algorithm [16]	2019	SBA	Single	Classification	94.6
Grasshopper algorithm [17]	2018	SBA	Single	Classification	96.49
Whale optimization [18]	2018	SBA	Single	Classification	95
Butterfly optimization [19]	2020	SBA	Single	Classification	75
GWO-PSO [20]	2019	SBA	Single	Classification	93
Salp optimization [8]	2020	EA	Multi-objective	Classification	96
Mayfly [21]	2020	EA	Single	Classification	99
GA [21]	2020	EA	Single	Classification	97
BPSO [21]	2020	SBA	Single	Classification	96

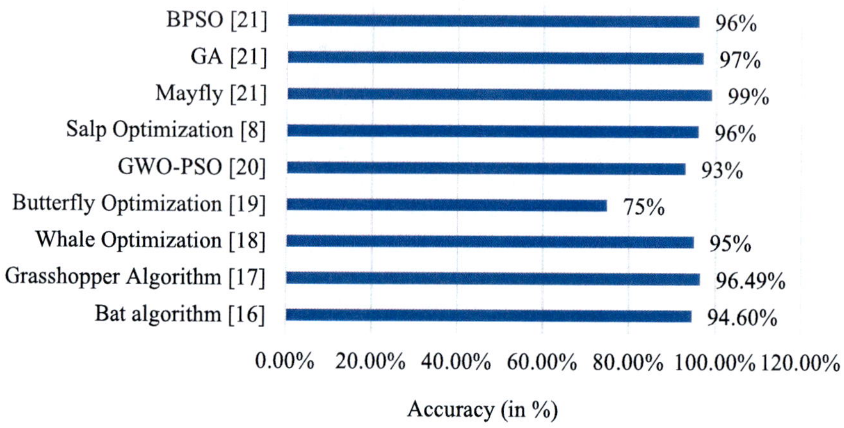

Fig. 2 Accuracy analysis of bio-inspired algorithms for feature selection

mostly they are designed for classification problems. Therefore, the main objective of this paper is to explore loopholes in current research works. There are very few contributions presented for multi-objective feature selection or feature reduction for time-series data or regression data. So, this can be adopted as future research direction.

5 Conclusion

This paper is dedicated to perform an analytical study on the feature reduction or feature selection methods for big data. A brief literature review is also presented that brings focus of researchers that bio-inspired optimization algorithm had contributed their effectiveness for feature reduction or selection as compared to conventional methods. Along with that this paper also presented accuracy comparison of some existing bio-inspired algorithms for feature reduction. From results, it has been seen that most of the researchers had contributed their efforts in designing swarm optimization algorithms. But from results, it can be observed that evolutionary optimization algorithm can outperform better. So, in future researchers can focus their efforts to design a hybrid optimization algorithm for feature reduction for both classification and time-series problems.

References

1. Ji B, Lu X, Sun G, Zhang W, Li J, Xiao Y (2020) Bio-Inspired feature selection: an improved binary particle swarm optimization approach. IEEE Access 8:85989–86002. https://doi.org/10.1109/ACCESS.2020.2992752
2. Valdez F, Castillo O, Melin P (2021) Bio-inspired algorithms and its applications for optimization in fuzzy clustering. Algorithms 14:122. https://doi.org/10.3390/a14040122
3. Bidgoli AA, Ebrahimpour-Komleh H, Rahnamayan S (2019) A many-objective feature selection algorithm for multi-label classification based on computational complexity of features. In: 14th international conference on computer science & education (ICCSE 2019), pp 85–91. https://doi.org/10.1109/ICCSE.2019.8845067
4. Larabi-Marie-Sainte S (2021) Outlier detection based feature selection exploiting bio-inspired optimization algorithms. Appl Sci 11(15):6769. https://doi.org/10.3390/app11156769
5. Akinyelu A (2020) Bio-inspired technique for improving machine learning speed and big data processing. In: 2020 international joint conference on neural networks (IJCNN), pp 1–8. https://doi.org/10.1109/IJCNN48605.2020.9206762
6. Shahmoradi MR, Ebrahimi M, Heshmati Z, Salehi M (2019) Multilayer overlapping community detection using multi-objective optimization. Futur Gener Comput Syst 101:221–235. https://doi.org/10.1016/J.FUTURE.2019.05.061
7. Amarjeet CJK (2017) Improving package structure of object-oriented software using multi-objective optimization and weighted class connections. J King Saud Univ—Comput Inf Sci 29:349–364. https://doi.org/10.1016/J.JKSUCI.2015.09.004
8. Elhariri E, El-Bendary N, Taie SA (2020) Using hybrid filter-wrapper feature selection with multi-objective improved-Salp optimization for crack severity recognition. IEEE Access 8:84290–84315. https://doi.org/10.1109/ACCESS.2020.2991968
9. Han M, Ren W (2015) Global mutual information-based feature selection approach using single-objective and multi-objective optimization. Neurocomputing 168:47–54. https://doi.org/10.1016/J.NEUCOM.2015.06.016
10. Gu S, Jin Y (2015) Generating diverse and accurate classifier ensembles using multi-objective optimization. IEEE SSCI 2014—2014 IEEE symposium series on computational intelligence, MCDM 2014—2014 IEEE symposium on computational intelligence in multicriteria decision-making, pp 9–15. https://doi.org/10.1109/MCDM.2014.7007182
11. Chen Z, Guo R, Lin Z et al (2020) A data-driven health monitoring method using multiobjective optimization and stacked autoencoder based health indicator. IEEE Trans Ind Informatics 17:6379–6389. https://doi.org/10.1109/TII.2020.2999323
12. Chen X, Shen Y, Cui Z, Ju X (2017) Applying feature selection to software defect prediction using multi-objective optimization. Proc—Int Comput Softw Appl Conf 2:54–59. https://doi.org/10.1109/COMPSAC.2017.65
13. Back T (1996) Evolutionary algorithms in theory and practice. Oxford University Press, Oxford
14. Bonabeau E, Dorigo M, Theraulaz G (1999) Swarm intelligence. Oxford University Press, Oxford
15. Yang XS (2009) Fire fly algorithm for multimodal optimization. International Symposium on Stochastic Algorithms, Cambridge, pp 169–178
16. Li J, Fong S, Liu L et al (2019) Dual feature selection and rebalancing strategy using metaheuristic optimization algorithms in X-ray image datasets. Multimed Tools Appl 7815(78):20913–20933. https://doi.org/10.1007/S11042-019-7354-5
17. Aljarah I, Al-Zoubi AM, Faris H et al (2018) Simultaneous feature selection and support vector machine optimization using the grasshopper optimization algorithm. Cogn Comput 103(10):478–495. https://doi.org/10.1007/S12559-017-9542-9
18. Tubishat M, Abushariah MAM, Idris N, Aljarah I (2018) Improved whale optimization algorithm for feature selection in Arabic sentiment analysis. Appl Intell 495(49):1688–1707. https://doi.org/10.1007/S10489-018-1334-8

19. Tubishat M, Alswaitti M, Mirjalili S et al (2020) Dynamic butterfly optimization algorithm for feature selection. IEEE Access 8:194303–194314. https://doi.org/10.1109/ACCESS.2020.3033757
20. Al-Tashi Q, Abdul Kadir SJ, Rais HM et al (2019) Binary optimization using hybrid grey wolf optimization for feature selection. IEEE Access 7:39496–39508. https://doi.org/10.1109/ACCESS.2019.2906757
21. Bhattacharyya T, Chatterjee B, Singh PK et al (2020) Mayfly in harmony: a new hybrid meta-heuristic feature selection algorithm. IEEE Access 8:195929–195945. https://doi.org/10.1109/ACCESS.2020.3031718

Chapter 7
A Machine Learning Approach for Predicting the Death Time and Mortality

Abhishek Sharma, Digbijoy Dasgupta, Shreya Bose, Udayan Misra, Ishita Pahari, Raktim Karmakar, and Sudipta Basu Pal

1 Introduction

The intensive care unit, commonly known as ICU, is the facility of the healthcare units for providing intensive treatment medicine for the patients who are suffering from severe illness or injury. Every ICU has a higher ratio of staff to the patient than regular wards, keeping the patients' data intact from one case to another and monitoring the patients comprehensively while generating massive healthcare data that is electronic-based and known as EHR. This EHR is helpful for the other patients to predict their status of the condition and how much health care they needed at the moment of ICU. Beth Israel Deaconess Medical Centre gathered at around 70,000–80,000 data from approximately 40,000 patients, who were in the ICU, from the year 2001 to 2012. This huge amount of collection of data is called The Medical Information Mart for Intensive Care III [1–4], in short, MIMIC III. The pros of this database are it is freely available, and it is a rich database because it is based on the real-time scenario of every patients. The MIMIC III database is built with two sub-databases; they are CareVue and MetaVision. CareVue provided the data from 2001 to 2008, while the MetaVision provided the data from 2008 to 2012. These two organizations have collected and stored the data by their own methods.

In the past years, the mortality prediction models are deployed using the rule-based method and data mining approach, and these models are providing good ICU scores [5, 6]. Among these, various frameworks were built using Acute Physiology and Chronic Health Evaluation (APACHE [7–9]), Simplified Acute Physiology Score (SAPS [10]), and Sepsis-Related Organ Failure Assessment Score (SOFA [11]). APACHE is used to provide the score which defines the morbidity of the patient to

A. Sharma · D. Dasgupta · S. Bose · U. Misra · I. Pahari · R. Karmakar · S. B. Pal (✉)
Computer Science and Engineering Department, University of Engineering & Management Kolkata, Kolkata, India
e-mail: sudipta_basu68@yahoo.com

L. Mandal et al. (eds.), *Proceedings of International Conference on Computational Intelligence, Data Science and Cloud Computing*, Algorithms for Intelligent Systems,
https://doi.org/10.1007/978-981-19-1657-1_7

predict the mortality of the patient. SAPS is a comparative analysis through which a patient's mortality can be determined, respected to other patients' data. SOFA is used to detect the possibility of organ failure of the patient during the ICU.

2 Literature Review

In this modern era of technologies and advancements of distributed computing and machine learning, a lot more better functionalities are there in the recent inventions. These models are basically based on the larger input features. Different types of reference databases [12–14] are necessary resources for researchers and developers of algorithms, who will investigate the various types of biomedical signals that need to verify the algorithms with realistic data and perform these tests frequently and reproducibly as algorithms clarifications are proposed. These databases also enrich medical science by providing adequate case studies of common and rare but clinically remarkable diseases. These databases will make it possible to formulate and answer numerous physiological questions without the need to evolve a new set of reference data at a high cost in each case [15] by making well-characterized clinical data available to researchers, developers, and users. In the past years, Desautels et al. [16] created a deep learning framework using the MIMIC III and the factors they have taken in the account are vitals, Glasgow Coma Score, age, and peripheral capillary oxygen saturation. They have predicted the sepsis very accurately. Next, Ghasemmi et al. [17] deployed the Latent Dirichlet Allocation using the MIMIC III database to make mortality prediction based on the free text of the hospitals. Hrayr et al. [18] deployed a machine learning model using this database and created a benchmark for providing four machine learning models which will predict the mortality based on the neural networks or NLP. Piracchio et al. [19] deployed the machine learning models based on the ensemble to predict the mortality more accurately. Awad et al. [20] used ensemble approach for prediction. Apart from these, a study shows how the HR and 30-day mortality models have a positive relationship [21]. Another thing is open science principle; open science principle is one of the most useful ways to make the research efficient [22]. From a study, we have seen that the patients survived while being in the ventilation is in between 48 h, not more than that [23]. Another study provided by the Angelina and co. about the quality of the data provided by the healthcare units as compared to the MIMIC database organization [24]. Steve and co. built a database network to work on various electronic healthcare datasets in a single machine and named it as CCHIC [25].

3 Problem Statement

In the past years, many works have been done using the MIMIC III. Still, they were designed for 24 h or 48 h after admitting to the ICU for providing the mortality prediction of the particular patient, and these models are accurate real-time prediction models until now. In this research paper, the main objectives of the authors are to create a machine learning model which has two phases. The first phase will provide the details of the prediction of the mortality of the ICU patient based on the individual factors. The second phase of the model will provide us with the information of the estimated death hours for the patient in the early stages of the admission in the ICU, i.e., 6-h time frame. Once a patient is declared dead in Phase 1 of the model, the model will also predict the death hour of that particular patient. Through this model, our main motive is to find out the high-risk patients who might be dead within some hours or a day after admission in the ICU in critical condition. For creating this model, we have extracted the data from the database as per our requirements so that the efficacy of the model. There are lots of models created based on this. After that, all the deployed models will be compared and based on the higher efficacy score. Finally, the best model will be specified as the final machine learning prediction for this particular dataset in the time frame of 6 h.

4 Methods and Solutions

4.1 Exploratory Data Analysis (EDA)

The whole MIMIC III dataset consists of 61,532 ICU records of distinct patients from 46,520 patients. To ensure the model to be accurate and efficient, we have removed the records of the patients which are having the staying time in the ICU less than one hour and also we have restricted the age limit from 16 to 89 years. By this, the huge 61 k dataset came down to 49,632 records with the unique patients count as 36,343. In this research work, we have deployed multiple machine learning models and have studied all of them based on the restricted dataset. Let us check out the summary of the dataset in Table 1,

The in-hospital mortality count is 5766. Now, we have to filter out the negative data in case of the mortality count, and after filtering those data, we have cut to the mark as 5718 in-hospital mortalities were resulted. After getting all those positive mortality data, we have calculated the average death time of the patients and the

Table 1 Variables based on those Phase 1 are deployed

Variables	Statistics
Age	Mean 62.61
Gender	Male 57.79%
Ethnicity	White 71.00%
Admission type	Emergency 82.31%
Number of ICU stays	Mean 1.37
In-hospital mortality ratio	11.62%

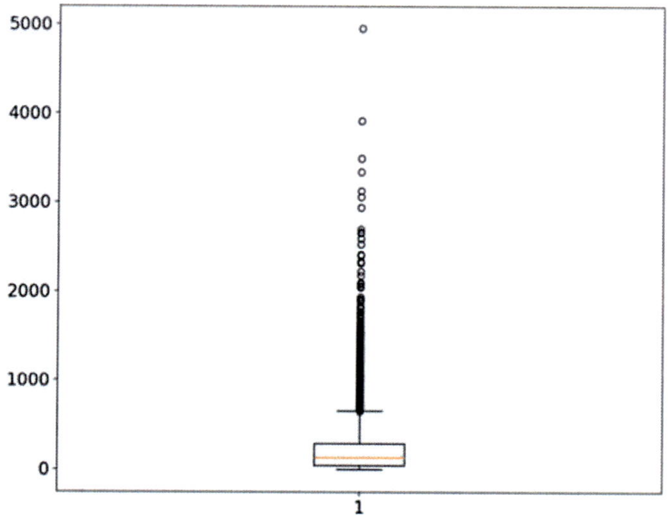

Fig. 1 Box plot to show the death time hours since the ICU admission

value is 9.57 days. The maximum death time of the records is 206.38 days, and the minimum is 0 days, which is showed in Fig. 1.

4.2 Feature Engineering from the Revised Dataset

This research work is based on the prediction of mortality and death time of the ICU patients. For our purposes, we have collected the data of 6 h, then 12 h, and lastly for 24 h since the patients had taken admission in the ICU. After that, we have extracted the feature values of the given time frame. Finally, we have got 123 features covering 5 static variables and 40 physiological variables.

Among all the variables, the static variables are like admission type, total number of previous and current stays in ICU, and other features like the demographic ones such as gender, age, and ethnicity of a particular patient. Now talking about the temporal data, there are physiological variables that include the vital information about the heart rate and blood pressure. In the laboratory testing system, we have also got the urine test results to judge the condition of the patient. Most of these variables and data are judged based on the maximum, minimum, and average during the specified time frame. Now, Table 2 will provide all the data about the variables and their corresponding data values.

Table 2 List of features those are extracted from the dataset

Category	Variables	Features
Demographic and static features	Gender, age, ethnicity, admission type, number of ICU stays	Not applicable
Significant signs	Heart rate, systolic blood pressure, diastolic blood pressure, mean blood pressure, respiratory rate, temperature, peripheral capillary oxygen saturation, glucose rate	Minimum Maximum Mean
Glasgow coma scale	Glasgow coma scale (GCS), GCS components (motor, verbal, eyes)	Minimum Maximum Mean
Blood samples and chemistry values	base excess, carboxyhemoglobin, methemoglobin, partial pressure of oxygen, partial of carbon dioxide, pH, ratio of partial pressure of oxygen to fraction of oxygen inspired, total carbon dioxide concentration	Minimum Maximum Mean
Lab results	Anion gap, albumin, immature band forms, bicarbonate, bilirubin, calcium, creatinine, chloride, hematocrit, hemoglobin, lactate, platelet, potassium, partial thromboplastin time, international normalized ratio, sodium, blood urea nitrogen, white blood cell count	Minimum Maximum Mean
Urine results	Urine output	Sum

4.3 Model Architecture

To build the model and implement that successfully, we have proposed a two-phase machine learning framework which will predict the in-hospital mortality and also their death time since the admission in the ICU. To implement the model of predicting the mortality of the patients, we have used 123 features those are extracted from the database and also shown in Table 2. And then to predict the death time of the patients, we have used a multiclass classifier on the same set of the features those are used in the Phase 1 of the prediction. We have also predicted the death time of the patients who were dead after the admission in the ICU in the hospitals.

For the Phase 1 model implementation, we have total of 49,632 data counts, and using that, we have created the model. We have used 80% of the data counts to train the model, and the rest 20% data count is used to test the model. For the predicting purpose, we have used random forest classifier as it is the effective one than the other classifiers implemented on the same dataset. The model is implemented in the time frame of 6 h, 12 h, and 24 h format. To make the model more efficient and accurate, we have tuned the model using the hyperparameters. The model performed well after the tuning procedure in each of the time frames of 6 h, 12 h, and 24 h.

Now in the Phase 2 of the model implemented, we have to predict the death time of the patients. For that, firstly we have obsolete the dead patients with negative death time given to them. After filtering them, we have 5718 data counts, and from these data counts, we have trained the model using the 80% data count and tested the model using 20% data count. Again, we have used the random forest classifier model to predict the death time of the patients, and after that using the hyperparameters, we have tuned the model (fivefold CV). Among all the classifier models implemented in this dataset, random forest classifier has the best response to all of them by the means of accuracy, efficacy, and efficiency.

4.4 Evaluation

The whole dataset was split into 80:20 ratio in case of training and testing the dataset. After that, hyperparameter tuning was used on the fivefold CV of the model and the test set of the dataset finally evaluated the whole thing. In terms of efficiency and accuracy, we have deployed various models and finally opted the random forest classifier model. Now to check the efficiency of the models, we have used the area under the curve parameter by which we can determine which model is more effective and accurate. ROC is basically the ratio of true and false positive rate. In a classifier, AUROC provides the single measure. Hence, we have used AUROC in the Phase 1 of the two models.

Now for Phase 2 of the model, we have micro-average and macro-average AUROC, and using these two parameters, we have to evaluate the model's performance. Macro-average basically takes all the classes equally and then calculates the metric value independently for each class, and after that, it takes the average of all the values. By this method, the model becomes more efficient and accurate than other parameters. And the micro-average takes all the aggregate values, then makes the average of them, and provides the metric. If in any model there is imbalance in class, anyone can use the micro-average without any hesitation. Apart from these two parameters micro-average and macro-average, we have used the confusion matrix, precision, recall, and F1 score through which we can determine the accuracy of the model during testing procedure of the model.

4.5 Implementation

This research paper and project consists of two models and two stages of implementation procedure. Firstly, we have used Apache Hive 2.1.0 on Microsoft Azure remote cluster and, secondly, the machine learning using Python 3.7 on a local server. As the data collected from the MIMIC III Web site is more than 50 GB of space we have used a big data tool as Apache Hive to store the information and data so that we can easily connect the database with our machine learning model server. Next, the building of the machine learning model. For that, we have used Anaconda Navigator and Jupyter Notebook. We have used Python 3.7 and various library packages, Pandas, Scikit-learn, and many more.

5 Result Analysis

During the research work, we have distinguished the data into two parts. The patient counts are divided into two parts, firstly the distinct ICU stay counts and, secondly, the distinct patients. The different ICU count provides the count of 61,532, and the specific patient count provides the count of 46,520. After that, we have eliminated all the null values, and the final data count for ICU stay becomes 49,632 and that of distinct patients becomes 36,343. Once the data is collected correctly, all the null values are eliminated from the dataset, and the models are deployed (Table 3).

In this research paper, we have deployed two models named Phase 1 and Phase 2. Phase 1 model shows the mortality prediction of the ICU patients. For the Phase 1 model, we have divided the time frame into three parts. They are 6, 12, and 24 h. Based on these time limits, we have deployed the models and calculated their accuracy and efficacy using two parameters. Firstly, the AUROC on fivefold CV and, secondly, AUROC on the test set. Table 4 shows the score of the models based on these two parameters.

From Fig. 2, we can determine that the model is very much eligible to predict the data values and efficient enough to predict the values correctly. However, it has been seen that the models with broader time frames have comparatively higher efficiency and accuracy. The accuracy score ranges from 0.88 to 0.92, which shows the error

Table 3 Distribution of data counts across the classes

Class	Description	Number of data
Class 0	Death time hours < 24	838
Class 1	$24 <$ death time hours $< 24 \times 7$	2517
Class 2	$24 \times 7 \leq$ death time hours	2363

Table 4 Accuracy score of the models

Models	AUROC on fivefold CV	AUROC on test
Random forest classifier model in the time frame of 6 h	0.883	0.883
Random forest classifier model in the time frame of 12 h	0.901	0.902
Random forest classifier model in the time frame of 24 h	0.920	0.919

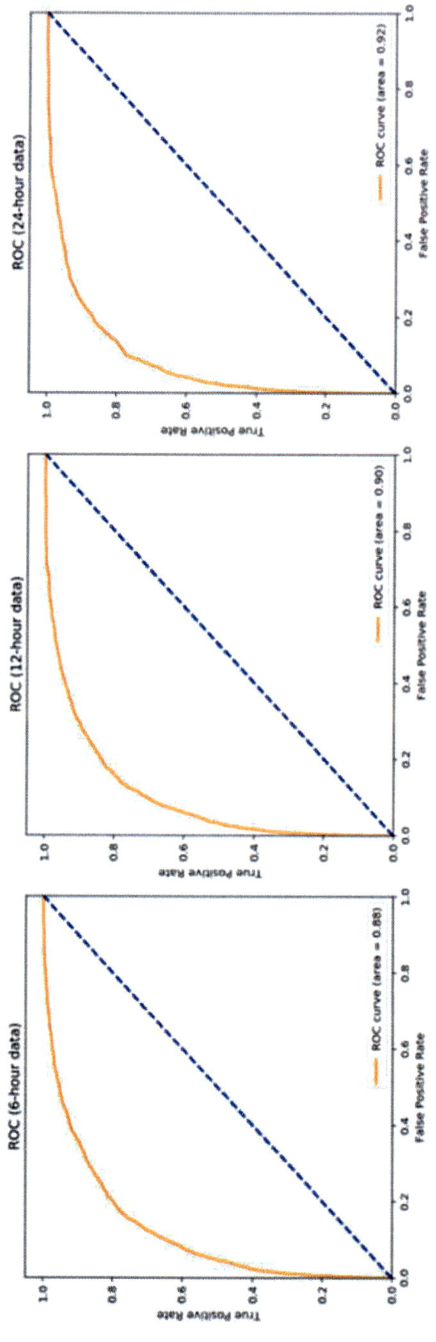

Fig. 2 ROC curves of Phase 1 using the random forest classifier

percentage is tiny while predicting the values. The time frame of 6 h provides a low score because of the data inefficiency compared to the 24-h data frame.

57% of the dataset is consisting records for male category, and 43% are of female category. In data counts, male patients' count is 21001, and the female patients' count is 15342. The next main feature is ethnicity. In this case, there are five categories, and they are white, not specified, black/African, Hispanic or Latino, and others. 70% of the dataset consists of white people, as the count suggests 25,577, not specified category includes 3497, black or African consists 2594, Hispanic or Latino consists 939, and other categories count 853. Now, based on the admission type, there are three categories, emergency, elective, and urgent.

82% of the data consists of emergency patients as the count suggests 29,915, after that elective patients are of 5267, and lastly the urgent patients are 1161 in count. The last main feature is in-hospital mortality, 11.61% of the dataset were dead in the admission of the hospital, and the rest of patients were alive as the count suggests 43,866 for alive patients and 5766 for dead patients.

In Phase 2 of the model, we have deployed the death time calculation procedure, for that the maximum death time since the ICU admission is 206.38 days and the lowest death time is 0 days. Average death time since the admission in the ICU is 9.57 days. In the deployment of Phase 2 model, we have trained the 80% of the data and rest of the data used in the test set. During the evaluation of the training set of the Phase 2 model, certain parameters scored certain values.

In Phase 2, we also deployed the random forest classifier to predict the situation and taken the micro-average of the AUROC to determine the score and accuracy of the models. Table 5 shows the comparative analysis of the three models of Phase 2. Based on those factors, we can determine the accuracy of the models that are deployed using the dataset.

As we have deployed three separate models for three different time frames, hence for each for the model, we have different ROC curves for three classes, Class 0 (death time hours < 24), Class 1 ($24 <$ death time hours $< 24 \times 7$), and Class 2 ($24 \times 7 \leq$ death time hours). Table 6 and Fig. 3 show the ROC curve score for each of the classes according to the models.

During the implementation of both the models, we have combined the demographic features with the dynamic features of the ICU patients such as physiological variables measured during the admission of the patients in the ICU. And combining

Table 5 Micro-average and macro-average AUROC score of different models

Models	Micro-average AUROC	Macro-average AUROC
Random forest classifier on 6 h	0.77	0.72
Random forest classifier on 12 h	0.79	0.74
Random forest classifier on 24 h	0.82	0.79

Table 6 ROC curve of different classes

Models	ROC curve value for Class 0	ROC curve value for Class 1	ROC curve value for Class 2
6-h time frame model	0.84	0.60	0.71
12-h time frame model	0.89	0.62	0.72
24-h time frame model	0.93	0.66	0.76

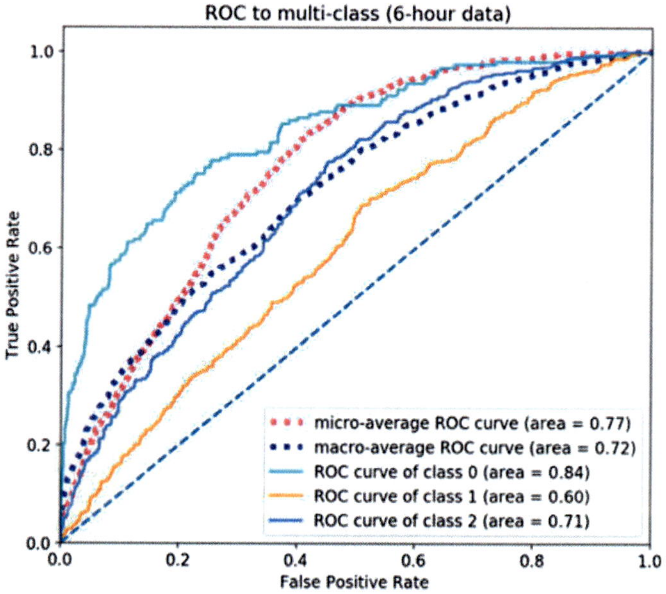

Fig. 3 ROC curve of 6-h prediction model

all those features provides us a compact way and easier way to deploy the model through which we predicted the in-hospital mortality of 6, 12, and 24 h, and also, we have predicted the death time of the ICU patients during their ICU based on 6, 12, and 24 h. The Phase 1 of the model is having the AUROC value ranging from 0.88 to 0.92 based on the three time frames. And the Phase 2 model which predicts the death time of the patients is having the micro-average AUROC ranging from 0.77 to 0.82.

After all those extraction and prediction of the model, we have come to the point that the critical patients died on the same day the graph shows that parameter. During the period of 100 days, more than 800 death cases are there while the patients were admitted on that day in the hospital. And the trend goes downwards as the days passed.

6 Conclusion and Discussion

In this research paper, authors have proposed a new two-phase model on the MIMIC III dataset. This two-phase model is built on the MIMIC III dataset, which combines both the static and dynamic data of ICU patients admitted to the hospital, stored in an electronic record system. The two-phase model shows a higher accuracy while they are treated in a larger time frame than 6 h. For Phase 1 of the machine learning framework, the 24-h time frame machine learning framework has higher accuracy than the 6-h machine learning framework. Similarly, this happens to Phase 2 of the model while predicting the death time of the patients in ICU. Thus, the machine learning framework will help the healthcare systems a lot, despite the lack of data and the improvements to be done.

Now, coming to the discussion part of this machine learning framework [26], the question arrives why we do not use the regression models? Why we went with the classification models? To answer this, we have deployed the regression model on the same dataset; but unfortunately, the model was not up to the mark as it has shown an accuracy of 20%. A model with an accuracy of 20% is a dumb model, and it cannot predict things accurately in these critical conditions. That is why we have gone with the classification models and tuned them by the hyperparameters to make them more accurate to predict the correct parameters in the critical conditions. Also, the model is treated with cross-validation to ensure the framework's efficacy, and it resulted in being a far better model than the regression ones. These are reasons we went with classification models over the regression models.

References

1. Johnson AEW, Pollard TJ, Shen L, Lehman L-WH, Feng M, Ghassemi M, Moody B, Szolovits P, Celi LA, Mark RG (2016) MIMIC-III, a freely accessible critical care database. Sci Data 3:160035
2. Goldberger AL, Amaral LAN, Glass L, Hausdorff JM, Ivanov PCh, Mark RG, Mietus JE, Moody GB, Peng C-K, Stanley HE (2000) PhysioBank, PhysioToolkit, and PhysioNet: components of a new research resource for complex physiologic signals. Circulation 101(23):e215–e220 [Circulation Electronic Pages; http://circ.ahajournals.org/content/101/23/e215.full]
3. Johnson AEW, Stone DJ, Celi LA, Pollard TJ (2017) The MIMIC code repository: enabling reproducibility in critical care research. J Am Med Inf Assoc, ocx084
4. Rogers P, Wang D, Lu Z (2021) Medical information mart for intensive care: a foundation for the fusion of artificial intelligence and real-world data. Front Artif Intell 4:691626. https://doi.org/10.3389/frai.2021.691626
5. Soares M, Fontes F, Dantas J, Gadelha D, Cariello P, Nardes F et al (2004) Performance of six severity-of-illness scores in cancer patients requiring admission to the intensive care unit: a prospective observational study. Crit Care 8(4):R194203. https://doi.org/10.1186/cc2870.PMC522839
6. Strand K, Flaatten H (2008) Severity scoring in the ICU: a review. Acta Anaesthesiol Scand 52(4):46778. https://doi.org/10.1111/j.1399-6576.2008.01586.x

7. KnausWA DEA, Wagner DP, Zimmerman JE (1985) APACHE II: a severity of disease classification system. Crit Care Med 13(10):81829. https://doi.org/10.1097/00003246-198510000-00009.PMID3928249 (This is the first published description of the APACHE II scoring system)
8. Knaus WA, Zimmerman JE, Wagner DP, Draper EA, Lawrence DE (1981) APACHE-acute physiology and chronic health evaluation: a physiologically based classification system. Crit Care Med 9(8):5917. https://doi.org/10.1097/00003246-198108000-00008
9. Knaus WA, Wagner DP, Draper EA, Zimmerman JE, Bergner M, Bastos PG, Sirio CA, Murphy DJ, Lotring T, Damiano A et al (1991) The APACHE III prognostic system. Risk prediction of hospital mortality for critically ill hospitalized adults. Chest 100(6):161936. https://doi.org/10.1378/chest.100.6.1619
10. Le Gall J-R, Lemeshow S, Saulnier F (1993) A new simplified acute physiology score (SAPS II) based on a European/North American multicenter study. JAMA 270:2957–2963 (This is the first published description of the scoring system)
11. Vincent JL, Moreno R, Takala J, Willatts S, De Mendona A, Bruining H, Reinhart CK, Suter PM, Thijs LG (1996) The SOFA (Sepsis-related Organ Failure Assessment) score to describe organ dysfunction/failure. On behalf of the working group on sepsis-related problems of the European Society of Intensive Care Medicine. Intensive Care Med 22(7):707–710
12. Moody GB, Mark RG (1990) The MIT-BIH arrhythmia database on CD-ROMand software for use with it. Comput Cardiol, 185–188
13. Moody GB, Mark RG (1996) A database to support development and evaluation of intelligent intensive care monitoring. Comput Cardiol, 657–660
14. Moody GB, Feldman CL, Bailey JJ (1993) Standards and applicable databases for long-term ECG monitoring. J Electrocardiol 26(suppl):151–155
15. Rockwell RC, Abeles RP (1998) Sharing and archiving data is fundamental to scientific progress. J Gerontol B Psychol Sci Soc Sci 53:S5–S8
16. Desautels T, Calvert J, Hoffman J, Jay M, Kerem Y, Shieh L, Shimabukuro D, Chettipally U, Feldman MD, Barton C, Wales DJ, Das R (2016) Prediction of sepsis in the intensive care unit with minimal electronic health record data: a machine learning approach. JMIR Med Inform 4(3):e28
17. Ghassemi M, Naumann T, Doshi-Velez F, Brimmer N, Joshi R, Rumshisky A, Szolovits P (2014) Unfolding physiological state: mortality modelling in intensive care units. In: Proceedings of the 20th ACM SIGKDD international conference on knowledge discovery and data mining, KDD 14. ACM, New York, NY, USA, p 7584
18. Harutyunyan H, Khachatrian H, Kale DC, Ver Steeg G, Galstyan A (2017) Multitask learning and benchmarking with clinical time series data. arXiv preprint arXiv:1703.07771
19. Pirracchio R, Petersen ML, Carone M, Rigon MR, Chevret S, van der Laan MJ (2015) Mortality prediction in the ICU: can we do better? Results from the Super ICU Learner Algorithm (SICULA) project, a population-based study. Lancet Respir Med 3(1):42–52. https://doi.org/10.1016/S2213-2600(14)70239-5
20. Awad A, Bader-El-Den M, McNicholas J, Briggs J (2017) Early hospital mortality prediction of intensive care unit patients using an ensemble learning approach. Int J Med Inform 108:185–195. https://doi.org/10.1016/j.ijmedinf.2017.10.002 Epub 2017 Oct 5
21. Guo Q, Li H, Ouyang H, Sun R, Wang J, Wu M, Pan Y, Wang J, Zhang Y (2021) Heart rate fluctuation and mortality in critically ill Myocardial infarction patients: a retrospective cohort study. Front Cardiovasc Med 14(8):577742. https://doi.org/10.3389/fcvm.2021.577742
22. Besançon L, Peiffer-Smadja N, Segalas C, Jiang H, Masuzzo P, Smout C, Billy E, Deforet M, Leyrat C (2021) Open science saves lives: lessons from the COVID-19 pandemic. BMC Med Res Methodol 21(1):117. https://doi.org/10.1186/s12874-021-01304-y
23. Reis AMD, Midega TD, Deliberato RO, Johnson AE, Bulgarelli L, Correa TD, Celi LA, Pelosi P, Gama De Abreu M, Schultz MJ, SerpaNeto A (2021) Effect of spontaneous breathing on ventilator-free days in critically ill patients—an analysis of patients in a large observational cohort. Ann Transl Med 9(9):783. https://doi.org/10.21037/atm-20-7901

24. Kurniati AP, Rojas E, Hogg D, Hall G, Johnson OA (2019) The assessment of data quality issues for process mining in healthcare using Medical Information Mart for Intensive Care III, a freely available e-health record database. Health Inf J 25(4):1878–1893. https://doi.org/10.1177/1460458218810760 Epub 2018 Nov 29PMID: 30488750
25. Harris S, Shi S, Brealey D, MacCallum NS, Denaxas S, Perez-Suarez D, Ercole A, Watkinson P, Jones A, Ashworth S, Beale R, Young D, Brett S, Singer M (2018) Critical care health informatics collaborative (CCHIC): data, tools and methods for reproducible research: a multi-centre UK intensive care database. Int J Med Inf 112:82–89. https://doi.org/10.1016/j.ijmedinf.2018.01.006. Epub 2018 Jan 31.PMID:29500026
26. Sharma A et al (2022) Covid-19—analysis and prediction—a case study using machine learning. In: Cyber intelligence and information retrieval. Lecture notes in networks and systems, vol 291. Springer, Singapore. https://doi.org/10.1007/978-981-16-4284-5_31

Chapter 8
Prediction of Crops Production Using Random Forest Regression

Abhirup Paria and Susovan Jana

1 Introduction

Machine learning (ML) showed as a promising field of an area over the last decagon by answering many complex and enlightened real-world problems. The solicitation sector included almost all the real-world problems like business application, health care, natural language processing, weather carving and voice and image processing. One of the most remarkable sectors of machine learning is forecasting; ML models are used to guide the upcoming route needed in many solicitation sectors as well as disease forecasting, stock market forecasting, disease diagnosis including weather forecasting [14].

Agriculture has been the crucial element in our daily life, taking the dramatic change in the health and nourishment of all individuals. The agronomist upgrades its fertility depending on the time administration, providing the right nourishment required and the actinic use which will guide in getting good crops and more yields. This will finally help the agronomists preserving more resources that can be used later when required [6].

The contemporary analysis centres attention on the results of various parameters like state name, district name, crop year, season, crop and irrigated areas of crops. The introductory scrutiny was done only with supervised algorithms because the training of the models is quite faster than the unsupervised and reinforcement algorithm. The multiple linear regression, support vector regression, random forest regression with

A. Paria (✉)
Department of Computer Science and Engineering (AIML), Haldia Institute of Technology, Haldia, India
e-mail: abhirup.paria@gmail.com

S. Jana
Department of Information Technology, Institute of Engineering and Management, Kolkata, India

L. Mandal et al. (eds.), *Proceedings of International Conference on Computational Intelligence, Data Science and Cloud Computing*, Algorithms for Intelligent Systems,
https://doi.org/10.1007/978-981-19-1657-1_8

stochastic habits tend to be the option amongst other algorithms. There are other choices like artificial neural network or deep neural network but that helps the operations slower because more calculations require examining the predicted output [8].

2 Literature Survey

Suvidha et al. utilise data mining to forecast future crop output, such as rice, wheat and maize, with regard to different observations made throughout time (1950–2013). This research only examined rain, temperature, land under irrigation, output, and yield. It found that for both the rice and wheat data set and the maize data set, multivariate adaptive regression splines (Earth) outperform multiple linear regression and random forest regression in terms of performance [8].

Fatin et al. show two distinct ML algorithms designed to evaluate agricultural production. Support vector regression (SVR) and linear regression (LR) are ideal for verifying the continuous variable parameters [5].

Abdipour et al. ran three models with ANNs, including RBF ANNs, component analysis and multi-linear regression models. The objective of the research is to differentiate the efficiency of agricultural seed yield predictions with a major influence on the error analyses, root mean squares error and mean absolute error components (MAE) [7].

Khosla et al. proposed two-step estimation of kharif crop yield first estimates rainfall using modular artificial neural network (MANN) and then predicts yield with support vector regression (SVR). Information for the years 2000–2016 were included in the experimental data set, and the result for the year 2018–2019 was calculated using that data [12].

Jheng et al. presented a new model with the scaling technique added to the data preparation stages. This design is a mix of conventional and SVR models. The hybrid SVR model offers improved agricultural production prediction reliability and sustainability; the model was assessed using root mean square error (RMSE) and correlation coefficient (CC) [9].

Aymen et al. discuss food shortages in rural Upper Egypt. The framework indicates how much grain the Egyptians will require. With WEKA, the prediction is built utilising an artificial neural network (ANN) and multilayer perception. This assists in forecasting the yearly amounts of key crops (rice, wheat and beans) up to 2020 [11].

Afan et al. examine deep learning techniques for weather prediction. The outcomes of the prediction models are compared by the RNN, conditional Boltzmann machine and convolution network (CN). The results are intended to support weather predictions for a range of applications, including aviation, agriculture and tourism [15].

Niketa et al. provide a comparison analysis of rice yields for the kharif season in India using the class algorithms, BayesNet and NaiveBayes. This is done by selecting 27 Maharashtra districts, taking into account parameters including minimum temperature, maximum temperature, average temperature, crop reference evapotranspiration, precipitation area, Kharif production and yield for a five-year time period from 1998 to 2002 (June to November). Finally, the conclusion is that BayesNet's performance was superior to NaiveBayes [4].

Niketa et al. compare classification methods to estimate the output of rice crops in the tropical wet and dry climate zones of India during the kharif season. The classification methods have been performed using the free-source tool WEKA, and experimental findings include precision, sensitivity, specificity, F1-score, mean absolute error and root average squared error. The classifications J48 and LADTree have the highest sensitivity, precision and specificity. The lowest sensitivity, accuracy and specificity values are shown in the LWL classification [3].

Rupanjali et al. provide several linear regression techniques to estimate the future yield prediction for tea production using the patterns seen in climate change throughout the period (1977–2006). The research chose criteria for the four areas (South Bank, North Bank, Upper Assam and Cachar) of Assam: temperature, rainfall, relative humidity, sunlight and evaporation. The result is that estimated tea production equal treatments have been verified for the future yield forecast for the four areas (2007, 2009 and 2010), and a model that has been created may be utilised to predict tea output precisely for each region [2].

3 Materials and Methods

It describes specifics of the data sets, machine learning methods, and prediction model construction.

3.1 Data Set

Data set was gathered from publicly accessible documents by the Indian government over a 18-year period, from 1997 to 2015. It includes state name, district name, crop year, season, crop, irrigated areas and production (see Table 1).

3.2 Machine Learning Techniques

Machine learning is useful in helping us answer questions about patterns in our data, whether we are discovering them or making predictions about them. The technique of data analysis machine learning is best described as a process that uses algorithms to create data models automatically.

Table 1 Data set description

Features	Description
State name	States name of India
District name	District name of each state
Crop year	Year starting from 1997 to 2015
Season	Seasons like Kharif and Rabi
Crop	Various crops like Maize, Wheat
Area	Area under irrigation
Production	Total production in each year

In this crop yield predicting research, best fitted regression models used:

- Random Forest Regression
- Support Vector Regression
- Multiple Linear Regression

Random Forest Regression One of the most common methods used for machine learning is the decision tree. Trees that are growing extremely deep have learned very abnormal behaviour; in other words, they are overfitting their training sets. Random forests are a technique for decreasing variation by averaging several deep decision trees, each trained on a different subset of the same training set, with the aim of generating a less diverse outcome. In order to maximise the performance of the final model, the increased bias and reduced interpretability must be accepted [17].

The random forest training algorithm uses the standard bootstrap method of aggregation or packaging to tree branches. Due to a training set $A = a_1, \ldots, a_n$ with answers $B = b_1, \ldots, b_n$ (C times), a random sample has to be selected to replace the training set and trees for the following samples are adapted:

For $c = 1, \ldots, C$:

- These are some training samples taken from A, B, which we will name A_c, B_c.
- Incorporate A_c, B_c into a classification or regression tree Z_c.

Following training, $\bar{a}$ may be taken by averaging the predictions from all the different regression trees on $\bar{a}$ for unspecified samples:

$$\hat{Z} = \frac{1}{C} \sum_{c}^{C=1} Z(x) \tag{1}$$

In the case of classification trees, or by obtaining a majority vote.

To make a better model, first conduct bootstrapping. The bootstrapping method lowers the variance of the model without increasing the bias. In other words, a single tree's predictions are very vulnerable to noise in the training set, whilst the result of averaging several trees is less impacted if the trees are mutually independent. The bootstrap method is a strategy for de-correlating the trees by giving them training methods sets [18].

Furthermore, the standard deviation of the predictions from all the different regression trees on $\bar{a}$ may be used to determine the uncertainty of the prediction.

$$\Delta = \sqrt{\frac{\sum_{c}^{C=1}(Z(x) - \hat{Z})^2}{C - 1}} \tag{2}$$

C is a free parameter, and there are an infinite number of trees to draw from. Depending on the size and type of the training set, the usual range is from several hundred to several thousand trees. A balance between the number of trees C and the prediction errors for each training example a_i may be obtained via cross-validation, or by looking at the out-of-bag error (the mean prediction error for each training example, calculated for all samples except the bootstrap sample). Once a certain number of trees have been measured, the training and test error should be equal [16].

3.3 Evaluation Parameters

In the current research, we assess each learning performance of a model as regards R^2 score, R^2_{Adjusted}, mean square error (MSE), mean absolute error (MAE), and root mean square error (RMSE).

Mean Absolute Error (MAE) The mean absolute error of the predicted results is the average size of the mistakes. This is an average for testing data between the forecasts of the model and real data, in which each difference is equal in importance. Its value range is from 0 to infinity, and lower scores indicate the quality of learning models, which is why they are sometimes termed negative values [19].

$$\text{MAE} = \frac{1}{m}\sum_{i=1}^{m}|z_i - \hat{z}_i| \tag{3}$$

Mean Square Error (MSE) Another method for evaluating regression models' effectiveness is via mean square error. The MSE removes the data points and squares them off the regression line. Squaring is needed because the negative sign is removed, and bigger discrepancies are considered more important. The lower the mean squared error indicates, the nearer the best fit [1]. The calculation of MSE is possible as:

$$\text{MSE} = \frac{1}{m}\sum_{j=1}^{m}(z_j - \hat{z_j})^2 \tag{4}$$

Root Mean Square Error (RMSE) The standard deviation of forecast errors may be defined as the root mean square error. The gap between the fitted line and the real-data points is also known as residual predictions. RMSE is thus a measure of how the actual data points are clustered across the best possible line. This is the following error rate shown by the MSE square root [19].

$$\text{RMSE} = \sqrt{\frac{1}{m}\sum_{j=1}^{m}(z_j - \hat{z_j})^2} \tag{5}$$

R^2 The R^2 value is a statistical metric used to assess regression models' effectiveness. The statistics indicate the variance per cent of the predictor variables, which determines the independent variable jointly. It assesses a comfortable 0–100% scale on the strength of the connection between the dependent variable and regression. We may assess the fitness of the proposed model using the R^2 score after training on the regression model. R^2 results from the dispersion of data points around the regression line, also known as the determination coefficient. Its rating always ranges from 0 to 100. 0 indicates that the response variable does not have the average variability of its model, whilst 100 means that the answer variable is based on its mean. The high R^2 value demonstrates the quality of the trained model. R^2 is a mathematical model which describes the variable's fluctuating percentage [13]. You may find it as:

$$R^2 = 1 - \frac{\text{Regression line squared sum error}}{\text{Mean line squared sum error}} \tag{6}$$

R^2_{Adjusted} The R^2_{Adjusted} is an updated R^2 form, which displays how the data sets fit to the curve as well as R^2. The main distinction between R^2 and R^2_{Adjusted} is whether the later adjustments are made to the amount of features in a forecasting model. For R^2_{Adjusted}, it may lead to the formation in new features if the new characteristics are beneficial for the prediction model. Nevertheless, the value has dropped if the additional features are worthless. The R^2_{Adjusted} is:

$$R^2_{\text{Adjusted}} = 1 - [(\frac{m-1}{m-p-1})(1 - R^2)] \tag{7}$$

4 Methodology

The research's primary goal is to discover the production over the irrigated area. The two main sequenced courses in ML are classification and regression; in other terms, a distinct and uninterrupted value is generated. In the process of assuming a numeric quantity, the regression function is necessary, and the template should be assessed using the chosen kernel function if classification is worked on. A data set is gathered in the first instance. Approximately, 246,091 data were directly collected from the Kaggle website [10] only with the right data to make the authenticity of actual information more valid. Then, each information is averaged by year.In this instance,

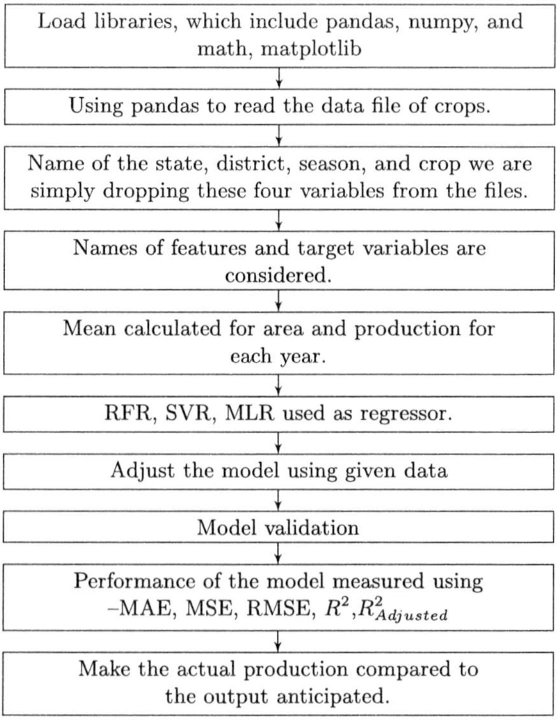

Fig. 1 Flow diagram of crop prediction

the random forest regression model is used to evaluate the values of crops that are spread across a broad variety of irrigated areas. Figure 1 shows how the process flows, which takes place once each stage is completed.The method will enable farmers to forecast the production quantity of crops by irrigated area. In this phase, data outlines are important in order to ensure that the information is worked and learned using various histograms and charts. Random forest regression was established to make it more effective for the real crop estimate. Validation is an essential step to improve the precision of the level.

In this instance, the MAE, MSE, RMSE, R^2, R^2_{Adjusted} values are given to offer a more accurate look at the model's prediction levels in the data sets. The evaluation and the visualisation are the phases where results are shown.

5 Experimental Results

In this part, we compare and analyse the crop prediction outcomes achieved using the three machine learning methods.

We utilised the Python programming language for implementation and testing. Sklearn is the package used to extract the algorithms, as well as obtain the data

Table 2 Actual versus predicted values of various machine learning algorithms

Year	Actual	RFR	SVR	MLR
1997	95,654.8	226,570.7	549,671.8	432,058.1
1998	517,254.5	437,370.6	549,671.8	445,119.6
1999	517,214.5	521,666.5	549,671.9	458,181.2
2000	549,672.3	542,323.4	549,672.3	471,242.8
2001	561,614.4	554,786.0	549,672.8	484,304.3
2002	465,466.5	502,035.3	549,673.4	497,365.9
2003	461,985.7	498,786.8	549,673.9	510,427.5
2004	590,955.5	568,534.2	549,674.5	523,489.0
2005	594,996.4	598,621.1	549,674.9	536,550.6
2006	621,201.5	621,201.5	549,675.3	549,612.2
2007	482,125.0	523,848.0	549,675.7	562,673.8
2008	542,306.2	519,647.0	549,676.0	575,735.3
2009	556,438.8	532,372.3	549,676.3	588,796.9

Table 3 Evaluation metrics

Evaluation metrics	RFR	SVR	MLR
R^2	0.74	6.17	0.096
R^2_{Adjusted}	0.69	3.12	0.042
MAE	67,341.88	146,820.90	144,278.86
MSE	13,505,706,155.00	5,297,4342,951.95	47,856,532,476.91
RMSE	116,214.05	230,161.55	218,761.35

and show the results of the algorithms. We utilised the SVR class from the support vector machines (SVMs) package to conduct binary and multi-class categorization. The random forest regression algorithm was taken from the ensemble library. The multiple linear regression algorithm was taken from the linear model library. The Pandas library was used to read the data file with the.csv extension, and the NumPy library was used to extract the data required for the methods. After executing all algorithms, the result for each algorithm was taken. The variation in values across the data is shown in Table 2, which contrasts the real data collected with the complicated findings produced.

Table 3 displays the results produced by the algorithms using the crop data set. According to the findings, RFR performed the best. Support vector and multiple linear regression both provide excellent results with minimal variations from SVM. Figure 2 shows a comparison of actual and predicted results for the machine learning methods to the crop data set.

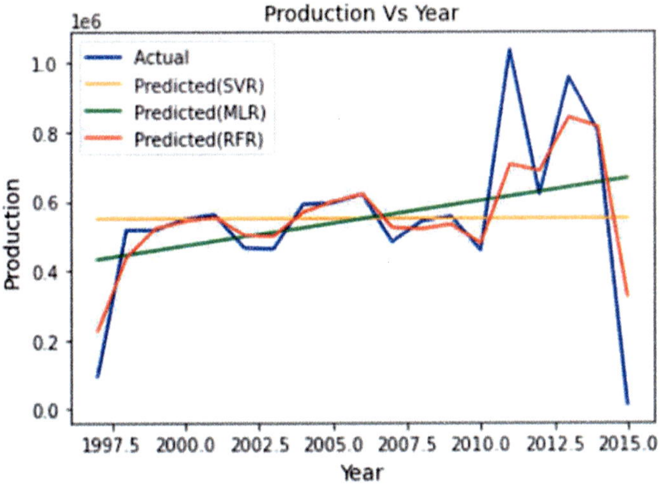

Fig. 2 Comparison between real and forecasted productions

6 Conclusion

The primary goal of this study was to examine machine learning methods in order to get the best crop forecast results. This study examined three machine learning algorithms: support vector machine, random forest and multiple linear regression. Five measures were used to evaluate these methods to determine which performed better: MAE, MSE, RMSE, R^2 and R^2_{Adjusted}. Crop data sets, a publicly accessible database, were used for the tests. The acquired findings were also compared to those of other research studies. According to the trial findings, the random forest method was the most efficient, followed by SVM and MLR. Random forest results on the crop data set are the best obtained. The algorithms discussed in this article are very effective and may be utilised for crop forecasts and for other predictions.

This research may be furthered by researching different machine learning methods, developing our own data set including information on various kinds of environmental factors such as temperature, humidity, rain fall and so on, and developing and implementing an application that is helpful in identifying the potential of production based on various environmental indicators.

References

1. Baran S, Nemoda D (2016) Censored and shifted gamma distribution based emos model for probabilistic quantitative precipitation forecasting. Environmetrics 27(5):280–292. https://doi.org/10.1002/env.2391, https://onlinelibrary.wiley.com/doi/abs/10.1002/env.2391

2. Baruah RD, Roy S, Bhagat RM, Sethi LN (2016) Use of data mining technique for prediction of tea yield in the face of climate change of Assam, India. In: 2016 international conference on information technology (ICIT), pp 265–269. https://doi.org/10.1109/ICIT.2016.060
3. Gandhi N, Armstrong LJ (2016) Rice crop yield forecasting of tropical wet and dry climatic zone of India using data mining techniques. In: 2016 IEEE international conference on advances in computer applications (ICACA), pp 357–363. https://doi.org/10.1109/ICACA.2016.7887981
4. Gandhi N, Armstrong LJ, Petkar O (2016) Predictingrice crop yield using bayesian networks. In: 2016 international conference on advances in computing, communications and informatics (ICACCI), pp 795–799. https://doi.org/10.1109/ICACCI.2016.7732143
5. Haque FF, Abdelgawad A, Yanambaka VP, Yelamarthi K (2020) Crop yield analysis using machine learning algorithms. In: 2020 IEEE 6th world forum on internet of things (WF-IoT), pp 1–2. https://doi.org/10.1109/WF-IoT48130.2020.9221459
6. Haque FF, Abdelgawad A, Yanambaka VP, Yelamarthi K (2020) Crop yield prediction using deep neural network. In: 2020 IEEE 6th world forum on internet of things (WF-IoT), pp 1–4. https://doi.org/10.1109/WF-IoT48130.2020.9221298
7. Huang X, Wang H, Luo W, Xue S, Hayat F, Gao Z (2020) Prediction of loquat soluble solids and titratable acid content using fruit mineral elements by artificial neural network and multiple linear regression. Sci Hortic 278. https://doi.org/10.1016/j.scienta.2020.109873
8. Jambekar S, Nema S, Saquib Z (2018) Prediction of crop production in india using data mining techniques. In: 2018 fourth international conference on computing communication control and automation (ICCUBEA), pp 1–5. https://doi.org/10.1109/ICCUBEA.2018.8697446
9. Jheng TZ, Li TH, Lee CP (2018) Using hybrid support vector regression to predict agricultural output. In: 2018 27th wireless and optical communication conference (WOCC), pp 1–3. https://doi.org/10.1109/WOCC.2018.8372729
10. Kaggle: agricuture-crops-production-in-India. https://www.kaggle.com/srinivas1/agricuture-crops-production-in-india (2021)
11. Khedr A, Kadry M, Walid G (2015) Proposed framework for implementing data mining techniques to enhance decisions in agriculture sector applied case on food security information center ministry of agriculture. Egypt. Procedia Comput Sci 65:633–642. https://doi.org/10.1016/j.procs.2015.09.007
12. Khosla E, Ramesh D, Sharma R (2020) Crop yield prediction using aggregated rainfall-based modular artificial neural networks and support vector regression. Environ Dev Sustain 22. https://doi.org/10.1007/s10668-019-00445-x
13. Lupón J, Gaggin HK, de Antonio M, Domingo M, Galán A, Zamora E, Vila J, Peñafiel J, Urrutia A, Ferrer E, Vallejo N, Januzzi JL, Bayes-Genis A (2015) Biomarker-assist score for reverse remodeling prediction in heart failure: the st2-r2 score. Int J Cardiol 184:337–343
14. Rustam F, Reshi AA, Mehmood A, Ullah S, On BW, Aslam W, Choi GS (2020) Covid-19 future forecasting using supervised machine learning models. IEEE Access 8:101489–101499. https://doi.org/10.1109/ACCESS.2020.2997311
15. Salman AG, Kanigoro B, Heryadi Y (2015) Weather forecasting using deep learning techniques. In: 2015 international conference on advanced computer science and information systems (ICACSIS), pp 281–285. https://doi.org/10.1109/ICACSIS.2015.7415154
16. Tibshirani GJDWTHR (2013) With applications in r. An introduction to statistical learning 1:316–321. https://doi.org/10.1007/978-1-4614-7138-7
17. Trevor H, Tibshirani R, Friedman J (2009) Data mining, inference, and prediction, 2nd edn. In: The elements of statistical learning, vol 2, pp 587–588. https://doi.org/10.1007/978-0-387-84858-7
18. Wikipedia: random forest. https://en.wikipedia.org/wiki/Random_forest (2021)
19. Willmott C, Matsuura K (2005) Advantages of the mean absolute error (MAE) over the root mean square error (RMSE) in assessing average model performance. Clim Res 30:79. https://doi.org/10.3354/cr030079

Chapter 9
Drowsy Driving Detection Based on Deep Neural Network for Accident Avoidance

Srijita Bandopadhyay, IshanJyoti Ray, Shalini Mondal, Subhrajyoti Manna, Sohini Mitra, Kyamelia Roy, Soumen Banerjee, and Sheli Sinha Chaudhuri

1 Introduction

Drowsy driving is one of the significant factors responsible for causing accidents in automobiles. Today's hectic lifestyle contributes in people becoming tired and fatigued which in turn leads to drowsiness during driving [1]. Implementing a risk alert system for detecting drowsiness is not only recommended but a necessity of the hour. This alert system should be able to awake the driver or autonomously slow down the car and even stop it in breakdown lane [2]. Several techniques have been implemented in this domain to measure driver drowsiness, viz., steering monitoring and acceleration, and breaking monitoring [3]. However, these techniques possess several limitations in that they can be used only in certain driving conditions and are not robust. The perspective for drowsiness detection to avoid accident could be segregated using domain analysis performed over target. One approach is to directly analyze the driver's behavior and to identify changes in it. This approach analyses facial elements such as eye and mouth using visual sensors or detects particular patterns in electrophysiological signals occurring when a driver is falling asleep. Other approaches indirectly infer a driver's state through analysis of signals extracted from the steering system. The most significant method adopted by authors in Babaeian et al. [4] and Yu et al. [5] included the use of bio-signal namely Electroencephalogram (EEG) and facial expression such as eye closure movements.

S. Bandopadhyay (✉) · I. Ray · S. Mondal · S. Manna · S. Mitra · K. Roy · S. Banerjee
Department of Electronics and Communication Engineering, University of Engineering and Management, Kolkata, West Bengal, India
e-mail: srijitabando30@gmail.com

S. S. Chaudhuri
Department of Electronics and Tele-Communication Engineering, Jadavpur University, Kolkata, West Bengal, India

L. Mandal et al. (eds.), *Proceedings of International Conference on Computational Intelligence, Data Science and Cloud Computing*, Algorithms for Intelligent Systems,
https://doi.org/10.1007/978-981-19-1657-1_9

In this paper, the authors presented a computer vision-based Deep Learning approach toward detection of drivers' drowsiness. Deep Learning is generally considered for its better efficiency and also the excellent feature performed from the raw data. Therefore, the proposed method yields higher accuracy with less features than other feature selection methods. In this model, images of the driver's face are taken as input at regular intervals and various CNN architectures are applied to classify the drowsiness behavior into two main classes, i.e., Open-Eyes and Closed-Eyes class. In this paper, comparative studies between two CNN architectures are presented and the detection is mainly done with Faster Region-Based Convolutional Networks (F-RCNN). The biggest advantage of the proposed model is that it can be compressed enough to be deployed on embedded board while preserving reasonable accuracy. To deploy a driver drowsiness detection system in daily use vehicles, a compressed model is significant. Therefore, to avoid accident a real-time drowsy detector is highly recommended.

2 Related Works

Various approaches toward drowsiness detection are reported in literature. To enhance the accuracy and to reduce the time complexity which are faced during compilation multiple methods are suggested by researcher in Dasgupta et al. [6], Zhao et al. [7], Sunagawa et al. [8], and Tanveer et al. [9]. Several machine learning approaches toward drowsiness detection are found in literature in which the latest additions are the approaches using deep learning techniques.

2.1 Drowsiness Detection Using Machine Learning

In [10], machine learning method was used to detect driver's drowsiness in which the concept of Hidden Markov's model was highlighted over support vector method [11] which is considered to the best model for classification. The authors in Lin et al. [12] proposed a drowsiness detection model where OpenCV functions were used to train the detector which included a 6-stage strong classifier comprising of 9–14-stage week classifiers at each stage. This model reflected average processing frame rates of up to 245 fps by a PC with 2.59 GHz operating frequency [13]. The detection rate of eye closure was 91.49% while wearing glasses and 95% without glasses.

2.2 Drowsiness Detection Using Deep Learning

In [14], the authors have designed a framework using CNN Architecture in which dlib's API was used for facial landmark detection and Percentage of Eyelid Closed

(PERCLOS) value was utilized to predict whether the driver was drowsy or not. This particular model showed a detection accuracy of 94%. In [5], a new model was suggested which takes video clips of Driver's state as input and correspondingly generated the condition of drowsiness as output based on facial expression. They adopted a fully connected neural network as there was possibility that given spatiotemporal representations have complex distributions and showed 76% accuracy in drowsiness detection. In [9], a different concept on drowsiness detection was proposed. In this model, input to the algorithm is the facial landmark coordinates, which were extracted from images using DLib Library as represented in Real-Time Drowsiness Detection Algorithm with CNN based on five layers. The model achieved an average of 83.33% of accuracy for all categories where its maximum size did not exceed 75 KB. In [15], two deep learning architectures were used—the first one to classify between drowsy/alert states, i.e., DNN, and the other one for detecting the strength of each channel. The CNN architecture was used for classification of color map images created from each time window for all channels in which the accuracy rate was 82.7%. The authors in [16], in 2019, proposed a dependable framework that accomplished 95.98% accuracy. The framework comprised a hereditarily weighted troupe of CNN. The module was trained on MRL Eye Dataset using Convolutional Neural network [17–20]. It consisted of three functions—(a). Data Preprocess for better performance and convergence, (b). Hyper parameter tuning for Fine-tuning, and lastly (c). Prediction on the test set accuracy of the model.

3 Methodology

The proposed system workflow adopted in this paper includes data extraction, model training, output classification, and final detection as shown in Fig. 1.

3.1 Dataset

The dataset used in this paper has been acquired from MRL Eye Dataset by Patil [16] as available in Kaggle. The dataset is of medium size and contains 4000 human eye images of two different eye positions, open eyes and closed eyes. Figure 2 clearly depicts some typical data set images.

3.2 Data Pre-processing

The data comprising of open and closed eye images are first segregated into two sets (folders) namely Open_Eyes and Closed_Eyes. Then, the images in each set are splitted into two categories, viz., training set and validation set consisting of

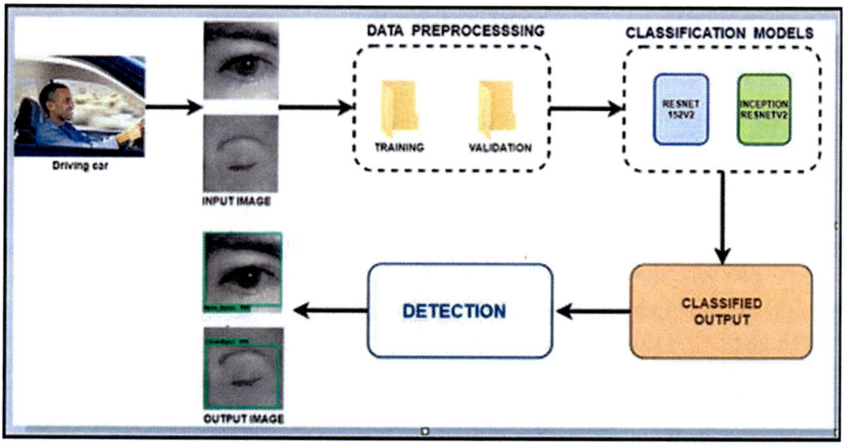

Fig. 1 System workflow

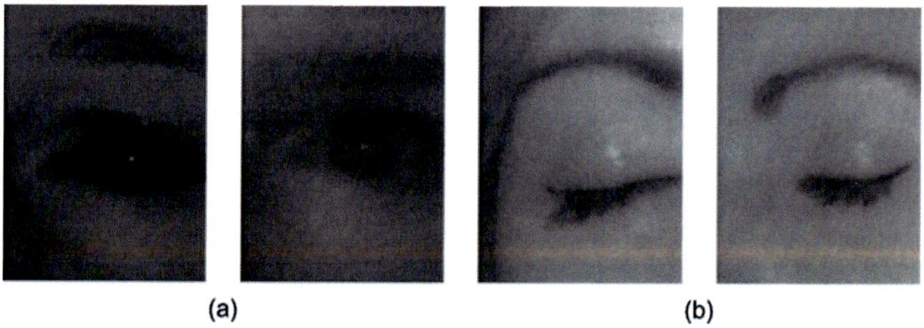

Fig. 2 Typical Data set images. **a** Opened eyes and **b** Closed eyes

70% and 30% of the total images, respectively. The images were then annotated and were labeled as open_eyes and closed_eyes. The annotations were used to generate TensorFlow records and label map files which were later used during the detection process.

3.3 Training The Models

The CNN models are trained using transfer learning for hyperparameter tuning and to get classification reports and accuracy of the models. The base architecture used for training are InceptionResNetv2 and ResNet152v2. The initial model consisted of 136 layers and latter of 152 layers. For training the model, sigmoid activation function is used as it takes real values as input and gives output in the form of 0s and 1s and hence is best suited for binary classification. For the optimizer, Adam Optimizer is used as it has the best accuracy for enhancing the CNN ability in classification and segmentation.

3.4 Classified Outputs

After training the models, classified outputs were acquired in the form of confusion matrix and classification report which consisted of parameters like precision, recall, and *F1-score* which were later used for a comparative study between the two base CNN models.

3.5 Detection

The last step is detection in which Faster Region-Based Convolutional Neural Networks is used with base architecture as InceptionResNetv2 from which desired detected output of closed and open eyes were obtained.

4 Experimental Results

The values for certain parameters, viz., precision, recall, $F1$-score, and specificity are calculated by observing the values of True Positive (TP), True Negative (TN), False Positive (FP), and False Negative (FN) from the Confusion Matrix. The formulas for the four parameters are as follows:

$$\text{Precision} = \text{TP}/(\text{TP} + \text{FP}) \tag{1}$$

$$\text{Recall} = \text{TP}/(\text{TP} + \text{FN}) \tag{2}$$

$$F1 - \text{Score} = 2 * \text{Precision} * \text{Recall}/(\text{Precision} + \text{Recall}) \tag{3}$$

$$\text{Specificity} = \text{TN}/(\text{TN} + \text{FP}) \tag{4}$$

The models InceptionResNetV2 and ResNet152V2 produced the best values of accuracies as 99.87% and 99.99% along with it promising values of precision, recall and $F1$-score. The other models are also compared and found to have lower values of accuracies as reported in Table 1. Adam optimizer is used the mentioned Classification models. The training and validation graphs depicted in Figs. 3 and 4 are

Table 1 Comparative study of CNN models

CNN models	Accuracy (%)	Precision	Recall	F1-score	Specificity
InceptionResNetv2	99.87	0.832	0.832	0.832	0.832
ResNet152V2	99.99	0.852	0.852	0.852	0.852
InceptionV3	99.30	0.80	0.80	0.80	0.80
ResNet101	95.24	0.75	0.751	0.759	0.75
VGG19	98.43	0.78	0.78	0.78	0.78
VGG16	98.25	0.778	0.778	0.778	0.778

Bold highlights the performance parameters of the current research work in comparison to the existing onesalready reported in literature

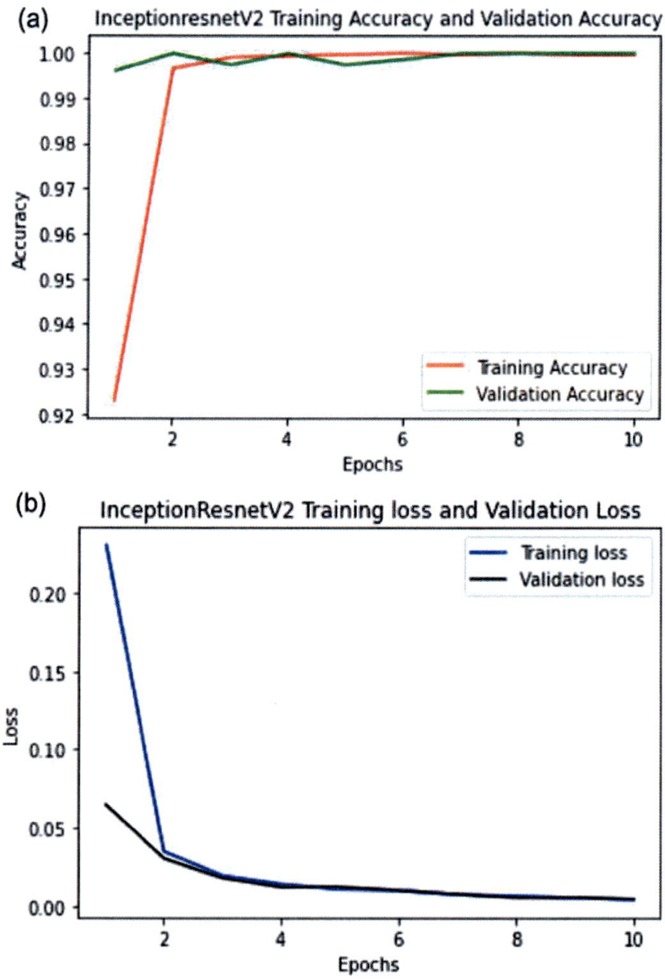

Fig. 3 Training versus Validation graph (InceptionResNet v2)

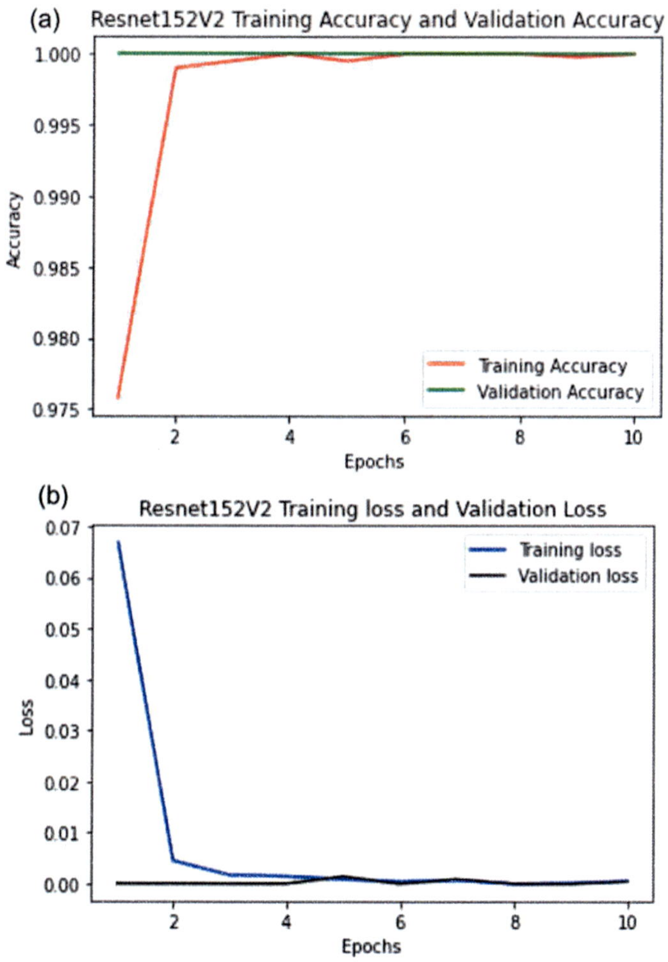

Fig. 4 Training versus Validation graph (ResNet152V2)

for the two CNN-based models (InceptionResNetV2 and ResNet152V2) used for detection purpose. After the training process, the detected output samples are shown in Figs. 5 and 6. A brief comparative analysis of the approaches implemented on MRL Eye Dataset [18] in different papers using CNN classifiers is reported in Table 2. From Table2, it can be witnessed that the models implemented in the present work give best results over existing models. The detection in this work is performed using Faster Region-Based Convolutional Neural Network (F-RCNN). Intersection over Union (IOU) of 0.8 is achieved with a threshold score of 0.7. F-RCNN gives better result over other detection results.

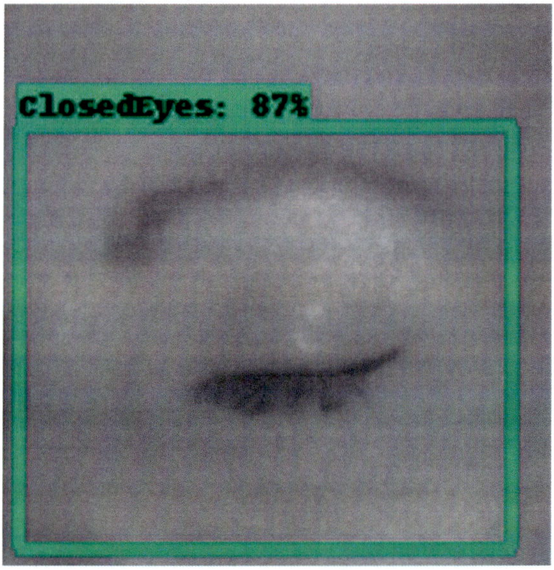

Fig. 5 Closed eye detected

Fig. 6 Open eye detected

5 Conclusion

In this paper, an algorithm for driver drowsiness detection in real time is proposed. This provides a solution which analyzes individual's eyes for detection of drowsiness. Here, the authors have presented the conception and implementation of a system for

Table 2 Comparative study of proposed approach with different other existing models using MRL eye dataset [16]

Papers	Classifiers	Accuracy (%)
Yeresime et al. [19]	CNN	86.05
Gordana et al. [20]	CNN (EAM)	90
Rahman et al. [13]	CNN (EyeNet)	99
Tran et al. [20]	CNN (Hybrid)	96
The proposed work	**CNN (InceptionResNetV2)**	**99.87**
The proposed work	**CNN (ResNet152v2)**	**99.99**

Bold highlights the performance parameters of the current research work in comparison to the existing ones already reported in literature

detecting drowsiness based on the vision that focuses to warn the individual if one is in a drowsy state. The framework portrayed in the present paper consists of a drowsiness detection technique based on Convolutional Neural Network (CNN) and this technique generated accuracies of 99.99% and 99.87%, respectively. The work was further validated using Faster RCNN for detection, and it generated promising results. Extensive training with different psychological behavior of driver can be beneficial in improving accuracy. This system can determine drivers' state under real day and night conditions. The main objective of this proposed work is to monitor the slight change in driver's eye state while driving a car. Thus, the approach adopted would be very beneficial to mankind for prevention of accidents occurring due to drowsy driving and would surely be helpful for saving human lives.

References

1. Rau PS (2005, June) Drowsy driver detection and warning system for commercial vehicle drivers: field operational test design, data analyses, and progress. In: 19th international conference on enhanced safety of vehicles, pp 6–9
2. Dwivedi K, Biswaranjan K, Sethi A (2014, February) Drowsy driver detection using representation learning. In: 2014 IEEE international advance computing conference (IACC). IEEE. pp 995–999
3. Reddy B, Kim YH, Yun S, Seo C, Jang J (2017) Real-time driver drowsiness detection for embedded system using model compression of deep neural networks. In: Proceedings of the IEEE conference on computer vision and pattern recognition workshops, pp 121–128
4. Babaeian M, Bhardwaj N, Esquivel B, Mozumdar M (2016, November) Real time driver drowsiness detection using a logistic-regression-based machine learning algorithm. In: 2016 IEEE Green Energy and Systems Conference (IGSEC). IEEE. pp 1–6
5. Yu J, Park S, Lee S, Jeon M (2018) Driver drowsiness detection using condition-adaptive representation learning framework. IEEE Trans Intell Transp Syst 20(11):4206–4218
6. Dasgupta A, Rahman D, Routray A (2018) A smartphone-based drowsiness detection and warning system for automotive drivers. IEEE Trans Intell Transp Syst 20(11):4045–4054
7. Zhao Z, Zhou N, Zhang L, Yan H, Xu Y, Zhang Z (2020) Driver fatigue detection based on convolutional neural networks using em-cnn. Comput Intell Neurosci

8. Sunagawa M, Shikii SI, Nakai W, Mochizuki M, Kusukame K, Kitajima H (2019) Comprehensive drowsiness level detection model combining multimodal information. IEEE Sens J 20(7):3709–3717
9. Tanveer MA, Khan MJ, Qureshi MJ, Naseer N, Hong KS (2019) Enhanced drowsiness detection using deep learning: an fNIRS study. IEEE access 7:137920–137929
10. Ngxande M, Tapamo JR., Burke M (2017, November) Driver drowsiness detection using behavioral measures and machine learning techniques: a review of state-of-art techniques. In: 2017 Pattern recognition association of south africa and robotics and mechatronics (PRASA-RobMech). IEEE, pp 156–161
11. Arefnezhad S, Samiee S, Eichberger A, Nahvi A (2019) Driver drowsiness detection based on steering wheel data applying adaptive Neuro-fuzzy feature selection. Sensors 19(4):943
12. Lin CY, Chang P, Wang A, Fan CP (2018, May) Machine learning and gradient statistics based real-time driver drowsiness detection. In: 2018 IEEE international conference on consumer electronics-Taiwan (ICCE-TW). IEEE, (pp 1–2)
13. Rahman A, Sirshar M, Khan A (2015, December) Real time drowsiness detection using eye blink monitoring. In: 2015 National software engineering conference (NSEC). IEEE, pp 1–7
14. Jabbar R, Shinoy M, Kharbeche M, Al-Khalifa K, Krichen M, Barkaoui K (2020, February) Driver drowsiness detection model using convolutional neural networks techniques for android application. In: 2020 IEEE international conference on informatics, IoT, and enabling technologies (ICIoT). IEEE, pp 237–242
15. Kusuma S, Udayan JD, Sachdeva A (2019, July) Driver distraction detection using deep learning and computer vision. In: 2019 2nd international conference on intelligent computing, instrumentation and control technologies (ICICICT), vol 1. IEEE, pp 289–292
16. https://www.kaggle.com/prasadvpatil/mrl-dataset
17. Suresh Y, Khandelwal R, Nikitha M, Fayaz M, Soudhri V (2021, October) Driver drowsiness detection using deep learning. In: 2021 2nd international conference on smart electronics and communication (ICOSEC). IEEE, pp 1526–1531
18. Jotanovic G, Jausevac G, Kostadinovic M, Damjanovic A, Brtka V (2021, March) Eye detection model for assessing the working capacities of employees. In: 2021 20th international symposium INFOTEH-JAHORINA (INFOTEH). IEEE, pp 1–5
19. Rahman MM, Islam MS, Jannat MKA, Rahman MH, Arifuzzaman M, Sassi R, Aktaruzzaman M (2020, February) EyeNet: An improved eye states classification system using convolutional neural network. In: 2020 22nd international conference on advanced communication technology (ICACT). IEEE, pp 84–90
20. Tran C, Namazi N (2021) Real-time detection of early drowsiness using convolution neural networks. Electron Imaging 2021(8):233–241

Chapter 10
An Intelligent Fire-Fighting Robotic System to Combat Fire in Coal Mines

Amlan Majumdar, Arka Das Chatterjee, Aniket Kirtania, Bidisa Das, and Piyali Mukherjee

1 Introduction

An intelligent fire-fighting robotic system can trace fire automatically and employing a bot it can outstretch to the focal point of fire and extinguish it by spraying CO_2 from the bot's manipulator pipe. Due to the high peril of catching fire inside the coal mine, this proposed advanced technique will help to blow out fire in a shorter period which in turn would reduce the chances for fire breakouts, thus resulting in lowering of losses due to death or property.

The main purpose of using robots inside the coal mine is that many times the human-defined rescue team reaches the firing spot much later when there already had been a fire outburst. This leads to an increased number of manpower being injured or even results in death cases. To minimize such occasions and lessen the risk, bots have proved to be an extremely friendly and powerful tool. The robot is modeled in such a way that it can reach the desired location and extinguish the fire automatically and swiftly.

Works in the field of robotics are a subject that is well documented in the literature. Robots have also been made to operate as firefighters in recent times as well. The design and implementation process of a safety firefighter robot with a dual controlling mechanism has been proposed by Rahat [1]. Additionally, an automatic fire extinguisher robot has been designed by Sampath [2]. All these systems are aimed primarily to extinguish the fire for domestic or industrial purposes. Very recently, researchers have shifted their focus to fight fire in coal mines as well. Zhao in 2017

A. Majumdar (✉) · A. Das Chatterjee · A. Kirtania · B. Das · P. Mukherjee
Department of Electronics & Communication Engineering, University of Engineering & Management, Kolkata, India
e-mail: amlanm980@gmail.com

A. Das Chatterjee
e-mail: daschatterjeearka@gmail.com

L. Mandal et al. (eds.), *Proceedings of International Conference on Computational Intelligence, Data Science and Cloud Computing*, Algorithms for Intelligent Systems,
https://doi.org/10.1007/978-981-19-1657-1_10

proposed a search-and-rescue robot system [3] for remotely sensing the underground coal mine environment. A few works in the related field have also been documented since then. This proposed work aims to design an intelligent system that can detect the source of fire as well as extinguish it without any human intervention. Moreover, the cost-effectiveness of the system would be immensely beneficial for its use in developing countries like India.

This paper introduces an overview of an advanced fire-fighting robotic system to fight a fire inside the coal mine. Section 2 describes the system design and introduces the working mechanism for the proposed system to work. In Sect. 3, the overview of the structure of the robot, which is the heart of the entire system, has been detailed. Section 4 describes the whole workflow of the designed system. Lastly, Sect. 5 concludes the proposed system with a light to the future scope of work that may be carried out in the future.

2 System Design

2.1 General System Architecture

In Fig. 1, the general system architecture has been proposed. This system is designed in such a way that it can trace out the origin of fire by itself and reach the spot automatically and extinguish the fire effectively. The sensory module, placed inside the mine chamber, consists of multiple sensors which can detect the changes in the different refractive index if any and detect the fire source. A Zigbee wireless connection has been set up between the sensory module and base station. Another

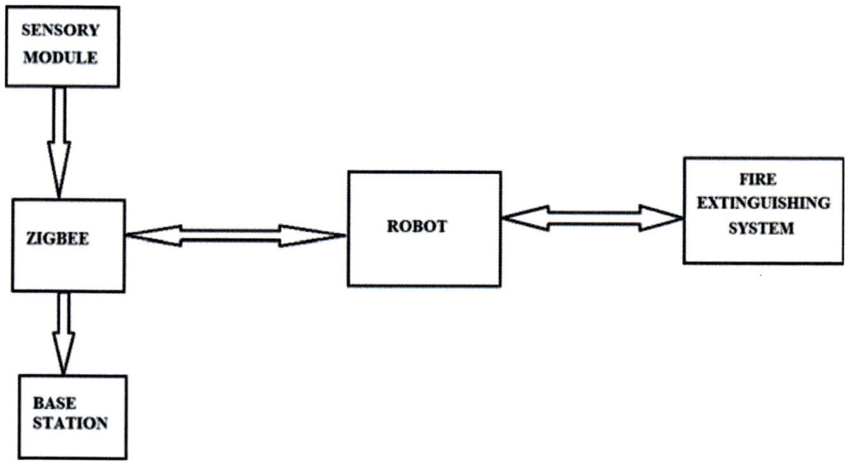

Fig. 1 General design architecture

Zigbee connection additionally connects the bot and base station. Depending on the sensor module, the bot is responsible for performing the fire extinguishing process. As per the function of the sensory module, the robot works efficiently and performs its operation.

2.2 *Sensory Module*

To keep a track of the changes in environmental information (e.g., temperature, humidity, etc.), a package of sensors are installed. The collection of air data from these sensors is to be sent to the microcontroller unit as its input. To trace the real-time data of the physical conditions within the chamber, this particular module is placed within it and is in turn responsible for transmitting those data to the base station via a Zigbee module. The changes in temperature, humidity, and presence of gases or flame, if any, within the mine are detected by these sensors. The individual sensors present in the proposed sensor module are described vividly.

Thermocouple Sensor: This sensor relies on the principle which employs two different metals which are co-joined at the terminals. On the occasion of heating or cooling, a voltage gets generated at the junction, thus depicting the correlation of temperature with the generated voltage. Thus, it provides an uncomplicated, vigorous, and cost-effective temperature sensor that can be used in a wide area of temperature measurement operation. The thermocouple sensor module shares its output with the aid of a SPI interface [4]. This sensor proves to be the best candidate for temperature measurement within the mine and hence is proposed to be employed in the system.

Humidity Sensor: This humidity sensor identifies the amount of water vapor in the gaseous environment and sends a report of the relative humidity of the air (RH). The humidity sensor operates in the voltage range around 3–5.5 DC volt and also gives an output of humidity approximately around 20–95% RH [5]. Substantially, the reason for using the humidity sensor inside the coal mine chamber is to identify the change of humidity and electrical component or the temperature of the air.

Flame Sensor: The flame sensor is responsive to flame or fire. The response of the flame sensor is faster, and the accuracy of this sensor is used in a wide area for the detection of flame or fire. It is also sensitive to the ordinary light source in the range of wavelength 760–1100 nm. It operates in the voltage range around 3.3–9 V DC. The flame detection of the sensor is greater than the temperature of 120 degrees approximately [6]. To avert the build-up of gas or catastrophic explosions, flame detector is being used. This sensor is sensitive to the flame and radiation within the mine and sends an immediate signal to the receiver.

Gas Sensor: To determine the concentration of gases in the environment, gas sensor is very indispensable in the system. The sensor is to be used as a gas sensor is cost-effective and can detect the presence of flammable gases inside the coal mine. Extensively, to measure the gas concentration and detect the poisonous gas inside the mine, it is used. The detecting range varies from 300 to 10,000 ppm [7]. Practically, it is responded faster to determine the proper output. To detect smoke, carbon monoxide, methane, and other flammable and poisonous gases, the gas sensor is very convenient in this system.

2.3 Data Transmission Employing Wireless Communication

Presently, the communication system which are used in the coal mine for security purposes are not much effective; thus, the innocent lives of coal miners might go into danger. Communication is the most important thing for safety purposes in the coal mine. But using a wired connection through a data transmission, cable might hamper the system during any kind of accident and may breakdown the whole communication process in a while. So, using wireless communication is better than wired communication inside the coal mine. Zigbee technology is one of the most convenient ways of transmission in wireless communication which is currently used in underground communication [8–10]. Zigbee technology is preferred over other wireless networks as this is cost-effective and also consumes less power. There is a variety of frequency range between 2.4 and 2.48 GHz (for lower ISM bands much other frequency range also can be applied). Due to its high reliability and short time delay response, it is a very effective network in the coal mine. Transceivers are used in multiple points in the systems which are described in this section. Transceivers are used to transmit and receive data (radio signals) for communication purposes using the combination of transmitters and receivers. Three transceivers connect the Zigbee module. One transceiver is connected to the sensor module that is placed within the mine; the sensors mainly include the flame sensor and the gas sensor that are used to detect the fire and also the presence of flammable gases in a particular place. Another transceiver is directly connected to the fire extinguisher robot, which is capable of extinguishing the fire with the help of the CO_2 sprays. The third transceiver is connected to the base station which has a fixed distance from the chambers of the mine. The base station will receive the signal in case of fire accidents through the wireless communication process using Zigbee. When the sensor module detects the fire or any toxic flammable gases inside its range, a signal is sent to the fire extinguisher robot and another signal is sent to the base station. After receiving the signal, the fire extinguisher robot will move to the desired location rapidly and will extinguish the fire using CO2 sprays. As soon as the fire is blown out, a signal is exchanged between the transceivers of the robot and the base station. This will thus lower the chances of spreading of fire within the mine and maintain the safety and security of humans as well as property.

2.4 *Fire Extinguisher System*

The proposed fire extinguishing system consists of a fire extinguisher cylinder filled with CO2. In fire extinguisher system, a vessel is attached to the cylinder to spray CO2 when an emergency arises. The direction of the fire extinguisher pipe is controlled by a robot manipulator that is connected to the cylinder vessel. At the time of fire accident when the robot reaches near a certain distance from the flame, a signal is sent to the microcontroller. With the help of the flame detector, the microcontroller sends a signal to the manipulator to fix the point to discharge the gas in the same direction where the flame detector detects the fire. After getting the signal from the microcontroller to the manipulator, the trigger of the cylinder gives pressure to the valve, and then, gaseous CO2 comes out through the pipe which was stored in high-pressure liquid form in the cylinder. After extinguishing the fire, the flame detector sends a signal to the microcontroller and the fire extinguishing system automatically turns off. CO2 is preferred over other gases because it can extinguish class *B* and class *C* fires [11].

3 Robot Architecture

3.1 *Overview of Robot Architecture*

The robot is designed to extinguish the fire in the coal mine very rapidly due to its higher mobility compared to humans. So, it is designed in such a way that it can move faster to the desired location, and with the help of the fire extinguishing manipulator, it can extinguish the fire. This intelligent robot can detect the source of fire all by itself.

The heart of the robot is a microcontroller, and the whole system needs a power supply to operate. Two lithium batteries are connected with the system to provide essential current for all the components of the system. Each battery of 12 V voltage and 2200 mA supplies the required voltage using linear voltage regulator ICs. The servo motors, wheels, microcontroller, and other electrical components of the system get the essential voltage from the batteries. The wheels of the robot system is based on caterpillar wheels concept. This type of robot wheel is used because it can move through uneven and rugged surfaces that are suitable for movement inside coal mines.

The transceiver is used to transmit and receive data simultaneously in an ordered manner. It is connected to the microcontroller to receive the signal from the transceiver module connected to the sensor package inside the coal chamber. Then, it transmits the signal to the microcontroller. On the other hand, when the fire is extinguished, it then transmits the signal to the base station to let know that the fire has been extinguished (Fig. 1).

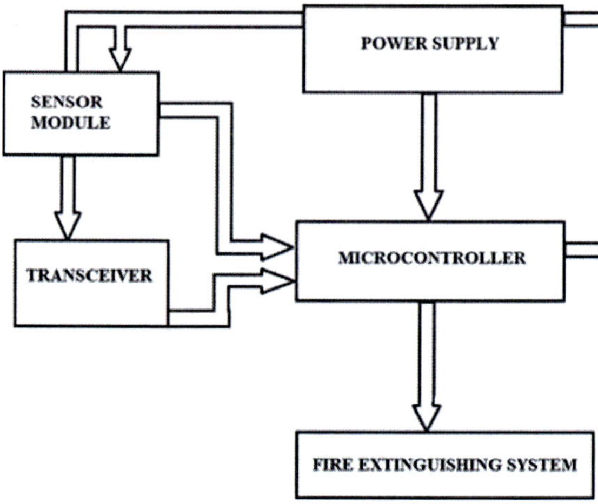

Fig. 2 Block diagram of robot architecture

3.2 *Microcontroller*

The robot comprises one Arduino Mega microcontroller. The specification of the microcontroller Arduino Mega 2560 is that it is an 8-bit board with 54 digital pins, 16 analog inputs, and 4 serial ports. There is a reset button and 4 serial ports (USART) so it can achieve speedy communication that is very much necessary in this system [12]. To protect from short and overcurrent, it has a resettable poly fuse. It is easy to code and has a user-friendly IDE and hence proves to be the best choice for its use in the proposed system. The microcontroller receives the input signal from the sensor module and the transceivers. The movement of the robot is controlled by a microcontroller, and also sensors are commanded by this controller unit.

3.3 *Internal Sensor Module*

The sensor module used in the robot is slightly different from the sensor module used in the coal mine chamber. This sensor module consists mainly of the two most vital sensors. These are flame sensors and proximity sensors that are to be associated with the robot.

Flame Sensor: The flame sensor used in the robot and the one that is placed in the mine are functionally the same, but the purpose of the use is different. The flame sensor is equipped in front of the robot. When the robot reaches the firing spot, by using the flame detector it can detect the flame and make the robot maintain a safe distance from the flame; otherwise, the robot itself may get burnt out. Additionally, the flame sensor will help to find the flame point so that the manipulator can discharge CO2 in the desired location.

Proximity Sensor: A proximity sensor is a commonly used sensor in automatic systems and robotics, which is mainly employed for detecting the nearby object and path planning motion. The sensor operates in the voltage of 5V, while the DC operating current is 15 mA. The sensor can detect a nearby object in the range of approximately 2cm-400cm based on the practical use [13]. Typically, it is installed with the robot body for the detection of the obstacles that may appear on the path covered by it. It conveys output to the microcontroller if there is any obstacle in its path.

3.4 Manipulator

A single manipulator is equipped with the body of the bot. A CO2 extinguisher cylinder is also equipped with the robot's body which is connected to manipulator. The servo motor, desired to be controlled by the microcontroller, is connected with the manipulator to move the gas pipe in a desired location and aid in the extinguishing process.

4 Workflow of The System

Coal mine chambers are mounted with a sensor module which includes a flame sensor, gas sensor, thermocouple sensor, and humidity sensor. The flame sensor detects flame inside the mine. After detection of flame, a signal is sent to the base station through the Zigbee connectivity module. Thereafter, signal transmission base controller gets notified about the throwaway. Subsequently, the robot system gets activated, starts moving to the chamber, and positions itself at close vicinity of the epicenter of the flare. A predefined path is programmed to the bot, and using the proximity sensor, it avoids the obstacles coming toward it or in its path. Thereafter, the robot embedded flame sensor also gets activated to easily find the epicenter of the fire. As soon as it detects the flame, the robot manipulator gets activated and starts emitting CO2 from the robot-mounted CO2 cylinder. When the fire finally gets extinguished, it will emit a signal to the base station with the help of the transceiver. The robot also returns to the base station. This is the overview of the detailed working of the proposed robot system which is defined to operate. A flowchart depicting the overall workflow has been depicted in Fig. 3.

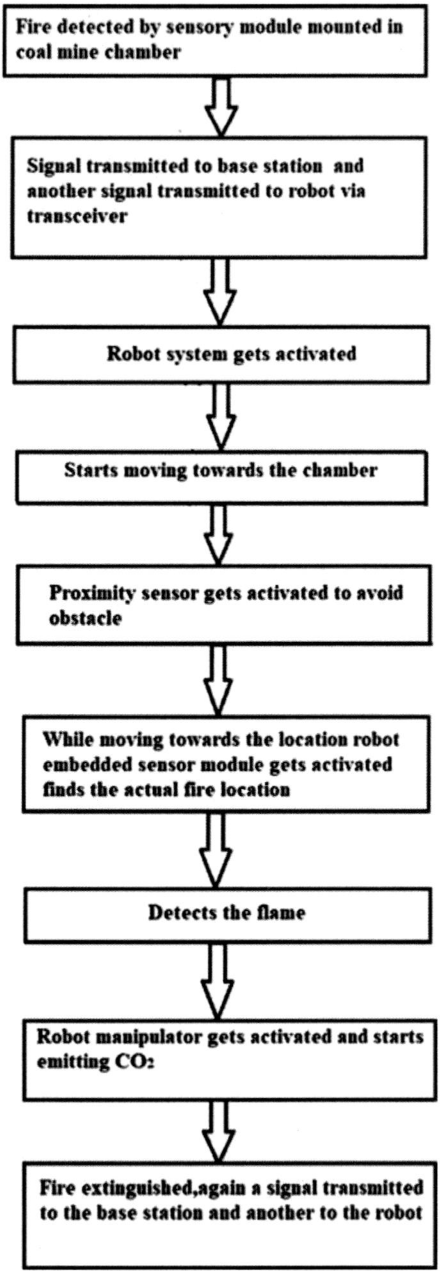

Fig. 3 Workflow of the system

5 Conclusion and Future Scope

In consolidation, the proper security and control system inside the mine are aimed to be replaced by an advanced technology employing an intelligent robot-based system that utilizes a wireless Zigbee module and fire brawling during catastrophe caused by an explosion or by flammable gas. In brief contention, the robot system module covers a wide area underground and it is capable of finding hazardous situations and is also efficient to rectify embryonic mishaps simultaneously. Moreover, this robot system module can be a game-changer to extinguish the fire in less amount of time to ensure the safety of the coal mine workers. This can be the finest augmentation of fire combatant. In the proposed security system, the sensor module monitors the humidity, temperature, and other flammable gases within the mine. Unexpectedly, if any flame gets detected, the bot immediately sends a signal to the transceiver after which the robot function would set to be in motion. In the robot framework, a sensor module has been additionally installed to detect the flame and as well to detect the obstacle in front of it to ensure that the bot is not damaged as well.

This system can be extensively used by implementing modest improvements. The crowded places like educational sectors, corporate sectors, industries, banks, etc. where this smart fire control and security system can be exerted. It is much reliable and will work faster until a rescue team reaches the spot and handle the situation if any embryonic mishaps occur.

The finest part of using this system is that the installation cost of the system is extremely low and hence beneficial for developing countries like India. Additionally, if some damages occur, only minimal parts of the robot have to be replaced not the entire system. Coating materials may also be used for resisting the external damages caused by the flames. The proposed system would thus prove itself to be the quantum leap for fire control and security system.

References

1. Rahat MRA (2020, June) Design and implementation of a safety fire fighter (SAFF) robot with dual controlling mechanism. In: 2020 IEEE region 10 symposium (TENSYMP)
2. Swetha Sampath B (2011, Nov) Automatic fire extinguisher robot. In: 8th international conference on ubiquitous robots and ambient intelligence (URAI 2011)
3. Zhao J (2017) A search-and-rescue robot system for remotely sensing the underground coal mine environment. Sensors 17:2426
4. https://en.wikipedia.org/wiki/Thermocouple
5. http://wiki.sunfounder.cc/index.php?title=Humiture_Sensor_Module
6. https://en.wikipedia.org/wiki/Flame_detector
7. https://en.wikipedia.org/wiki/Gas_detector
8. Gao D, Weiwei Li, Dai K (2016) Department of automation engineering, Tangshan Polytechnic College, Tangshan, 063200 Design of Coal Mine Intelligent Monitoring System based on ZigBee Wireless Sensor Network. In: International conference of mechanics, materials and structural engineering (ICMMSE 2016)

9. Yuana Y, Shena Z, Quan-fua W, Pei S a School of Information and Electrical Engineering, China University of Mining and Technology, Xuzhou 221008, China b Yan Zhou Mining Group Company, Jining 272100, China, Long distance wireless sensor networks applied in coal mine. In: 6th international conference on mining science & technology
10. Longkanga W, Baishenga N, Ruminga Z, Shengruia Z, Hailong L China University of Mining and Technology, Beijing 100086, ZigBee-based positioning system for coal miners. In: First international symposium on mine safety science and engineering
11. https://www.umb.edu/ehs/fire_safety/fire_extinguishers/carbon_dioxide
12. http://eprints.polsri.ac.id/4598/8/File%20VIII%20%28Lampiran%29.pdf
13. https://en.wikipedia.org/wiki/Proximity_senso

Chapter 11
Predictive Analysis of COVID-19 Using LSTM and ARIMA Models with Various Orders in India

Subhalaxmi Chakraborty, Arindam Chakraborty, and Pritam Kundu

1 Introduction

In India, in January, 2019, the first phase of COVID-19 was reported. According to statistics, India has the greatest number of COVID-19 confirmed cases in Asia [1]. The most challenging but essential part in this scenario is to predict the cases that may come under the attack of COVID-19. In Italy, Malato [2] predicted the spread of coronavirus using mathematical models and compared the logistic and exponential models. Chen and Yu [3] employed a second model to predict the coronavirus in China with cumulatively diagnosed cases in the first two months. Moreover, various deep learning approaches such as long short-term memory (LSTM), auto-regressive integrated moving average (ARIMA) can play a pivotal role to predict various aspect of different domain of science and technology [4–6]. Initially, LSTM was used to study, and forecasting of stock prices in China was done by Chen et al. [4]. Later, Wang et al. [5] used an upgraded model based on LSTM to predict spread of coronavirus. Deep learning with the rolling update mechanism has been used by them on the data given by Hopkins University. Shahid et al. [6] proposed models such as ARIMA, support vector regression (SVR), LSTM and Bidirectional LSTM which are used for forecasting of confirmed, deaths and recoveries in ten countries affected by pandemic. Forecasting is the most important but challenging part of time series data analysis. The time series analysis first uses linear regression and then moving average. Such technique is ARIMA which follows linear regression. ARIMA models perform suitably for predicting for the shorter period of time that is it works for short-term forecasts but fails to perform reasonably for long-term prediction. In India, this model has been used to forecast cases. It is a linear time series technique which has two parts as auto-regressive (AR) and moving average (MA). ARMA is formed by

S. Chakraborty (✉) · A. Chakraborty · P. Kundu
Department of Computer Science, University of Engineering and Management, Kolkata, India
e-mail: subhalaxmi2008@gmail.com

L. Mandal et al. (eds.), *Proceedings of International Conference on Computational Intelligence, Data Science and Cloud Computing*, Algorithms for Intelligent Systems,
https://doi.org/10.1007/978-981-19-1657-1_11

joining AR and MA, and it holds the properties of both of them. Hoong et al. [7] used a linear time series approach with the ARMA model so that network traffic on bit-torrent applications can be predicted. They collected data from a point-to-point bit-torrent network. Simple Network Prediction Studio (SNPS) follows ARMA to predict the bit-torrent traffic patterns. The outcome has been evaluated by MSE. This has also been mentioned that ARMA achieved better accuracy in short-term predictions. The integrated ARMA model is called the ARIMA model. Shu et al. [8] introduced the ARIMA algorithm and the model. The MAPE is used to evaluate predictions. This result concludes that the ARIMA model gives a more accurate result than the ARMA model. Traditionally, the ARIMA-based model was used for time series forecasting in fields of economics and finance [9, 10]. Meyler et al. [11] created ARIMA to predict Irish inflation. In their research work, they mainly focused on optimizing forecast performance while giving more attention to minimizing out-of-sample forecast errors. As ARIMA-based models work reasonably for short-term forecasts, it can be easily used to predict the total number of people getting infected with COVID-19 infections in the future. Contreras et al. [12] used ARIMA to predict electricity prices for next days. ARIMA and LSTM have been trained with data made with the data collected from the 'Worldometer' statistical website [13]. In this paper, analysis of two deep learning approaches, LSTM and ARIMA (with different orders), are used to estimate the COVID-19 cases in India. Moreover, in case of various ARIMA models: (1,1,0), (0,1,1), (2,1,0), (2,1,1), (0,2,1), various parameters namely RMSE, MAPE, MAE are measured. ARIMA model with various orders is analyzed to achieve better results in terms of the above-mentioned parameters. It is observed that ARIMA model with order (2,1,1) is recommended as it produced more accurate results.

2 Methodology Used

LSTM is one kind of recurrent neural network (RNN) [14]. This time RNN is betokened to evade long-term dependence quandaries and is opportune for processing and presaging time series. The LSTM model was made by Schmidhuber Jurgen and Hochreiter Sepp in 1997 [15]. The LSTM has a set of recollection cells which are unique that supersede the obnubilated neurons of the RNN, and the key is the recollection cells' state. The LSTM model filters information through the gate to maintain and upgrade the recollection cells' state. The LSTM have input, forgotten and output gates. There are three sigmoid layers, one tanh layer in each cell. Figure 1 is the structure of LSTM cells.

The forgotten gate determines the state information of the cell to be discard [15]. The memory cell accepts the previous output through ht-1 and the external information and combines the previous output and external information in a long vector through the σ transition function in Fig. 1. The general equation is:

$$F_t = \sigma(W_f * [h_{t-1}, X_t] + b_f) \quad (1)$$

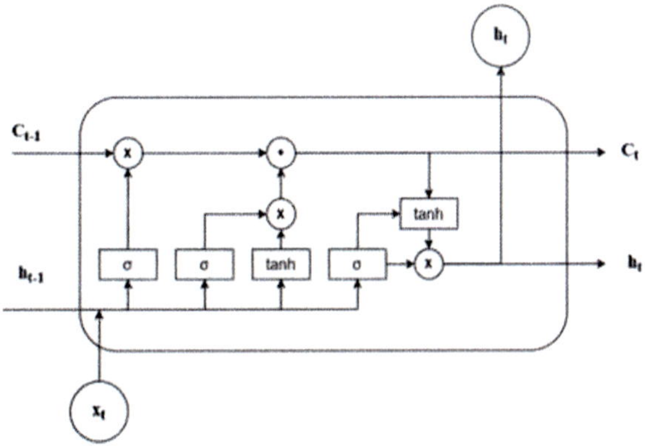

Fig. 1 LSTM recollection cell showing the layers and inputs

In the Eq. (1), *Wf* = Weight matrix, *bf* = Forgotten gate's bias, σ = Sigmoid function. This forgotten gate gives the output in terms of 0 and 1. Here, 0 indicates that the cell state information has been discarded completely, and the 1 indicates that the cell state information has been reserved completely. The input gate determines how much of the current time network input xt is reserved into the cell state *Ct*, which obviates non-essential content from entering the recollection cells. It has two functions. One, the cells' state to update, the value to be upgraded by the sigmoid, as in Eq. (2). Other updates the information to be updated to the state of the cell. An incipient candidate vector is engendered through the tanh layer to control how much incipient information is integrated, as in Eq. (3). Determinately, Eq. (4) is utilized to update the cell state of the recollection cells:

$$i_t = \sigma(W_f * [h_{t-1}, X_t] + b_i) \tag{2}$$

A new vector $\hat{C}$ is created through the tanh to control the added information. The vector function and the memory cell's final state function are as below in Eq. (3) and Eq. (4):

$$\hat{C} = \tanh(W_c * [h_{t-1}, X_t] + b_c) \tag{3}$$

$$C_t = f_t * C_{t-1} + i_t * \hat{C}_t \tag{4}$$

The output gate discards some portion of the current cell's state. The final output is given in Eq. (5) and Eq. (6):

$$O_t = \sigma(W_\sigma * [h_{t-1}, X_t] + b_O) \tag{5}$$

$$h_t = O_t * \tanh(C_t) \tag{6}$$

Figure 2 shows the steps to be followed to train the LSTM. The data is given into the proposed model. Then, it is trained based on the values chosen from the dataset. The training error and the validation error are evaluated to justify the accuracy of the model. The model is used for high accuracy. The model is trained for low accuracy. The corresponding RMSE, MAE, MAPE are also evaluated to measure the accuracy of the model. For modeling the time series, ARIMA can be the best choice because these models are linear as future values are cramped to be linear functions of the past given training data. Kurawarwala and Matsuo [16] calculated the seasonal demand. ARIMA has p, d, q where $p =$ Auto-regressive no.; $d =$ Differences no.; $q =$ Moving averages no. *Yt* can be expressed given by Eq. (7).

$$Y_t = \alpha Y_{t-1} + €_t \tag{7}$$

Literally, each visual examination consists of a desultory component (desultory shock, ε) and a linear coalescence of the anterior visual examinations. α_1 is self-regression coefficient. Short-term values may fluctuate around the averaged value, and the level of the series remains unchanged over a long span of time. A differentiation having order 1 defines the constant differences between the two successive values of Y. An integrated process can be defined as,

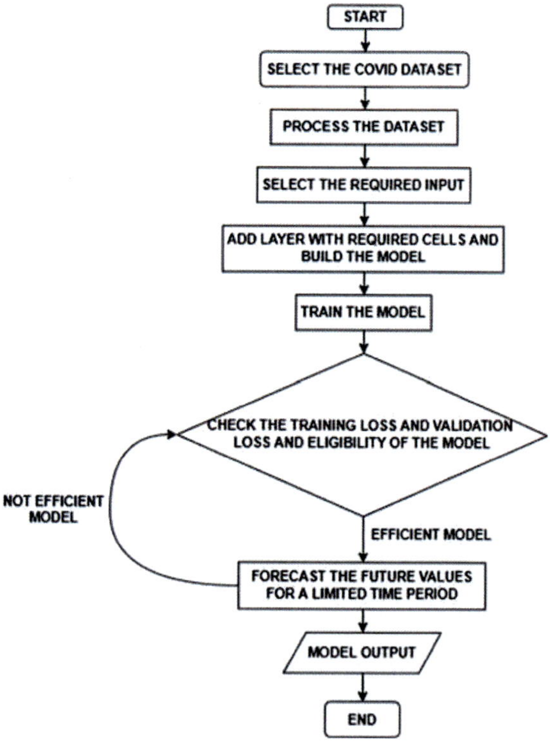

Fig. 2 Proposed flowchart of LSTM model

$$Y_t = Y_{t-1} + €_t \tag{8}$$

Here in Eq. (8), random perturbation $€t$ defines a white noise. The moving average order denotes the periods number in the current value. Moving average is defined by Eq. (9).

$$Y_t = €_t - \theta_1 €_{t-1} \tag{9}$$

Yule [18], Wold [19] founded Jenkins [17] and Box to perform ARIMA. It consists of three steps [20], parameter estimation with identification of model. Conclusively, we peregrinate to the diagnostic checking of model adequacy. In this last step, we ascertain that the hypothesis we made about the errors is contended. If the model fails to give the desired result, we have to change the parameters followed by the model validation. Diagnostic information can avail us come up with incipient models. Box-Jenkins model constitutes a process approach to follow and reiterate till reaching a high degree of contentment about the model and having truncated errors. Consequently, there are some limits for utilizing the ARIMA model. Nevertheless, once we apply the ARIMA model, we reach a high-quality in diametrical time series models.

Figure 3 shows the steps to train model. Different values of p, d, q are taken to train it. The various values of p, d and q are used to select the particular value of

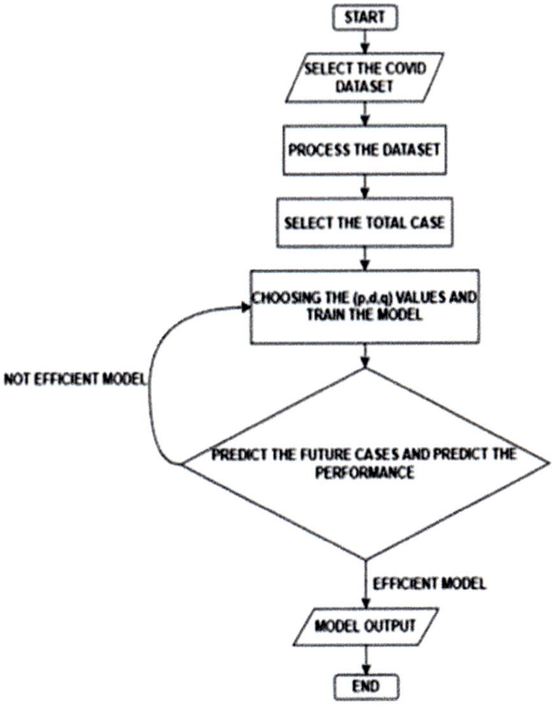

Fig. 3 Proposed flowchart of ARIMA model

these variables by evaluating the model against the training dataset where the model has lowest AIC value. The performance is calculated by RMSE, MAE, MAPE [21].

Formula for RMSE:

$$\sqrt{\frac{1}{n}\sum_{t=1}^{n} e_t^2} \tag{10}$$

Formula for MAE:

$$\frac{1}{n}\sum_{t=1}^{n} \left|e_t^2\right| \tag{11}$$

Formula for MAPE:

$$\frac{100}{n}\sum_{t=1}^{n} e_t^2 \tag{12}$$

In Eq. (10), (11), (12), e_t is the difference between the predicted value and observation.

3 Result and Analysis

In this paper, the models have been implemented using Python3 language and Jupyter Notebook. The model has two LSTM layers having 10 and 4 cells, respectively. The first layer has input shape (7,1), and the return sequence is true. The activation of both the LSTM layers is 'ReLU'. After the first two layers, a third dense layer has been added. While compiling, 'Adam' optimization is used and the loss is MSE. The forecasting of COVID-19 cases has been performed by LSTM model with different epochs, 25, 50, 75, 100, 125, 150 [22]. The corresponding training loss and validation loss are given in the Figs. 4a, b, c, d, e, f, respectively.

In Figs. 4a, b, c, d, e, f, the blue represents training loss, while the orange represents validation loss. The validation loss and the training loss have merged to a point for epochs 25, 50, 75, 100 and 150. So, for these epochs, we can get a valid model which may predict the total COVID-19 cases in India. For epoch 125, the validation loss and the training loss do not merge. So, for this case, the model cannot give the desired prediction.

In Figs. 5a, b, c, d, e, f, the graphs show prediction for LSTM. In the case of epoch 125, as that training error and validation error do not merge in Fig. 4, the graph cannot give the result. Because, it can be seen that the total COVID-19 case

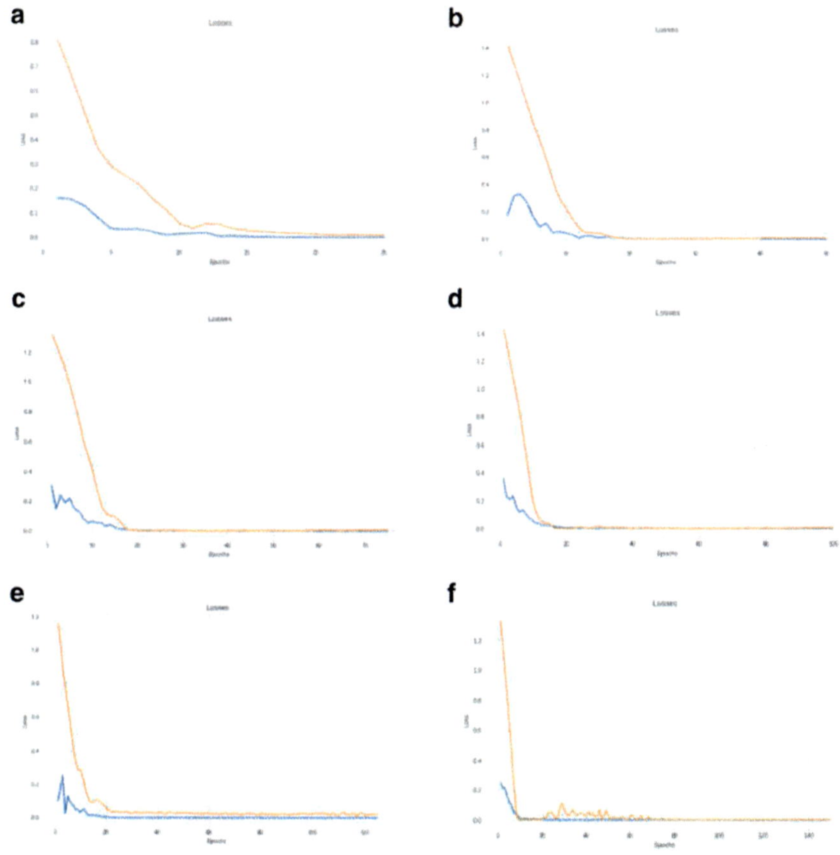

Fig. 4 **a** Graphical result of training loss and validation loss for epoch 25. **b** Graphical result of training loss and validation loss for epoch 50. **c** Graphical result of training loss and validation loss for epoch 75. **d** Graphical result of training loss and validation loss for epoch 100. **e** Graphical result of training loss and validation loss for epoch 125. **f** Graphical result of training loss and validation loss for epoch 150

predicted graph is decreasing in nature which is supposed to be increasing order. According to the valid graphs, the COVID-19 cases would have reached between 1.44 crore and 1.9 crore. While prediction of COVID-19 cases, using ARIMA model has been performed by taking different orders [22] of *p*, *d*, *q* and the accuracy has been measured in terms of RMSE, MAE, MAPE.

In case of Fig. 6a, it is observed that the value of MAPE is 0.0006897460234913194. It can predict that within two month, the total COVID-19-infected case will reach up to 35 crores. In case of Fig. 6b, it is observed that the value of MAPE is 0.004369794165458522. It can predict that within two month, the total COVID-19-infected case will not increase further.

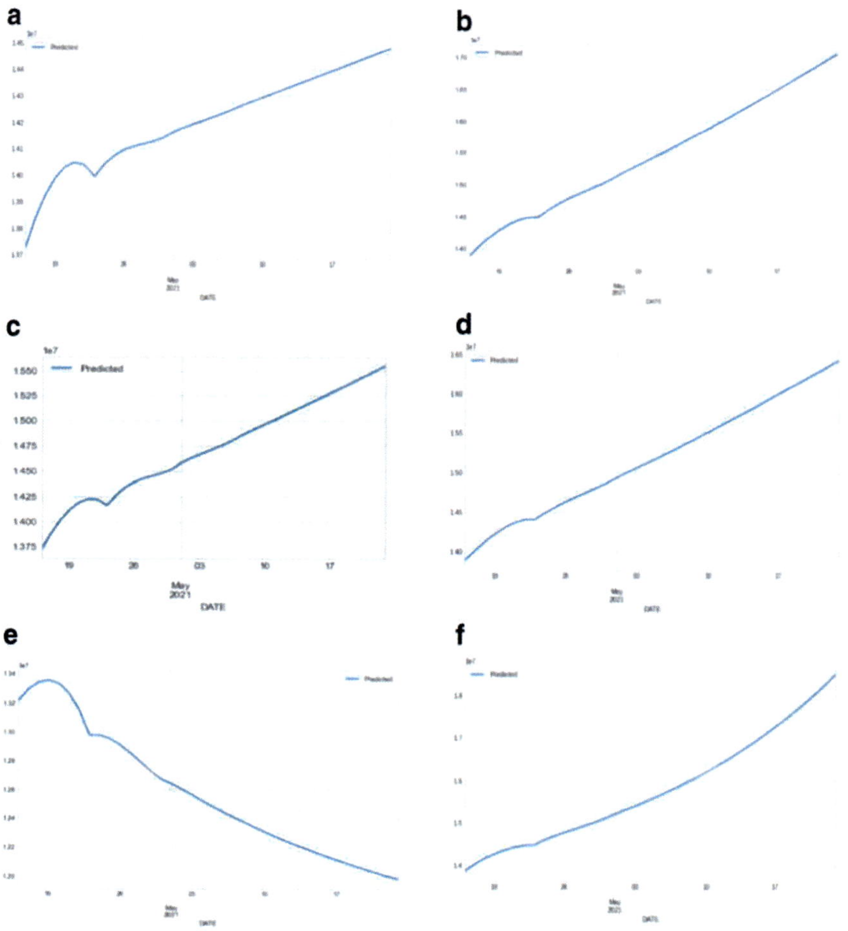

Fig. 5 **a** Graphical result predicted total case for epoch 25. **b** Graphical result predicted total case for epoch 50. **c** Graphical result predicted total case for epoch 75. **d** Graphical result predicted total case for epoch 100. **e** Graphical result predicted total case for epoch 125. **f** Graphical result predicted total case for epoch 150

In case of Fig. 6c, it is observed that the value of MAPE is 0.000693319121315967. It can predict that within two month, the total COVID-19-infected case will reach up to 35 crores.

In case of Fig. 6d, it is observed that the value of MAPE is 0.000659180090 0373274. It can predict that within two month, the total COVID-19-infected case will reach up to 70 crores.

In case of Fig. 6e, it is observed that the value of MAPE is 0.000659310389402987. It can predict that within two month, the total COVID-19-infected case will reach up to 70 crores.

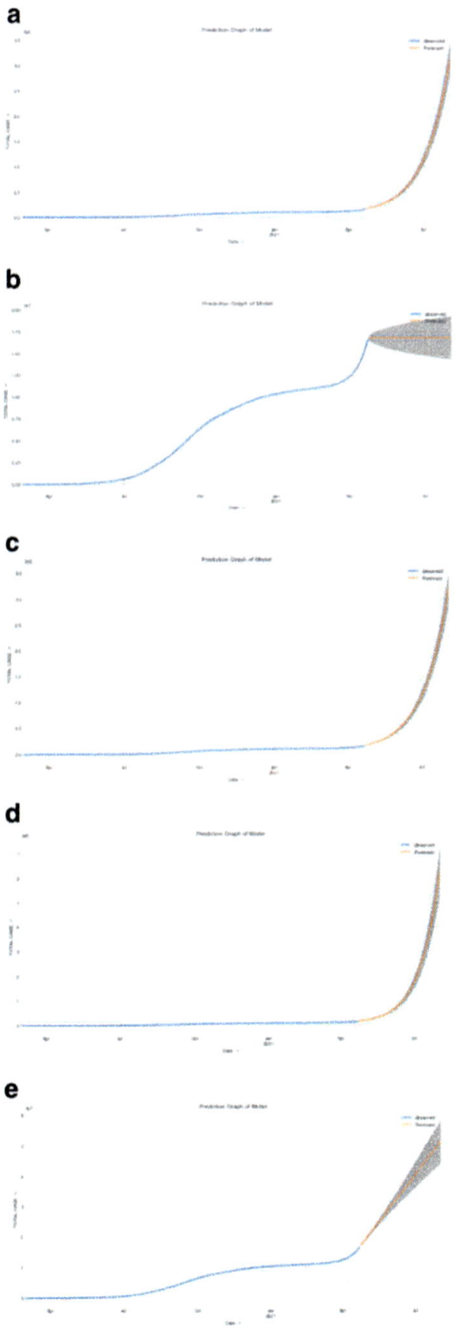

Fig. 6 **a** Total case prediction generated by ARIMA (1,1,0). **b** Total case prediction generated by ARIMA (0,1,1). **c** Total case prediction generated by ARIMA (2,1,0). **d** Total case prediction generated by ARIMA (2,1,1). **e** Total case prediction generated by ARIMA (0,2,1)

Table 1 Variation of RMSE, MAE, MAPE of ARIMA models of various orders (1,1,0), (0,1,1), (2,1,0), (2,1,1), (0,2,1)

Sl. no.	(*p*,*d*,*q*) values	RMSE	MAE	MAPE
1	1,1,0	6481.578536995546	4715.517425121931	0.0006897460234913194
2	0,1,1	40,380.38846040409	29,162.928628725473	0.004369794165458522
3	2,1,0	6475.601236132582	4732.3695921034205	0.000693319121315967
4	2,1,1	6231.700696636397	4413.200770891955	0.0006591800900373274
5	0,2,1	6929.070169182649	4773.767826495757	0.000659310389402987

Table 1 demonstrates the RMSE, MAE, MAPE of the ARIMA models. These values describe the error rate of the model. The model having the lowest RMSE value is most accurate.

The ARIMA (2,1,1) was used to develop COVID-19 case prediction model, and it has the lowest RMSE value. So, the model is considered to be more accurate.

4 Conclusion

Presently, deep learning approaches happen to be very effective in predicting various parameters in various fields. Apart from various forecasting in the areas, namely stock price, traffic networks, these approaches play a major role in predicting the coronavirus outbreak so that the precautions can be taken earlier. It has been seen that deep learning models like ARIMA series models are very effective in predicting and any kind of forecasting work. The main moto of this paper is to build an effective model using the ARIMA and LSTM. Applying different orders of ARIMA model and using different epochs, cells and layers of LSTM, a decision is inferred to choose which model can perform considerably in COVID-19 prediction. In case of LSTM, it is observed that except epoch 125, all other epochs (25, 50, 75, 100 and 150) can build a valid model to predict total COVID-19 cases in long term as the training loss and validation loss have merged to a point. Moreover, ARIMA model of various orders, (1,1,0), (0,1,1), (2,1,0), (2,1,1), (0,2,1) are evaluated for short-term prediction. It is found from the variation of RMSE, MAE, MAPE parameters of the above-mentioned ARIMA model that the model (2,1,1) has the capability to predict more accurate result as all the parameters produce lowest value in this case. These prediction models can be applicable to other health and allied fields to calculate the upcoming wave of COVID-19 and thus prepare themselves for serving better so that no life is lost during the worst situation. In the future, there is a scope to include the model in the webpage and make CPU and memory utilization more effective so that if we have a large data in case of different countries, the model can perform well and predict the cases more accurately.

References

1. WHO Coronavirus (COVID-19) Dashboard I WHO Coronavirus (COVID-19) Dash- board with vaccination data situation by country territory or area
2. Bertolaccini L, Spaggiari L (2020) The hearth of mathematical and statistical modelling during the coronavirus pandemic:801–802
3. Chen X, Yu B (2020) First two months of the 2019 coronavirus disease (COVID-19) epidemic in China: real-time surveillance and evaluation with a second derivative model. Glob Health Res Policy 5(1):1–9
4. Chen K, Zhou Y, Dai F (2015) A LSTM-based method for stock returns prediction: a case study of China stock market. In: 2015 IEEE international conference on big data (big data). IEEE
5. Wang, P et al (2020) Prediction of epidemic trends in COVID-19 with logistic model and machine learning technics. Chaos Solitons Fractals 139(2020):110058
6. Shahid F, Zameer A, Muneeb M (2020) Predictions for COVID-19 with deep learning models of LSTM, GRU and Bi-LSTM. Chaos Solitons Fractals 140:110212
7. Hoong PK, Tan IK, Keong CY (2012) Bittorrent network traffic fore- casting with ARMA. arXiv preprint arXiv:1208.1896
8. Shu Y et al (2005) Wireless traffic modeling and prediction using seasonal ARIMA models IEICE Trans Commun 88(10):3992–3999
9. Stellwagen E, Tashman L (2013) ARIMA: the models of Box and Jenkins. Foresight Int J Appl Forecast 30(2013):28–33
10. Ariyo AA, Adewumi AO, Ayo CK (2014) Stock price prediction using the ARIMA model. In: 2014 UKSim-AMSS 16th international conference on computer modelling and simulation. IEEE
11. Meyler A, Kenny G, Quinn T (1998) Forecasting irish inflation using ARIMA models:1–48.
12. Contreras J et al (2003) ARIMA models to predict next-day electricity prices. IEEE Trans Power Syst 18(3):1014–1020
13. India COVID: 33,953,475 Cases and 450,621 Deaths World Meter Website, cases of India
14. LeCun Y, Bengio Y, Hinton G (2015) Deep learning. Nature. Google Scholar Google Scholar Cross Ref Cross Ref 521(7553):436–444
15. Abedinia O, Amjady N, Zareipour H (2016) A new feature selection technique for load and price forecast of electrical power systems. IEEE Trans Power Syst 32(1):62–74
16. Kurawarwala AA, Matsuo H (1998) Product growth models for medium- term forecasting of short life cycle products. Technol Forecast Soc Chang 57(3):169–196
17. Box GEP, Jenkins GM (1970) Control. Halden-Day, San Francisco
18. Yule GU (1926) Why do we sometimes get nonsense-correlations between Time- Series?--a study in sampling and the nature of time-series. J Royal Stat Soc 89(1):1–63
19. Wold H (1938) A study in the analysis of stationary time series. Diss. Almqvist & Wiksell
20. Ghosh S (2020) Forecasting of demand using ARIMA model. American J Appl Mathematics Comput 1(2):11–18
21. Yang, CT et al (2019) Implementation of a real-time network traffic monitoring service with network functions virtualization. Future Generation Comput Syst 93(2019):687–701
22. Moghar A, Hamiche M (2020) Stock market prediction using LSTM recurrent neural network. Procedia Comput Sci 170:1168–1173

Chapter 12
DeceptionTime: Predicting the Movement of Shares Using Momentum Indicators

Kushal Kapoor

1 Introduction

The democratising of data in the stock market has led to a lot of research in this domain. Stock market data with daily intervals have become a popular way for people to learn about data science and machine learning. That data follow a specific trend and are generally easy to predict. However, the shorter you reduce the period, you increase the volatility of the model. This increase in volatility makes such data hard to interpret for the algorithm. This causes a loss of accuracy.

Noisy data are a common problem for any machine learning engineer working for models to apply in the real world. Noisy data can often lead to negative results [1]. Normally, the data would be processed to remove such unwanted information from the data, but this unwanted data can contain valuable information that might give the model a better result. This model aims to reduce the dependency on data pre-processing to reduce noise, and we hand over more control to the model in deciding what is noise and what is not. This way, we do not need to artificially inject noise into the model, and during actual occurrences of noise, our model will be able to filter them out better. The only removal of data that we have done is to remove post-market data, which are faulty data, for our model.

Classification data that are biased towards one class can lead to mishaps in the prediction of new data, by overemphasising one class in the results or this case only predicting one class. The problems with the current InceptionTime make it hard for it to classify biased data and only predicts one class [see Results].

We have proposed a new architecture to solve this problem. We have used a lot of elements that were created to withstand volatile data. The model was built using InceptionTime [2] as an inspiration. Most of the changes made in the architecture

K. Kapoor (✉)
Kolkata, India
e-mail: kushalk173@gmail.com

L. Mandal et al. (eds.), *Proceedings of International Conference on Computational Intelligence, Data Science and Cloud Computing*, Algorithms for Intelligent Systems,
https://doi.org/10.1007/978-981-19-1657-1_12

were from InceptionTime. Here, we show that an analytical approach to creating the architecture of the model by adding and removing layers can give us better results.

Outline: The remainder of the paper is as follows. Section 2 talks about the existing works that have been published in the stock recognition field, Sect. 3 highlights the model's architecture and its unique features, Sect. 4 highlights the data, the causes of the problems and data augmentation, and Sect. 5 tabulates and compares the result.

2 Past Works

2.1 Research Conducted on Convolutional Filters

The deep neural and convolutional networks [3] have been used to show unmatched accuracy when testing on data from the stock markets. Such networks use activation functions and multiple hidden layers to replicate the structure of the dynamic stock markets.

For modern deep learning, one of the key contributions has been extracting features and representing them as a part of the learned model. The creation of CNN [4] is a prime example of this. The filters are tuned automatically to the utility function for the entire network to optimise. The application of CNNs in multiple domains, from object tracking [5] to segmentation [6], has been very successful. There have been a few published works that adopt CNNs to analyse financial microstructure data, but the existing architectures lack features that give them the flexibility to analyse data that might seem impossible to analyse.

2.2 Research Conducted on The Link Between Finance and Stock Market

Prediction of the stock markets has been an important topic of research for a very long time [7, 8]. Even after varied opinions, there are ample widely studies that show that stock markets are predictable [9–12]. There are two major ways that researchers have used to analyse the markets—traditional statistical models and machine learning data-driven models [13]. The traditional statistical models work under the assumption of the generation of time series is a parametric process [14]. There have been more studies that now show that stock market behaviour is complex which is highly nonlinear [15, 16]. Many machine learning techniques are capable of capturing and compiling such arbitrary nonlinear information [17, 18].

Deep learning has been previously used to analyse time series data in the share markets with their momentum [19]. Univariate time series forecasting [20] and sectoral stock selection [21] have been a couple of techniques where researchers have used deep learning for analysing stocks. Long short-term memory (LSTM) has a popular model to use for this task [22] with extensive research done on this and using multiple data sources and markets [23].

2.3 *Inception*

Inception network was created by Google Inc. [24]. The network was considered a landmark paper in deep learning. They created a model by stacking multiple CNNs and going deep in the convolutions. They did this to analyse the connections deeper than what they have done before. Inception networks have.

The use of inception modules has been seen before in financial markets [18]. They have been used to infer local interactions over different time horizons.

2.4 *Gaps in Existing Models*

The existing works either look at all the indicators separately or take a bunch into account. According to our research, no other study uses such a diverse set of momentum indicators in ultra-short-term data. Certain momentum indicators are good at predicting the outcomes in certain cases, and using multiple of them helps our model to create an accurate prediction, by using a combination of 12 different indicators. The lack of studies on the use of deep learning models on Indian stock markets makes this paper unique. In the context of the broader theme of artificial intelligence, the handling of noisy and biased data in classification makes our architecture unique. The usage of transpose convolutional filters has been virtually unexplored, and the uniqueness of the architecture makes this paper different. We have also used a new dataset for time series classification.

3 Proposed Model

The model comprises four modules—a modulated InceptionTime module, a convolutional module, a series of gated recurrent units and a parallel echo-state network. Deception modules with convoluted and transpose convoluted filters are used to automate feature extraction. This process is quite difficult in financial data, as it is noisy with a low signal-to-noise ratio. The weights are learned during the inference, and extra features are learned from the dataset. The GRUs help to capture the time dependencies produced from the resulting features extracted from the temporal data. The ESNs are used to compute the initial data for time series analysis.

3.1 An Overview

The input data are passed into three streams. The first stream of data goes into a deception module, a convolution module and an ensemble of echo-state networks. The inputs from the first deception module and the convolutional module are concatenated, and an activation function and dropout are added to the result. The new matrix is passed as inputs into another deception module and another convolutional block. The same process continues. The final matrix is passed through a series of three GRUs and then concatenates with the results from each of the echo-state networks. The final result is passed through a single dense layer, to give us a classification (Fig. 1). In the figure, The dark blue boxes represent the Decpetion Module, which is describled in detail in Sect. 3.1.5.

3.1.1 Reduction in Number of Layers

When you compare this model to the inception network, you notice the simplicity of this model. The reduction in the number of blocks helped reduce the complexity of the model. The marginal increase in accuracy compared to the increase in time needed to train the model showed us that two inception blocks are the best trade-off and provide the maximum value, compared to the number of blocks.

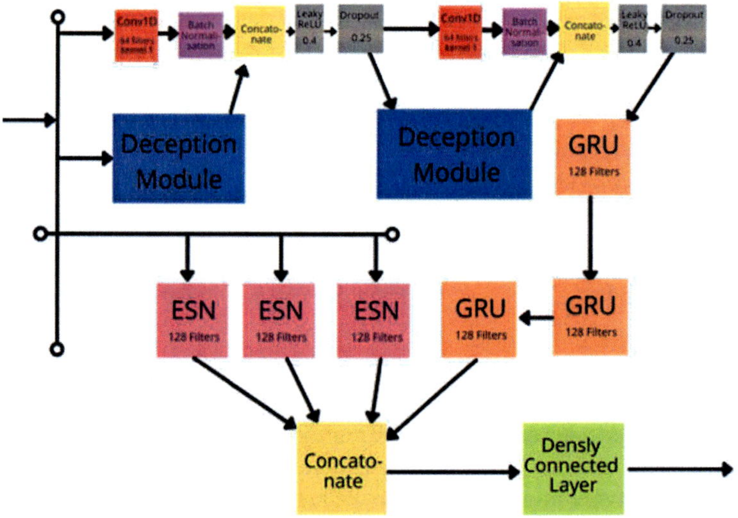

Fig. 1 Architecture of the proposed model

3.1.2 Echo-state Networks (ESN)

ESNs are the best RNNs used for noisy data. Echo-state networks have also been proven very effective to predict stock prices in the past [25]. The structure of ESNs makes them good for analysing and predicting noisy data. This fact is proven by the individual results, where the ESN outperformed the LSTM and GRU by 45% and 30%, respectively. Hence, in the architecture of the model, we have used ESNs running directly from input data streams [26]. I used echo-state networks paired with hyperbolic tangent activation function.

3.1.3 Gated Recurrent Units (GRU)

The data coming out of the final convolutional block do not account for time. Hence, we pushed the output into a system of GRUs in a series connection [see Fig. 1]. The reason for choosing GRU over LSTM came down to the core structure. LSTM has three gates and GRU has two. This gives the LSTM the ability to process longer series of time; GRU, on the other hand, gives lesser relevance to the information from the previous streams of data. This is due to the extra dropout layer in LSTMs which reduces the amount of data coming in from the new inputs. Since we needed a greater emphasis on more recent price movements in our model, the choice of GRUs was made over LSTM. The GRU was comprised of 64 units and followed by the hyperbolic tangent activation function.

3.1.4 The Convolutional Module

The convolutional block is made up of three layers: a convolutional layer, a batch normalisation and a concatenate layer. The convolutional layer is made on a part of the input and comprises a kernel of size 1. This addition to the network is a primary feature of the model and helps comprise.

3.1.5 The Deception Module

The deception module is at the core of this architecture. The data flow into a 1D convolutional block with kernel size 1 and then flow out into nine streams. The first stream of data goes through an average pooling layer. This average pooling layer again sends the resulting matrix into another 1D convolutional block with kernel size one. We then apply LeakyReLU with a constant of 0.4 and dropout with a value of 0.25. This result gives us the first matrix. The next eight streams of data are passed into 1D convolutional and 1D transpose convolutional layers. The kernel sizes of the layers are [1, 2, 4, 8] for the transpose convolutional layers and [16, 24, 32, 34] for the convolutional layers. The result from each of these layers is passed through a LeakyReLU (0.4) and a dropout (0.25). The result from all the nine layers is concatenated and returned as the result from the deception module.

Transpose Convolutional Networks

Transpose convolutional networks work just opposite to the convolutional networks. Convolutional neural networks are used to analyse patterns from a given dataset. Transpose convolutional filters are used here to help expand the dataset, but I have used these filters with the smallest weights and applied a convolutional layer to that.

Transpose convolutional networks do not compile results but create data that do not exist in original data. Such transpose convolutional networks are used as noise generators for our model. This helps us to make our model more resilient to noise. We have not used any extra addition of noise but have run it parallelly and have mixed the resulting matrix with the original data and the output from the convolutional filters. This addition of noise coupled with the usage of dropout layers has been a very important factor in our results.

The tests run on CNN, and InceptionTime derives that alone convolutional CNNs are not enough to give us accurate predictions. Since both models have only predicted one class in the tests, CNNs alone have been very disadvantageous on this category of data. Hence, the proposition to use transpose convolutional networks. The tests running with the addition of transpose convolutional networks give me the results that can distinguish between different classes (Fig. 2).

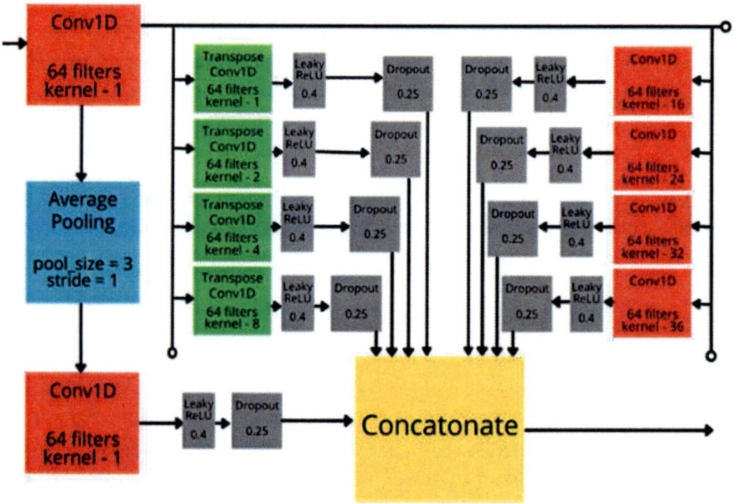

Fig. 2 Architecture of the deception module

3.2 How Does the Model Work?

The model takes in the input of the values of the current input data—the volume, price, momentum indicators, etc. and tries to compute the direction of a price change that we would see in the next second. Depending on the current position that the model holds, we get three different directions. Buy, Sell and Hold. The trades that the model recommends are squared off the next instant. The model squares off any old positions before taking a new position, if we get a buy signal, then the model buys the security. If we get a sell signal, the model shorts the security, and if we get a hold signal, the model maintains the current position. The hold position is required, because, in the original data, there are data points deciding between a buy and a sell which become so difficult that the model either chooses to dropout of the trade or repeat the last position, depending on how the user wants to set the programme.

4 Data

4.1 Why is This Data Volatile?

Stock market data in the short term, when each time step is held up independently, can be very noisy due to the fluctuations in price that initially seem illogical. The ad hoc movement of the shares driven by the traders adds another level of complexity to the model, where there is no clear indication of what will happen in the future. This behaviour exhibited by the shares can reflect the complex tasks of nature [27].

4.2 The Data

I wanted to test the model on varying levels of noise and levels of noise and bias. Each share has a unique composition. The reason for the selection of each share is to get a better idea on the different kinds of securities that trade on the Stock Market, with varying levels of activity in each. Table 1 is used to list each of the Securities used, and gives a link on how to access that data. Table 2 sorts each of the eight stocks in varying levels of noise and bias based on the activity. As a means of segregation, Datasets in which the Daily traded value in indian rupees was less than 60,000 Lakh or 6,000 Million Indian Rupees, were assumed as having low levels of noise and with a higher value, were assumed as having high levels of noise. Datasets where the split percentage between the classess (See sect. 4.4) was below 37% for all three classes were assumed as having low levels of bias, and the rest were assumed to have high levels or bias.

4.3 Momentum Indicators

The use of momentum indicators to predict movements has been around for quite some time. Momentum in the stock market is somewhat similar to its counterpart in physics. Linear momentum in physics aims to calculate the product of the mass of the object with its velocity. Similarly, momentum in the stock markets is a way of analysing past trends in the prices of the stock, and the velocity of price changes to predict the future price movement of the stock. See Table 3 for a list of which momentum indicators have been used, and how.

4.4 Creation of Classes

To create predictions, I created three classes—Buy, Sell, Hold. The goal of this model was to execute a trade every second, squaring off the previous trade before a new position is taken. The exception is the hold, where the model would hold the trade for another second.

4.5 Data Augmentation

To standardise the data, we normalised the data using a z-score without any emendation. The formula we used was- $\mathrm{Z} = (\mathrm{X} - \mu)/\sigma$

4.6 Data Split

The data was split into training and validation data. The training data were the first 66.67% of the dataset, and the remaining 33.33% were in the validation set. The testing data were taken out from 1.5.21 to 31.5.21. The data in both sets had between 350,000 and 400,000 data points, depending on the share we used.

5 Results and Conclusions

I tested the model, on data from 1.5.21 to 30.5.21. The back test was also run on 8 common machine learning models, the previous InceptionTime network, a fully connected CNN network, An ESN, GRU and LSTM model.

Due to discrepancies in the results, InceptionTime and CNNs have not been listed in the results tally. They have only predicted one class and ignored the other two, even though they received a higher accuracy score than the mod.

We henceforth conclude, from the table the benefit of this model. The usage of transpose convolutional layers and GRUs has proved useful. Further research in the use of transpose convolutional networks is warranted from the conclusions.

The data have been split into two tables. Table 4 has datasets 1–4, which contain experimental states for all different permutations of the level of bias and volatility. Table 5 includes another example of the permutations. These results help prove that the accuracy of DeceptionTime improves with the increase in volatility and accuracy.

Table 1 Details of shares used to test the model

	Name of share	Number of data points	API link
Dataset 1	Tata steel ltd	400,096	http://froggly.io/get_history/tatasteel_tick/2021-04-1/2021-04-30/demo
Dataset 2	Adani ports and special economic zone ltd	393,370	http://froggly.io/get_history/adaniports_tick/2021-04-1/2021-04-30/demo
Dataset 3	Cipla ltd	363,298	http://froggly.io/get_history/cipla_tick/2021-04-1/2021-04-30/demo
Dataset 4	Reliance industries ltd	401,314	http://froggly.io/get_history/reliance_tick/2021-04-1/2021-04-30/demo
Dataset 5	Adani enterprises ltd	380,346	http://froggly.io/get_history/adanient_tick/2021-04-1/2021-04-30/demo
Dataset 6	SRF ltd	283,346	http://froggly.io/get_history/srf_tick/2021-04-1/2021-04-30/demo
Dataset 7	United breweries ltd	217,228	http://froggly.io/get_history/ubl_tick/2021-04-1/2021-04-30/demo
Dataset 8	Infosys ltd	397,467	http://froggly.io/get_history/infy_tick/2021-04-1/2021-04-30/demo

Table 2 Classification of data into varying levels of noise and bias

Dataset	Data split of classes			Average number of shares traded per day	Daily Traded Value in inr	Volatility or noise level	Level of bias
Dataset 1	34.15%	34.28%	31.56%	21,234,858	111,834.30 Lakh	High	Low
Dataset 2	34.07%	32.92%	33.0%	7,521,213	53,069.68 Lakh	Low	Low
Dataset 3	27.3%	26.83%	45.86%	3,310,964	30,196.65 Lakh	Low	High
Dataset 4	26.33%	25.96%	47.7%	4,836,812	107,777.23 Lakh	High	High
Dataset 5	33.11%	33.52%	33.35%	9,853,280	104,532 Lakh	High	Low
Dataset 6	33.36%	31.70%	34.95%	1,018,333	24,105 Lakh	Low	Low
Dataset 7	24.26%	23.45%	52.27%	3,78.237	6,422 Lakh	Low	High
Dataset 8	25.82%	25.99%	48.17%	13,438,109	237,950 Lakh	High	High

Table 3 Momentum indicators used in the model and their purpose

Name of indicator	Definition	Purpose	Window of data
Simple moving averages	The moving average of the past *n* days	To determine the flow of previous prices	5, 10, 15 and 20 time steps
Exponential moving average	Moving of the average of the past *n* days, in which each data point is multiplied by a weight	To determine the way the previous prices are working with an emphasis on the most recent works	9,10 and 20 time steps
Moving average convergence divergence	The difference between two different moving averages	To determine the direction of change of price by finding out the strategy to use with the moving averages	
Bollinger bands	A pair of upper and lower bands aim to judge the state of the stock, based on previous prices	To judge if a stock is overbought or oversold based on the momentum and the price volatility of the stock	10 time steps
Chaikin oscillator	Measure the accumulation-distribution line of MACD	To predict oscillations of the price around the accumulation-distribution line	10 time steps

(continued)

Table 3 (continued)

Name of indicator	Definition	Purpose	Window of data
Force Index	An indicator that aims to measure the amount of force being applied on the stock which is changing the price	They can accurately measure the significant changes in the forces that affect the market	10 time steps
Rate of change	The velocity at which the price of the stock has changed	To identify the direction of change of price of the stock using the velocity of its price change	10 time steps
Commodities channel index	The difference between the current price and historical prices	To determine if a stock is overbought or oversold based on its historic levels	10 time steps
Volume weighted average price	The average price of a stock by taking into account the change in volume	To determine the average price of a stock for a period by only taking the trend and not sudden volume spurts	10 time steps
Relative strength index	The measurement of a stock's recent price changes	To determine if a stock is overbought or oversold based on its price changes	10 time steps
Stochastic oscillator	The comparison of the price of a security to its historic levels	To determine if a stock is overbought or oversold while also predicting reversal in momentum	10 time steps
Mass index	The comparison of the historic range between the highest and lowest trading price of a stock	To predict reversals of trends in prices and momentum	26 time steps

Acknowledgements Special thanks to Professor Sanjoy Kumar Saha of Jadavpur University, Kolkata, and Mr. Aishik Pyne for mentoring me through the process.

Table 4 Results of the model on datasets 1–4

Name of model	Dataset 1		Dataset 2		Dataset 3		Dataset 4	
	Accuracy	Loss	Accuracy	Loss	Accuracy	Loss	Accuracy	Loss
Extreme gradient boosting	0.3817	1.1268	0.3429	1.228	0.2512	1.4638	0.3839	1.1964
Logistic regression	0.3896	1.0809	0.3418	1.0963	0.5376	1.0107	0.4683	1.2204
Multinomial Naive Bayes	0.3614	1.0933	0.2934	1.1088	0.5358	1.0416	0.4674	1.058
K-nearest neighbours	0.354	11.2752	0.3029	12.8467	0.2652	15.6675	0.3349	11.6485
Decision tree	0.3652	21.9251	0.3561	22.2396	0.3121	23.7583	0.3454	22.6075
Random forest	0.4007	1.0848	0.4017	1.0902	0.3857	1.0922	0.4292	1.0627
Gaussian Naive Bayes	0.3363	1.38422	0.4332	1.8501	0.5156	1.1679	0.4612	1.6458
Quadratic discriminant analysis	0.3826	1.6415	0.4412	1.5835	0.2269	6.1882	0.2746	5.786
Linear discriminant analysis	0.3702	1.1228	0.3697	1.088	0.5419	0.9834	0.4705	1.1599
LSTM	0.245	3.9489	0.2908	7.041	0.3994	1.0928	0.3823	1.073
ESN	0.355	1.0806	0.3052	1.0907	0.3343	1.1095	0.3048	1.0927
GRU	0.2738	6.6694	0.3952	3.6886	0.1707	5.2588	0.3011	1.1099
DeceptionTime	0.4208	1.0914	0.4391	1.0714	0.5515	1.0345	0.5208	1.0014

Table 5 Results of the model on datasets 5–8

Name of model	Dataset 5		Dataset 6		Dataset 7		Dataset 8	
	Accuracy	Loss	Accuracy	Loss	Accuracy	Loss	Accuracy	Loss
Extreme gradient boosting	0.3202	1.1935	0.4153	1.2661	0.4486	1.1123	0.5645	0.9553
Logistic regression	0.3636	1.0872	0.3243	1.1103	0.5322	1.0117	0.6008	0.9918
Multinomial Naive Bayes	0.2765	1.1307	0.3144	1.09967	0.5296	1.0171	0.6119	1.0015
K-nearest neighbours	0.2892	14.8279	0.344	10.369	0.3596	9.6083	0.4119	9.1086
Decision tree	0.3519	22.3863	0.3414	22.7458	0.3726	21.6711	0.4	20.7248
Random forest	0.334	1.166	0.3763	1.1104	0.4585	1.044	0.5481	0.966
Gaussian Naive Bayes	0.4131	1.2792	0.3775	1.4564	0.4112	3.7656	0.6091	1.305
Quadratic discriminant analysis	0.2895	2.0516	0.3054	2.4347	0.457	8.8813	0.2021	3.0305
Linear discriminant analysis	0.4573	1.066	0.3457	1.1035	0.5372	1.0093	0.6096	0.9764
LSTM	0.2016	1.1403	0.2784	1.0798	0.2389	1.029	0.2013	0.992
ESN	0.3029	1.1367	0.2784	1.0908	0.4279	1.5979	0.3154	1.0699
GRU	0.2794	1.1279	0.1503	1.0167	0.5277	1.0213	0.1337	0.9008
DeceptionTime	0.4319	1.1204	0.3975	1.0954	0.5972	0.9987	0.6392	0.9962

References

1. Rani S, Anurag, Rao S (2019) Study and analysis of noise effect on big data analytics 8:5841–5850
2. Ismail Fawaz H, Lucas B, Forestier G et al (2020) InceptionTime: finding Alexnet for time series classification. Data Min Knowl Disc 34:1936–1962. https://doi.org/10.1007/s10618-020-00710-y
3. Prater A (2017) Comparison of echo state network output layer classification methods on noisy data. Int Joint Conf Neural Netw (IJCNN) 2017:2644–2651. https://doi.org/10.1109/IJCNN.2017.7966179
4. LeCun Y, Bengio Y et al (1995) Convolutional networks for images, speech, and time series. Handb Brain Theory Neural Netw 3361(10):1995
5. Wang N, Yeung DY (2013) Learning a deep compact image representation for visual tracking. Adv Neural Inf Proc Syst:809–817
6. Long J Shelhamer E, Darrell T (2015) Fully convolutional networks for semantic segmentation. In: Proceedings of the IEEE conference on computer vision and pattern recognition, pp 3431–3440
7. Ang A, Bekaert G (2006) Stock return predictability: Is it there?" The review of financial studies. J Latex Class Files 20(3):651–707
8. Bacchetta P, Mertens E, Van Wincoop E (2009) Predictability in financial markets: What do survey expectations tell us? J Int Money Financ 28(3):406–426
9. Bollerslev T, Marrone J, Xu L, Zhou H (2014) Stock return predictability and variance risk premia: statistical inference and international evidence. J Financ Quant Anal 49(3):633–661
10. Ferreira MA, Santa-Clara P (2011) Forecasting stock market returns: the sum of the parts is more than the whole. J Financ Econ 100(3):514–537
11. Mandelbrot B, Hudson RL (2007) The misbehavior of markets: a fractal view of financial turbulence. Basic books
12. Mandelbrot BB (2008) How fractals can explain What's wrong with wall street. Sci Am 15(9):2008
13. Agrawal J, Chourasia V, Mittra A (2013) State-of-theart in stock prediction techniques. Int J Advanced Res Electr Electron Instrum Eng 2(4):1360–1366
14. Cavalcante RC, Brasileiro RC, Souza VL, Nobrega JP, Oliveira AL (2016) Computational intelligence and financial markets: a survey and future directions. Expert Syst Appl 55:194–211
15. Cao Q, Leggio KB, Schniederjans MJ (2005) A comparison between Fama and French's model and artificial neural networks in predicting the Chinese stock market. Comput Oper Res 32(10):2499–2512
16. Sirignano J, Cont R (2018) Universal features of price formation in financial markets: perspectives from deep learning. arXiv preprint arXiv:1803.06917
17. Atsalakis GS, Valavanis KP (2009) Surveying stock market forecasting techniques–Part II: Soft computing methods. Expert Syst Appl 36(3):5932–5941
18. Zhang Z, Zohren S, Roberts S (2019) DeepLOB: deep convolutional neural networks for limit order books. IEEE Trans Sig Process 67(11):3001–3012. https://doi.org/10.1109/TSP.2019.2907260
19. Girshick R, Donahue J, Darrell T, Malik J (2014) Rich feature hierarchies for accurate object detection and semantic segmentation. In: Proceedings of the IEEE conference on computer vision and pattern recognition, pp 580–587
20. Vengertsev D (2014) Deep learning architecture for univariate time series forecasting. Technical report, Stanford University
21. Heaton JB, Polson NG, Witte JH (2017) Deep learning for finance: deep portfolios. Appl Stoch Model Bus Ind 33(1):3–12. https://doi.org/10.1002/asmb.2209
22. Lee MC, Liao JS, Yeh SC, Chang JW (2020, January) Forecasting the short-term price trend of taiwan stocks with deep neural network. In: Proceedings of the 2020 11th international conference on e-education, e-business, e-management, and e-learning. Academic Press, pp 296–299

23. Srivastava P, Zhang J, Eachempati P (2021) Deep neural network and time series approach for finance systems: predicting the movement of the indian stock market. J Organ End User Comput 33:204-226. https://doi.org/10.4018/JOEUC.20210901.oa10
24. Szegedy C, Liu W, Jia Y, Sermanet P, Reed S, Anguelov D, Erhan D, Vanhoucke V, Rabinovich A (2014) Going deeper with convolutions
25. Dan J, Guo W, Shi W, Fang B, Zhang T (2014) Deterministic echo state networks based stock price forecasting. Abstr Appl Anal 2014:1–6. https://doi.org/10.1155/2014/137148
26. Szegedy C et al (2015) Going deeper with convolutions. IEEE Conf Comput Vision Pattern Recognit (CVPR) 2015:1–9. https://doi.org/10.1109/CVPR.2015.7298594
27. Chowhan PK, Shukla V (2003) Volatility in Indian stock markets. Available at SSRN: https://ssrn.com/abstract=325041 or http://dx.doi.org/https://doi.org/10.2139/ssrn.325041

Chapter 13
Strategy for Margin Trading Based on Algorithm

Partha Sarathi Paul and Indranil Basu

1 Introduction

The stock price movements are stochastic in nature, so it is impossible to make correct prediction of stock price every time, however sophisticated the algorithm used may be. However, using real-time online brokers, who also allow operation in multiple exchanges simultaneously, trading on margin can be carried out on several selected stocks in many exchanges, with the purpose of diversifying risk and increasing profits.

The aim of the proposed algorithm is to take decision to BUY a stock if it predicts an upward movement of the stock beyond a set threshold in the next time step or take SELL decision if it is predicted to be going down beyond a certain set threshold or HOLD if the predicted value is within a threshold value. After running several iteration of the algorithm on actual stock price movement, it is shown that a significant advantage may accrue by trading in margin, over a period of time by frequent algorithmic trading.

2 Literature Review

Several publications were studied on stock price prediction to get an overview of the state-of-the art methods.

P. S. Paul (✉) · I. Basu
Institute of Engineering and Management, Kolkata, India
e-mail: parthasarathi.paul@iemcal.com

I. Basu
e-mail: indranil.basu@iemcal.com

L. Mandal et al. (eds.), *Proceedings of International Conference on Computational Intelligence, Data Science and Cloud Computing*, Algorithms for Intelligent Systems,
https://doi.org/10.1007/978-981-19-1657-1_13

Jin et al. [1] have proposed a scheme and theoretical basis for models with sentiment analysis, decomposition of the complex sequence of stock price by adopting empirical modal decomposition (EMD) and an improved LSTM algorithm.

Hagströmer et al. [2] have studied high-frequency trading (HFT) using data from NASDAQ-OMX and shown that about 70% of trading volume comes from HFT. They have also shown that HFT activity mitigates intraday price volatility.

Conegundes et al. [3] have observed in their study that by using the DRL-2 (deep reinforcement learning) method, best cumulative return and the best annual return may be achieved, which can beat the benchmarks.

Hansen [4] in his research has studied the relationship between and interaction dynamic of humans and models in finance by applying machine-learning algorithms for investment management and trading.

Zhou et al. [5] have studied the drivers and deterrents of computerized trading based on literature and regulatory investigations to assess the potential for future growth of computerized trading in Asia–Pacific economies. A comparison of algorithmic and high-frequency trading in Asia–Pacific region with global empirical and theoretical literature has been made.

Adebiyi et al. [6] use autoregressive integrated moving average-based stock price predictive model. Results show that ARIMA has a strong potential for short-term prediction and can compete favorably with other existing techniques for stock price prediction.

Maji et al. [7] employ machine learning (ML) on historical stock price data to ascertain future trend. Since ML has the potential to discover insights not humanly possible to decipher, it can be used to make accurate predictions. Long short-term memory (LSTM) is used here with company's future growth prospects for analysis.

However, none of the papers address the practical issue of how to trade online, with the help of existing online trading systems. Keeping this in view, a practical approach is presented in this paper which may be developed into an online trading system, for common investors who do not have enormous resources of a high-frequency trader and yet have a reasonable chance of succeeding in the stock market.

3 Basics of Margin Trade and Use of Algorithm

Efficient market hypothesis (EMH) says that stocks always trade at fair market price, and no undervaluation or overvaluation is possible. At any point of time, the stock price reflects all information so next movement up or down is random, hence impossible to predict according to a theory known as Random Walk theory.

There are others who believe, contrary to EMH, that it is possible to beat the market and that stocks deviate from their fair market price. This paper tests this, with practical stock price movements based on prices from stock exchange on a sample basis, and the results show that the latter is probably true. This opens up opportunity of making a profit from frequent trading.

Algorithmic trading is responsible for around 60–73% of all U.S. equity trading. According to SelectUSA, financial markets in the U.S. are the largest and most liquid in the world. In *India, algorithmic trading* is still less than 50%, and firms are relatively small in size.

There are essentially three types of algorithmic trading, depending on frequency of trade. In low-frequency trading, trading transactions occur in intervals of days, weeks, or months. In intraday trading, the trading transactions occur several times a day. In high-frequency trading (HFT), transactions can take place within a fraction of a second. HFT is carried out with high power computers that follow complex algorithms to analyze multiple markets and executes trades based on market conditions. It relies solely on price movements and market making processes with faster mechanism to make potential positive returns. Financial institutions develop proprietary algorithms for this purpose.

3.1 Artificial Intelligence (AI) and Machine Learning (ML) is Used in Trading in the Following Ways

- Pattern Formation
- Predictive Trading (Sentiment Based)
- Increased Trading Speed.

Pattern Formation

Artificial intelligence is used to analyze numerous data points to identify trading patterns and take rapid trading decisions, whereas humans cannot identify and build patterns with such speed.

Market Sentiment-Based Predictive Trading

Artificial intelligence is used to analyze news headlines, social media comments, and other sources to forecast the stock price movements. Company news in electronic text format is available from commercial providers like Bloomberg [10], public new Web sites, and twitter feeds. Automated systems can identify company names and keywords and make news-based trades before human traders can process the information.

Increased Trading Speed

AI facilitates trading in millisecond interval in today's fast-paced, technology-oriented functioning of stock markets.

Normal equity trading in stock market involves buying and stocks which you hold, by attempting to buy low and sell high (after holding the stock for some time, normally months or years). Here, we are talking of buying and selling at frequent intervals (may be several times in a day) using margin trading to increase leverage. Margin trading is allowed in India (in selected stocks) where one can trade stocks without holding them. The key purpose of the algorithm is to predict the direction of movement of a stock and then take BUY, SELL, or HOLD decision accordingly. The idea is to get a few predictions right and make large amount of money and not lose too much money on the wrong predictions made. Automated **limit orders** and **stop-loss** mechanisms have to be used to control losses. Margin trading allows chance of gaining whether the stock price goes up or down. Changes in stock price generally will be small, particularly, if intraday trading is considered. However, since the trading on margin is done in multiple of minimum lot sizes of 500 or 1000 stocks at a time, the possibility of significant gain is always there.

This proposal is primarily for trading with high leverage using margin trading, not for ordinary equity stock trading. Generally, transaction frequency is high for increasing return on investment. Some 200 large cap and mid-cap stocks are allowed in India, for margin trading or futures trading. There is also the requirement of minimum lot size for futures trading. Say HCL stock futures can be traded in lots of 500. In terms of money, the value would be of the order of 3–5 lacs for each trade. Of course, only say 10% percent of that amount has to be kept with the broker as margin money, at the time of trading. Futures margin list is available from service providers of DMAT account or brokers. Short selling is possible only in futures, not in normal stock. You need to exit on expiry (maximum 3 months), irrespective of whatever happens. Intraday trading is possible. In futures, profit or loss is accounted for on a daily basis called MTM or market-to-market settlement. For normal delivery, it happens on designated settlement day, i.e., T+2 day.

In Table 1, result of trading *xyz* stock on margin is shown. Say, the minimum lot size is 500 and the current price is 638 with margin money of 15%. That means approximately 50,000/- have to be maintained as margin with the broker to trade. Margins have a typical expiry date of 3 months beyond which margin has to be recalculated based on current stock price. It is to be noted that positive margin has to be maintained at all times, and if it goes negative, the situation is known as "margin call" and the trader has to make up from his bank account or by bringing in fresh funds. "Match" in the table below means that the algorithmic prediction/decision has

Table 1 Explains margin trades on a lot size of 500 with algorithmic trading decisions

Time step	Lot size	*xyz* stock price EOD	MTM	Algorithmic decision	Comment
1	500	638		BUY	Match
2	500	648	5000	BUY	Wrong bet
3	500	628	10,000	BUY	Match
4	500	658	15,000	HOLD	No trade
...	500	...	...	HOLD	No trade
$N-1$	500	928		BUY	Match
N	500	938	5000	HOLD	No trade
Total MTM			15,000		

gone in the correct direction (UP for BUY or DOWN for SELL) as predicted by the algorithm. HOLD means no transaction is made.

In the example below, after *N* transactions, which may be in a period of one/two/six/twelve months, the trader makes 15,000/- on a margin of 50,000/-. The trader makes 4 BUY decisions out of which 1 goes wrong and he loses money. In other time steps, the trader does not transact. In margin trade, the trader can make money even when the stock value is going down, by short selling. The table would also be similar for trading decisions for SELL decision, not included in the table.

3.2 Building a Reasonable Algorithmic Trading System

- A cost-effective algorithmic trading system that operates on the market every 15 mins (not in nanoseconds as in HFT) and yields benchmark beating returns may be possible to build. For reasonable ROI, intraday trading at 15-min interval would be sufficient.
- Using Google cloud for developing and running the algorithm. The algorithm may be shared within a group of individual investors and traders. Keeping the algorithm in cloud would also make implementation of new strategies and updates easier to deploy among users. This may not be as good as HFT, which uses FPGA-based hardware but should be adequate for trading every 15 mins.
- Start by using app-based software from home PC and try futures and options trading with much higher leverage than normal equity.
- Using a broker who operates in multiple exchanges in India and overseas to spread risk. Interactive Brokers [9] allows integrating third party software for trading.
- Costs to be considered
  - Broker fees (a broker operating in multiple stock exchanges in India and overseas may be selected to diversifying the portfolio. Indian government allows foreign trades in stocks up to 250,000 USD). Full service brokers, like ICICI or Axis Bank, are more expensive than no-frills brokers like Interactive Brokers [9]. Interactive Brokers charge typically 1% and lower for larger margin loans. It is also possible for several traders to pool-in and operate one account to avail volume discount.
  - What does it cost to get real-time data from exchange? (Often it is included in the fees charged by broker).
  - Cost of developing and maintaining the algorithm which may be based on simple indicator-based strategies to complex reinforcement neural network-based algorithms.
  - Costs of using Google cloud platform or other cloud services. The cost of which may be included in broker fees. It may be necessary to use cloud platform to be able to trade on multiple exchanges simultaneously.

  - Dedicated hardware allowing trading at 15-min intervals (may even be a PC at home which make it possible for manual intervention in your portfolio, if required at times).

- Algorithm has to back-tested to see performance. Confusion matrix on historical data can be checked. How many false positives and false negatives? The ROI must beat a benchmark like Sensex, after all costs are taken into account, before trying it out live.
- Thumb rule estimates show that even if it is possible to predict and ride an upward or downward movement in a stock, say 50 or 60 times in a year (with the provision for trading in the portfolio every 15 mins), enough profit may accrue. Even if the algorithm does not yield the right prediction and act on it, the resulting losses can be controlled by **stop-loss** mechanism provided by the online trading system. Considering gains and losses over a year, should yield much higher ROI and beat the benchmark. If after testing and validating the algorithm on historical data, it is found satisfactory, live deployment may be done.

4 Methods

Two methods of stock price prediction were tried and results analyzed before employing the method which gave slightly better mean square error (MSE) on a sample of actual stock prices from publicly available data from finance.yahoo.com [8].

In the first method, LSTM method was used. In this model, 50 LSTM units were used in the hidden layer and one output layer to predict a single numerical value. It uses the efficient Adam version of stochastic gradient descent, and it is optimized using mean square error (MSE) loss function. Once the model is defined, it was fitted on the training dataset. MSE was found out on the sample data, and the result is reported in Results section.

A manually operated algorithm was used next, and the operation and the steps are shown below along with the flowchart in Fig. 1.

- Check present stock price.
- Calculate volatility from last 15 prices taken at 1-min interval (can be seen on https://in.finance.yahoo.com/chart/).
- Calculate trend from last 15 prices.
- Simulate 10 or more scenarios, with trend and volatility as input parameters and get an estimated value of future stock price for each scenario. Take an average of all scenarios. If the average value is more than a certain threshold, BUY on margin, if the average value is less than threshold value, SELL on margin, and for intermediate values HOLD. Basically, every 15 mins takes this BUY, SELL, or HOLD decision.
- Cumulative normal inverse function has been used here; however, some other suitable functions which yield good MSE may be used.

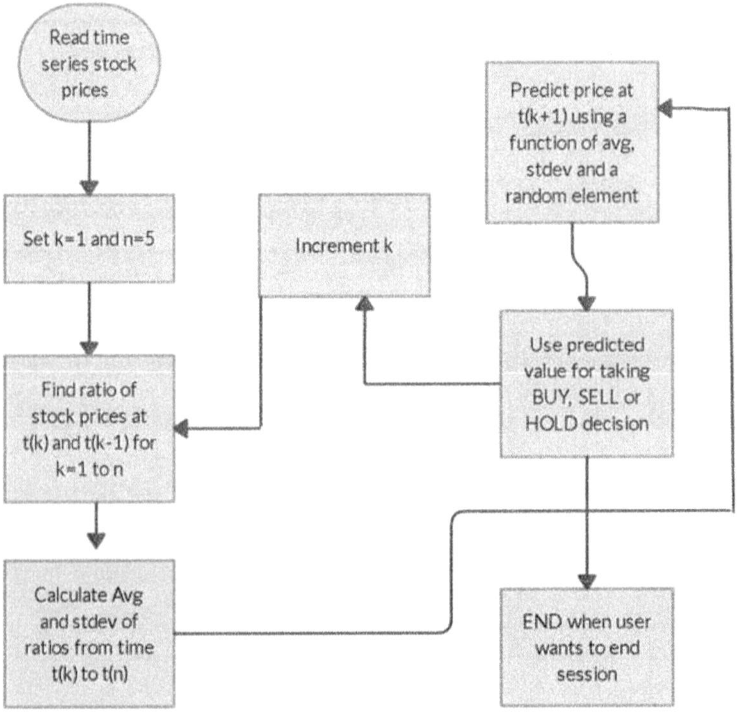

Fig. 1 Flowchart of algorithm used for stock price prediction

- Stop loss and limit order features may be used to minimize losses, if the bet goes the wrong way.

5 Results

As described in the Method section, the two methods resulted in the following MSE figures:

LSTM method MSE: 0.000586

Manual method MSE: 0.000493

Since the manual method yielded slightly better MSE, it was used to predict and take action on predicted values. The results of the simulation were checked with actual values and tabulated in Table 2. "Match" in the table means that the algorithm prediction or decision has gone in the correct direction (UP for BUY or DOWN for SELL), as predicted by the algorithm. HOLD means no transaction is made.

Table 2 Result of simulation on sample of stock prices from actual public data

Results						
Epoch	Match	Loss	HOLD	Total steps	Correct (%)	Incorrect (%)
1	14	7	38	59	67	33
2	13	8	38	59	62	38
3	13	4	42	59	76	24
4	17	12	30	59	59	41
5	17	7	35	59	71	29
6	14	5	40	59	74	26
7	18	6	35	59	75	25
8	13	9	37	59	59	41
9	15	3	41	59	83	17
10	18	11	30	59	62	38
11	16	9	34	59	64	36
12	12	7	40	59	63	37
13	19	5	35	59	79	21
14	12	9	38	59	57	43
15	13	8	38	59	62	38
16	15	9	35	59	63	38
17	11	6	42	59	65	35
18	17	7	35	59	71	29
19	17	6	36	59	74	26
20	12	10	37	59	55	45
21	10	11	38	59	48	52
22	17	6	36	59	74	26
23	13	8	38	59	62	38
24	15	9	35	59	63	38
25	13	7	39	59	65	35

6 Conclusion

It can be seen that after many simulations, correct decisions are almost always much higher than incorrect ones. This signifies that if the algorithm is employed for stock trading, there is a fair chance of making profit.

7 Future Scope of Work

- To develop an algorithm which can work on an online trading system, e.g., Interactive Broker.
- To back-test algorithm on many samples of data from different exchanges. Make modifications depending on the margin trading rules followed in that exchange.
- Get an online connection to Interactive Broker or equivalent brokers and get real-time stock market data. Use the data and see off-line how the predictions and recommended actions work.
- Once confidence id developed on the use of the algorithm, pay margin money as required and use the system in real time.
- Trading based on sentiments and news to be incorporated in algorithm, to make it sensitive to rapid movements of stock price.

References

1. Jin Z et al (2020) Stock closing price prediction based on sentiment analysis and LSTM. Neural Comput Applic 32:9713–9729. https://doi.org/10.1007/s00521-019-04504-2
2. Nordén L et al (2013) The diversity of high-frequency traders. J Fin Markets 16(4):741–773. http://dx.doi.org/https://doi.org/10.1016/j.finmar.2013.05.009
3. Conegundes L et al (2020) Beating the stock market with a deep reinforcement learning day trading system. http://dx.doi.org/https://doi.org/10.1109/IJCNN48605.2020.9206938
4. Hansen KB (2020) The virtue of simplicity: on machine learning models in algorithmic trading. journals.sagepub.com/home/bdshttps://doi.org/10.1177/2053951720926558
5. Zhou H, Kalev PS (2018) Algorithmic and high frequency trading in Asia-Pacific, now and the future. Pacfin. https://doi.org/10.1016/j.pacfin.2018.10.006
6. Ayodele AA et al (2014) Stock price prediction using the ARIMA model. In: 2014 UKSim-AMSS 16th international conference on computer modelling and simulation. Doi:https://doi.org/10.1109/UKSim.2014.67
7. Maji G et al (2019) Stock price prediction using LSTM on Indian share market. In: Proceedings of 32nd international conference on computer applications in industry and engineering, vol 63. Doi: https://doi.org/10.29007/qgcz
8. URL: https://finance.yahoo.com/?guccounter=1
9. URL: https://www.interactivebrokers.co.in/en/index.php?f=41254
10. URL: Bloomberg News - Wikipedia

Chapter 14
Variational Autoencoder-Based Imbalanced Alzheimer Detection Using Brain MRI Images

Soumyajit Maity, Rohit Prasan Mandal, Saranya Bhattacharjee, and Sankhadeep Chatterjee

1 Introduction

Alzheimer's disease (AD) is the most prevailing type of dementia. Approximately, 5.8 million people in the USA age 65 and older live with Alzheimer's disease. Of those, 80% are 75 years old and older [1]. Out of the approximately 50 million people worldwide with dementia, between 60 and 70% are estimated to have Alzheimer's disease. This disease destroys brain cells causing people to lose their memory, mental functions, and ability to continue daily activities. Initially, Alzheimer's disease affects the part of the brain that controls language and memory. As a result, Alzheimer's disease patients suffer from memory loss, confusion, and difficulty in speaking, reading or writing, reasoning, making judgments. Ultimately, Alzheimer's disease destroys the part of the brain controlling breathing and heart functionality which leads to death.

To detect AD in earlier stage from brain MRI report, it needs a medical expert. Now biomedical images are well-studied for machine learning and digital image processing [12]. Magnetic resonance imaging (MRI) is used to analyze the anatomical structures of the brain due to its high spatial resolution and ability to contrast soft tissue. Magnetic resonance has empowered neuroscience with a tool to investigate the inner structure and workings of the central nervous system. This work is pertaining to MRI over PET/CT scan because MRI that provides makes it easy for machine learning models to spot abnormalities in the brain [19].

S. Maity · S. Bhattacharjee
Department of Computer Science and Engineering, University of Engineering and Management, Kolkata, West Bengal, India

R. P. Mandal · S. Chatterjee (✉)
Department of Computer Science and Technology, University of Engineering and Management, Kolkata, West Bengal, India
e-mail: chatterjeesankhadeep.cu@gmail.com

L. Mandal et al. (eds.), *Proceedings of International Conference on Computational Intelligence, Data Science and Cloud Computing*, Algorithms for Intelligent Systems,
https://doi.org/10.1007/978-981-19-1657-1_14

Alzheimer's disease classification requires a large amount of data for accurate diagnose that is why it is challenging for clinicians to analyze large and complex MRI datasets and to extract explicit information manually. When the dataset is imbalanced, then the dataset itself becomes a problem statement. That is why variational autoencoder (VAE) comes into play. Using VAE we can easily handle the class imbalance problem for digital images [17].

In the rest of the chapter, we have discussed the background of variational autoencoders and latent vector formation from the original imbalance dataset. Then we have discussed the latent vector resampling techniques, and finally, the result and conclusion have been reported.

2 Proposed Method

In this section, we have described our proposal to perform the experiments in the mentioned steps: (1) we built an variational autoencoder (VAE) to generate the latent vector from the original imbalanced dataset. (2) To cope with the imbalance problem, we resampled the data using some popular sampling techniques. (3) Measure the performance of classifiers using some performance metrics. Figure 1 represents the flow of our proposed experimental methods.

2.1 Autoencoder

Autoencoder is an unsupervised machine learning algorithm which we have used to reduce the dimension of our original imbalanced dataset. Autoencoders are consist of two parts, encoder and decoder [5]. Encoder helps to reduce the dimension of the original data and convert it into a latent vector form. The decoder helps to reconstruct the encoded image and gives the output as close as input image. The encoder generates latent vector from the original image data this latent vector can be defined as:

$$z = \sigma(Wx + b) \tag{1}$$

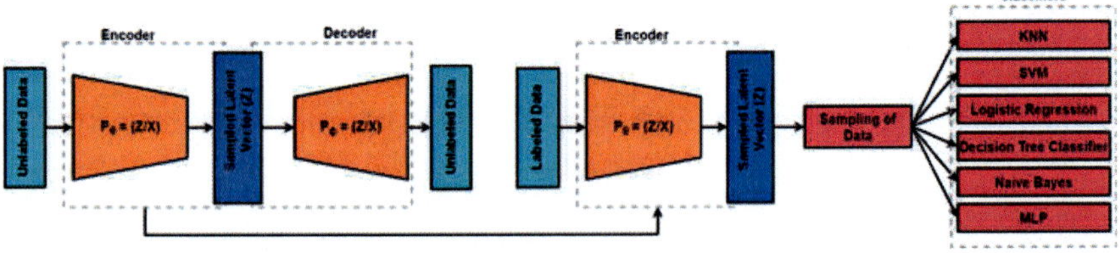

Fig. 1 Variational autoencoder is trained, and encoder is later used to obtain latent vectors

where z representes a latent vector form, σ is a activation function, W is a weight matrix, x is the input image, and b is a bias vector.

2.2 Variational Autoencoder

In the case of autoencoders, the latent vector we get from the encoder part only contains latent attributes but not in a probabilistic fashion. That is why we consider variational autoencoder (VAE) [10] for our experiment. VAE provides the latent vector which contains latent attributes as a probability distribution. But in this consent, a question may arise why not we are using principal components analysis (PCA) for the dimensionality reduction. The solution is, in the case of dimensionality reduction PCA only performs linear dimensionality reduction but VAE performs large-scale nonlinear dimensionality reduction also VAE can reduce the dimensionality fast and accurately without losing much information. The VAE model the input data as follows:

$$p_\theta(x|z) = f(x; z, \theta) \quad p(z) = N(z|0, I) \tag{2}$$

where f is a posterior probability function that uses deep neural network to perform a nonlinear transformation with z parameters. The exact computation of the posterior $p_\theta(z|x)$ in this model is not mathematically feasible. Instead, a distribution $q_\phi(z|x)$ [10] is used to approximate the true posterior probability. This inference network $q_\phi(z|x)$ is parameterized as a multivariate normal distribution as shown below:

$$q_\phi(z|x) = N(z|\mu_\phi(x), \text{diag}(\sigma_\phi^2(x))) \tag{3}$$

where both $\sigma_\phi^2(x)$ and $\mu_\phi(x)$ represent the vectors variance and means, respectively.

2.3 Latent Vector Resampling Techniques

Till now we have got latent vector by feeding the original imbalance data to our proposed VAE. Figure 2 depicts how our proposed VAE is generating latent vector from input image data. In this section, we are mainly focusing on the training biasing problem due to the imbalance in the dataset. In order to resolve this imbalance problem, we are applying some popular resampling techniques including oversampling methods (synthetic minority oversampling technique (SMOTE) [6], adaptive synthetic sampling (ADASYN) [9], borderline SMOTE [8] (for our experiment we used BorderlineSMOTE-1) and SVM SMOTE [13]) followed by some undersampling techniques (Cluster Centroid [20], Tomek's Links [16], Edited Nearest Neighbor (ENN) [18], Neighborhood Cleaning Rule (NHC Rule) [11] and All KNN [15]) and two hybrid sampling method, viz. SMOTE ENN [2] and SMOTE Tomek [3].

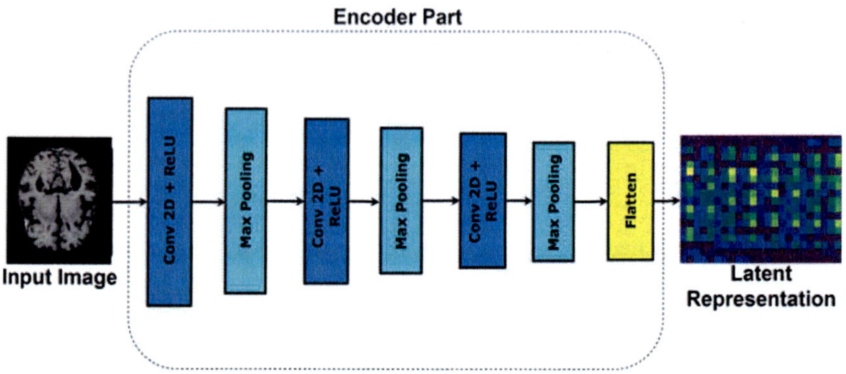

Fig. 2 Proposed encoder architecture

In the following, we are briefly describing the resampling techniques used in our proposed experimental method:

2.3.1 Oversampling Method

In the aspect of oversampling, the most primary oversampling method is random oversampling which randomly copies the minority class data and increases the no of samples to balance the data which can increase the probability of overfitting. So, we are using the oversampling methods, viz. SMOTE, ADASYN, Borderline SMOTE-1 & SVM SMOTE. In SMOTE, it calculates k-nearest neighbor (KNN) for minority class data and generates synthetic samples for minority class. Next the ADASYN, it also uses interpolation to generate the data samples for minority class but it mainly focused on generating samples based on the next original data samples which are wrongly classified using KNN. The borderline SMOTE represented its method by name itself. It is an option of the original SMOTE technique but it considers the borderline samples to generate the synthetic data. And another variant of SMOTE we have used, that is SVM SMOTE, it uses a SVM classifier to find the support vectors and then considers that it generates the new synthetic data. We used these two variants of SMOTE because they generate the data in the opposite direction of the nearest neighbor by considering the borderline data samples of the optimal decision functions which is more optimal oversampling solution than the previous methods.

2.3.2 Undersampling Method

In order to balance the data, if we consider the undersampling methods, the most simple random undersampling method randomly deletes samples from majority class and balances the data which can cause underfitting. Thus, we used the following controlled undersampling techniques which also help to clean the data more efficiently.

Cluster centroid is a method which replace the clusters of samples by using K-means algorithm. We used Tomek's Links, it removes the samples that overlapped between classes until all the samples are minimally distanced in a specific nearest neighbor. ENN reduces the samples of the dataset by editing the dataset, and it applies the nearest neighbor algorithm and removes the samples which are not agreeing enough with its neighbor. All KNN is nothing but the repetitive implementation of ENN but there is a difference ENN only considers the majority class in case of implementing nearest neighbor algorithm, but all KNN considers both the classes, and thus, the number of neighbors in the internal nearest neighbor increases in each iteration. Then NHC, it gives the priority to clean the data without considering the data. It uses the samples which are rejected by ENN and the output of three nearest neighbors. Basically, it uses ENN and KNN to remove the noise of the dataset.

2.3.3 Combination of Oversampling and Undersampling (Hybrid Sampling) Method

As hybrid sampling method, we have used SMOTE ENN and SMOTE Tomek which itself represents its working principle. It is already proved that the hybrid sampling method performs better than single oversampling or undersampling [3]. SMOTE + ENN describe that SMOTE is used to oversample the minority class data with synthetic data, and ENN is used to clean the noise from the data. And SMOTE + Tomek also uses SMOTE to generate synthetic data and uses Tomek Links to clean the noise from the dataset. It is how these hybrid sampling methods are working so efficiently. These techniques also increase the efficiency of the dataset to perform over the classifiers.

3 Results and Discussion

In this section, we evaluated different performance metrics of various validation classifiers on original imbalance data and synthetic data obtained by applying sampling methods on imbalance latent vector. A description of dataset is provided followed by the implementation of our experiment. Our simulations are implemented with NVIDIA GeForce GTX 1650, Ryzen 5-3550H, 8 GB RAM, Windows 10 Home 21H1, TensorFlow 2.5.0.

3.1 Experimental Setup

We have used a public dataset [7] for our experiment. This dataset contains of four classes, viz. '*Very Mild Demented*,' '*Mild Demented*,' '*Non-Demented*,' and '*Moderate Demented*.' In the training set, 'Non-Demented' class contains 2560 samples, 'Very Mild Demented' contains 1792 samples, 'Mild Demented' contains 717 samples, and 'Moderate Demented' contains 52 samples. So it is clear from the description of the dataset is, it is a multiclass imbalance problem where 'Non-Demented' is the most majority class and 'Moderate Demented' is the minority class. Dataset contains every image of size 96×96, and each image is a 1-channel grayscale image. Figure 3 represents some data samples of our used dataset.

In order to get the latent vector firstly, we trained the VAE by feeding training samples along with the test samples of our dataset. During this unsupervised training, ReLU activation function and RMSprop optimizer have been used to take leverage of nonlinear and nonsaturating characteristics. The value of learning rate is set to 0.001. All input images were normalized to [0, 1] before training the VAE. The trained VAE is then used to bring the input images down to a latent vector of size 12. However, even after transforming the images into latent vectors, it remains imbalanced.

To be confident about the performance improvement of validation classifiers, we applied six different classifiers, viz. K-nearest neighbor (KNN), support vector machine (SVM), decision tree (DT), logistic regression (LR), naïve Bayes (NB), and multilayer perceptron (MLP). Model hyperparameters are decided by applying grid search [4]. The classifiers are implemented using *scikit-learn* package [14]. We applied KNN classifier by setting number of neighbors to 10. SVM is used with the

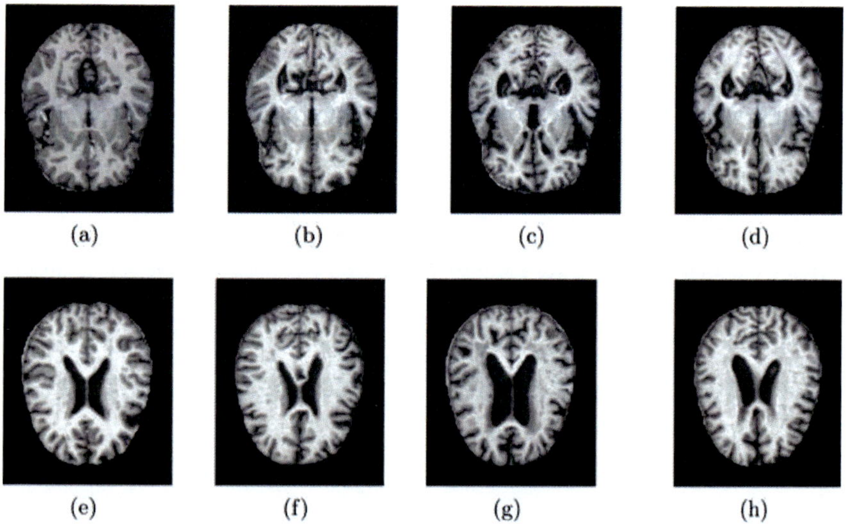

Fig. 3 Brain MRI images of non, very mild, mild, and moderate demented Alzheimer's cases

radial basis function (RBF) as kernel. Logistic regression is implemented by setting the 'lbfgs' solver and maximum iteration of 100. Decision tree classifier is applied with 'best' splitter. We set the variance smoothing attribute at 1e-09 for Naïve Bayes classifier, and in case of MLP classifier, we used RELU activation function, Adam solver also set learning rate at a constant value of 0.001 and maximum iteration of 200.

Performance of the classifiers is checked to check the performance of each synthetic data along with imbalance data over each validation classification method, and we applied some evolution metrics, viz. accuracy score, precision score, recall score, and area under curve (AUC) score. In order to ensure the stability of our results, we implemented tenfold cross-validation in every validation classifier. For every experiment, we divide all the data into 80% for training purpose and 20% for testing purpose.

3.2 Comparison of Classifiers

Table 1 represents the accuracy score obtained over various classification algorithms for different sampling strategies including imbalance data. The column *imbalance* represents the accuracy score of the data which we got in latent vector form by feeding the original imbalance data to our proposed VAE. As it can be shown the accuracy score of the imbalance data is very undesirable, they obtain a score of 0.58, 0.57, 0.57, 0.62, 0.54, and 0.60 on the classifiers, viz. KNN, SVM, LR, DT, NB, and MLP, respectively. But after applying the oversampling using SMOTE, the performance of the data has been improved. And the other sampling methods, viz. ADASYN, Borderline SMOTE, SVM SMOTE, Cluster Centroid, Tomek Links, ENN, NHC Rule, All KNN, and SMOTE Tomek also performed so well in the similar fashion. It is noticeable from Table 1 that for all the classifiers, viz. KNN, SVM, LR, DT, NB, and MLP the hybrid sampling method called SMOTE ENN performed best. For all the classifiers, it acquired an accuracy score of 0.98, 0.99, 0.99, 0.98, 0.97, and 0.98, respectively.

Table 2 depicts the precision score we got for every sampling method over different classifiers. In the same manner, Tables 3 and 4 also show the recall score and AUC score, respectively. From Tables 2, 3, and 4, we can observe that the imbalance data in every aspect obtained very poor score which is not expected as well. For the classifiers, viz. KNN, SVM, LR, DT, NB, and MLP, the *imbalance data* attain a **Precision** score of 0.58, 0.56, 0.57, 0.63, 0.59, 0.57, respectively; a **Recall** score of 0.58, 0.57, 0.57, 0.63, 0.54, 0.59, respectively, and a **AUC** score of 0.48, 0.54, 0.45, 0.57, 0.50, 0.52, respectively. Then after applying oversampling method using SMOTE, the **Precision**, **Recall**, and **AUC** score improved drastically. For all the classifiers, after applying SMOTE the overall **Precision** score has been improved by 60%, overall **Recall** score improved by 41% and overall **AUC** score has been improved by 74%. It can be seen from Tables 2, 3, and 4 that after applying the sampling methods, the performance improved over every classifier unexpectedly.

Table 1 Comparative analysis in terms of accuracy

	Imbalance	SMOTE	ADASYN	Borderline SMOTE	SVM SMOTE	Cluster centroid	Tomek links	ENN	NHC rule	ALL KNN	SMOTE ENN	SMOTE Tomek
KNN	0.58	0.89	0.89	0.88	0.87	0.88	0.89	0.92	0.91	0.9	0.98	0.91
SVM	0.57	0.93	0.93	0.94	0.93	0.94	0.94	0.95	0.94	0.93	0.99	0.95
LR	0.57	0.91	0.91	0.91	0.91	0.91	0.91	0.94	0.93	0.92	0.99	0.92
DT	0.62	0.9	0.91	0.91	0.91	0.91	0.91	0.92	0.91	0.92	0.98	0.92
NB	0.54	0.88	0.88	0.87	0.87	0.88	0.88	0.89	0.89	0.89	0.97	0.89
MLP	0.6	0.93	0.91	0.92	0.92	0.91	0.92	0.95	0.93	0.92	0.98	0.94

Table 2 Comparative analysis in terms of precision

	Imbalance	SMOTE	ADASYN	Borderline SMOTE	SVM SMOTE	Cluster centroid	Tomek links	ENN	NHC rule	ALL KNN	SMOTE ENN	SMOTE Tomek
KNN	0.58	0.94	0.93	0.9	0.96	0.93	0.93	0.98	0.97	0.96	1	0.95
SVM	0.56	1	0.97	0.96	0.99	0.99	0.99	0.99	0.96	0.97	1	0.99
LR	0.57	0.9	0.91	0.88	0.94	0.92	0.94	0.96	0.94	0.91	0.99	0.94
DT	0.63	0.93	0.87	0.85	0.91	0.9	0.96	0.93	0.93	0.93	1	0.93
NB	0.59	0.88	0.87	0.88	0.93	0.85	0.9	0.93	0.9	0.95	1	0.95
MLP	0.57	0.94	0.89	0.91	0.96	0.93	0.97	0.98	0.95	0.97	1	0.97

Table 3 Comparative analysis in terms of recall

	Imbalance	SMOTE	ADASYN	Borderline SMOTE	SVM SMOTE	Cluster centroid	Tomek links	ENN	NHC rule	ALL KNN	SMOTE ENN	SMOTE Tomek
KNN	0.58	0.77	0.83	0.82	0.87	0.81	0.78	0.89	0.86	0.85	0.98	0.88
SVM	0.57	0.84	0.9	0.88	0.92	0.86	0.88	0.94	0.89	0.89	0.99	0.93
LR	0.57	0.85	0.9	0.88	0.92	0.88	0.89	0.93	0.89	0.9	0.99	0.94
DT	0.63	0.88	0.94	0.91	0.92	0.93	0.91	0.96	0.92	0.9	0.99	0.93
NB	0.54	0.81	0.86	0.87	0.91	0.86	0.83	0.91	0.87	0.86	0.99	0.9
MLP	0.59	0.85	0.94	0.91	0.94	0.9	0.9	0.95	0.91	0.9	0.99	0.94

Table 4 Comparative analysis in terms of AUC

	Imbalance	SMOTE	ADASYN	Borderline SMOTE	SVM SMOTE	Cluster centroid	Tomek links	ENN	NHC rule	ALL KNN	SMOTE ENN	SMOTE Tomek
KNN	0.48	0.86	0.88	0.87	0.92	0.88	0.86	0.93	0.92	0.9	0.99	0.92
SVM	0.54	0.92	0.94	0.93	0.96	0.94	0.94	0.96	0.92	0.92	0.99	0.96
LR	0.45	0.88	0.91	0.89	0.93	0.9	0.91	0.94	0.91	0.89	0.98	0.94
DT	0.57	0.91	0.89	0.89	0.91	0.92	0.93	0.93	0.92	0.91	0.99	0.92
NB	0.5	0.85	0.87	0.88	0.92	0.86	0.86	0.9	0.87	0.9	0.99	0.92
MLP	0.52	0.9	0.91	0.92	0.95	0.92	0.93	0.96	0.93	0.93	0.99	0.96

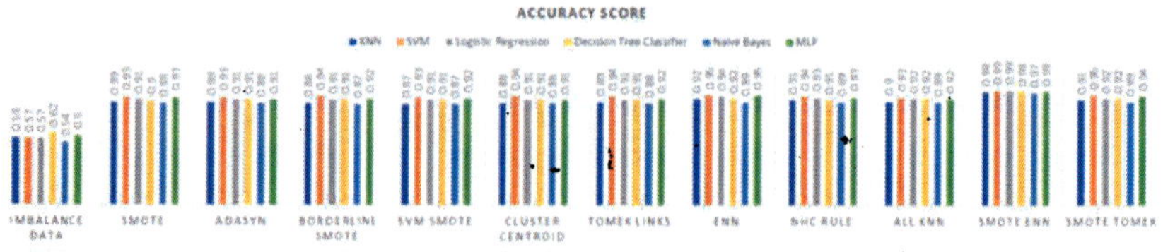

Fig. 4 Performance analysis in terms of accuracy score

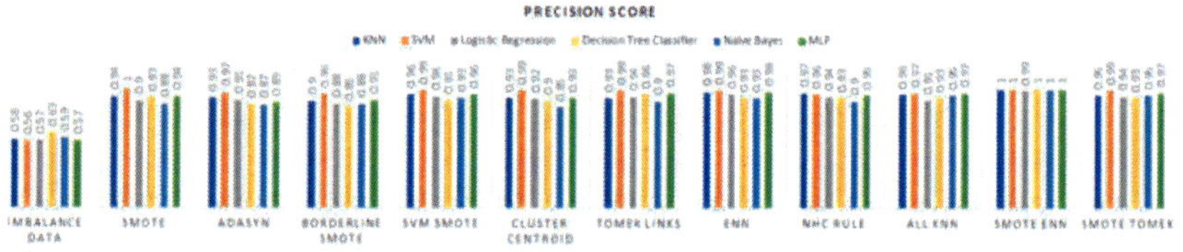

Fig. 5 Performance analysis in terms of precision score

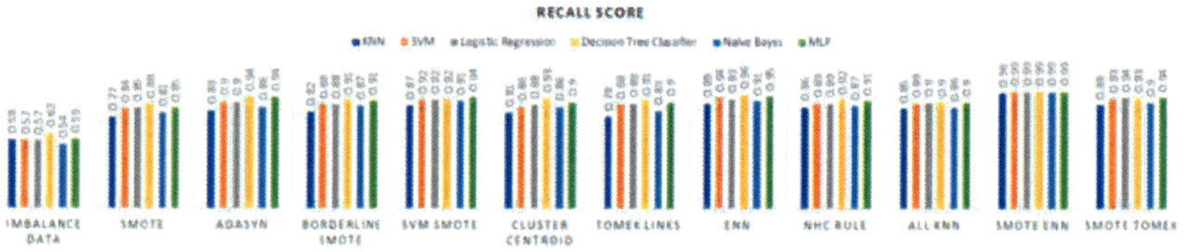

Fig. 6 Performance analysis in terms of recall score

And the hybrid sampling technique SMOTE ENN acquired the best performance score for all the performance metrics, viz. precision, recall, and AUC.

From Tables 1, 2, 3, and 4, it is evident that the imbalance data we used in our experiment which we get by feeding the original imbalance image data to our proposed VAE performs worse on every classifier. But after resample the data using some well-established oversampling, undersampling, and hybrid sampling algorithm, the synthetic data performs far better on every scenario. As it can be shown in Tables 1, 2, 3, and 4, the all results obtained for every performance metrics computation all the synthetic data performed much better than the imbalance data and there is a consistency over the result of each data over all the classifier.

It is obvious from Figs. 4, 5, 6, and 7 that the performance of all the classifiers to compute performance metrics used in our experiment on imbalance data has been improved after resampling the data. As shown in the figures that the validation classifier performance over some cases has been improved remarkably by applying SMOTE ENN hybrid sampling technique. In the other words, we can say that the sampling methods on our imbalance dataset work efficiently so that the drastic performance improvement can be seen over the sampled data in comparison with imbalance data.

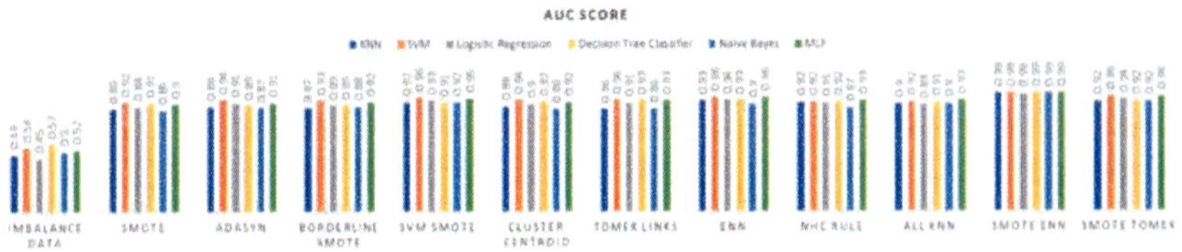

Fig. 7 Performance analysis in terms of AUC

4 Conclusion

In this chapter, we have focused on the detection of Alzheimer's disease using imbalance brain MRI image dataset. By using these raw brain MRI images, we have trained our proposed variational autoencoder to obtain the latent vector including latent attributes in a probabilistic manner. Due to imbalanced data, the latent vector we extract also has biasing problem. In order to cope with that we resampled the biased latent vector using resampling methods, viz. oversampling, undersampling, and hybrid sampling algorithms. Then, the balanced data is used to train and test different classifiers. The experimental result also reflects that the proposed method using VAE followed by resampling strategies has drastically improved the performance of classifiers. Another observation comes from our experiment, the SMOTE ENN hybrid sampling method has performed best among all the sampling methods we applied.

References

1. Arvanitakis Z, Shah RC, Bennett DA (2019) Diagnosis and management of dementia. JAMA 322(16):1589–1599
2. Batista GEAPA, Bazzan ALC, Monard MC et al (2003) Balancing training data for automated annotation of keywords: a case study. In: WOB, pp 10–18
3. Batista GEAPA, Prati RC, Monard MC (2004) A study of the behavior of several methods for balancing machine learning training data. ACM SIGKDD Explor Newslett 6(1):20–29
4. Bergstra J, Bengio Y (2012) Random search for hyper-parameter optimization. J Mach Learn Res 13(2)
5. Charte D, Charte F, García S, del Jesus MJ, Herrera F (2018) A practical tutorial on autoencoders for nonlinear feature fusion: taxonomy, models, software and guidelines. Inf Fusion 44:78–96
6. Chawla NV, Bowyer KW, Hall LO, Philip Kegelmeyer W (2002) Smote: synthetic minority over-sampling technique. J Artif Intell Res 16:321–357
7. Dubey S (2019) Alzheimer's dataset (4 class of images). https://www.kaggle.com/tourist55/alzheimers-dataset-4-class-of-images
8. Hui H, Wang W-Y, Mao B-H (2005) Borderline-smote: a new over-sampling method in imbalanced data sets learning. In: International conference on intelligent computing. Springer, pp 878–887
9. He H, Bai Y, Garcia EA, Li S (2008) Adasyn: adaptive synthetic sampling approach for imbalanced learning. In: 2008 IEEE international joint conference on neural networks (IEEE world congress on computational intelligence). IEEE, pp 1322–1328

10. Kingma DP, Welling M (2013) Auto-encoding variational bayes. arXiv preprint arXiv:1312.6114
11. Laurikkala J (2001) Improving identification of difficult small classes by balancing class distribution. In: Conference on artificial intelligence in medicine in Europe. Springer, pp 63–66
12. Lundervold AS, Lundervold A (2019) An overview of deep learning in medical imaging focusing on mri. Z Med Phys 29(2):102–127
13. Nguyen HM, Cooper EW, Kamei K (2011) Borderline over-sampling for imbalanced data classification. Int J Knowl Eng Soft Data Paradigms 3(1):4–21
14. Pedregosa F, Varoquaux G, Gramfort A, Michel V, Thirion B, Grisel O, Blondel M, Prettenhofer P, Weiss R, Dubourg V et al (2011) Scikit-learn: machine learning in python. J Mach Learn Res 12:2825–2830
15. Tomek I et al (1976) An experiment with the edited nearest-nieghbor rule
16. Tomek I et al (1976) Two modifications of CNN
17. Wan Z, Zhang Y, He H (2017) Variational autoencoder based synthetic data generation for imbalanced learning. In: 2017 IEEE symposium series on computational intelligence (SSCI). IEEE, pp 1–7
18. Wilson DL (1972) Asymptotic properties of nearest neighbor rules using edited data. IEEE Trans Syst Man Cybern 3:408–421
19. Yamanakkanavar N, Choi JY, Lee B (2020) Mri segmentation and classification of human brain using deep learning for diagnosis of alzheimer's disease: a survey. Sensors 20(11):3243
20. Yen S-J, Lee Y-S (2009) Cluster-based under-sampling approaches for imbalanced data distributions. Expert Syst Appl 36(3):5718–5727

Chapter 15
Breast Cancer Classification Using Deep Convolutional Neural Networks

Monalisa Dey, Anupam Mondal, Sainik Kumar Mahata, and Darothi Sarkar

1 Introduction

1.1 Motivation

In the recent past, breast cancer has claimed an alarming number of lives, both in India and abroad. In the early days traditional methods were used to identify and classify cancerous tissue. This process was workforce intensive and required human intervention to a large extent. Now with the advent of robust computer vision algorithms, the process has become simpler.

Cancer that develops from breast tissue is known as breast cancer. Computer vision used traditional techniques like line detectors, filters, and edge detectors for classification in the early 90s. With the advent of convolutional neural networks, we have come a long way in the twenty-first century. Convolutional neural networks are presently being used to accurately identify microscopic pictures of breast tissue in order to assess whether or not the tumor is malignant and what form of cancer it is. It has become important to find a more robust, fast, and accurate method for detection of cancer. Convolutional neural networks have been observed to perform well for such problems with added preprocessing techniques.

M. Dey (✉) · A. Mondal · S. K. Mahata · D. Sarkar
Institute of Engineering and Management, Kolkata, India
e-mail: monalisa.dey@iemcal.com

A. Mondal
e-mail: anupam.mondal@iemcal.com

S. K. Mahata
e-mail: sainik.mahata@iemcal.com

D. Sarkar
e-mail: darothi.sarkar@iemcal.com

L. Mandal et al. (eds.), *Proceedings of International Conference on Computational Intelligence, Data Science and Cloud Computing*, Algorithms for Intelligent Systems,
https://doi.org/10.1007/978-981-19-1657-1_15

In the current medical scenario, breast cancer is claiming an alarming number of lives. So, it has become absolutely essential to be able to identify a tumor as benign or malignant with almost perfect accuracy as there is no room for error.

Deep convolutional neural networks with suitable preprocessing has immense potential in simplifying and automating this process of classification and hence we have used it in the current work.

2 Related Works

According to world cancer report, among women, breast cancer has high morbidity and mortality. For identifying breast cancer, the histopathological diagnosis based on light microscopy is a gold standard. To perform breast cancer diagnostics, the materials retrieved in the operation room are first formalin-treated and then embedded in paraffin. The tissue is then cut by a high precision instrument and mounted on glass slides. After that, the slides are dyed with haematoxylin and eosin (HE) to make the nuclei and cytoplasm visible. Then, the pathologists visually inspect the histopathological slides under the microscope to perform diagnosis. The pathologists who are inspecting the slide need to have strong professional background and rich experience [1]. Though this method is traditional, but, manual diagnosis needs intense workload and diagnostic errors may occurs due to prolonged work of pathologists [2]. The above problems can be solved by designing intelligent diagnostic algorithms.

Automatic breast cancer histopathology image recognition schemes are designed by extensive pieces of literature. The algorithms of the literature typically can be classified into two categories. In the first of the two said categories, first, nuclei segmentation is performed and then handcrafted features, such as texture and morphological features are extracted from the segmented nuclei. The generated features are finally put into classifiers for automatic image type decision. In this work [3], the authors offer a huge, publicly accessible and annotated dataset made up of 7909 clinically relevant microscopic pictures of breast tumor tissue gathered from 82 individuals. Six handcrafted features, including LBP [4] and LPQ [5], as well as four standard classifiers, including 1-nearest neighbor (1-NN) and support vector machines (SVM), have been thoroughly assessed. To build acceptable features for this sort of technique, substantial effort and effective specialist topic knowledge are usually necessary.

Recent research shows that CNN-based algorithms outperform the best traditional machine learning method and achieve promising results. To increase diagnosis efficacy, the authors in [6] introduce deep learning. This improves the analysis of the histopathology slide and concludes that it holds great promise. The use of deep max-pooling CNN to detect mitosis, which is an important indicator of breast cancer, has been well written about in a paper [7]. The ICPR 2012 mitotic detection competition was won by the presented approach. To reduce training time, the DeCAF features are retrieved using a pre-trained CNN, and then a classifier is learned for the new classification job. To classify breast cancer histopathology images [8], both single-task CNN and multi-task CNN architectures are proposed. The generalization ability is insufficient in the second category as most of the CNN-based schemes adopt one single model to recognize cancer. In one of the research papers [9], the authors trained different patch-level CNNs and merged these CNN models to predict the final image label based on an improved existing CNN, and this method was observed to achieve great results on the large public breast cancer dataset [6].

The approaches based on CNN produce better outcomes than the first category. But because of high number of model parameters and higher computing burden in the inference stage, they are more complex than the traditional scheme. More layers and parameters are present in the recently designed networks, such as Reset [10] has more than 100 layers and 60 million parameters. This will cause several problems. The problem of classifying images can be addressed in two ways. Either the images can be considered as a whole or the images can be divided into patches of equal dimensions and then these patches are classified. We categorized the photos using patch-wise classification, which separated each image into patches and classified each of these patches separately.

3 Proposed Framework

3.1 Overview

Dataset Used

The dataset that have we used for this project is the breast cancer histology (BACH) dataset. The BACH dataset is publicly available and consists of 400 H&E (haemotoxylin and eosin) stain images of size 2048 $\times$ 1536 pixels. Color model: RGB. All the images are digitally collected with the same acquisition conditions to maintain parity, with a magnification of 200X and pixel size of 0.42 μm $\times$ 0.42 μm. Each image is assigned to one of four balanced classes: normal, benign, in situ, or invasive, with class according to the image's main cancer type. Two medical specialists annotated the photographs, and any images in which there was dispute were eliminated.

Data Preprocessing

Several preprocessing techniques were used, as shown in Fig. 1, to make our model learn features easily. Grayscaling is done so that the image sharpness increases also the light and dark areas are easily identifiable. Dataset normalization is done so that all the values of the data lie at zero-centered mean and a variance of one. To increase

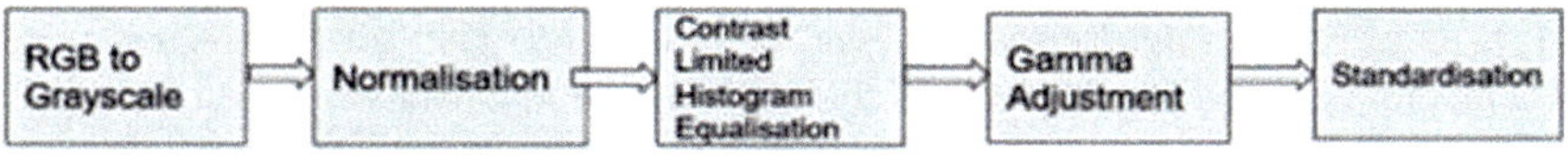

Fig. 1 Preprocessing steps

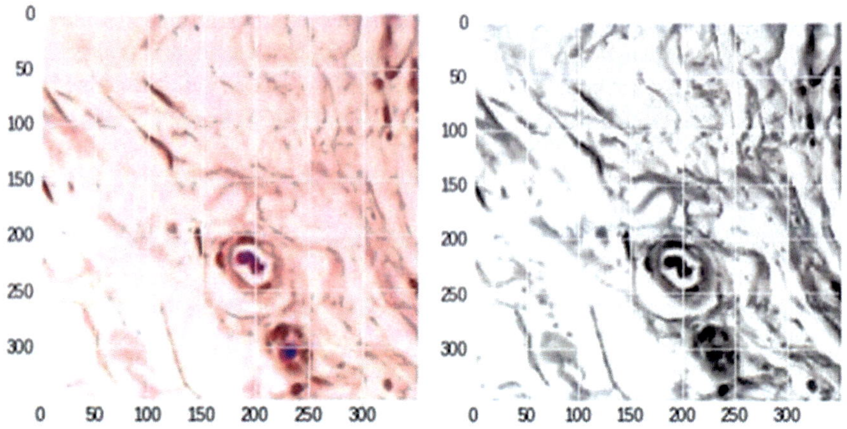

Fig. 2 Before and after preprocessing

the overall contrast and making the highlights look brighter, CLAHE is used. Gamma correction is again used for brightness control and to make sure the images do not become too dark or bleached out. Figure 2 shows patch of image before and after preprocessing.

Model Training

A convolutional neural network (ConvNet/CNN) is a deep learning algorithm which is one of the most influential innovations in the field of research in computer vision. It works on the concept of filters.

Our model architecture uses repeating blocks of convolution followed by max-pooling and batch normalization. The parameters of convolution block have a kernel size 3, stride of 1 and uses same padding. ReLU is applied after each convolution operation. ReLU is a nonlinear activation function which increases nonlinearity. Pooling layer gives translation invariance. With a pool size of 2 and a stride of 2, max-pooling is employed here. Batch normalization is then used which reduces internal covariant shift and regularizes the network. The initial block had 32 convolution filters then the second block had 64 convolution filters and the third block had 128 convolution filters. The fourth block and the fifth block had 256 convolution filters.

The sixth block and seventh block had 512 convolution filters after that the resulting output was attuned to a dense layer with hidden units of 512 and activation of ReLU. Final output layer consists of one hidden unit and an activation of sigmoid was used. The final output layer predicts whether the image is cancerous or not. Cancerous tissues are labeled 1 and non-cancerous tissues are labeled as 0.

The main challenge of our research was the low amount of training data. We know that to train a neural network, we require large amount of data. There were 100 images of each type. To address the challenge of low data, we used techniques of patch-wise classification coupled with data augmentation.

The CNN architecture used is shown in Fig. 3 and Fig. 4 shows the various training graphs.

The augmentation techniques applied are random rotation and horizontal and vertical flip augmentation. During our initial experimentation, we found out that when a single classifier was used to classify all four classes of tissues it was not able to differentiate between them so we had to use three separate binary classifiers each trained on two classes which significantly performed better independently and a combination-based inference strategy was developed.

The flowchart showing the steps leading to the final classification is shown in Fig. 5.

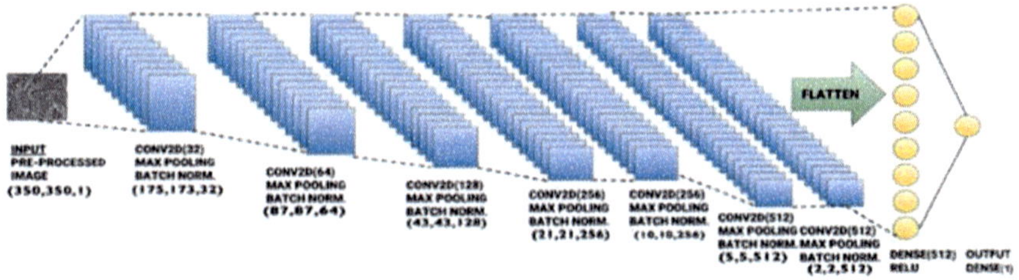

Fig. 3 CNN architecture

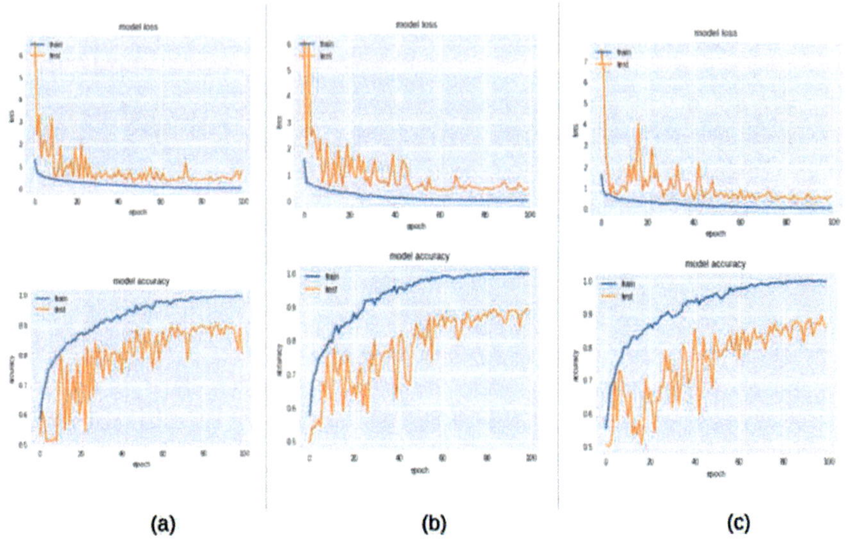

Fig. 4 Training graphs for **a** cancerous versus non-cancerous, **b** in situ versus invasive, **c** normal versus benign

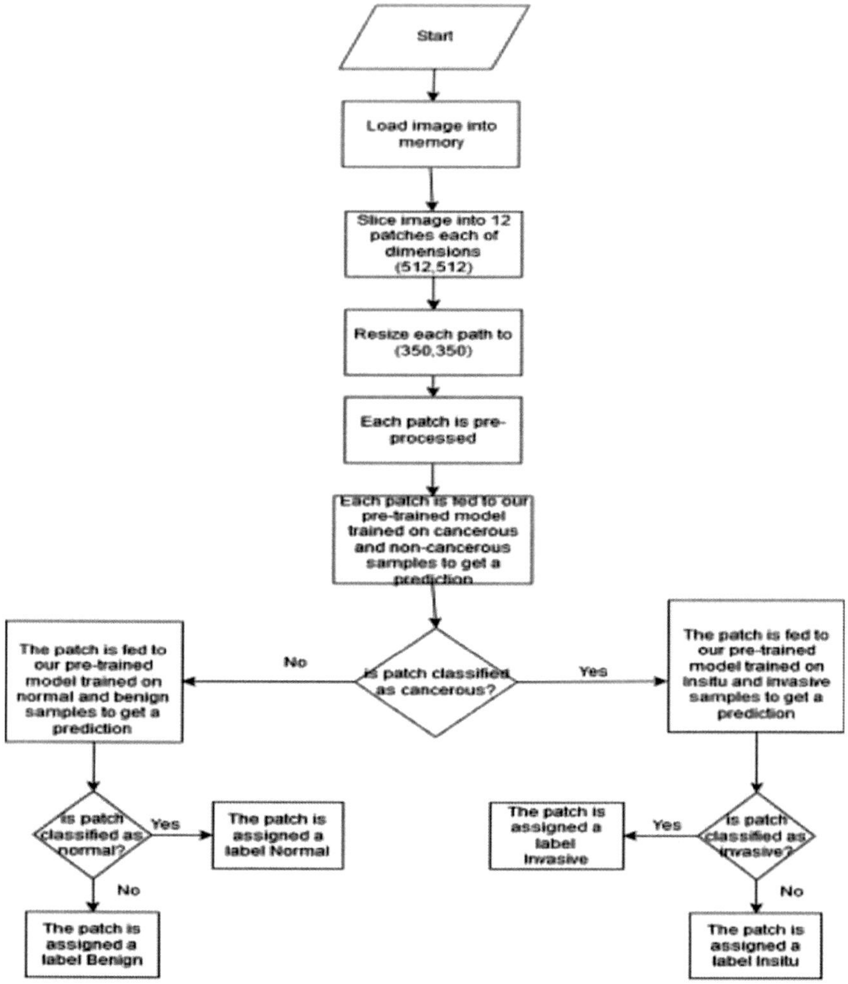

Fig. 5 Classification flowchart

4 Experimental Results

Accuracy, specificity, and specify are performance measures that may be used to evaluate the aforementioned methodology's performance. False negatives (FN), true positives (TP), false positives (FP), and true negatives (TN) are the four main parameters needed to generate the above metrics, as given in Table 1.

Accuracy: The proportion of the total number of Pulsar Stars that are exactly recognized by the classifier = (TP + TN)/($P + N$) is the accuracy of a classifier on a certain collection of Pulsar Stars.

Table 1 Performance of different classifiers

Model	Accuracy	Sensitivity	Specificity
Not cancerous versus cancerous	89.27	93.91	84.36
Benign versus normal	89.16	91.25	87.08
In situ versus invasive	88.33	85.83	90.83

Sensitivity: Defined as the ratio of true positives to total positives. It determines the proportion of positives correctly identified out of the total number of positives = TP/*P*.

Specificity: Defined as the ratio of genuine negatives to total negatives. The proportion of the total number of negatives that are appropriately detected is calculated as TN/*N*.

Figure 6 shows the confusion matrix for different classifiers.

5 Conclusion

In today's world, breast cancer is one of the leading causes of fatality. The accurate and quick diagnosis with correct medication can control the aggravated and complicated situation. Currently, the analysis of breast tissue histopathology images are required to find out various anomalies and by taking help of artificial intelligence to assist pathologists is the future of medical diagnosis. In this experiment, we propose a novel methodology to detect cancer in breast tissues using a combination of preprocessing and a combination of three deep convolutional neural networks that can be deployed as a Web application or a mobile application. In the future, we are required to deal with the problem of a smaller dataset for training and we also hope to get better computational resources so that training can be done on larger batches of images. To help pathologists more comprehensively, there is room for detection of the damaged areas of the tissue more precisely.

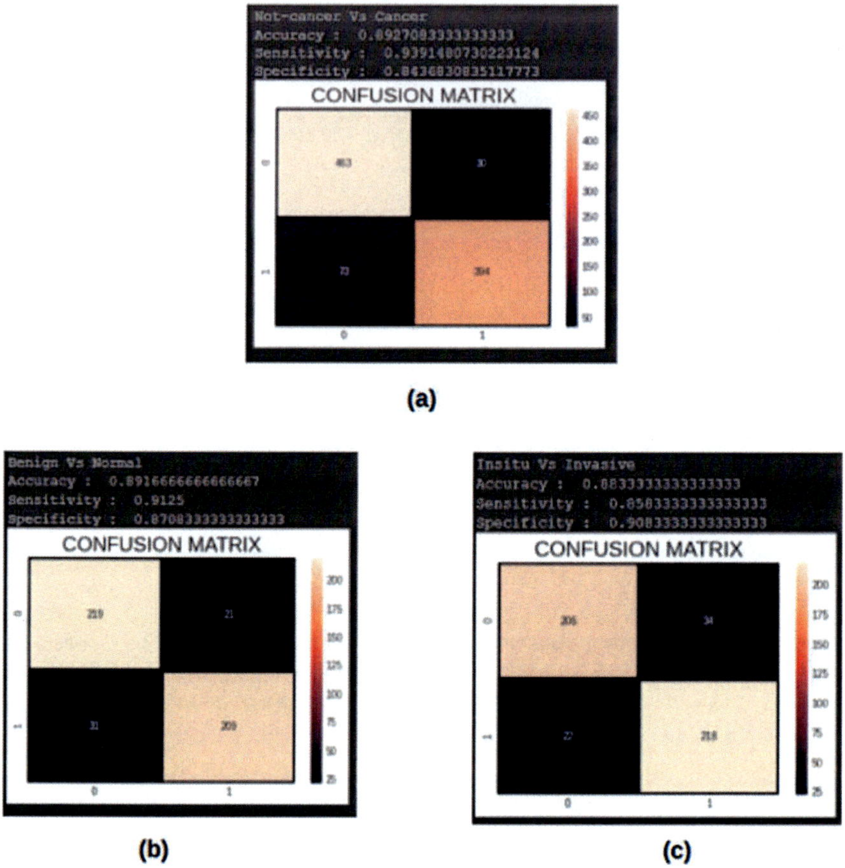

Fig. 6 Confusion matrix and calculated metrics for **a** cancerous versus non-cancerous, **b** in situ versus invasive, **c** normal versus benign

References

1. Han Z, Wei B, Zheng Y, Yin Y, Li K, Li S (2017) Breast cancer multi-classification from histopathological images with structured deep learning model. Sci Rep 7(1):4172
2. Zhu C, Song F, Wang Y et al (2019) Breast cancer histopathology image classification through assembling multiple compact CNNs. BMC Med Inform Decis Mak 19:198
3. Spanhol FA, Oliveira LS, Petitjean C, Heutte L (2016) A dataset for breast cancer histopathological image classification. IEEE Trans Biomed Eng 63(7):1455–1462
4. Ojala T, Pietikainen M, Maenpaa T (2002) Multiresolutiongray-scale and rotation invariant texture classification with local binary patterns. IEEE Trans Patt Anal Mach Intell 24(7):971–987
5. Ojansivu V, Heikkilä J (2008) Blur insensitive texture classification using local phase quantization. In: International conference image signal proceedings. Springer, Berlin, pp 236–43
6. Litjens G, Sánchez CI, Timofeeva N, Hermsen M, Nagtegaal I, Kovacs I, Hulsbergen-Van De Kaa C, Bult P, Van Ginneken B, Van Der Laak J (2016) Deep learning as a tool for increased

accuracy and efficiency of histopathological diagnosis. Sci Rep 6:26286
7. Cireşan DC, Giusti A, Gambardella LM, Schmidhuber J (2013) Mitosis detection in breast cancer histology images with deep neural networks. In: International conference on medical image computing and computer-assisted intervention. Springer, Berlin, pp 411–418
8. Bayramoglu N, Kannala J, Heikkilä J (2016) Deep learning for magnification independent breast cancer histopathology image classification. In: Pattern recognition (ICPR), pp 2440–2445
9. Spanhol FA, Oliveira LS, Petitjean C, Heutte L (2016) Breast cancer histopathological image classification using convolutional neural networks. In: Neural networks (IJCNN). IEEE, Vancouver, pp 2560–2567
10. He K, Zhang X, Ren S, Sun J (2016) Deep residual learning for image recognition. In: Proceedings of the IEEE conference on computer vision and pattern recognition. IEEE, Las Vegas, pp 770–778

Chapter 16
Uncertainty Principle in the Light of Quantum Information Science

Saptarshi Sahoo, Amit Kumar Mandal, Nirmalya Misra, and Vishal Mandal

1 Introduction

Heisenberg's uncertainty principle [1, 2] is one of the important aspects of quantum physics. Uncertainty is the inherent nature of quantum mechanical systems; it does not depend on the degree of preciseness of experimental apparatus.

If we assume, the uncertainty of momentum of a quantum mechanical particle moving along +x direction to be Δp_x and uncertainty in position be Δx, then according to uncertainty principle, $\Delta p_x \Delta x \geqslant \frac{\hbar}{2}$. In a naive way, it can be stated that if the location of a quantum mechanical particle is precisely known (i.e., uncertainty in location is infinitesimally small), then we will not have any information about its momentum (i.e., uncertainty in momentum is very very large). This notion is also true for energy and time. In fact, this principle holds true for any two canonically conjugate variables.

Although the main idea of uncertainty comes from Werner Heisenberg (1927), this formal inequality was proposed by Kennard [3] and Weyl [4]. The generalized uncertainty principle was given by Robertson [5].

This article mainly discusses uncertainty principle and attempts to derive the following associated inequality on the light of quantum information science [6–9, 29]. The necessary mathematics required to understand the text is also provided in the mathematical preliminaries section. We hope not only this article will give a good introduction about uncertainty principle to newcomers to this field but also provide a amusable read to experienced readers.

S. Sahoo (✉) · A. K. Mandal · N. Misra · V. Mandal
Institute of Engineering and Management, Salt Lake Electronics Complex, Y-12, Gurukul, Street Number 18, EP Block, Sector V, Bidhannagar, Kolkata, West Bengal 700091, India
e-mail: uzumaki.sahoo@gmail.com

A. K. Mandal
e-mail: amit.mandal@iemcal.com

L. Mandal et al. (eds.), *Proceedings of International Conference on Computational Intelligence, Data Science and Cloud Computing*, Algorithms for Intelligent Systems,
https://doi.org/10.1007/978-981-19-1657-1_16

Before plunging into the core topic, we would like to bring the reader's attention to double-slit experiment which not only laid a strong foundation for quantum mechanics but also demonstrates the uncertainty principle.

2 Double-Slit Experiment

In the classic double-slit experiment [10], light waves passing through two parallel slits create a characteristic light and dark patterns on a screen positioned behind the slits [11]. The patterns correspond to the points on the screen where the peaks and troughs of the waves diffracting out from the two slits (S_1, S_2) combine with one another either constructively or destructively. This happens due to the interference of photons with one another (Fig. 1).

However, if a detector is placed to determine through which slit the particle passed, interference pattern is destroyed. This is a version of Werner Heisenberg's uncertainty principle, which states that it is not possible to precisely measure both the position and the momentum of a photon simultaneously.

These similar interference pattern can also be observed from experiment conducted with electrons [12], atoms, and even some molecules [13–15], including buckyballs [16], thereby justifying the particle-wave nature of these aforementioned particles and the uncertainty principle.

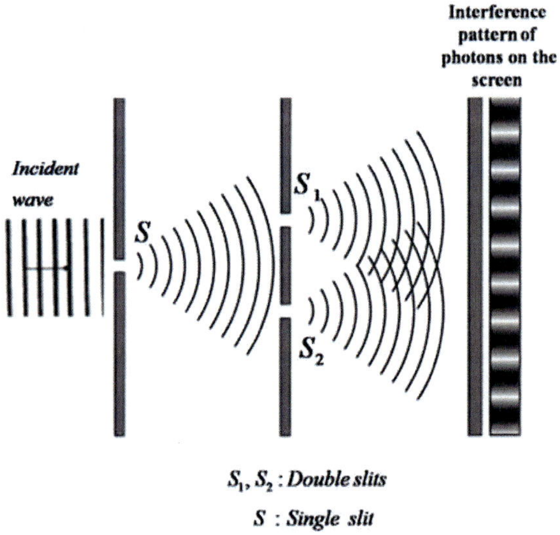

Fig. 1 Young's double-slit experiment

3 Mathematical Preliminaries

In this section, we have discussed some mathematical preliminaries with which one should be conversant to step into the generic form of uncertainty principle. Quantum state resides in Hilbert space ($\mathcal{H}$) [17, 18]. The complete information about a quantum mechanical system can be extracted from its wave function.

3.1 Hilbert Space and Observable

The elementary mathematical space in which the wavefunction is situated called Hilbert space [19]. Let $|\psi\rangle \in \mathcal{H}$ be a quantum state and $A \in \mathcal{L}(\mathcal{H})$(the space of all linear operator on the Hilbert space $\mathcal{H}$) be a hermitian operator ($A = A^\dagger$), which acts as observable. Unlike classical mechanics, in quantum mechanics, the physical observables, position, momentum, energy, etc. do not have a precise value in a particular state. The spectral decomposition [20] of A:

$$\begin{aligned} A &= \sum_i \lambda_i |\lambda_i\rangle\langle\lambda_i| \\ &= \sum_i \lambda_i P_i \qquad [\text{Let, } P_i = |\lambda_i\rangle\langle\lambda_i|] \end{aligned} \tag{1}$$

where $\{\lambda_i \in \mathbb{R}, |\lambda_i\rangle\}$ are the eigen value and eigenvector. $\mathcal{H}$ is the concerned Hilbert space and $\mathcal{L}(\mathcal{H})$ is the set of linear operator on $\mathcal{H}$ [18, 21].

3.2 Born's Rule

When measuring in basis of operator A, $|\psi\rangle$, can be represented in the basis of operator's eigen states as following:

$$|\psi\rangle = \sum_i \langle\lambda_i|\psi\rangle\, |\lambda_i\rangle \tag{2}$$

Therefore, probability for the $|\psi\rangle$ to collapse into a particular eigen state $|\lambda_k\rangle$ with eigen value λ_k after measurement is:

$$
\begin{aligned}
Pr(A = \lambda_k|\ |\psi\rangle) &= |\langle\lambda_k|\psi\rangle|^2 \\
&= \langle\lambda_k|\psi\rangle\langle\psi|\lambda_k\rangle \\
&= \langle\psi|\lambda_k\rangle\langle\lambda_k|\psi\rangle \\
&= \langle\psi|(|\lambda_k\rangle\langle\lambda_k|)|\psi\rangle \\
&= \langle\psi|P_k|\psi\rangle \qquad [\text{Let, } P_k = |\lambda_k\rangle\langle\lambda_k|]
\end{aligned} \tag{3}
$$

Above is the formal definition of Born's rule [18, 21].

3.3 *Expectation Value* ($E_\psi[A]$)

Expectation of quantum mechanical operator is useful while doing physical experiments as well as in theoretical derivation. This will help in finding out the standard deviation of an operator ($\Delta_\psi[A]$) which is directly linked with uncertainty during measurement. Expectation [22] of an operator can be defined as [18] follows [18]:

$$
\begin{aligned}
E_\psi[A] &= \sum_i \lambda_i Pr(A = \lambda_i||\psi\rangle) \\
&= \sum_i \lambda_i \langle\psi|P_i|\psi\rangle \\
&= \langle\psi|\sum_i \lambda_i P_i|\psi\rangle \\
&= \langle\psi|A|\psi\rangle \qquad [\text{by Spectral Theorem (1)}]
\end{aligned} \tag{4}
$$

3.4 *Variance* ($V_\psi[A]$) *and Standard Deviation* ($\Delta_\psi[A]$) *of an Operator*

Using the expressions derived in the above Sect. (3.3), we will be looking into the variance of an operator thereby standard deviation of the operator. Variance [22] of an operator can be written down as follows:

$$
\begin{aligned}
V_\psi[A] =& \sum_i (\lambda_i - E_\psi[A])^2 Pr(A = \lambda_i | \, |\psi\rangle) \\
=& \sum_i (\lambda_i^2 + E_\psi[A]^2 - 2\lambda_i E_\psi[A]) Pr(A = \lambda_i | \, |\psi\rangle) \\
=& \sum_i (\lambda_i^2 \langle\psi|P_i|\psi\rangle + \langle\psi|P_i|\psi\rangle E_\psi[A]^2 - 2\lambda_i \langle\psi|P_i|\psi\rangle E_\psi[A]) \\
=& (\sum_i \lambda_i^2 \langle\psi|P_i|\psi\rangle + \sum_i \langle\psi|P_i|\psi\rangle E_\psi[A]^2 - 2\sum_i \lambda_i \langle\psi|P_i|\psi\rangle E_\psi[A]) \\
& [\because A^2 = \sum_i \lambda_i^2 P_i \, , \, \sum_i P_i = \mathbb{I} \text{ and from the expectation expression (4)}] \\
=& (\langle\psi|A^2|\psi\rangle + 1 \cdot E_\psi[A]^2 - 2E_\psi[A] \cdot E_\psi[A]) \\
=& \langle\psi|A^2|\psi\rangle + E_\psi[A]^2 - 2E_\psi[A]^2 \\
=& \langle\psi|A^2|\psi\rangle - (\langle\psi|A|\psi\rangle)^2 \\
=& E_\psi[A^2] - E_\psi[A]^2 \\
=& (\Delta_\psi[A])^2
\end{aligned}
\tag{5}
$$

where $\Delta_\psi[A]$ is the standard deviation (S.D). We would like to point out the fact that this S.D is our required uncertainty in the measurement by a arbitrary operator on a arbitrary quantum state [18, 21].

3.5 *Cauchy-Schwarz Inequality*

This inequality is an integral aid in deriving the uncertainty principle. For any $|\theta\rangle, |\psi\rangle \in \mathcal{H}$, following inequality [23] can be written as follows:

$$
\begin{aligned}
|\langle\phi|\theta\rangle|^2 \leq \langle\phi|\phi\rangle \cdot \langle\theta|\theta\rangle = ||\phi||^2||\theta||^2 \\
\implies |\langle\phi|\theta\rangle| \leq ||\phi||||\theta||
\end{aligned}
\tag{6}
$$

where $|| \cdot ||$ is a 2-norm on Hilbert Space($\mathcal{H}$).

3.6 *Commutator and Anticommutator*

The whole theme of uncertainty boils down to whether the operators involved in simultaneous measurement commute. Hence, the commutator and anticommutator plays a vital role in the current topic of discussion. Let $A, B \in \mathcal{L}(\mathcal{H})$, then commutator [21] is defined as:

$$[A, B] = AB - BA \tag{7}$$

and anticommutator [21] is defined as :

$$\{A, B\} = AB + BA \tag{8}$$

If both A and B are Hermitian then following result can be written down:

$$\begin{aligned} AB &= \frac{\{A, B\}}{2} + i\frac{[A, B]}{2i} \\ &= X + iY \qquad \text{(Let.)} \end{aligned} \tag{9}$$

Here, both $X, Y \in \mathcal{L}(\mathcal{H})$ are Hermitian. Above result can be proved just by substituting the values for $[A, B]$ and $\{A, B\}$ in the above equation.

4 Derivation of Generic Form of Heisenberg-Robertson's Uncertainty Principle

We will use all the facts and equations developed in the previous section of mathematical preliminaries (3) to derive the generalized uncertainty principles in full glory. Let $A, B, \widetilde{A}, \widetilde{B} \in \mathcal{L}(\mathcal{H})$ hermitian. Say we did a simultaneous measurement using the operator A and B. Let them be related as follows:

$$\begin{aligned} \widetilde{A} &= A - E_{\psi}[A] \cdot \mathbb{I} \\ \widetilde{B} &= B - E_{\psi}[B] \cdot \mathbb{I} \end{aligned} \tag{10}$$

4.1 *Expectations of* $\widetilde{A}$ *and* $\widetilde{B}$

$$\begin{aligned} E_{\psi}[\widetilde{A}] &= \langle\psi|A - E_{\psi}[A]\mathbb{I}|\psi\rangle \\ &= \langle\psi|A|\psi\rangle - E_{\psi}[A]\langle\psi|\psi\rangle \\ &= E_{\psi}[A] - E_{\psi}[A] \cdot 1 \\ &= 0 \end{aligned} \tag{11}$$

Therefore from above, we get, $E_{\psi}[\widetilde{A}] = 0$, and similarly, we can calculate $E_{\psi}[\widetilde{B}] = 0$.

4.2 Variance of $\widetilde{A}$ and $\widetilde{B}$

$$
\begin{aligned}
V_{\psi}[\widetilde{A}] &= E_{\psi}[\widetilde{A}^2] - E_{\psi}[\widetilde{A}]^2 \\
&\quad \text{[Substituting the expectation of}\widetilde{A}\text{from (11)]} \\
&= \langle\psi|\widetilde{A}^2|\psi\rangle - 0 = ||\widetilde{A}|\psi\rangle||^2 \\
&\qquad [\because \forall A \text{ are hermitian, } \langle\psi|A^2|\psi\rangle = \Big(\langle\psi|A\Big)\Big(A|\psi\rangle\Big) = ||A|\psi\rangle||^2] \\
&\quad \text{[Again by substituting the value of } \widetilde{A} \text{ by (10)]} \\
&= \langle\psi|(A - E_{\psi}[A]\mathbb{I})^2|\psi\rangle \\
&= \langle\psi|(A^2 + E_{\psi}[A]^2\mathbb{I} - 2A \cdot E_{\psi}[A])|\psi\rangle \\
&= \langle\psi|A^2|\psi\rangle + E_{\psi}[A]^2\langle\psi|\psi\rangle - 2\langle\psi|A|\psi\rangle \cdot E_{\psi}[A] \\
&= \langle\psi|A^2|\psi\rangle - E_{\psi}[A]^2 \\
&= (\Delta_{\psi}[A])^2
\end{aligned} \tag{12}
$$

Therefore, we can write from above (12) $V_{\psi}[\widetilde{A}] = V_{\psi}[A] = (\Delta_{\psi}[\widetilde{A}])^2 = (\Delta_{\psi}[A])^2 = ||\widetilde{A}|\psi\rangle||^2$. Similarly for B & $\widetilde{B}$ it can be written that $V_{\psi}[\widetilde{B}] = V_{\psi}[B] = (\Delta_{\psi}[\widetilde{B}])^2 = (\Delta_{\psi}[B])^2 = ||\widetilde{B}|\psi\rangle||^2$.

4.3 Relation Between $[A, B]$ and $[\widetilde{A}, \widetilde{B}]$

We can prove $[\widetilde{A}, \widetilde{B}] = [A, B]$ like following:

$$
\begin{aligned}
[\widetilde{A}, \widetilde{B}] &= \widetilde{A}\widetilde{B} - \widetilde{B}\widetilde{A} \\
&= (A - E_{\psi}[A]\mathbb{I}) \cdot (B - E_{\psi}[B]\mathbb{I}) - (B - E_{\psi}[B]\mathbb{I}) \cdot (A - E_{\psi}[A]\mathbb{I}) \\
&= AB - \cancel{E_{\psi}[A]B} - \cancel{E_{\psi}[B]A} + \cancel{E_{\psi}[A]E_{\psi}[B]\mathbb{I}} \\
&\quad - BA + \cancel{E_{\psi}[B]A} + \cancel{E_{\psi}[A]B} - \cancel{E_{\psi}[A]E_{\psi}[B]\mathbb{I}} \\
&= AB - BA \\
&= [A, B] \qquad \text{Hence Proved.}
\end{aligned} \tag{13}
$$

4.4 Assembling the Previous Sections

By Cauchy-Schwarz inequality (6) we have:

$$
|\langle\psi|\widetilde{A}\widetilde{B}|\psi\rangle|^2 \leq ||\widetilde{A}|\psi\rangle||^2||\widetilde{B}|\psi\rangle||^2 \tag{14}
$$

Above can be written by assuming $|\phi\rangle = \widetilde{A}|\psi\rangle$ and $|\theta\rangle = \widetilde{B}|\psi\rangle$. From the commutator and anticommutator section, we can write the following (9):

$$\begin{aligned}\widetilde{A}\widetilde{B} &= \frac{\{\widetilde{A}, \widetilde{B}\}}{2} + i\frac{[\widetilde{A}, \widetilde{B}]}{2i} \\ &= X + iY \qquad \text{(Let.)}\end{aligned} \tag{15}$$

Therefore, we can write:

$$\begin{aligned}|\langle\psi|\widetilde{A}\widetilde{B}|\psi\rangle|^2 &= |\langle\psi|X + iY|\psi\rangle|^2 \\ &= |\langle\psi|X|\psi\rangle + i\langle\psi|Y|\psi\rangle|^2\end{aligned}$$

[Next step is just like taking a modulus of a complex number and squaring it]

[Also, $\forall X$ which are hermitian operator,$\langle\psi|X|\psi\rangle \in \mathbb{R}, \forall|\psi\rangle \in \mathcal{H}$]

$$\begin{aligned}&= \langle\psi|X|\psi\rangle^2 + \langle\psi|Y|\psi\rangle^2 \\ &= \langle\psi|X|\psi\rangle^2 + \langle\psi|Y|\psi\rangle^2 \geq \langle\psi|Y|\psi\rangle^2 \text{[Implicit]}\end{aligned} \tag{16}$$

Now using the above two inequality (14 and 16) and using the definitions of the S.D (12), we can further write down as:

$$\Longrightarrow |\langle\psi|\widetilde{A}\widetilde{B}|\psi\rangle|^2 \geq \frac{1}{4} \cdot |\langle\psi|[\widetilde{A}, \widetilde{B}]|\psi\rangle|^2 \quad \text{[Substituting } Y = \frac{[\widetilde{A}, \widetilde{B}]}{2i} \text{ by (15)]}$$

[By using the Cauchy-Schwarz inequality(14) previously stated]

$$\Longrightarrow ||\widetilde{A}|\psi\rangle||^2||\widetilde{B}|\psi\rangle||^2 \geq |\langle\psi|\widetilde{A}\widetilde{B}|\psi\rangle|^2 \geq \frac{1}{4} \cdot |\langle\psi|[\widetilde{A}, \widetilde{B}]|\psi\rangle|^2$$

$$\Longrightarrow ||\widetilde{A}|\psi\rangle||^2||\widetilde{B}|\psi\rangle||^2 \geq \frac{1}{4}|\langle\psi|[\widetilde{A}, \widetilde{B}]|\psi\rangle|^2$$

$$\Longrightarrow ||\widetilde{A}|\psi\rangle||\,||\widetilde{B}|\psi\rangle|| \geq \frac{1}{2}|\langle\psi|[\widetilde{A}, \widetilde{B}]|\psi\rangle|$$

[By using the definition of S.D stated before we can further write]

$$\therefore \Delta_\psi[A]\Delta_\psi[B] \geq \frac{1}{2}|\langle\psi|[A, B]|\psi\rangle| \tag{17}$$

Hence, the generalized uncertainty principle is derived.

5 Typical Examples of Two-State Quantum Mechanical Systems

In nature, we have an abundance of multi-level quantum systems [17]. Although, if we look closely around us, we find rather very common and daily used two-level systems [24–26]. We will try to provide two of such examples:

- Spin-$\frac{1}{2}$ particles; most often electrons.
- Photon Polarization.

5.1 Spin-$\frac{1}{2}$ Particles

Spin is an intrinsic property of elementary particles, just like mass, charge, etc. The spin state of an object with spin-$\frac{1}{2}$ (like an electron) can be either an up($|\uparrow\rangle$) or down($|\downarrow\rangle$) along an axis described in $\mathbb{R}^3$-space. Refer to Fig. 2 for better understanding. They are expected to obey properties of superposition and entanglement of quantum mechanics. An example of superposition would be:

$$|\psi\rangle = a|\uparrow_z\rangle + b|\downarrow_z\rangle \tag{18}$$

In Eq. (18) $|\psi\rangle \in \mathcal{H}^2$(A two dimensional Hilbert Space). $a, b \in \mathbb{C}$ and $|a|^2 + |b|^2 = 1$. An example of entanglement would be:-

$$|\text{singlet}\rangle = \frac{|\uparrow_z\rangle_A|\downarrow_z\rangle_B - |\downarrow_z\rangle_A|\uparrow_z\rangle_B}{\sqrt{2}} \tag{19}$$

Here, (19) $|\text{singlet}\rangle \in \mathcal{H}_{AB} = \mathcal{H}_A \otimes \mathcal{H}_b$(A four dimensional Hilbert space).

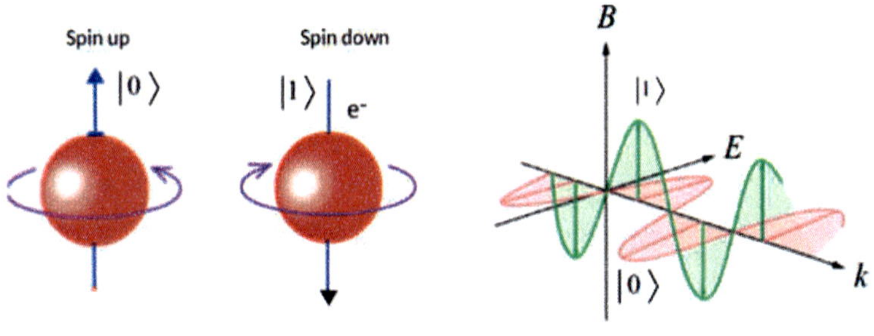

(a) States of a spinning electron (b) States of a polarized photon

Fig. 2 Typical examples of two-state quantum systems

A standard definitions of the the basis along three axes are given:

$$|\uparrow_z\rangle = \begin{bmatrix}1\\0\end{bmatrix} \; |\uparrow_x\rangle = \begin{bmatrix}\frac{1}{\sqrt{2}}\\ \frac{1}{\sqrt{2}}\end{bmatrix} = \frac{|\uparrow_z\rangle+|\downarrow_z\rangle}{\sqrt{2}} \; |\uparrow_y\rangle = \begin{bmatrix}\frac{1}{\sqrt{2}}\\ \frac{i}{\sqrt{2}}\end{bmatrix} = \frac{|\uparrow_z\rangle+i|\downarrow_z\rangle}{\sqrt{2}}$$

$$|\downarrow_z\rangle = \begin{bmatrix}0\\1\end{bmatrix} \; |\downarrow_x\rangle = \begin{bmatrix}\frac{1}{\sqrt{2}}\\ -\frac{1}{\sqrt{2}}\end{bmatrix} = \frac{|\uparrow_z\rangle-|\downarrow_z\rangle}{\sqrt{2}} \; |\downarrow_y\rangle = \begin{bmatrix}\frac{1}{\sqrt{2}}\\ -\frac{i}{\sqrt{2}}\end{bmatrix} = \frac{|\uparrow_z\rangle-i|\downarrow_z\rangle}{\sqrt{2}}$$

The observables along these axes are described by the Pauli matrices:

$$\sigma_1 = \sigma_x = \begin{bmatrix}0 & 1\\1 & 0\end{bmatrix} \; \sigma_2 = \sigma_y = \begin{bmatrix}0 & -i\\i & 0\end{bmatrix} \; \sigma_3 = \sigma_z = \begin{bmatrix}1 & 0\\0 & -1\end{bmatrix}$$

For a much more generalized observable in spherical coordinate in $\mathbb{R}^3$:

$$\sigma_{\theta,\phi} = \sin\theta\cos\phi \cdot \sigma_x + \sin\theta\sin\phi \cdot \sigma_y + \cos\theta \cdot \sigma_z \tag{20}$$

In (20), θ is the polar angle and ϕ is the azimuthal angle.

5.2 *Photon Polarizations*

Photons are spin-1 particles. Hence, the two-state system of photon is provided by two independent polarizations. Refer to Fig. 2b for better grasping the idea. Generally, they are horizontally polarized ($|H\rangle$) and vertically polarized ($|V\rangle$). These states can be represented as follows:

$$|V\rangle = \begin{bmatrix}0\\1\end{bmatrix} \; |H\rangle = \begin{bmatrix}1\\0\end{bmatrix}$$

Just like earlier, we can see that they participate in quantum properties like superposition and entanglement.

6 Qubit: Unit of Quantum Information

A qubit [27] is defined as an unit of quantum information modeled by a two-state(orthogonal) quantum mechanical system. This is a more abstract encapsulation of the Sects. 5.1 and 5.2. A qubit can be written in the computational basis (σ_z basis) as:

$$|\psi\rangle = a|0\rangle + b|1\rangle \tag{21}$$

In (21) $|\psi\rangle \in \mathcal{H}$(which is in this isomorphic to $\mathbb{C}^2$ under the standard Eucledian metric) and $a, b \in \mathbb{C}$. The most appreciated way to visualize pure state qubits is through the Bloch sphere [27]. It describes the qubits as a point on an unit sphere in $\mathbb{R}^3$ space and operations on qubit are represented as motion of the state $|\psi\rangle$ on the surface of the unit sphere. In a very hand wavy way, we can say that Bloch sphere is a restricted mapping from $\mathbb{C}^2 \to \mathbb{R}^3$. The restrictions are as follows:

- Unit length of the state ($|a|^2 + |b|^2 = 1$).
- Global phase does not matter while measuring the qubit. But the relative phase between $|0\rangle$ and $|1\rangle$ matters.

Due to the above constraints, the qubit can be further simplified into the below expression:

$$\begin{aligned} |\psi\rangle &= \cos\frac{\theta}{2}|0\rangle + e^{i\phi}\sin\frac{\theta}{2}|1\rangle \\ &= \begin{bmatrix} \cos\frac{\theta}{2} \\ e^{i\phi}\sin\frac{\theta}{2} \end{bmatrix} \end{aligned} \tag{22}$$

where $0 \leq \theta \leq \pi$ and $0 \leq \theta < 2\pi$. Hence, it can be seen that, by using the above expression, we can map on the Bloch sphere. It is very intuitive that $|0\rangle$ is at the north pole of the unit sphere and $|1\rangle$ is at the south pole of the unit sphere.

Following states are plotted using Qiskit [28] for better clarity:

As it is perceived, Bloch sphere can give a quite good representation of qubit. But there is a catch, when representing a larger number of qubits with Bloch sphere, it can be cumbersome and clumsy. Also Bloch sphere does not serve much use when we work with mixed quantum state.

6.1 Uncertainty in the Light of Quantum Information Science

In this section, we will show how uncertainty principle can be applied to a qubit, which is a generalization of two-state quantum system. Here is another related work [29]. We will show the example using the Pauli matrices. It is can be easily proved that (Figs. 3 and 4):

$$\sigma_a\sigma_b = \delta_{ab}\mathbb{I} + i\varepsilon_{abc}\sigma_c \tag{23}$$

Thereby using (23) we can also prove,

$$[\sigma_a, \sigma_b] = i2\delta_{ab}\sigma_c \tag{24}$$

In (23) and (24), the δ_{ab} is the kronecker delta [30], ε_{abc} is the Levi-Civita symbol [31] and $\mathbb{I}$ is the 2×2 identify matrix.

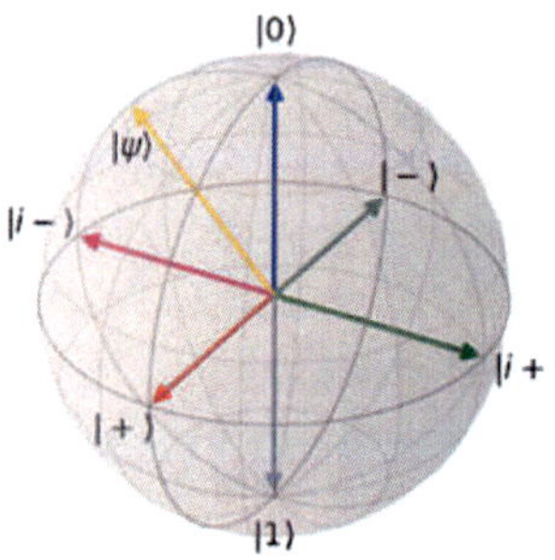

Fig. 3 Bloch sphere

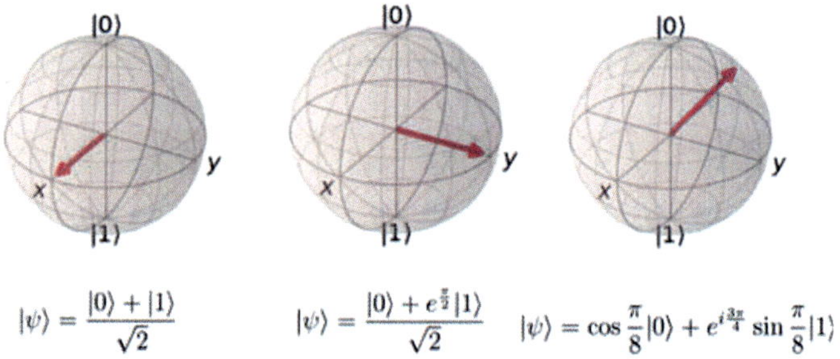

Fig. 4 Bloch sphere examples

Say, we have a quantum state as $|\psi\rangle \in \mathcal{H}$, and we want to state the uncertainty principle of σ_a and σ_b:

$$\begin{aligned} &\Delta_\psi[\sigma_a]\Delta_\psi[\sigma_b] \geq \frac{1}{2}|\langle\psi|[\sigma_a, \sigma_b]|\psi\rangle| \quad \text{[by (17)]} \\ &\implies \Delta_\psi[\sigma_a]\Delta_\psi[\sigma_b] \geq \frac{1}{2}|\langle\psi|i2\sigma_c|\psi\rangle| \quad \text{[by (24)]} \\ &\implies \Delta_\psi[\sigma_a]\Delta_\psi[\sigma_b] \geq |\langle\psi|\sigma_c|\psi\rangle| = |E_\psi[\sigma_c]| \end{aligned} \tag{25}$$

One thing to note that in (25) uncertainty in two orthogonal directions is directly depend on the expectation value of the wavefunction in the third orthogonal direction. We can further choose to decompose σ_c using spectral theorem (1) as $\sigma_c = |c_0\rangle\langle c_0| - |c_1\rangle\langle c_1|$[+1 and −1 are always the eigen values of the Pauli matrices.]. Hence, $|\psi\rangle$ can be written as $|\psi\rangle = a_{c_0}|c_0\rangle + a_{c_1}|c_1\rangle$, in the eigen basis of σ_c; thereby, we can compute $E_\psi[\sigma_c]$:

$$
\begin{aligned}
E_\psi[\sigma_c] &= \langle\psi|\sigma_c|\psi\rangle \\
&= \langle\psi|\Big(|c_0\rangle\langle c_0| - |c_1\rangle\langle c_1|\Big)\Big(a_{c_0}|c_0\rangle + a_{c_1}|c_1\rangle\Big) \\
&= \langle\psi|\Big(a_{c_0}|c_0\rangle - a_{c_1}|c_1\rangle\Big) \\
&= \Big(\langle c_0|a^*_{c_0} + \langle c_1|a^*_{c_1}\Big)\Big(a_{c_0}|c_0\rangle - a_{c_1}|c_1\rangle\Big) \\
&= |a_{c_0}|^2 - |a^2_{c_1}|
\end{aligned}
\tag{26}
$$

Combining (26) and (25), we can present the following statement:

$$
\Delta_\psi[\sigma_a]\Delta_\psi[\sigma_b] \geq |\langle\psi|\sigma_c|\psi\rangle| = |E_\psi[\sigma_c]| = ||a_{c_0}|^2 - |a^2_{c_1}|| \tag{27}
$$

Above is the uncertainty principle for the Pauli matrices on qubits. It is to be also noted that, if we want to calculate uncertainty in angular momentum of a spin-1/2 particle, then we can use the same mathematical statement (27) just we need to attach a $\frac{\hbar}{2}$ on the right side of the equation.

6.2 *Position and Momentum of a Particle*

We can also derive the most common form (uncertainty due to momentum and position of a particle) of Heisenberg's uncertainty principle. The process goes as follows:

Let,

$A = \hat{x} = x$ (Position operator along x-axis)

$B = \hat{p}_x = -i\hbar\frac{\partial}{\partial x}$ (Momentum operator along x-axis)

Now,

$$
\begin{aligned}
|\langle\psi|[A, B]|\psi\rangle| &= |\langle\psi|\Big(x(-i\hbar)\frac{\partial}{\partial x} + (i\hbar)\frac{\partial}{\partial x}x\Big)|\psi\rangle| \\
&= |\langle\psi|\Big(-i\hbar x\frac{\partial|\psi\rangle}{\partial x} + i\hbar\frac{\partial x|\psi\rangle}{\partial x}\Big)| \\
&= |\langle\psi|\Big(-i\hbar x\frac{\partial|\psi\rangle}{\partial x} + i\hbar x\frac{\partial|\psi\rangle}{\partial x} + i\hbar|\psi\rangle\frac{\partial x}{\partial x}\Big)| \\
&= |i\hbar\langle\psi|\psi\rangle| \\
&= \hbar = \frac{h}{2\pi}
\end{aligned}
\tag{28}
$$

Using the above expressions (17), (28), we can write the most common form of uncertainty principle which encompasses the uncertainty in measurement of position and momentum along abscissa:

$$\Delta_{\psi}\hat{x}\Delta_{\psi}\hat{p}_x \geq \frac{h}{4\pi} \qquad \text{Hence Proved.} \tag{29}$$

This Eq. 29 can also be written along y-axis and z-axis too as:

$$\Delta_{\psi}\hat{y}\Delta_{\psi}\hat{p}_y \geq \frac{h}{4\pi}$$
$$\Delta_{\psi}\hat{z}\Delta_{\psi}\hat{p}_z \geq \frac{h}{4\pi}$$

6.3 *Making an Analogy with Position and Momentum*

Let us try to give an analogy between the two examples given above (27), (29) but first we will take a specific example for (27) and then compare with the 29.

Suppose,

$$|\psi\rangle = \frac{|0\rangle - i|1\rangle}{2}$$

$$\sigma_a = \sigma_x$$

$$\sigma_b = \sigma_z$$

$$\sigma_c = \sigma_y$$

Therefore,

$$\begin{aligned} &\Delta_{\psi}[\sigma_x]\Delta_{\psi}[\sigma_z] \geq |\langle\psi|\sigma_y|\psi\rangle| \\ &\implies \Delta_{\psi}[\sigma_x]\Delta_{\psi}[\sigma_z] \geq 1 \end{aligned} \tag{30}$$

By comparing (29) and (30), we can see that the uncertainty in position is analogous to uncertainty in measuring the qubit in the x-direction(in the Bloch sphere), and in a similar way, the uncertainty in momentum is analogous to uncertainty of measuring the qubit in the y-direction.

6.3.1 Establishment of Uncertainty Relation in Measuring the Qubit in Two Basis States

Let us define the spread of the quantum state, $|\Psi\rangle$, in $|0\rangle, |1\rangle$ basis as $|\Psi_b\rangle$ and in $|+\rangle, |-\rangle$ basis as $|\Psi_s\rangle$.

$$|\Psi_b\rangle = a_0|0\rangle + a_1|1\rangle \tag{31}$$
$$|\Psi_s\rangle = a_+|+\rangle + a_-|-\rangle \tag{32}$$

Now, let us introduce two parameters:

$\Delta\Psi_b = |a_0| + |a_1|$, spread or uncertainty in $|0\rangle, |1\rangle$ basis.
$\Delta\Psi_s = |a_+| + |a_-|$, spread or uncertainty in $|+\rangle, |-\rangle$ basis.

When we come to know the bit values with perfect accuracy, $\Delta\Psi_b = 1$.
Whereas, when we do not know the bit values accurately, $\Delta\Psi_b = \frac{1}{\sqrt{2}} + \frac{1}{\sqrt{2}} = \sqrt{2}$

In the similar manner, when sign values are accurately known, $\Delta\Psi_s = 1$. For uncertain situation, $\Delta\Psi_s = \frac{1}{\sqrt{2}} + \frac{1}{\sqrt{2}} = \sqrt{2}$
So, the product of $\Delta\Psi_b$ and $\Delta\Psi_s$ for any qubit,

$$\Delta\Psi_b \Delta\Psi_s \geq \sqrt{2} \tag{33}$$

That means both $\Delta\Psi_b$ and $\Delta\Psi_s$ can not be simultaneously one.

7 Conclusion

Uncertainty principle has always drawn the attention of researchers in the field of sub-atomic physics [32]. Even if the uncertainty principle is innate nature in the parlance of measurement in quantum mechanical systems, its interpretation is bit ambiguous [33]. In this paper, we have represented uncertainty principle in the light of qubit, and in future, it can be used in quantum networks.

One may think that this imprecision is due to the fault imparted by the observer, but this is not due to the observer effect [34], [35, p.3]. This is inherent wave-like systems, and it pops in quantum mechanics because of the particle-wave duality on which the whole system resides on.

In recent years, people have come up with clever tricks to bypass this inherent principle [36]. In [37], instead of elementary particle, the authors did experiment on two vibrating drumheads one-fifth of the width of a human hair. These drumheads were configured to behave in a quantum mechanical way. This technique can in future used in quantum networks.

Acknowledgements We acknowledge the use of IBM quantum services for this work. The views expressed are those of the authors and do not reflect the official policy or position of IBM or the IBM Quantum team.

References

1. Heisenberg W (1985) Über den anschaulichen inhalt der quantentheoretischen kinematik und mechanik. In: Original scientific papers Wissenschaftliche Originalarbeiten. Springer, pp 478–504
2. Sen D (2014) The uncertainty relations in quantum mechanics. Curr Sci 203–218
3. Kennard EH (1927) Zur quantenmechanik einfacher bewegungstypen. Z für Phys 44(4):326–352
4. Weyl H (1927) Quantenmechanik und gruppentheorie. Z Phys 46(1–2):1–46
5. Robertson HP (1929) The uncertainty principle. Phys Rev 34(1):163
6. Berta M, Christandl M, Colbeck R, Renes JM, Renner R (2010) The uncertainty principle in the presence of quantum memory. Nat Phys 6(9):659–662
7. Girdhar P, Doherty AC (2020) Testing generalised uncertainty principles through quantum noise. New J Phys 22(9):093073
8. Ozawa M (2012) Mathematical foundations of quantum information: measurement and foundations. arXiv preprint arXiv:1201.5334
9. Narasimhachar V, Poostindouz A, Gour G (2016) Uncertainty, joint uncertainty, and the quantum uncertainty principle. New J Phys 18(3):033019
10. Wikipedia, Double slit experiment
11. Green B (2005) The fabric of the cosmos: space, time, and the texture of reality. J Chem Educ 82(6):822–823
12. Donati O, Missiroli G, Pozzi G (1973) An experiment on electron interference. Am J Phys 41(5):639–644
13. Nairz O, Brezger B, Arndt M, Zeilinger A (2001) Diffraction of complex molecules by structures made of light. Phys Rev Lett 87(16):160401
14. Nairz O, Arndt M, Zeilinger A (2003) Quantum interference experiments with large molecules. Am J Phys 71(4):319–325
15. Arndt M, Nairz O, Vos-Andreae J, Keller C, Van der Zouw G, Zeilinger A (1999) Wave-particle duality of c 60 molecules. Nature 401(6754):680–682
16. Al-Khalili J (2012) Quantum: a guide for the perplexed. Hachette
17. Griffiths DJ (2010) Introduction to quantum mechanics, 2nd edn. Prentice Hall
18. Hall BC (2013) Quantum theory for mathematicians, vol 267. Springer
19. Kreyszig E (1991) Introductory functional analysis with applications, vol 17. Wiley, New York
20. Axler S (1997) Linear algebra done right. Springer Science and Business Media
21. Hayashi M, Ishizaka S, Kawachi A, Kimura G, Ogawa T (2014) Introduction to quantum information science. Springer
22. Guttorp P (1992) The art of probability for scientists and engineers
23. Wu H-H, Wu S (2009) Various proofs of the cauchy-schwarz inequality. Octogon Math Mag 17(1):221–229
24. Feynman R (1979) Feynman lectures on physics, vol 3, quantum mechanics. le cours de physique de feynman, vol 3. mecanique quantique
25. Landau LD, Lifshitz EM (2013) Quantum mechanics: non-relativistic theory, vol 3. Elsevier
26. Cohen-Tannoudji C, Diu B, Laloe F (2009) Quantum mechanics, vol 1. 4. rev. upd. ed.; quantenmechanik. bd. 1
27. Nielson MA, Chuang IL (2000) Quantum computing and quantum information
28. IBMQuantum (2021) https://quantum-computing.ibm.com/
29. Kak S (2002) Uncertainty in quantum computation. arXiv preprint quant-ph/0206006
30. Wikipedia, Kronecker Delta
31. Wikipedia, Levi-Civita symbol
32. Furuta A (2012) One thing is certain: Heisenberg's uncertainty principle is not dead. Sci Am
33. Ozawa M (2003) Universally valid reformulation of the heisenberg uncertainty principle on noise and disturbance in measurement. Phys Rev A 67
34. Dent EB (2005) The observation, inquiry, and measurement challenges surfaced by complexity theory. Managing the complex: philosophy, theory and practice. IAP, Greenwich, pp 253–268

35. Dirac P (1967) The principles of quantum mechanics, 4th edn (revised), vol 159. Clarendon, Oxford
36. Suh J, Weinstein A, Lei C, Wollman E, Steinke S, Meystre P, Clerk AA, Schwab K (2014) Mechanically detecting and avoiding the quantum fluctuations of a microwave field. Science 344(6189):1262–1265
37. de Lépinay LM, Ockeloen-Korppi CF, Woolley MJ, Sillanpää MA (2021) Quantum mechanics-free subsystem with mechanical oscillators. Science 372(6542):625–629

Part II
Image and Natural Language Processing

Chapter 17
Extraction of Appearance-Based DUS Characteristics of Okra Stem, Flower, and Seed Using Image Processing

Gopinath Bej, Abhra Pal, Tamal Dey, Sabyasachi Majumdar, Amitava Akuli, Alokesh Ghosh, and Nabarun Bhattacharyya

1 Introduction

For enhanced crop variety, Research and Development utilizing traditional breeding methods and approaches to biotechnology require considerable investments in both scientific and financial resources. Plant breeding programs, which provide sufficient incentives for breeding breeders, must be protected against misuse and should encourage investment, growth and development. This would not only safeguard new variations from unlawful commercial exploitation but also promote the development of new species, employing efficient plant variety protection (PVP) rule. [1]. The PPVFR act, which incorporates measures to safeguard both rights of breeders and farmers [1, 2], was applied in India during 2001. This legislation safeguards both

G. Bej (✉) · A. Pal · T. Dey · S. Majumdar · A. Akuli · A. Ghosh · N. Bhattacharyya
Centre for Development of Advanced Computing (C-DAC), Plot E-2/1, Block-GP, Sector-V, Salt Lake, Kolkata 700091, India
e-mail: gopinath.bej@cdac.in

A. Pal
e-mail: abhra.pal@cdac.in

T. Dey
e-mail: tamal.dey@cdac.in

S. Majumdar
e-mail: sabyasachi.majumdar@cdac.in

A. Akuli
e-mail: amitava.akuli@cdac.in

A. Ghosh
e-mail: alokesh.ghosh@cdac.in

N. Bhattacharyya
e-mail: nabarun.bhattacharya@cdac.in

L. Mandal et al. (eds.), *Proceedings of International Conference on Computational Intelligence, Data Science and Cloud Computing*, Algorithms for Intelligent Systems,
https://doi.org/10.1007/978-981-19-1657-1_17

innovative and established varieties and farmers. For 15 years, the rights of plant breeding agents allowed the breeder (or his heirs, contractors, or licensees) to ban any person from producing, selling, promoting, distributing, exporting, or importing protected varieties. A breeder may also use a protected species following the law for research reasons. Under the "Protection of Plant Varieties and Farmers' Rights Act," a new plant variety can be registered and protected for a set period, which is 15 years for annuals and 18 years for vines and trees. A variety can only be registered and protected if it fulfills the distinctness, uniformity, and stability (DUS) standards. [2, 3]. The term "distinctness" refers to the ability of a variety to be differentiated from other variations by one or more essential characteristics. If the variety's important features are sufficiently uniform, the variety is deemed uniform, subject to the variance that may be expected owing to the population's unique characteristics. It is deemed stable if the variety's key traits do not change after repeated propagation. Crop-specific DUS testing guidelines [4] were created to describe ideas and procedures for various aspects of DUS testing of new crop varieties. Plant material needs, test processes, methodologies and observations, variety grouping, characteristics and symbols, a table of characteristics, literature, and a technical questionnaire are all included in the guideline. The DUS study almost always needs at least two independent growth cycles [3–5].

The needs for distinctness, homogeneity, and stability are evaluated using characteristics. The characteristics are either a general attribute of the plant or a characteristic of a specific portion of the plant. These characteristics might be morphological, biochemical, or any other kind. The table of characteristics given by authorities is the most important component of test recommendations and DUS testing [6–8].

It also contains special instructions for conducting the DUS test for Okra indicating the table of features of the okra plant elements (e.g., leaf, stem, flower, fruit, and seed). The morphological and color properties specified in Table 1 are presented in DUS guidelines for Okra stem, flower, and seed. The manual measurement of various field properties might lead to mistaken data. Furthermore, the subjective, error-prone, and manual measurement depend on so many of the observer's characteristics (e.g., age, eye, sex, mental state, etc.) [9, 10]. This study proposes a machine vision technique for reducing human mistakes and objectively deriving attributes [9–14].

After collecting the Okra stem, flower, and seed samples, an imaging setup consisting of a NIKON DSLR camera and an illuminating arrangement was used to capture the image of each sample. To limit the interference of ambient light, the imaging equipment was put in a dark room. To determine the DUS-defined physiological and color characteristics of Okra stem, flower, and seed, algorithms based on digital image processing were developed. After the characteristics were extracted using the developed algorithms, they were validated against the result of manual measurement, and the accuracy was optimistic.

2 Literature Survey

Machine vision has a lot of benefits when it comes to analyzing features for DUS tests [8–10]. Curvature, area, perimeter, surface roughness, and other metrics that cannot be assessed with standard measuring equipment are more routinely measured using image-based approaches [6, 7, 11, 13]. Furthermore, the vision system can have a far greater resolution than human eyesight, allowing for the observation and analysis of minute features utilizing a machine vision system [9, 10, 12, 14].

3 Materials and Methods

3.1 Sample Collection

In NBPGR, New Delhi, the candidate variety was grown in three replications according to DUS standards. Samples were collected from the field at the specific stage of maturity as mentioned in DUS guidelines. For this experiment, a total of 131 samples were collected and investigated. Out of 131, 26 samples were of the stem, 55 samples were of flower, and 50 samples were of seed. These were used for derivation and validation of characteristics 1 to 8 as per Table1. The number of stem samples was reduced since the plant was cut to obtain stem samples, destroying plants.

3.2 Acquire Image

The images of stem, flower, and seed were captured following a designed protocol. The images were captured using an imaging setup consisting of a NIKON DSLR camera and an illumination system. For improved image clarity, the camera's distance from the item was set at 30 cm for stem and 20 cm for flower and seed. For all the samples, illumination was provided from the top.

3.3 Image Processing, Analysis and Feature Extraction

Applying digital image processing and analysis techniques, algorithms were developed to extract specified DUS features of Okra stem, flower, and seed from the acquired images. In a nutshell, Figs. 1, 2, 3 and 4 are showing the specified DUS features of stem, flower, and seed, respectively. From the image of the Okra stem, color, the diameter of the stem, and the intensity of green color were determined (serial no. 1, 2, and 3 from Table1). Images of the whole flower and the separated petal were captured and considered for extraction of DUS-defined features as shown

Table 1 DUS characteristics of Okra stem, flower, and seed

Sl No	Characteristics	Definition	Stage of recoding
1	Stem: color	The visible skin color of the stem. (green/red)	Active vegetative growth before flowering
2	Stem: intensity of green color	Measurement of green color intensity. (light/medium/dark)	Active vegetative growth before flowering
3	Stem: diameter (at 10 cm above ground level)	The thickness of the stem	About 50% of fruits have reached the turning stage (physiological maturity). Advanced seed filling
4	Flower: petal color	Visible color of the flower petal. (cream/yellow/purple)	50% of plants in flowering
5	Flower: petal base color (purple)	Visible color of the flower petal base. (Inside only/both sides)	50% of plants in flowering
6	Flower: length (cm)	Length of the flower	50% of plants in flowering
7	Flower: diameter (at the top of flower)	Diameter of the flower	50% of plants in flowering
8	Seed color	The visible skin color of the seed. (green/brown)	Seed maturity
9	Seed hairiness	Presence of hair around the placenta of the seed	Seed maturity

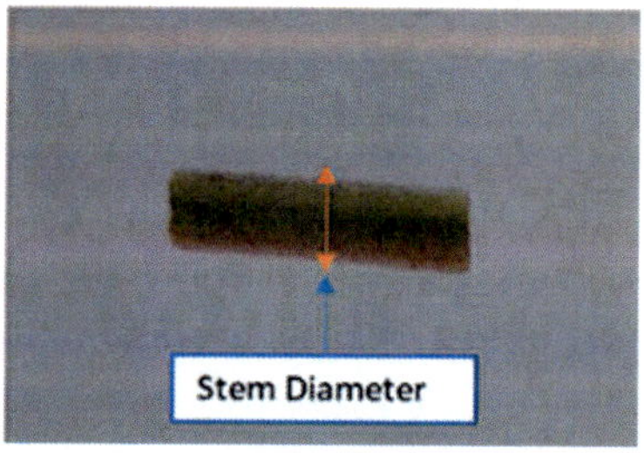

Fig. 1 DUS-defined features of Okra stem

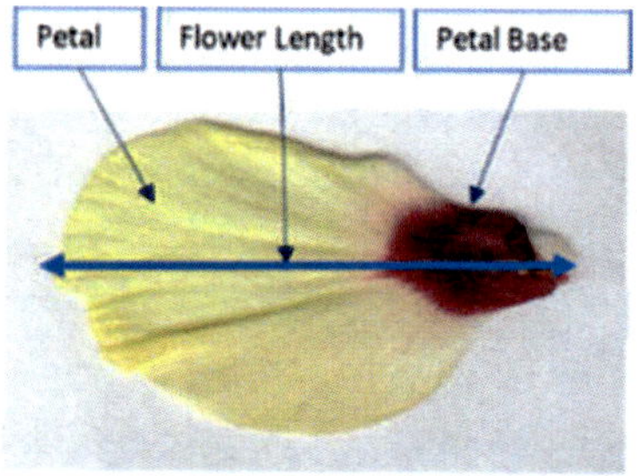

Fig. 2 DUS-defined features of Okra flower petal

Fig. 3 DUS-defined features of Okra flower

Fig. 4 DUS-defined features of Okra seed

in Figs. 4 and 3, respectively. The image of the petal was used to derive the petal color, petal base color, and flower length (serial no. 4, 5, and 6 from Table1); whereas, the image of the whole flower was used to derive the width as defined in serial no 7 in Table1. For seed samples, two images from one seed were acquired, e.g., testa view (placenta in rear side) and placenta view (placenta in the front side) as shown in Figs. 5 and 6, respectively. Testa view of seed image was used to evaluate the seed color (serial no. 8 from Table1) and placenta view was used to derive the seed hairiness parameter (serial no. 9 from Table1).

Algorithms related to the extraction of DUS features of okra stem, flower, and seed have been described below.

Segmentation of Okra Stem. The process begins with segmenting the Okra stem from the background regions. Figure 7 illustrates the steps involved in segmenting the stem. As shown in Fig. 7, dynamic thresholding was applied based on histogram analysis of the Blue channel. After thresholding, the noise in the image was reduced

Fig. 5 Okra seed testa view

Fig. 6 Okra seed placenta view

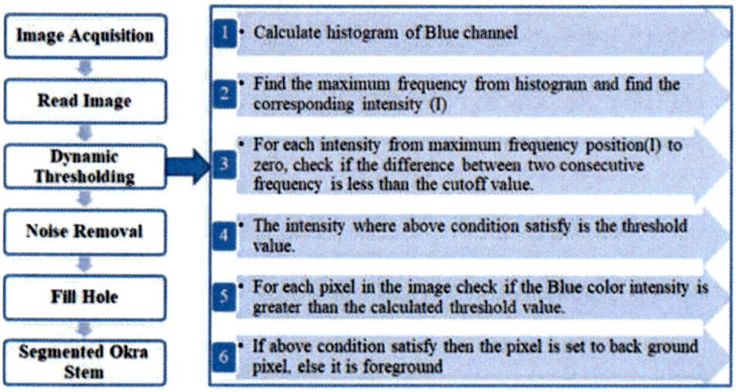

Fig. 7 Segmentation of Okra Stem

using area filtering, which involved setting relatively tiny foreground components to background pixels. The fill hole operation was used to fill the tiny holes that formed within the foreground regions areas as a result of the thresholding process. These processes produced a segmented picture of the Okra stem. Following segmentation, DUS-defined features, i.e., color and diameter of the stem were extracted.

Derive Color of Stem. The color properties of the Okra stem were extracted from the captured picture using RGB color space. The array of the extracted RGB color was compared with standard Green (0, 255, 0) and Red (255, 0, 0) colors using Mahalanobis distance [10]. The minimum distance between the calculated distance from Green and Red colors was determined for classifying the color of the stem in terms of Green/Red.

Derive Intensity of Green Color. The green channel image of the segmented colored Okra stem image was extracted. The average intensity of the green channel was calculated to represent the intensity of the green color. This average value was classified into light, medium, and dark based on the determined range of intensity values.

Derive Diameter of Stem. The diameter of the stem was calculated from the segmented image of the Okra stem. The boundary pixels of the stem were extracted by applying the chain code algorithm. From these boundary pixels, two pixels were

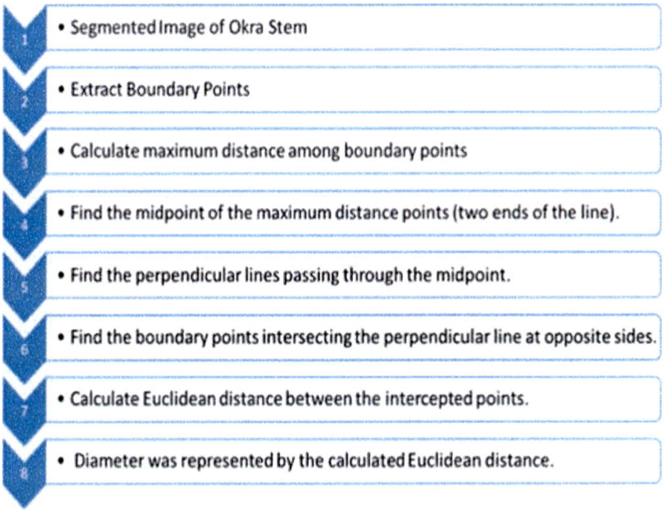

Fig. 8 Steps to derive diameter of Okra stem

found out which represents the maximum distance among the boundary pixels. These two pixels were the furthest points concerning each other. A line was drawn between these two pixels. The midpoint of the line was calculated by averaging the obtained X and Y coordinates of the two ends of the line. The perpendicular line passing through the midpoint of the line intercepts two boundary points at opposite ends. The diameter was calculated by calculating the Euclidean distance between the two intercepted boundary points. Detailed steps are shown in Fig. 8.

Segmentation of Petal. For segmentation of petals, the algorithm mentioned in Fig. 7 was used. After applying the segmentation, the fill hole operation and area-based noise removal were applied to get the whole portion of the petal. This segmented petal image was used to determine the color of the petal, the color of the petal base region, and flower length. The block diagram of the developed algorithm is shown in Fig. 9.

Derive Petal Color. After segmenting the whole petal, another threshold operation was applied to segment the petal only excluding the petal base. This threshold was applied on HSV color space ($H > 35°$ and $H < 75°$ and $V > 75$). This segmented image was a colored image of the petal. The color of the segmented petal was extracted in RGB color space. The array of RGB color was compared with standard Cream (255, 253, 208), Yellow (255, 255, 0), and Purple (128, 0, 128) colors using Mahalanobis distance [10]. The minimum distance among the three calculated distances for Cream, Yellow, and Purple was determined. The color of the petal was classified into Cream/Yellow/Purple based on the shortest distance rule.

Derive Petal Base Color. From the whole petal, another threshold was applied in HSV color space to extract the purple petal base only. This threshold ($H < 10°$ or $H > 335°$ and $V < 70$) extracts only the purple part from the image. The resultant image

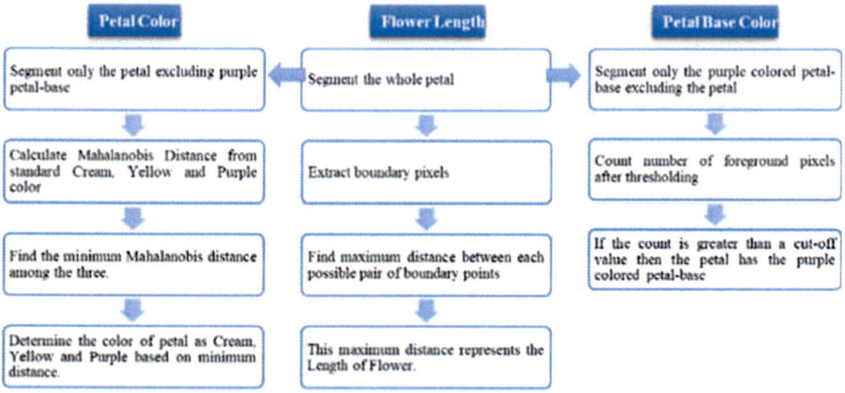

Fig. 9 Steps to derive the color of the petal, flower length, and the presence of purple-colored petal base

after this threshold operation was a binary image. The number of foreground pixels counted after this threshold was used to determine the presence of purple color in the petal base.

Derive Flower Length. The flower length can be represented by the length of the whole petal. First, the boundary pixels of the whole petal was determined using the chain coding algorithm. The maximum distance between the boundary points was calculated by comparing the Euclidean distance between each possible pair of points. The calculated maximum distance was the length of the flower.

Segmentation of the Whole Flower. The whole flower was first segmented using the algorithm mentioned in Fig. 7. Fill the hole and noise removal was applied to extract the foreground object as only the whole flower.

Derive Flower Width. From the segmented image of the whole flower, another threshold operation was applied to remove the green portion consisting of sepal, peduncle, and the receptacle from the whole Okra flower. This threshold was also applied in HSV color space ($H > 60°$). The resultant image was having the petal and petal base excluding the green peduncle side. From this segmented image, boundary pixels were determined using the chain code algorithm. The maximum distance between the boundary points was determined and eventually, the two furthest points were determined. A straight line was drawn between those two points. The pixels lying on this line were traversed and for each pixel, a perpendicular line was drawn concerning the drawn line representing the maximum distance. The two ends of each perpendicular line touch the perimeter of the okra flower. The length of each perpendicular line was determined by calculating the Euclidean distance between the two endpoints touching the boundary. The maximum length of these perpendicular lines was the width of the okra flower.

Derive Seed Color. Seed color was determined from the testa view image of the seed. The image segmentation of Fig. 7 was applied to extract the testa view of the seed only. After segmentation, noise removal and fill hole operations, the color

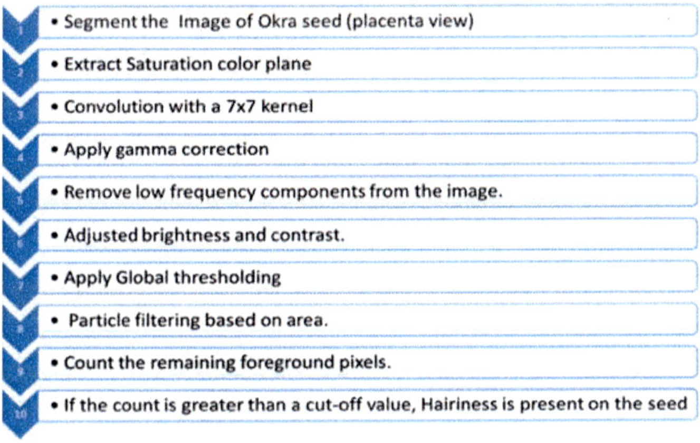

Fig. 10 Steps to determine the presence of hairiness on seed

of the seed was determined using the RGB color space. The extract array of RGB color was compared with Green (0, 255, 0) and Brown (165, 42, 42) colors using the Mahalanobis distance [10]. Further, the minimum distance was calculated between the determined Mahalanobis distances. The color of the seed was classified into Green/Brown based on the determined minimum distance.

Derive Seed Hairiness. Hairiness was determined to form the images of seed placenta view. The saturation color plane was extracted, and a convolution filter was applied by a 7×7 kernel. The kernel was having all -1(minus one) except the center pixel as $+$ 60(sixty). Gamma correction (0.44) was applied to make the hair more prominent. FFT was applied and low-frequency components were removed. The brightness and contrast of the resultant image were adjusted and subsequently segmented using global thresholding in gray-scale. Then particle filter was applied based on the area and calculated length of each foreground component. The calculated foreground area of the resultant image represents hairiness. A fixed area threshold was used to determine the presence/absence of hairiness. The flowchart is shown in Fig. 10.

3.4 Manual Measurement by Expert Panel

At the National Bureau of Plant Genetic Resources (NBPGR) in New Delhi, an expert panel of eight agriculture specialists was formed. This panel was using manual methods to measure the DUS trait. The generated data using the developed algorithms were validated against the manual measurement data.

3.5 Calibration and Implementation

In image processing, the pixel is the measuring unit. The human perception of length, on the other hand, is based on quantities like centimeters and millimeters. As a result, units like pixels have to be converted to centimeters or millimeters. The manual measurement was compared to the computerized measurement to identify one factor. The factor was determined by utilizing (1).

$$\text{Factor} = \text{Length Measured in Centimeter/Length Measured in Pixel} \quad (1)$$

After computing this factor, the needed centimeter value was obtained by multiplying it by pixel values. Similarly, the area was determined by multiplying the factor's square with the pixel area, as illustrated in (2).

$$\text{Length (in cm)} = \text{Length in Pixel} * \text{Factor}$$

$$\text{Area}\left(\text{in cm}^2\right) = \text{Area in Pixel} * \text{Factor}^2 \quad (2)$$

All of the aforementioned methods were implemented in software using Microsoft VC++ and the MYSQL database. All of the samples' DUS characteristics were recorded using this developed application.

4 Result and Validation

The proposed algorithms for deriving DUS-defined characteristics of Okra stem, flower, and seed were validated using a total of 131 samples. 26 okra stem samples, 55 okra flower samples, and 50 okra seed samples were evaluated. All of the samples were manually measured for DUS-defined morphological and color characteristics after they were imaged. The mutually agreed-upon DUS-specified characteristics were determined by an expert group of eight scientists. The findings of manual measurement were compared to those obtained utilizing the algorithms devised. The color of the stem was defined in two classes, namely Green and Red. When comparing with manual measurement of stem color, the accuracy of the developed algorithm was in the tune of 100% as shown in Fig. 11. As these two were prominent colors (i.e., Green and Red), it was not overlapping in the results. The intensity of the green color of the stem was classified into light, medium, and dark in DUS protocols. The comparison result between the manual measurement and software measurement was plotted in the confusion matrix in Table 2 for the intensity of green color. Also the performance metrics, e.g., precision (TP/(TP + FP)), pecall (TP/(TP + FN)), F1-score ((2*TP)/(2*TP + FP + FN)), specificity (TN/(TN + FP)), and accuracy ((TP + TN)/(TP + TN + FP + FN)) were derived from the confusion matrix [15–17]. The extracted diameter (in cm) of the stem using the developed algorithms was plotted

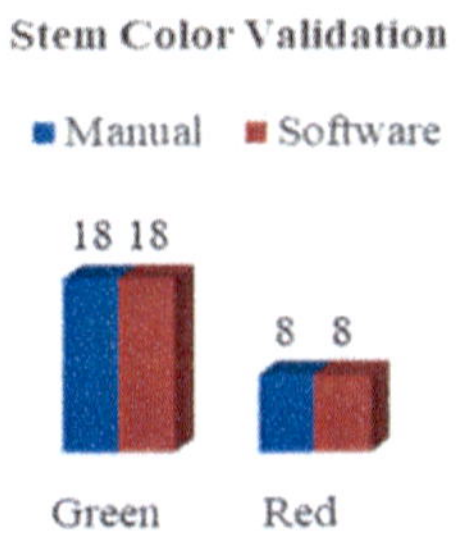

Fig. 11 Stem color validation (Software vs. Manual)

Table 2 Validation of the measurement of intensity of green color of stem

Actual/predicted	Light	Medium	Dark
Light	8	0	0
Medium	2	13	0
Dark	0	1	2
True positive	8	13	2
True negative	16	10	23
False positive	2	1	0
False negative	0	2	1
Precision	0.80	0.93	1.00
Recall	1.00	0.87	0.67
F1-score	0.89	0.90	0.80
Specificity	0.89	0.91	1.00
Accuracy	0.92	0.88	0.96

against the manual measurement in scatter charts as shown in Fig. 12. The calculated correlation coefficient (R2) for diameter validation of stem was around 96%.

The color of the petal was classified into three color groups, i.e., Cream/Yellow/Purple as per DUS guidelines. The obtained class of color of the petal using the developed algorithm was determined and compared with manual measurements. The comparison result is plotted in the confusion matrix form in Table 3. The performance metrics are also derived and given in the table. The whole petal consists of a purple base also for a few varieties. As per the DUS guidelines, it was

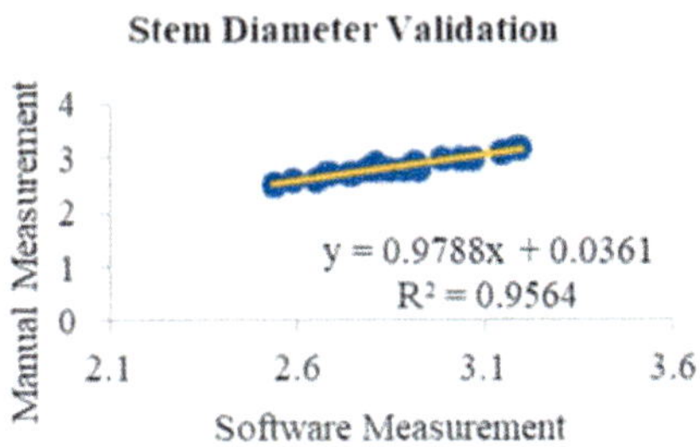

Fig. 12 Stem diameter measurement validation

Table 3 Validation of the measurement of petal color

Actual/predicted	Cream	Yellow	Purple
Cream	42	1	0
Yellow	5	7	0
Purple	0	0	0
True positive	42	7	–
True negative	7	42	–
False positive	5	1	–
False negative	1	5	–
Precision	0.89	0.88	–
Recall	0.98	0.58	–
f1score	0.93	0.70	–
Specificity	0.58	0.98	–
Accuracy	0.89	0.89	–

required to determine the presence of purple color in the petal base. The developed algorithm declared the presence of purple color in the petal base using the count of purple pixels in the petal. The software result was validated with manual results and it was having 100% accuracy for determining the presence of purple color in the petal base. The derived length (in cm) and diameter (in cm) of the Okra flower were plotted against the manual measurement in Fig. 13 and Fig. 14, respectively. The calculated correlation coefficient (R2) for length validation was around 96% and for the diameter, validation was around 94%.

The color of the seed was classified into two color groups, namely Green/Brown. The classification results obtained from the developed algorithm were compared with manual results, and the accuracy was determined as 96%. Two seeds were found for which the result of the developed algorithm for seed color derivation was conflicting with manual assessment. As per DUS, the presence of hairiness on the seed was also a defined parameter. The developed algorithm was calculating the area of hairiness and based on some threshold value it was declared absent/presence. This threshold value was essential to reduce the misclassification due to the uneven surface properties of

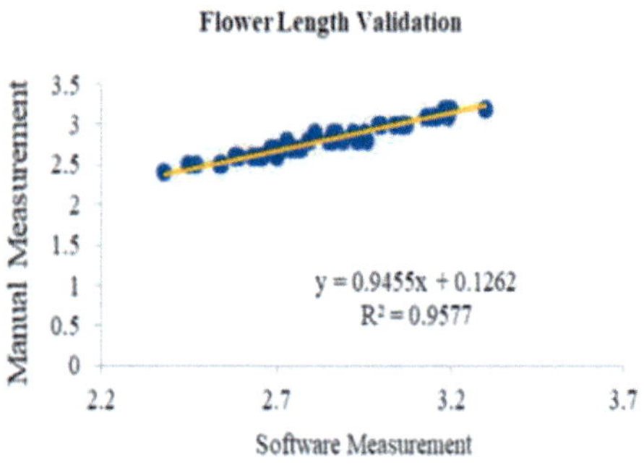

Fig. 13 Flower length measurement validation

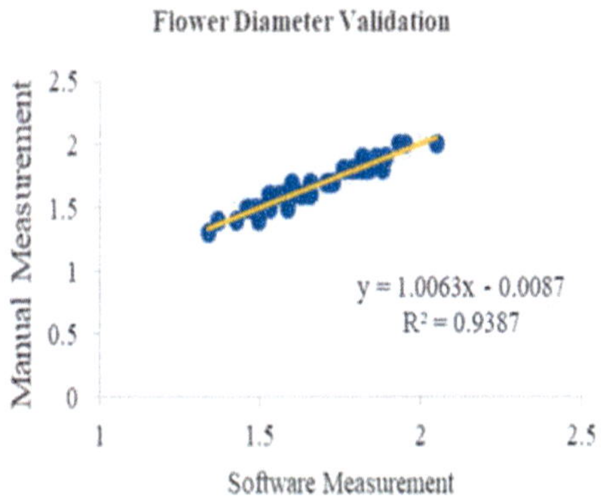

Fig. 14 Flower diameter measurement validation

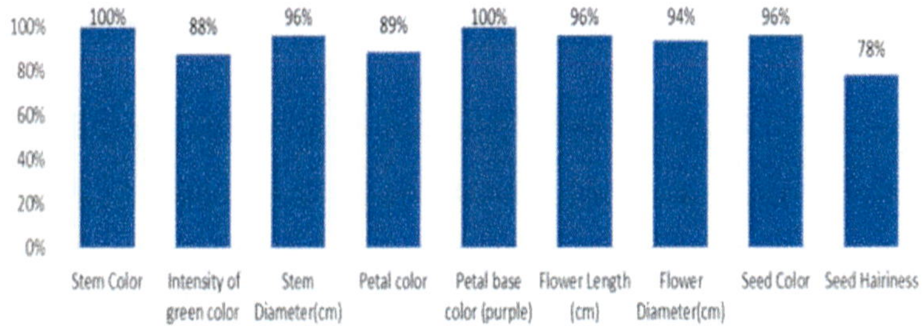

Fig. 15 Accuracy using the proposed methodology

the seed. Out of 50 samples, 11 samples were misclassified as present of hairiness; whereas, manual results were defined as absent of hairiness. Thus, the accuracy of the developed algorithm was calculated as 78% for estimation of seed hairiness.

The overall accuracy from each confusion matrix was calculated using the ratio of total true positive and the total no. of samples. The overall accuracy for the intensity of the green color of the stem was calculated as 88% ($\frac{23}{26} * 100$), for petal color, it was calculated as 89% ($\frac{49}{55} * 100$). The accuracy achieved using the developed algorithms for each selected DUS-defined feature is shown in Fig. 15.

5 Conclusion

DUS testing may be made easier using the created algorithm. When human specialists collect data, there is a chance that some of it will be incorrect. Furthermore, the manual data collecting procedure is time-consuming, and the data collected by specialists may differ. During DUS testing, this machine vision technology might be a valuable tool for this error-prone human data acquiring procedure. This technique may give scientists with DUS characteristic measurement findings in digital format, allowing them to process DUS tests more quickly and accurately. To build an AI/ML-based automated model for DUS testing, more research on the data provided by the established algorithms is required [9, 18].

Acknowledgements This project was funded by the National Fund for Basic, Strategic and Frontier Application Research in Agriculture (NFBSFARA), Indian Council of Agricultural Research, New Delhi. The authors are grateful to the funding agency for providing support and guidance. The authors are also grateful to Dr. Nachiket Kotwaliwale (CIAE, Bhopal, India) and Dr. K.K. Gangopadhyay (NBPGR, New Delhi, India) for their constant support.

References

1. Protection of plant varieties and farmers' rights authority, India.http://www.plantauthority.in/. Accessed on 27 Apr 2020
2. Training Manual on DUS Test in Cotton with reference to PPV & FR legislation, 2001, 19th to 25th January, 2005
3. Crop DUS Guidelines. https://www.plantauthority.gov.in/crop-dus-guidelines, Accessed on: second September 2021.
4. Jyoti B, Usha PS (2015, December) Role of DUS testing in registration of plant varieties under PPVV & FR Act, 2001. Rashtriya Krishi 10(2). e ISSN–2321–7987
5. Guidelines for the conduct of test for Distinctiveness, Uniformity, and stability on Okra/Lady's Finger (*Abelmoschusesculentus (L.) Moench.*): protection of plant varieties and Farmer's rights authority, Government of India: SG/35/2009, November 3, 2009. https://www.plantauthority.gov.in/sites/default/files/fokra.pdf
6. Ranjan P (2009) Recent developments in India's plant variety protection, seed regulation and linkages with UPOV's proposed membership. J World Intellect Property 12(3):219–243. https://doi.org/10.1111/j.1747-1796.2009.00365.x
7. Yu J-K, Chung Y-S (2021) Plant variety protection: current practices and insights. Genes 12:1127. https://doi.org/10.3390/genes12081127.
8. Yang CJ, Russell J, Ramsay L et al (2021) Overcoming barriers to the registration of new plant varieties under the DUS system. Commun Biol 4:302. https://doi.org/10.1038/s42003-021-01840-9
9. Kotwaliwale N, Singh K, Chakrabarty SK, Joshi MA, Kalne A, Tiwari M, Gangopadhyay KK, Bhattacharya n, Akuli A, Aggarwal D, Bej G (2019) Machine vision for characterization of some phenomic features of plant parts in distinguishing varieties–a review. Int J Bio-Inspired Comput (IJBIC) 14(4). https://doi.org/10.1504/IJBIC.2019.103960
10. Bej G et al (2021) Comparison of different color models for priority based color matching of plant parts used in DUS testing. In: Balas VE. Hassanien AE, Chakrabarti S, Mandal L (eds) Proceedings of international conference on computational intelligence, data science and cloud computing. Lecture notes on data engineering and communications technologies, vol 62. Springer, Singapore. https://doi.org/10.1007/978-981-33-4968-1_37
11. Dhaliwal I et al (2020) DUS characterization of some released varieties and advanced breeding lines of Soybean (*Glycine max L.*) under Punjab agroclimatic conditions. Int J Curr Microbiol App Sci 9(8):2124–2135. https://doi.org/10.20546/ijcmas.2020.908.242
12. Palakshappa MG et al (2020) DUS testing of sesame (*Sesamum indicum L.*) Accessions using morphological descriptors and evaluation for foliar diseases of sesame. Int J Curr Microbiol App Sci 9(1):1837–1852. https://doi.org/10.20546/ijcmas.2020.901.206

13. Temam N, Mohammed W, Aklilu S (2021) Variability assessment of Okra (*Abelmoschus esculentus* (*L.*) *Moench*) genotypes based on their qualitative traits", 2021 ,International Journal of Agronomy,DOI: https://doi.org/10.1155/2021/6678561
14. Matthew O, Ohwo Uo, Osawaru M (2018) Morphological characterization of Okra (Abelmoschus [Medik.]) accessions. Makara J Sci. 22. https://doi.org/10.7454/mss.v22i2.9126
15. Al-jabery KK, Obafemi-Ajayi T, Olbricht GR, Wunsch II DC, 9—data analysis and machine learning tools in MATLAB and Python. In: Al-jabery KK, Obafemi-Ajayi T, Olbricht FGR, Wunsch II DC (2020) Computational learning approaches to data analytics in biomedical applications. Academic Press, pp 231–290. ISBN 9780128144824, https://doi.org/10.1016/B978-0-12-814482-4.00009-7
16. Liu Y-f et al (2013) Statistic analysis on quantitative characteristics for developing the DUS test guideline of *Ranunculus asiaticus L.* J Integr Agric 12(6): 971-978. https://doi.org/10.1016/S2095-3119(13)60317-2
17. Manliguez C (2016) Generalized confusion matrix for multiple classes. https://doi.org/10.13140/RG.2.2.31150.51523
18. Santhy V, Rathinavel K, Saravanan M, Meshram C, Priyadharshini M (2020) Genetic diversity assessment of extant cotton varieties based on principal component analysis (PCA) and cluster analysis of enlisted DUS traits. Electron J Plant Breed 11(2):430–438. Retrieved from http://www.ejplantbreeding.org/index.php/EJPB/article/view/3386

Chapter 18
A Typical Hybrid Optimization-Based Image Quality Enhancement Technique

Vineeta Singh and Vandana Dixit Kaushik

1 Introduction

The need of picture combination is filling in as of late in picture preparing applications because of the enormous measure of securing frameworks. Combination of pictures is characterized as an arrangement of essential Information from assorted sensors utilizing different numerical strategies to create a solitary fused picture. Integrating pictures to utilize in coordination of corresponding multi-view, multi-focus as well as multi-sensor Information to a single combined picture leveraged with good quality moreover via maintaining the trustworthiness of significant highlights. This was taken as an imperative pre-handling level for applications such as robot vision, satellite imaging, clinical imaging as well as a robot/vehicle direction. Picture combination alludes to separating and afterward joining the most significant data from various source pictures, expects to create a solitary picture that is more educational and helpful for ensuing applications. Multi-focus picture blend is an improvement method to make completely focused pictures that may address the depth of-field limit in imaging of optical central focuses. Multi-focus picture combination is extracting focused areas out of various pictures for making single focused everywhere melded picture. The central issue is that solitary articles inside the depth of-field possess a clearest visual in the picture, whereas other may seem to be obscured. The progressions in robotized demonstrative apparatuses permit scientists to acquire increasingly more data from clinical pictures. As of late, to get more educational clinical pictures, multi-modality pictures have been utilized. These pictures have fundamentally more data when contrasted with customary clinical pictures. In any case, the development of multi-modality pictures is certifiably not a simple undertaking.

V. Singh (✉) · V. D. Kaushik
Department of Computer Science and Engineering, Harcourt Butler Technical University Kanpur, Kanpur, India
e-mail: vineeta.singh.cs@gmail.com

L. Mandal et al. (eds.), *Proceedings of International Conference on Computational Intelligence, Data Science and Cloud Computing*, Algorithms for Intelligent Systems,
https://doi.org/10.1007/978-981-19-1657-1_18

2 Related Work

Mostly developed techniques produce the choice guide to acknowledge multi-focus picture combination, which as a rule lead to detail misfortune because of misclassification, particularly related to focused/defocused regions. For overcoming this issue in [1] researchers presented another generative antagonistic organization with versatile as well as slope joint limitations for melding multi-focus pictures. Here researchers proposed a model where a versatile choice square is acquainted with decide if input pixels are focused/ partially focused relied on the distinction of rehashed obscure. Devised model is unaided without requirement of ground-truth pictures to generate fused pictures. Moreover, another dataset exhibiting 120 multi-focus picture sets for benchmark assessment were given. Test results demonstrated the prevalence of devised technique in terms of visualization as well as quantitative measurements [1].

The involvement of deep learning has advanced colossal advancement in picture combination, and the amazing element extraction and reproduction abilities of neural organizations make the melded results promising. As of late, a few most recent deep learning advances have made picture combination detonate, e.g., generative ill-disposed organizations, autoencoder, and so on. Be that as it may, a complete audit and investigation of most recent deep learning techniques in various combination situations are inadequate. To this end and in this overview, we initially present the idea of picture combination, and arrange the strategies according to the viewpoints of the deep designs took on and combination situations. Then, at that point, we audit the cutting edge on the utilization of deep learning in different sorts of picture combination situations, including the computerized photography picture combination, the multi-modular picture combination and the honing combination. Thusly, the assessments for some agent strategies in explicit combination undertakings are performed subjectively and quantitatively. Also, authors momentarily presented a few run of the mill uses of picture combination, including photography perception, RGBT object following, clinical analysis, and far off detecting observing. At long last, they given the end, featured the difficulties in picture combination, and anticipated potential future exploration bearings [2].

In [3] a novel technique for integrating multi-focus pictures was proposed relied on unsupervised deep learning. They trained an encoder–decoder network in an unaided way for gaining deep highlights of info pictures. Further at that point, they used spatial recurrence, an angle-based strategy to gage sharp variety out of these deep highlights, for reflecting action levels. They applied some consistency check strategies for changing the choice guide and draw out the combined outcome. Developed technique breaks down sharp appearances in deep highlights rather than unique pictures, which can be viewed as another example of overcoming adversity of solo learning in picture handling. Trial results show that the proposed strategy accomplishes best in class combination execution contrasted with 16 combination strategies in unbiased and abstract evaluations, particularly in slope-based combination measurements [3].

Sickness determination is the most basic medical care work. On the off chance that an ailment is analyzed before the typical or arranged period it can save individuals' lives. Order strategy for AI can be valuable to help the clinical branch by conveying solid and moment sickness finding. Henceforth the helpful time for the two doctors and patients since coronary illness is one of the world's most super risky and perilous sicknesses today, because of the trouble to analyze the infection. Inside this paper we incorporate an audit of the grouping techniques for AI and picture combination that have been shown to help medical services experts distinguish coronary illness. Here researchers started with the AI brief and summed up depictions of the predominantly utilized arrangement procedures for diagnosing infections of heart. Then, at that point, further audited and exhibited some work on the utilization of arrangement procedures for AI and picture combination around there [4].

The combination of multi-modality clinical pictures makes a critical commitment to clinical conclusion and investigation since it permits indicative imaging experts to make a more precise determination. As per our present information, there are a few constraints of current clinical picture combination draws near. The main restriction is that the utilization of a weighted normal guideline for combining low-recurrence segments. This impediment prompts a lessening in the power of the brilliance of the intertwined picture. The subsequent impediment was to use combination techniques for high-frequency parts was not actually ideal. This is probably going to bring about the deficiency of point by point data in the melded picture. In this paper, an original methodology involving two calculations has been proposed to resolve the previously mentioned shortcomings. The principal calculation depended on the Grasshopper enhancement calculation (GOA) to discover ideal boundaries fully intent on melding low-recurrence segments. This permits the intertwined picture to have great difference. The subsequent calculation depends on the Kirsch compass administrator to make a productive guideline for the combination of high-recurrence parts. This permits the intertwined picture to altogether safeguard subtleties moved from input pictures. Exploratory outcomes show that the proposed approach not just successful in upgrading essentially the differentiation of the intertwined picture yet in addition safeguarding edge data conveyed from input pictures to the composite picture [5].

To safeguard the design data and play out the definite data of source pictures, a novel picture combination conspire dependent on picture animation surface decay and meager portrayal was proposed in [6]. Here source multi-methodology pictures were decayed into animation and surface parts. For animation parts an appropriate spatial-based strategy was introduced for morphological design protection. A fusion mechanism relied on energy was utilized to safeguard structure data of every input picture. In case of surface parts, a scanty portrayal-based strategy was proposed. A word reference with solid portrayal capacity was prepared for the proposed scanty portrayal-based combination strategy. At long last, as indicated by the surface upgrade combination rule, the intertwined animation and surface parts were incorporated. The experimentation results have demonstrated better results for visual as well as quantitative metrics [6].

In [7] scientists proposed a newly developed multiscale change disintegration strategy for multi-focus picture combination for improving intertwined execution. The non-subsampled contourlet change (NSCT) was right off the bat consumed to decay input multi-focus pictures into low-frequency as well as a few high-frequency groups to isolate two fundamental qualities of input pictures such as head data as well as edge subtleties. The "normal" as well as "max-outright" combination mechanisms were performed on low-as well as high-frequency parts, individually to produce a fused picture. Moreover here the distinction pictures between the fundamental intertwined picture and the source pictures were determined, and the energy of the inclination (EOG) of contrast pictures were used to refine the essential melded picture by incorporating normal channel and middle channel. Visual and quantitative investigation showed the better presentation of the proposed combination method [7]. In [8] the devised model, at first, decays the picture into subgroups utilizing a non-subsampled contourlet change (NSCT) area. From that point, an outrageous version of the Inception (Xception) was utilized to include extraction of the input pictures. The multi-destination differential advancement was utilized for selecting ideal components. From that point, the coefficient of assurance as well as the energy misfortune allied combination capacities were utilized for getting intertwined coefficients. Finally the combined picture was generated via incorporating the inverse NSCT. Proposed approach outflanked the other multi-focus picture fusion approaches [8].

As the quantity of space debris has expanded quickly lately, it represents a significant danger to the well-being of spacecraft in space as well as harm appraisal for space flotsam and jetsam hypervelocity sway is vital. To all the more exhaustively and precisely depict the harm absconds in the infrared investigation information gathered by infrared warm imaging innovation, in [9] creators proposed a super high speed sway harm location calculation dependent on infrared reproduction picture combination. The calculation first preprocesses the acquired infrared warm reaction picture arrangement and isolates the harm components, and afterward applies a multi-target transformative advancement calculation to separate the average transient warm reaction, and afterward reproduces the element infrared pictures. At long last, picture combination dependent on directed separating was executed on the harm recreation infrared pictures. Further a few remaking pictures that address various sorts of effect harm imperfection are consolidated to further develop the discovery capacity of the calculation. Infrared identification investigates harmed examples acquired from real hypervelocity impacts check the adequacy of the proposed calculation [9].

The pixel-based combination technique will in general lose nearby and spatial data as the connections between pixels are not thought about fittingly, which has a lot of influence over the nature of the combination outcomes. For resolving aforementioned issue, a district-based multimodal clinical picture combination structure has been proposed in [10] dependent on superpixel division and a post-handling advancement strategy. In the proposed work, the normal picture of the source clinical pictures is right off the bat acquired via a weighted average strategy. For acquiring viably homogeneous areas as well as safeguarding the total data of picture subtleties, the quick straight unearthly bunching (LSC) superpixel calculation is done to portion the normal picture as well as obtain superpixel marks. Here for every locale of the clinical pictures, log-Gabor channel, i.e., LGF as well as average changed laplacian were embraced for extricating surface component and difference include for the estimation of district significance. The main districts are chosen and the choice guide is created by examination. Besides, to get a more precise choice guide, another post-handling improved technique dependent on genetic algorithm has been yielded. A weighted method was incorporated to the extricated highlights as well as weighting variable may be adaptively changed via GA. The devised technique justified its viability via directing investigations on eight sets of clinical pictures out of assorted modalities [10].

The track of clinical imaging has seen a few headways somewhat recently. A few clinical imaging modalities have showed up somewhat recently involving Computed Tomography, i.e., CT, X-beam, Emission Tomography, i.e., PET, Magnetic Resonance, i.e., MR, ultrasound imaging as well as Positron Single-Photon Emission Computed Tomography abbreviated as SPECT. By and large, clinical pictures are utilized for the finding reason. Each kind of procured pictures has a few benefits and limits. To boost clinical pictures use with the end goal of conclusion, clinical imaging combination pattern has showed up as a current examination area. Different clinical modality images were combined for yielding final pictures with reciprocal data. In [11] a study investigation of clinical imaging modalities as well as features of them has been illustrated. Moreover, unique clinical picture combination draws near and their fitting quality measurements are introduced. The fundamental point of this exhaustive study investigation is to contribute in the progression of clinical picture moves toward that can help for better finding of various illnesses [11].

In [12–21] researchers have demonstrated in detail about variety of fusion techniques covering different application domains involving medical imaging, infrared as well as visible images, multi-focus images, etc.

3 Proposed Fusion Technique

In our proposed fusion technique, initially two source pictures of the single scene were taken which were focused at different locations. The main aim was to generate a solitary image by blending both of them. Initially the contourlet transform [22] was imposed to both of the input pictures, as a result of which four frequency subgroups were generated of each of the input picture. All the four corresponding subgroups were joined via proposed fusion rule which is based on optimization strategy. Here two fusion factors were taken into consideration, suppose named as α and β. Let us consider an expression $\alpha A + \beta B$, here A and B refers to the weighted subgroups while α and β are the fusion factors. Here first fusion factor is based on holo-entropy. Holo-entropy [23] is calculated for source pictures as well as for fused picture. If fused picture possess higher value that means there is some presence of redundant information which may degrade the fused image quality. So this will control the redundant information infusion. Second fusion factor is determined relied on optimization strategy known atom search optimization technique along with sine cosine algorithm. Optimization techniques are best known for convergence property and are widely used to solve real life engineering issues. Here the atom search optimization technique [24] is substituted in the update position of sine cosine algorithm [25]. In this devised approach further two objective functions were used. First one utilized the combination of holo-entropy and joint holo-entropy while other fitness function utilized mutual information, structural similarity index measurement as well as peak signal to noise ratio for determining optimum solution via help of fitness objective maximization procedure. On the basis of both objective functions the optimal solutions were selected and least value for the fitness function was taken as optimal solution and update positions were updated for Taylor series [26] then for hybrid of ASO and SCA and finally for Taylor-based ASO and SCA accordingly. These steps were repeated until the best optimal solution was obtained. You may refer the following sequence of steps.

- **Start**
- **Input source images (two focused at different locations)**
- **Apply contourlet transform**
- **Generation of frequency subbands**
- **Join respective frequency subbands as per proposed fusion mechanism**
- **Apply inverse contourlet transform**
- **Obtain final resultant image**
- **Stop**

4 Experimental Work and Simulation Details

The simulation work executed on MATLAB and dataset utilized was Lytro multi-focus dataset. The proposed fusion technique for blending pictures utilized three performance evaluators SSIM, MI, and PSNR. We have demonstrated here four sample image pairs with respective fused images generated by proposed fusion strategy (please refer Figs. 1, 2, 3 and 4). Refer Table 1 to have a look on experimental related results and performance assessment parameters, where proposed method has

Input image 1 Input image 2 Fused image

Fig. 1 Experimental result using image pair 1

Fig. 2 Experimental result using image pair 2

Fig. 3 Experimental result using image pair 3

Fig. 4 Experimental result using image pair 4

been comparatively analysed with State-of-the-Art techniques. Comparative results demonstrated here the better performance for proposed method.

Table 1 Comparative key performance values for proposed method

Images	Key indicators	[27]	[28]	Proposed method
1	MI	1.15	1.21	**1.34**
1	SSIM	0.88	0.92	**0.93**
1	PSNR	20.88	21.75	**23.9**
2	MI	1.16	1.24	**1.37**
2	SSIM	0.86	0.92	**0.93**
2	PSNR	19.61	23.71	**25.19**
3	MI	1.17	1.25	**1.40**
3	SSIM	0.85	0.89	**0.92**
3	PSNR	20.01	23.24	**26.51**

Bold values depict the proposed method with best performing values for the enlisted performance indicators

5 Conclusion and Future Scope

Our proposed model outperformed other state-of-the-art fusion models evaluated on certain key indicators such as MI, SSIM, PSNR with a value of 1.40, 0.93, 26.51 dB, respectively. In future we may extend this work further on more performance key indicators as well as datasets for wider diverse application of the proposed work.

References

1. Zhang H, Le Z, Shao Z, Xu H, Ma J (2021) MFF-GAN: An unsupervised generative adversarial network with adaptive and gradient joint constraints for multi-focus image fusion. Inf Fusion 66:40–53
2. Zhang H, Xu H, Tian X, Jiang J, Ma J (2021) Image fusion meets deep learning: a survey and perspective. Inf Fusion 76:323–336
3. Ma B, Zhu Y, Yin X, Ban X, Huang H, Mukeshimana M (2021) SESF-fuse: an unsupervised deep model for multi-focus image fusion. Neural Comput Appl 33(11):5793–5804
4. Diwakar M, Tripathi A, Joshi K, Memoria M, Singh P (2021) Latest trends on heart disease prediction using machine learning and image fusion. Mater Today Proc 37:3213–3218
5. Dinh PH (2021) A novel approach based on grasshopper optimization algorithm for medical image fusion. Expert Syst Appl 171:114576
6. Zhu Z, Yin H, Chai Y, Li Y, Qi G (2018) A novel multi-modality image fusion method based on image decomposition and sparse representation. Inf Sci 432:516–529
7. Li L, Ma H, Jia Z, Si Y (2021) A novel multiscale transform decomposition based multi-focus image fusion framework. Multimedia Tools Appl 80(8):12389–12409
8. Kaur M, Singh D (2021) Multi-modality medical image fusion technique using multi-objective differential evolution based deep neural networks. J Ambient Intell Humaniz Comput 12(2):2483–2493
9. Tan X, Huang X, Yin C, Dadras S, Cheng YH, Shi A (2021) Infrared detection method for hypervelocity impact based on thermal image fusion. IEEE Access 9:90510–90528

10. Duan J, Mao S, Jin J, Zhou Z, Chen L, Philip Chen CL (2021) A novel GA-based optimized approach for regional multimodal medical image fusion with superpixel segmentation. IEEE Access 9:96353–96366
11. Faragallah OS, El-Hoseny H, El-Shafai W, El-Rahman WA, El-Sayed HS, El-Rabaie E-SM, Abd El-Samie FE, Geweid GGN (2020) A comprehensive survey analysis for present solutions of medical image fusion and future directions. IEEE Access 9:11358–11371
12. Kaur H, Koundal D, Kadyan V (2021). Image fusion techniques: a survey. Arch Comput Method Eng 1–23
13. Javan FD, Samadzadegan F, Mehravar S, Toosi A, Khatami R, Stein A (2021) A review of image fusion techniques for pan-sharpening of high-resolution satellite imagery. ISPRS J Photogramm Remote Sens 171:101–117
14. Bhat S, Koundal D (2021) Multi-focus image fusion techniques: a survey. Artif Intell Rev 1–53
15. Amolins K, Zhang Y, Dare P (2007) Wavelet based image fusion techniques—an introduction, review and comparison. ISPRS J Photogramm Remote Sens 62(4):249–263
16. Hermessi H, Mourali O, Zagrouba E (2021) Multimodal medical image fusion review: theoretical background and recent advances. Signal Process 108036
17. Laimer G, Schullian P, Bale R (2021) stereotactic thermal ablation of liver tumors: 3D planning, multiple needle approach, and intraprocedural image fusion are the key to success—a narrative review. Biology 10(7):644
18. Ezimand K, Chahardoli M, Azadbakht M, Matkan AA (2021) Spatiotemporal analysis of land surface temperature using multi-temporal and multi-sensor image fusion techniques. Sustain Cities Soc 64:102508
19. Plooij JM, Maal TJ, Haers P, Borstlap WA, Kuijpers-Jagtman AM, Bergé SJ (2011) Digital three-dimensional image fusion processes for planning and evaluating orthodontics and orthognathic surgery. A systematic review. Int J Oral Maxillofac Surg 40(4):341–352
20. Zhang X (2021) Benchmarking and comparing multi-exposure image fusion algorithms. Inf Fusion 74:111–131
21. Perez J, Mazo C, Trujillo M, Herrera A (2021) MRI and CT fusion in stereotactic electroencephalography: a literature review. Appl Sci 11(12):5524
22. Li B, Peng H, Wang J (2021) A novel fusion method based on dynamic threshold neural P systems and nonsubsampled contourlet transform for multi-modality medical images. Signal Process 178:107793
23. Venkatesh M, Prasad KS (2021) High efficient video coding through improved Holo entropy encoding with hybrid grey wolf optimization. Ann Rom Soc Cell Biol 2370–2380
24. Zhao W, Wang L, Zhang Z (2019) Atom search optimization and its application to solve a hydrogeologic parameter estimation problem. Knowl-Based Syst 163:283–304
25. Mirjalili S (2016) SCA: a sine cosine algorithm for solving optimization problems. Knowl-Based Syst 96:120–133
26. Tang Q, Yang J, He X, Jia W, Zhang Q, Liu H (2021) Nighttime image dehazing based on Retinex and dark channel prior using Taylor series expansion. Comput Vis Image Underst 202:103086
27. Zhao W, Wang D, Lu H (2018) Multi-focus image fusion with a natural enhancement via a joint multi-level deeply supervised convolutional neural network. IEEE Trans Circ Syst Video Technol 29(4):1102–1115
28. Singh V, Kaushik VD (2021) Renyi entropy and atom search sine cosine algorithm for multi focus image fusion. Signal Image Video Process 1–10

Chapter 19
A Comparative Study on Reassembly of Image Fragments

Susovan Jana

1 Introduction

The fragment reassembly problem [1] deals with the task of joining multiple parts of an image to make the complete original image. The positioning of the fragments is most important at the time of reassembly [2]. This is a very interesting as well as a challenging task in the field of image processing. This problem is also treated as Jigsaw Puzzle solving problem in computer vision. The fragment reassembly problem is a common problem for multiple applications like art recovery, archaeology, digital forensics [3], reconstructive surgery, map stitching [4], etc. Manual reassembly is a very time-consuming task when the number of fragments is very large. This problem can be broadly categorized into two types. The first type is that the information of the complete image is available for the fragments to be reassembled. The second type is obviously for fragment reassembly where no prior information is available about the complete image. The second type is a very difficult one.

The shape, colour, and texture features of fragments play a very important role in fragment reassembly. The accurate extraction of fragment boundary improves the accuracy of fragment reassembly. It is very difficult to reassemble the fragments, where the shape of all the fragments are similar. Then the dependency on colour and texture features increases in absence of shape. In this paper, the prior approaches are categorized into three types, i.e. geometry-based reassembly, content-based reassembly, and reassembly using machine learning.

The rest of the papers has been organized as follows. Section 2 discusses the prior approaches based on geometry-based reassembly. Section 3 describes the

S. Jana (✉)
Department of Information Technology, Institute of Engineering & Management, Kolkata, India
e-mail: jana.susovan2@gmail.com

L. Mandal et al. (eds.), *Proceedings of International Conference on Computational Intelligence, Data Science and Cloud Computing*, Algorithms for Intelligent Systems,
https://doi.org/10.1007/978-981-19-1657-1_19

prior content-based reassembly techniques. Section 4 presents the machine learning-based approaches to address the fragment reassembly problem. Section 5 brings the experimentation and results on this problem. Section 6 concludes this study.

2 Geometry-Based Reassembly

The shape of the fragment boundary is the most important attribute for reassembling image fragments. It has been observed that only boundary curves are sufficient to reassemble correctly if the image was fragmented in random shape and the number of fragments is not very large. Hence, shape-based reassembly is very popular. It has limitations also. The fragment boundary is less important if all the fragments are similar in shape, i.e. rectangular, square, etc.

McBride et al. proposed a technique to reassemble the fragments of archaeological marbles [5]. The authors have divided their approach into two distinct stages. In the first stage, local pairwise affinity was calculated. The adjacency candidates are not only dependent on the number of corners but also dependent on a novel normalized energy which is determined by how long the similarities exist between sub boundaries. In the second stage, global optimal arrangements are identified using the best-first strategy. Multiple solutions are identified as backtracking is expensive. Global confidence is determined to select the best match. It takes around 7 min to reassemble only 15 fragments.

Xu et al. proposed a technique to reassemble image fragments in polygon shape [6]. At first, the corners are detected for a fragment using Hough transform. The length of a side is measured by the distance between two corner points. Then the matching fragment is identified from the length of a side. The contour of the matching fragment is rotated for merging with a matching angle. The grey correlation of matching sides is used to find the appropriate pair of matching fragments if the number of matching fragments is more than 2 based on the distance of matching sides. The execution time is too high because of computing Hough transform for all the corners. But the success of this technique mostly depends on the appropriate detection of corners. So, there is a high chance of performance degradation by a small amount of noise.

Zhang et al. proposed a graph-based optimization technique for addressing the challenges of fragment reassembly problems [7]. The authors adopted a three-step approach to reducing the effect of local optima. In the first step, pairwise matching has been identified based on the shape of boundary and colour information. The possible pair of matching fragments are identified by this step. The probability of matching error is mostly caused due to noise and local optima. In the second step, a graph-based global searching is done to find out the group-wise matching fragments. This global reassembly reduces the probability of the redundant matching of the first step. In the last step, a fine-tuning of group-wise reassembly is done by using a graph optimization algorithm. The algorithm takes 2 to 3 min for assembling 10 fragments and 15 min for assembling 26 fragments.

Kamran et al. proposed a three-stage fragment reassembly technique [8]. At first, the boundary was extracted from the scanned fragments. The boundary matching between all pair fragments is determined by considering it as a Longest Common Subsequence problem. The global optimal matching of fragments is done with the help of the greedy technique. The experimentation has been done on printed hand-torn images of various resolutions.

3 Content-Based Reassembly

The content of the image plays a very important role in the fragment reassembly of an image. One of the most difficult tasks in image processing is the analysis of image content. The content of an image is mostly analysed by colour and texture information. This is the reason that the prior researchers utilized colour and texture features for finding the adjacent or matching fragments. Sometimes the colour and texture are less important in the case of synthetic images, flags, etc. The shape of estimation of the object becomes vital in the case of an image that represents only an object.

Tsamoura et al. proposed a colour-based reassembly technique for fragmented images and paintings [9]. The proposed technique has been divided into four stages. In the first stage, a set of probable adjacent fragments are identified using the content-based image retrieval technique. This step reduces the computational burden for the next few steps. In the second stage, the colour information of the fragment contour is used to retain only one fragment from the set of probable adjacent fragments. In the third stage, the alignment of each fragment is identified by using different image registration techniques. The alignment angle will act as a novel feature in the final step. The final step does a global fragment reassembly based on the alignment angle. The success of the proposed technique is dependent on the performance of each step as these stages are arranged sequentially. The experimentation has been done on scanned torn pieces of a painting.

Youguang et al. proposed colour and contour-based image fragment reassembly techniques [10]. The fragments are scanned on a white background. In the first step, the image was converted into HSV colour space. The histogram for each of the channels is divided into 12 intervals. The Euclidean distance was computed for all the intervals. The probable adjacent fragments are identified based on the distance of histograms. Then each of the fragments is binarized using the Otsu thresholding technique. The curvature for each of the boundary points is computed. The absolute curvature difference is used to find the inflection points for a fragment. The contour of the fragments is also divided into multiple segments. The potential adjacent fragment list is refined based on the contour match using the LCS algorithm. Finally, the fragments are joined based on the refined adjacency list.

4 Reassembly Using Machine Learning

Machine learning has a wide range of application areas. Machine learning has already solved many challenging problems. Some researchers have already tried to address this fragment reassembly problem using machine learning in the last few years. There is a huge scope of research in solving this problem using machine learning.

Paumard et al. proposed an image reassembly technique using deep learning and shortest path algorithms [11]. They have divided the image into multiple square-shaped fragments. They have applied a deep learning-based technique to predict the relative position of the fragments. The prediction is mainly done based on the content of the image. Then the shortest path graph algorithm was used to reassemble the fragments. The authors contributed an image dataset to the community of fragment reassembly problems. The dataset contains 14,000 images, which have been captured from the Metropolitan Museum of Art (MET). The experimentation was done on their dataset as well as on the popular ImageNet dataset.

Abitbol et al. [12] proposed a machine learning-based approach to reassemble papyrus fragments. A convolutional neural network has been considered for finding pairwise local matching. The authors preferred recall over precession for this phase to cover all the potential matches. A classifier has been trained from the extracted features of the previous phase to decide for the elimination of false matches and aligning candidate fragments for all the positions. The proposed approach reduces a huge number of possible matches in the first stage.

5 Experimentations and Results

Some experiments have been done to explore the challenges. The dataset has been prepared by collecting images from the Internet. All the images of the dataset are resized to a uniform size, i.e. 600×800. At first, the images were converted to a grayscale to keep it simple. Figure 1 shows a sample image from the dataset after conversion into grayscale. The images are fragmented horizontally and vertically based on the number of partitions provided as input for both the direction. This means all the fragments will be of the same size. Then the set of fragments is shuffled. Some distortion has been applied to all the fragments to make it like a scanned image. Here, histogram equalization has been applied to each fragment. Figure 2 shows the fragments after shuffling and applying histogram equalization. The fragments are rotated arbitrarily to add more distortion. Figure 3 shows the appearance of each fragment after applying arbitrary rotations. The reassembly algorithms can be applied to the fragments shown in Fig. 3.

Tewari et al. [13] proposed a fragment reassembly approach where there is prior information about the actual image. At first, a median filter has been applied to remove noise. Then SURF features have been applied to orient the fragments into a proper angle. Then the absolute difference has been computed from the original fragments

Fig. 1 Sample image from dataset

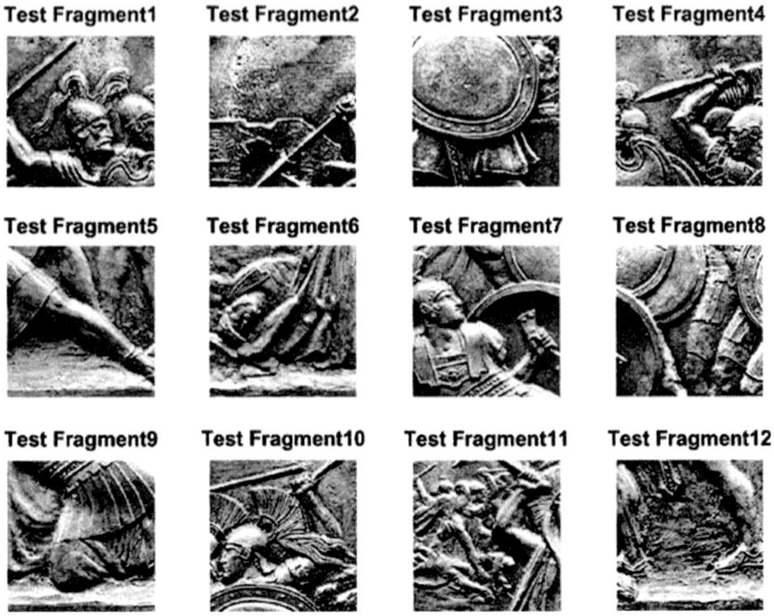

Fig. 2 After shuffling the fragments

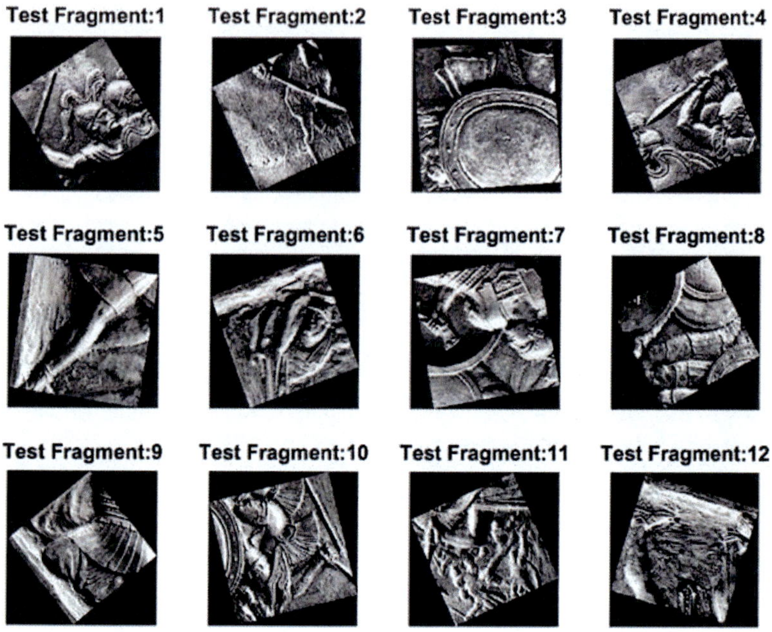

Fig. 3 After application of arbitrary rotation

to all the candidate fragments. The fragment which has minimum difference is fixed for the particular position. Figure 4 shows the outcome of this approach on the set of fragments in Fig. 3.

6 Conclusions

This paper contains a comparative study on different varieties of fragment reassembly problems as well as different types of solutions for these problems. The fragment reassembly can be divided into two types based on the availability of prior information, i.e. (a) reassembly with prior information and (b) reassembly without prior information. The fragment reassembly can also be categorized as (a) reassembly of equal shaped fragments and (b) reassembly on arbitrarily shaped fragments. There are three types of approaches to address those problems, i.e. (a) geometric approaches (b) content-based approaches, and (c) machine learning approaches. It has been observed that the fragment reassembly using boundary shape is time-consuming when the number of fragments is high. The content-based approaches are very sensitive to noise. Machine learning adds a new dimension to this problem. There is a huge scope of research on this problem. Some challenges are not touched by prior approaches, e.g. (a) reassembly of synthetic images and (b) a large number of fragments of similar shape, etc.

Fig. 4 After reassembly using Tewari et al. [13] approach

References

1. Papaodysseus C, Panagopoulos T, Exarhos M, Triantafillou C, Fragoulis D, Doumas C (2002) Contour-shape based reconstruction of fragmented, 1600 bc wall paintings. IEEE Trans Signal Process 50(6):1277–1288
2. Liu H, Cao S, Yan S (2011) Automated assembly of shredded pieces from multiple photos. IEEE Trans Multimedia 13(5):1154–1162
3. Memon N, Pal A (2006) Automated reassembly of file fragmented images using greedy algorithms. IEEE Trans Image Process 15(2):385–393
4. Jana S, Chattopadhyay M (2015) An event-driven university campus navigation system on android platform. In: 2015 applications and innovations in mobile computing (AIMoC). IEEE, pp 182–187
5. McBride JC, Kimia BB (2003) Archaeological fragment reconstruction using curve-matching. In: 2003 conference on computer vision and pattern recognition workshop, IEEE
6. Xu G, Xian Y (2009) A fast reassembly methodology for polygon fragment. In: 2009 international conference on computer engineering and technology. IEEE, pp 495–499
7. Zhang K, Li X (2014) A graph-based optimization algorithm for fragmented image reassembly. Graph Models 76(5):484–495
8. Kamran H, Zhang K, Li M, Li X (2018) An LCS-based 2D fragmented image reassembly algorithm. In: 2018 13th international conference on computer science & education (ICCSE). IEEE, pp 1–6
9. Tsamoura E, Pitas I (2009) Automatic color based reassembly of fragmented images and paintings. IEEE Trans Image Process 19(3):680–690
10. Youguang W, Li X, Li M (2013) Color and contour based reconstruction of fragmented image. In: 2013 8th international conference on computer science & education. IEEE, pp 999–1003

11. Paumard MM, Picard D, Tabia H (2018) Image reassembly combining deep learning and shortest path problem. In: Proceedings of the European conference on computer vision (ECCV), pp 153–167
12. Abitbol R, Shimshoni I, Ben-Dov J (2021) Machine learning based assembly of fragments of ancient papyrus. J Comput Cultural Heritage (JOCCH) 14(3):1–21
13. Tewari M, Jana S, Parekh R (2017) An automated system for image reconstruction from distorted image fragments. In: 2017 international conference on computer, electrical & communication engineering (ICCECE). IEEE, pp 1–6

Chapter 20
A Survey on Medical Image Diagnosis Systems: Problems and Prospects

Krishnendu Ghosh, Tamesh Halder, Maitri Roy, Chandranath Biswas, Rintu Kumar Gayen, and Debashish Chakravarty

1 Introduction

Modern medical imaging facilities have entirely reformed the healthcare industry in the last few decades. Physicians once had to rely on their ability to introspect inside the human body to determine a possible abnormality. With the latest diagnostic images, these physicians can see how well the inner organs function or the joints move. The installation of various medical imaging devices and their improved performances have amassed considerable biomedical data over time. With recent progress in devising performant artificial intelligence (AI)-based algorithms and high computational abilities of the associated machine learning (ML)-based techniques, these vast archives of medical images have been analyzed for various applications such as medical image processing, interpretation, segmentation, retrieval, analysis, multimodal fusion computer-aided diagnosis, image-guided therapy, and so on [17].

K. Ghosh
Center of Excellence in Artificial Intelligence (AI), Indian Institute of Technology Kharagpur, Kharagpur 721302, India

T. Halder · D. Chakravarty
Department of Mining Engineering, Indian Institute of Technology Kharagpur, Kharagpur 721302, India
e-mail: haldertamesh@iitkgp.ac.in

M. Roy (✉)
Department of Computer Science & Engineering, Institute of Engineering and Management, Kolkata 700002, India
e-mail: maitriroy122@gmail.com

C. Biswas
Indian Institute of Technology Kharagpur, Kharagpur 721302, India

R. K. Gayen
Department of Electronics Engineering, Institute of Engineering and Management, Kolkata 700002, India

L. Mandal et al. (eds.), *Proceedings of International Conference on Computational Intelligence, Data Science and Cloud Computing*, Algorithms for Intelligent Systems,
https://doi.org/10.1007/978-981-19-1657-1_20

Table 1 Typology of medical imaging modalities and associated organs

CT	X-rays	Ultrasound	MRI	PET	Histology
Abdomen	Radiography	Transrectal	Neuroimaging	Cardiology	Epithelium
Appendix	Mammography	Breast	Cardiovascular	Infected tissues	Endothelium
Bladder	Fluoroscopy	Doppler	Liver	Small animal	Mesenchyme
Breast	Dexa scan	Abdominal	Functional	Imaging	Placenta
Brain	Contrast radiography	Transabdominal	Oncology	Musculoskeletal	Blood cells
Cervix		Cranial	Phase contrast	Oncology	Germ cells
Chest	Arthography	Gallbladder		Pharmaco kinetics	Neuron
Kidney	Discography	Spleen			

Although these techniques have transformed all three stages of traditional clinical practice: diagnosis, prognosis, and treatment, the present paper focuses on the diagnosis perspective pertaining to different medical images. Medical image diagnosis systems are primarily deployed to analyze the diagnostic tasks and, consequently, ease the prognosis and treatment processes.

These diagnostic medical images can be categorized into six major classes on the basis of their modalities: computer tomography (CT), X-rays, ultrasound, magnetic resonance imaging (MRI), positron emission tomography (PET), and histology [6]. Images of different modalities relate to various anatomical and functional aspects of various body organs [14]. For example, X-rays are primarily used to determine skeletal abnormalities but can be used in mammography to detect and diagnose breast cancers, too. MRI is used to detect nerve injuries, tumors, stroke, or even the cause of a headache. Table 1 mentions different body organs associated with these six modalities of medical images.

Medical image diagnosis systems apply ML-based techniques to analyze images and identify possible diseases. These approaches typically begin with determining the image features that could be of importance for determining the results. After extracting, the best combination of such features is chosen for computing the desired metric with accuracy. Although we noted several features to render image properties and metrics to analyze the associated algorithm's performance, accurate feature selection and appropriate metrics can be challenging. In the absence of expensive feature extraction and selection processes, deep learning approaches preprocess its data (images) and feed its network. This network is trained by automatically identifying the features, and the final predictions are obtained. Among various diseases diagnosable with AI techniques lie: detecting neurological abnormalities (such as brain-related injuries or blood clotting), screening common cancers, diagnosing infections in the kidney/liver, detecting brain tumors, and identifying dental/bone abnormalities.

With the latest technologies, medical images can offer precise details, rich variables, and higher dimensions [17]. While it is possible to examine different abnormalities and their level of severity from these images, it is also challenging to develop good-quality medical imaging systems. The first hurdle in the course of implementing AI algorithms is the availability of medical image data. In the absence of standard data, researchers need to collect databases handling potential biases. Although hassle of data availability can be addressed using approaches like federated learning, interactive reporting, and synoptic reporting, annotation and curation tasks pose substantial barriers [22]. Due to the skewed nature of medical data, it is also not ardent to assign an unsupervised approach for such tasks. In the present scenario, working with medical data is also complicated due to legal and privacy issues. Medical data also suffers from compatibility issues, as data from different devices or similar devices of different makes often render variable signatures. Existing works primarily work on separate databases instead of developing standard databases or experimenting on other's data. Apart from these, handling medical data is challenging as they often relate to a critical case. The associated diagnosis process is susceptible to human errors, and the prognosis tasks are subject to disagreements between physicians. Considering the precarious nature of these tasks, it is not advisable to rely on further prognosis prescribed by a system trained over deep learning-based approaches. Another crucial aspect of medical image diagnosis with AI is the lack of reliable validation techniques for proposed algorithms. According to Kim et al. [9], essential characteristics of a proper validation methodology are shaped with diagnostic cohort design, inclusion of multiple institutions, and prospective data collection for external validation. Unfortunately, most existing studies did not follow these external validation tasks.

The current paper aims to provide a comprehensive survey on medical image diagnosis systems. This discussion starts with common challenges, exiting approaches to overcome those challenges, present prospects, and ends by depicting the current scenario to aid future research. We presented the basics of the state-of-the-art approaches followed by a summary stating their pros and cons. However, we emphasized (i) developing diagnosis systems to determine new abnormalities, (ii) developing hybrid systems to determine multiple abnormalities, and (iii) developing recommendation systems for the associated patients. This medical field of image processing is introduced in the first section, while existing works' methodologies are hierarchically illustrated in Sect. 2. In Sect. 3, we mentioned the research gaps of existing diagnosis systems followed by scopes for future research. Section 4 concludes the present task with directions for future works.

2 Medical Image Diagnosis

Medical image diagnosis systems are primarily realized over two modules, namely image segmentation and abnormality classification.

2.1 Medical Image Segmentation

Medical images are segmented into several objects such that segments with similar attributes could be analyzed easily. Segmentation methods identify objects based on either of two approaches: detecting discontinuities or detecting similarities. Different image segmentation methods are elaborated below.

2.1.1 Segmentation by Edge Detection

This approach looks for sharp changes in intensity in an image, and such differences between different parts create edges. These edges are linked to develop boundaries [16] that draw a binary image [26]. There are two edge detection-based segmentation approaches: gray histogram and gradient-based method [8].

Detecting edges is a well-known problem in image processing. Boundaries of regions and edges have a relationship due to the sharp adjustment in intensity. However, this boundary can often be disconnected due to a lack of information from neighboring pixels. Although edge detection is one of the segmentation tasks with closed region boundaries, it detects edges as boundaries between objects from an image. Segmentation can also be done on edges that are seen by some other local detectors. Lindeberg and Li suggested an object recognition approach that segments edges into straight and curved edge segments based on a minimum description length (MDL) criterion [11]. This criterion is optimized by a split-and-merge-like method. Candidate breakpoints are obtained from complementary junction cues to receive similar points at which partitions are to be considered.

2.1.2 Segmentation by Thresholding

Thresholding differentiates an object of interest from a background based on a threshold. Pixels with intensities above the threshold are considered as part of the object of interest. Such approaches perform well for images with high contrast and sharp edges but fail otherwise. While a constant threshold value cannot address images having noise, dynamic thresholds prove to be effective. Different thresholding techniques are elaborated below:

- **Simple Thresholding**: This method involves replacing the pixels in an image with either black or white. If the concerned pixel's intensity is less (more) than the threshold, it is replaced with black (white).
- **Otsu's Binarization**: This method searches for an optimal threshold that minimizes intra-class variance (i.e., the weighted sum of two different class variances) [13]. This technique segments the image into light and dark patches, denoting the object of interest and background. The optimal threshold is determined by minimizing the summation of the weighted group variances. These weights represent

individual groups' probabilities. [24]. This method is widely used for scanning documents, recognizing patterns, or removing unwanted colors from documents.
- **Adaptive thresholding**: Using this method, the threshold value is extracted separately for each pixel with neighborhood pixel statistics. Thus, it handles the variations in illumination.

2.1.3 Region-based Segmentation

Region-based segmentation is more robust to noise and simpler than edge detection-based techniques [25]. This methods segment images into regions using preset rules [7, 16] in following approaches:

- **Region Growing**: Region growing techniques assimilate the pixels or subregions inside a larger region based on predefined rules. It begins with an individual pixel and ends up with the segmented area. This technique suffers from leakage issues as it is difficult to determine a threshold that confines an actual object [23]. The difference in practice between thresholding and region growing lies in the locality of the extracted surface. Threshold segmentation extracts the whole set of labeled voxels and is suitable when the objective is to extract everything we notice colored in the threshold range. Region growing is used when we need to extract only a subregion.
- **Region Splitting and Merging**: This technique splits images into numerous segments and merges them based on predefined rules [26]. This method is commonly used for quadtree data.

2.1.4 Segmentation by Clustering

In clustering-based segmentation techniques, constituent pixels are classified by identifying a set of clusters [4]. Being unsupervised, clustering methods do not need any training step. When the number of classes is known ahead of time, clustering is performed. Pixels are compared using a similarity criterion [26], and similar pixels are clustered together. Centroids of these clusters are determined by reducing inter-class variability while increasing intra-class variability. Such methods depend on the similarity measures or assigned metrics. K-means, fuzzy clustering, and hard clustering are among the most common clustering methods.

2.1.5 Hybrid Segmentation Models

A popular puzzle in segmentation is creating homogenous segments that result in a heterogeneous segment when two neighbors are combined. Histogram-based segmentation techniques are prone to fewer errors (i.e., probability distribution function of intensity values of any image). Edge detection with differential filter is used to get the image gradient or Laplacian and further processed into contours analysis. Region-based techniques segment images into a set of homogeneous regions and merge them according to specific decision rules [20]. The actual image is realized by a Markov or Gibbs random field with a distribution function in the Markov random field-based segmentation technique. Edge-based and region-based approaches are integrated with hybrid segmentation techniques. In this method, images are first partitioned into areas, then merged using the split and merge process, and finally, the contours were recognized using the edge-based technique.

2.2 *Medical Image Classification*

After segmenting images, classification tasks are carried out to identify abnormalities out of the segmented images. Popular algorithms and their application in image classification are briefly described below.

2.2.1 K-Nearest Neighbor

K-nearest neighbor (KNN) algorithm is a classic instance-based or lazy learning algorithm. It classifies the images based on the most frequent class among its k nearest training instances. The nearness is measured using distance metrics such as Euclidean distance.

In a task for detecting abnormality and automatic medical image classification using k-nearest neighbor [15], this algorithm obtained an 80% classification rate. The abnormal images are then processed to detect the region of interest (ROI).

2.2.2 Support Vector Machines

Support vector machines (SVM) determine a hyperplane that separates the feature vectors belonging to different classes by the biggest possible margin. The hyperplane is selected such that test noises are not classified erroneously.

An SVM-based classifier is used for detecting microcalcifications in [5]. This task is evaluated using 76 clinical mammograms comprising 1120 microcalcifications. A free-response receiver operating characteristic curve was utilized to measure detection performance, and the findings were compared to other current approaches. Finally, this SVM classifier achieved an astounding accuracy of 94%.

2.2.3 Linear Discriminant Analysis

Texture analysis of ultrasound pictures is used to determine the stage of chronic renal disease [1]. Tissue heterogeneity causes local variations in intensity, which is termed texture. Texture analysis can be both structural and structural. For ultrasound images, the best texture-based feature extraction is done using statistical texture analysis. The most popular statistical texture analysis technique examines gray level co-occurrence matrix (GLCM) in the feature extraction step. GCLM tabulates how often different combinations of a gray level co-occur in an image. An image is made up of pixels, each with an intensity (a specific gray level). The contents of GCLM are then used to provide a measure of variation in intensity at the pixel level of interest. In the final classification step, a linear discriminant analysis (LDA) classifier is employed.

2.2.4 Neural Networks

Each neuron in the network adds the input data and then applies the activation function to the summed data before sending the output to the next layer. As a result of the hidden layers' ability to capture nonlinear relationships, adding more hidden layers allows a system to deal with increasingly complicated challenges. Deep neural networks are the name given to these types of neural networks (DNNs).

Cao et al. [2] used CNN to select viable embryos according to morphological characteristics for transfer. Existing methods like logistic regression, SVM, and Bayesian methods required handcrafted feature selection, which is difficult. Image standardization [18] and ROI extraction was followed by classification. The network has ten layers for classification. Two repeated convolutional layers and max-pooling layers make up the first four layers. Two repeated convolutional and max-pooling layers make up the middle four layers. The final two layers are fully connected. The other layers use ReLU activation, except for the last layer, which uses the sigmoid activation.

In another paper, deep learning and machine learning techniques are used to classify a healthy lung from a diseased one, and further processing is done to classify the lung disease into one of the existing 15 categories (14 diseases and one for no findings) [21]. Image processing techniques used are based on spatial transform [1] to get the preliminary information for the image. An accuracy of 69.3% was observed using CNN + VGG + STN. Further improvements could be made by training with more epochs and making the dataset better by removing noise.

Li et al. [10] solved the problem of acquiring annotated data by following a unified approach where disease identification and localization are made with the same model on chest X-rays. The proposed a CNN to first process the input image. Patch slicing layer used max-pooling or bilinear interpolation to shrink the convolutional features from the CNN. A fully convolutional recognition network is then applied to these regions. Two images are created using multi-instance learning assumptions during training. This trained model predicts both labels and class-specific localizations in the testing phase.

Roychowdhury et al. suggested a commonly performed diagnostic test known as optical coherence tomography for identifying retinal disorders [19]. This method uses a novel active learning framework to reduce the training complexity by reducing the number of images per class.

SimCLR is a recent model proposed for unsupervised visual representation learning [17]. The SimCLR model is trained with a batch size of 128 for 120 epochs. The model is stopped at epoch 75 with an accuracy of 99.77%. The training objective is to maximize the similarity between images and their augmented views (positive pair) while repulsing the similarity values with other differentiating images.

3 Limitations

AI- and ML-based imaging applications in the healthcare domain engulf numerous tasks ranging from abnormality detection and monitoring to recommendation of customized treatments. However, there are scopes for devising new tools to analyze and convert all medical images into meaningful representations. Apart from the factors affecting standard machine learning-based models, other limitations mainly restrict the medical image diagnosis systems. The first inhibiting factor is access to large high-quality labeled datasets for training. Data collection is particularly challenging due to expensive annotation tasks and standardization issues of medical hardware. Labeling datasets require inputs from multiple medical experts to master the issues of disagreement and human errors. The presence of rare diseases with indistinguishable symptoms complicates the labeling tasks and often skews the training data. Generative adversarial network (GAN) is considered in recent works to address the problem of data scarcity. On the other hand, imbalanced data could be partially addressed by transfer learning. Medical data suffers from another issue of interoperability and standardization. Images from different hardware or medical devices often carry different signatures, and as a result, researchers could not study data collected from different devices or sources. With widely accepted standards for data formats and medical devices, it is possible to pursue collaborative studies. Compared to other real-world data, sharing medical data is intricate due to privacy and legal issues. Data privacy has both societal and technical components that must be addressed together from both viewpoints. The black box problem in AI affects the application of deep learning in medical picture analysis when the inputs and outputs are known. However, the internal representations are not well-understood.

In areas where radiologists are unavailable, AI will aid in increasing access to care. AI can make healthcare accessible to anyone with a smartphone which is a necessity, especially in rural areas. A recent study revealed after a survey on existing literature that the existing radiological diagnostic image provides a lot of secondary benefits (not related to classification) that in turn might aid in the process of image classification accurately [12]. Secondary benefits include promoting education, communication, and community engagement. Another advantage of AI is that it improves the accuracy of disease detection and diagnosis. The most challenging hurdle for AI in many healthcare fields is assuring its acceptance in daily clinical practice, not whether the technologies are capable enough to be helpful. Although libraries like pandas and scikit-learn are the ones that come to mind for machine learning tasks, there are a bunch of other useful libraries that can prove to be very useful for all the extensive image preprocessing, segmentation, and classification tasks. Libraries like SimpleCV, Mahotas, SimpleITK, pgmagick, Pycaro, OpenCV-Python to name a few. AI systems must be approved by regulators, integrated with EHR systems, standardized to the point that similar products work similarly, taught to doctors, paid for by public or private payer organizations, and updated over time in the field for widespread adoption to occur. These obstacles will be overcome in the end, but they will take considerably longer than the maturation of the technologies themselves. As a result, we expect to see limited AI usage in clinical practice over the next five years and more widespread use within the next decade [3].

4 Conclusion

The scope of artificial intelligence in the sphere of medical image diagnosis is enormous. The algorithm can automatically detect diseases, lesions, and tumors in an ideal situation and highlights their locations in the large pile of images. In the real world, there are many medical exceptions that the training data may not capture, leading to misclassification. Though it is nowhere near to the human expertise of physicians, especially when it comes to detecting any abnormal patterns, AI can become the most reliable savior when it comes to tedious repetitive tasks involved in the diagnostic procedure. Also, especially in medical image analysis, we would want elaborate validation techniques as even one faulty classification might lead prove critical. Thus, collaboration of clinicians and medical organizations is required on a large scale to develop accurate and elaborate datasets. However, there are several challenges like addressing data complexity, data access, and a variety of clinical scenarios while ensuring patient privacy, data curation, lack of validation, and testing. Through continuous efforts, reliable, and cost-effective solutions to these challenges could be formulated, and AI can become the next most effective tool to save lives by accelerating the diagnostic process.

References

1. Ahmad R, Mohanty BK (2021) Chronic kidney disease stage identification using texture analysis of ultrasound images. Biomed Signal Process Control 69:102695
2. Cao Q, Liao SS, Meng X, Ye H, Yan Z, Wang P (2018) Identification of viable embryos using deep learning for medical image. In: Proceedings of the 2018 5th international conference on bioinformatics research and applications, pp 69–72
3. Davenport T, Kalakota R (2019) The potential for artificial intelligence in healthcare. Future Healthcare J 6(2):94
4. Dehariya VK., Shrivastava, S.K., Jain, R.: Clustering of image data set using k-means and fuzzy k-means algorithms. In: 2010 International conference on computational intelligence and communication networks. IEEE, pp 386–391

5. El-Naqa I, Yang Y, Wernick MN, Galatsanos NP, Nishikawa RM (2002) A support vector machine approach for detection of microcalcifications. IEEE Trans Med imaging 21(12):1552–1563
6. Heidenreich A, Desgrandschamps F, Terrier F (2002) Modern approach of diagnosis and management of acute flank pain: review of all imaging modalities. European Urol 41(4):351–362
7. Kaganami HG, Beiji Z (2009) Region-based segmentation versus edge detection. In: 2009 Fifth international conference on intelligent information hiding and multimedia signal processing. IEEE, pp 1217–1221
8. Kang WX, Yang QQ, Liang RP (2009) The comparative research on image segmentation algorithms. In: 2009 First international workshop on education technology and computer science, vol 2, IEEE, pp 703–707
9. Kim DW, Jang HY, Kim KW, Shin Y, Park SH (2019) Design characteristics of studies reporting the performance of artificial intelligence algorithms for diagnostic analysis of medical images: results from recently published papers. Korean J Radiol 20(3):405–410
10. Li Z, Wang C, Han M, Xue Y, Wei W, Li LJ, Fei-Fei L (2018) Thoracic disease identification and localization with limited supervision. In: Proceedings of the IEEE conference on computer vision and pattern recognition, pp 8290–8299
11. Lindeberg T, Li MX (1997) Segmentation and classification of edges using minimum description length approximation and complementary junction cues. Comput Vis Image Understand 67(1):88–98
12. Ndenga T, Ferreira M, Burbridge B (2014) Radiology: an underutilized resource for the teaching of anatomy in Angolan medical schools. In: Proceedings of the association for medical education (AMEE) annual conference
13. Otsu N (1979) A threshold selection method from gray-level histograms. IEEE Trans Syst Man Cybern 9(1):62–66
14. Rahman MM, Desai BC, Bhattacharya P (2008) Medical image retrieval with probabilistic multi-class support vector machine classifiers and adaptive similarity fusion. Comput Med Imaging Graph 32(2):95–108
15. Ramteke R, Monali KY (2012) Automatic medical image classification and abnormality detection using k-nearest neighbour. Int J Adv Comput Res 2(4):190
16. Rastgarpour M, Shanbehzadeh J (2011) Application of AI techniques in medical image segmentation and novel categorization of available methods and tools. In: Proceedings of the international multiConference of engineers and computer scientists 2011, vol I, IMECS 2011, 16–18 Mar 2011, Hong Kong. Citeseer
17. Razzak MI, Naz S, Zaib A (2018) Deep learning for medical image processing: overview, challenges and the future. In: Classification in BioApps, pp 323–350
18. Rocha JC, Passalia FJ, Matos FD, Takahashi MB, Maserati MP Jr, Alves MF, De Almeida TG, Cardoso BL, Basso AC, Nogueira MFG (2017) Automatized image processing of bovine blastocysts produced in vitro for quantitative variable determination. Sci Data 4(1):1–9
19. Roychowdhury S, Tang KS, Ashok M, Sanka A (2021) SISE-PC: Semi-supervised image subsampling for explainable pathology. arXiv:2102.11560
20. Singh KK, Singh A (2010) A study of image segmentation algorithms for different types of images. Int J Comput Sci Issues (IJCSI) 7(5):414
21. Tripathi S, Shetty S, Jain S, Sharma V (2021) Lung disease detection using deep learning. Int J Innov Technol Explor Eng **10**(8) (2021)
22. Willemink MJ, Koszek WA, Hardell C, Wu J, Fleischmann D, Harvey H, Folio LR, Summers RM, Rubin DL, Lungren MP (2020) Preparing medical imaging data for machine learning. Radiology 295(1):4–15
23. Yadav B, Jatti A (2014) Segmentation techniques for medical image analysis. Int J Sci Res Dev
24. Yousefi J (2011) Image binarization using Otsu thresholding algorithm. University of Guelph, Ontario, Canada
25. Zhang H, Fritts JE, Goldman SA (2008) Image segmentation evaluation: a survey of unsupervised methods. Comput Vision Image Understand 110(2):260–280
26. Zhang YJ (2006) An overview of image and video segmentation in the last 40 years. In: Advances in image and video segmentation pp 1–16

Chapter 21
An Analysis of Abstractive Text Summarization Using Pre-trained Models

Tohida Rehman, Suchandan Das, Debarshi Kumar Sanyal, and Samiran Chattopadhyay

1 Introduction

Given a long text, humans have a natural tendency to remember its most important points in a summary form. The volume of data around us is growing to the point that we need to find a solution that will deliver accurate and timely summary information. It requires a tool or approach for extracting an accurate summary from a large amount of data. Automatic text summarization is an example of a technology that may be used to achieve this. Broadly, there are two approaches for summarization [1, 2]: *extractive* and *abstractive*. *Extractive methods* [2–4] generally copy whole sentences from the input source text and combine them into a summary, throwing away the unimportant sentences from the input. *Abstractive methods* [5] can generate novel words during summarization similar to how a human being does the job, that is, first reads the whole document, understands it, then summarizes by inducing new and appropriate words. The main contributions of this paper are: (1) Evaluating the performance of three different pre-trained models, namely google/pegasus-cnn-dailymail, T5-base, facebook/bart-large-cnn on CNN-dailymail [6, 7], SAMSum [8], BillSum [9] datasets. We record significant results using the three pre-trained models. Performance evaluation of the pre-trained models has been done with ROUGH [10] and BLEU [11] metrics.

T. Rehman (✉) · S. Das · S. Chattopadhyay
Jadavpur University, Kolkata, India
e-mail: tohida.rehman@gmail.com

D. K. Sanyal
Indian Association for the Cultivation of Science, Kolkata, India

S. Chattopadhyay
Institute for Advancing Intelligence (IAI), TCG CREST, Kolkata, India

L. Mandal et al. (eds.), *Proceedings of International Conference on Computational Intelligence, Data Science and Cloud Computing*, Algorithms for Intelligent Systems,
https://doi.org/10.1007/978-981-19-1657-1_21

2 Literature Review

For abstractive summarization, Nallapati et al. [7] have proposed a model based on attentional encoder–decoder recurrent neural networks for an abstractive text summarization. Performance of basic encoder and decoder model has been improved through Bahdanau et al. [12]. Luong et al. [13] proposed attention mechanism: a global approach which always attends to all source words, and a local one that only looks at a subset of source words at a time. See et al. [14] offer a detailed study of numerous abstractive text summarization models using pointer-generator networks. Sutskever et al. [15] offer a multilayer LSTM-based end-to-end solution to sequence learning; here, the input for the encoder is a fixed length of text. Lin et al. [16] have proposed a global encoding mechanism of abstractive text summarization. This paper uses GRU-based encoder and decoder with one extra attention layer. Shi et al. [17] have proposed a seq2seq model to improve performance. This has improved fluency and human readability, generated high-quality summaries and captured the salient information. Aksenov et al. [18] have proposed a new method called BERT windowing. This method helps to process a long text (whose length is longer than BERT window) in a chunk-wise manner. Most of these techniques differ in one of these three categories: network structure, parameter inference, and decoding/generation nature.

3 Datasets

Text summarization can be applied to a variety of datasets. Three different datasets are used in this paper: CNN-dailymail [6, 7], SAMSum [8], and BillSum [9].

CNN-dailymail dataset is an English language dataset with little over 300,000 unique news items published by CNN and Daily Mail writers. Version 3.0.0, which can be used to train both abstractive and extractive summarization, was employed in this study. News stories and highlight sentences make up the dataset pair. The articles are utilized as the context in the question-answering setting, and entities are concealed one at a time in the highlight sentences, resulting in cloze-style problems in which the model correctly identifies which entity in the context has been hidden in the highlight. The highlight sentences are concatenated in the summarizing setting to produce an article summary. The model generated summary's are compared to the written highlights to find out the ROUGE [10] score and BLEU [11] score for a specific article.

SAMSum dataset consists of about 16k messenger-like chats with summaries. Linguists conversant in English created and recorded the conversations. The statistic represents the percentage of real-life issues. It has 16369 talks that are evenly split between four groups. Data field consists of text of dialogue or conversation strings, summary form of the dialogue or conversations written by human and unique id of each example. In split, there are 14,732 instances in the train set, 818 in the validation set, and 819 in the test set. The summaries are annotated, with the expectation that they will be (1) brief, (2) extract key bits of information, (3) include interlocutors' names, and 4) written in the third person. There is only one summary reference in each dialogue or conversation. Samsung R&D Institute, Poland prepared and published this dataset for non-commercial research purposes.

BillSum dataset summaries state bills in the United States Congress and California. It has the following attributes: (1) text: the text of the bill (2) synopsis/summary: a synopsis/summary of the legislation, (3) title: the bills titles, (4) text len is the number of character in the text, and (5) sum len is the number of characters in the summary. A text string, summary string, and a title string make up the data field. Legislation bills have complex sentence structure and nature. Hence, it is a challenging, important, and useful dataset for text summarization.

For comparison of the results of the three different pre-trained models, we have selected a set of 2000 test examples from each of the above datasets.

4 Pre-trained Models

Zhang et al. [19] introduced the Pegasus model in **PEGASUS**: "*Pre-training with Extracted Gap-sentences for Abstractive Summarization.*" Pegasus pre-training task is designed to be comparable to summarization: key sentences from an input document are removed/masked, and the remaining phrases are combined into one output sequence, similar to an extractive summary. Pegasus achieves good summary performance as evaluated by ROUGE and BLEU metrics in all 12 downstream datasets.

Raffel et al. [20] introduced the **T5-base** model in "*Exploring the Limits of Transfer Learning with a Unified Text-to-Text Transformer.*" **T5-base** model is an encoder–decoder model that has been pre-trained on a multi-task combination of unsupervised and supervised workloads, with each task transformed into text-to-text. T5-base works well on a range of jobs right out of the box by prefixing each task's input with a distinct prefix. T5 uses relative scalar embeddings. Encoder input padding can be done on the left and on the right. T5 comes in different sizes, here we used t5-base only.

Lewis et al. [21] introduced the **BART** (Bidirectional and Auto-Regressive Transformers) **model**: "*Denoising Sequence-to-Sequence Pre-training for Natural Language Generation, Translation and Comprehension.*" BART is a transformer encoder–encoder (seq2seq) model with a bidirectional (BERT(bidirectional encoder representations from transformers)-like) encoder and an autoregressive (GPT(generative pre-trained transformer)-like) decoder. It is implemented as a sequence-to-sequence model with a bidirectional encoder over corrupted text and a left-to-right autoregressive decoder [21]. The pre-training job entails rearranging the sequence of the original phrases at random and using a new in-filling method that replaces text spans with a single mask token. When fine-tuned for text production, BART is especially successful in summarization as well as other comprehension tasks.

4.1 Pre-trained Model Comparison

In this subsection, we compare several pre-trained summarizers. We choose google/pegasus-cnn-dailymail, T5-base, and facebook/bart-large-cnn models after going through the abstractive text summarization models on the hugging face website[1,2,3]. Rather than looking for models that were trained on the same set of data, we selected models that were trained on different text corpora. We are interested to analyze how these models perform on a common set of test documents.

To compare these models effectively, they were all initialized and put into a Python dictionary for subsequent use. The models were given the same articles from each of the datasets. The pre-trained models have character restrictions for that each and every model has some predefined number of accepted tokens in a sequence like 512 tokens only. When attempting to produce summaries for each article in the datasets, errors are generated if the restrictions are not obeyed. To address it, we have used the concept of BERT windowing [18]. Some BERT-based pre-trained models in the literature condense a lengthy text by eliminating everything except what it considers to be the most important sentences. We followed the same approach to reduce the input size; after an article has been reduced, an abstractive summary is created using the pre-trained models.

5 Evaluation

After the pre-trained models have summarized all the articles, the summaries are evaluated using ROUGE and BLEU metrics. When comparing the produced summaries to the reference summaries used for assessment, ROUGE determines the precision, recall, and f-measure for each model. Here, TP-True positive, FP-False positive, FN-False Negative naming conventions used.
Recall (r) is defined as:

$$r = \mathrm{TP}/(\mathrm{TP} + \mathrm{FN}) \tag{1}$$

Precision (p) is defined as:

$$p = \mathrm{TP}/(\mathrm{TP} + \mathrm{FP}) \tag{2}$$

F-measure (f) values are calculated using the formula:

$$f = 2 * ((r * p)/(r + p)) \tag{3}$$

We have used another metric called BLEU to evaluate the performance of the models. Bilingual assessment understudy (BLEU) is a method to assess the type of text

[1] https://huggingface.co/google/pegasus-cnn_dailymail.

[2] https://huggingface.co/t5-base.

[3] https://huggingface.co/facebook/bart-large-cnn.

Table 1 Observations of ROUGE and BLEU performance using CNN-dailymail dataset

	ROUGE									BLEU
	ROUGE-1			ROUGE-2			ROUGE-L			
	R	P	F1	R	P	F1	R	P	F1	
Google/pegasus-cnn-dailymail	40.22	32.15	34.60	17.64	13.98	14.97	37.51	30.06	32.32	55.09
T5-base	33.96	30.35	31.10	13.37	11.67	12.00	31.43	28.10	28.79	55.70
Facebook/bart-large-cnn	**41.76**	**33.69**	**36.45**	**18.67**	**14.60**	**15.92**	**38.84**	**31.36**	**33.92**	**55.71**

Table 2 Observations of ROUGE and BLEU performance using SAMSum dataset

	ROUGE									BLEU
	ROUGE-1			ROUGE-2			ROUGE-L			
	R	P	F1	R	P	F1	R	P	F1	
Google/pegasus-cnn-dailymail	38.26	25.12	28.39	10.69	7.24	8.05	33.76	22.63	25.36	24.59
T5-base	33.11	26.44	27.88	9.16	7.66	7.72	29.38	23.60	24.82	**26.54**
Facebook/bart-large-cnn	**40.63**	**30.57**	**33.02**	**13.16**	**10.10**	**10.58**	**36.38**	**27.69**	**29.78**	25.95

produced by a model. BLEU is a metric that for ensuring a link to human quality choices, and it is still one of the most well-known automated and minimum cost estimates today. It represents how much close the predicted summary is to the reference summary.

Using these measures, we are able to assess how well each model summarized the data. Despite the lack of GPU support in our computers, we have been able to test the three pre-trained models using three different dataset. We have successfully compared and assessed all the three models.

The observation for **CNN-dailymail** dataset is shown in Table 1. Figure 4 shows the graphical representation of their outputs. The observation for **SAMSum** dataset is shown in Table 2. Figure 5 shows the graphical representation of their outputs. The observation for **BillSum** dataset is shown in Table 3. Figure 6 shows the graphical representation of their outputs. (Note that in Figs. 4, 5 and 6 where the X-axis contains the labels ROUGE*1f* (f-measure for ROUGE-1), ROUGE*2f*, ROUGH*Lf*, ROUGH*1r* (recall for ROUGE-1), ROUGH2*r*, ROUGH*Lr*, ROUGE*1p* (precision for ROUGE-1), ROUGE2*p*, ROUGE*Lp*, and BLEU, and the color codes of the vertical bars are as follows: blue: google/pegasus-cnn-dailymail, orange: T5-base, green: facebook/bart-large-cnn.). Best results are highlighted in bold in Tables 1, 2 and 3.

Table 3 Observations of ROUGE and BLEU performance using BillSum dataset

	ROUGE									BLEU
	ROUGE-1			ROUGE-2			ROUGE-L			
	R	P	F1	R	P	F1	R	P	F1	
Google/pegasus-cnn-dailymail	**26.30**	**48.41**	**30.95**	**11.59**	24.82	**13.99**	**22.95**	42.78	**27.25**	53.81
T5-base	21.71	39.45	26.29	7.31	17.71	9.73	18.77	34.56	22.91	54.11
Facebook/bart-large-cnn	21.43	46.65	27.82	9.20	**24.99**	12.52	19.53	**42.84**	25.47	**58.41**

5.1 Sample Output

Figure 1 illustrates the effectiveness of the three pre-trained models on CNN-dailymail dataset. Figure 2 illustrates the effectiveness of the three pre-trained models on SAMSum dataset. Figure 3 illustrates the effectiveness of the three pre-trained models on BillSum dataset. The google/pegasus-cnn-dailymail model and the facebook/bart-large-cnn model generate better summary on CNN-dailymail dataset and SAMSum dataset while the google/pegasus-cnn-dailymail model generates better summary on BillSum dataset.

6 Conclusion

In this paper, we observed outputs from different pre-trained models using different datasets. While using CNN-dailymail dataset and SAMSum dataset, google/pegasus-cnn-dailymail model and facebook/bart-large-cnn model provides better results than T5-base model as per the f-score of ROUGE-1, ROUGE-2, and ROUGE-L performance metrics. While using BillSum dataset google/pegasus-cnn-dailymail model provides better results than T5-base model and facebook/bart-large-cnn model as per the f-measure of ROUGE-1, ROUGE-2, and ROUGE-L performance metrics. In the future, we aim to fine-tune the pre-trained models on the given datasets and examine if the performance is improved. We also plan to apply these models to documents in scholarly domain as an extension to our work in [22].

Article(truncated): CNN Emergency operators get lots of crazy calls but few start like this. Caller Hello I'm trapped in this plane and I called my job but I'm in this plane. Operator You're where? Caller I'm inside a plane and I feel like it's up moving in the air. Flight 448 can you please tell somebody to stop it. The frantic 911 call came just as the Alaska Airlines flight had taken off from Seattle Tacoma International Airport on Monday afternoon. The caller was a ramp agent who fell asleep in the plane's cargo hold. The cell phone call soon broke up but the man was making himself known in other ways as the crew and passengers reported unusual banging from the belly of the Boeing 737. The pilot radioed air traffic control and said he would make an emergency landing. There could be a person in there so we're going to come back around he told air traffic control. The ramp agent who took the untimely nap and caused all the fuss is an employee of Menzies Aviation a contractor for Alaska Airlines that handles loading the luggage. He'll no longer have the option of dozing aboard one of the airline's planes. The Menzies employee has been permanently banned from working on Alaska Airlines planes said Bobbie Egan a spokeswoman for the airline. Flight 448 which was on its way to Los Angeles only spent 14 minutes in the air. Other than being scared the agent never was in any real danger. The cargo hold is pressurized and temperature controlled the airline said. The passengers knew something wasn't right almost as soon as the plane took off. All of a sudden we heard all this pounding underneath the plane and we thought there was something wrong with the landing gear Robert Higgins told CNN affiliate KABC. Not everyone heard the banging but it was soon clear this wasn't a normal flight. We just took off for L.A. regular and then ... about five minutes into the flight the captain came on and said we were going back and we'd land within five to seven minutes and we did passenger Marty Collins told affiliate KOMO. When we landed was when all the trucks and the police and the fire trucks surrounded the plane. I think it's scary and really unsafe too Chelsie Nieto told affiliate KCPQ. Because what if it's someone who could have been a terrorist? The employee started work at 5 a.m. and his shift was scheduled to end at 2 30 p.m. just before the flight departed. The agent was off the two days prior to the incident and had taken a lunch break and a break in the afternoon before making his way into the cargo hold according to a source familiar with the investigation. The man had been on a four person team loading baggage onto the flight. During a pre departure huddle the team lead noticed the employee was missing. The team lead called into the cargo hold for the employee and called and texted the employee's cell phone but did not receive an answer. His co workers believed he finished his shift and went home the airline's blog said. It's believed he was hidden by luggage making it difficult for the rest of his team to see him the source said. All ramp employees have security badges and undergo full criminal background checks before being hired according to the airline. After the delay the flight with 170 passengers and six crew members on board made it to Los Angeles a couple of hours late. CNN's Dave Alsup Joshua Gaynor and Greg Morrison contributed to this report.

Human-written summary: The ramp agent fell asleep in the plane's cargo hold . He can no longer work on Alaska Airlines flights .

Predicted summary with google/pegasus-cnn-dailymail: The frantic 911 call came just as the Alaska Airlines flight had taken off .<n>The caller was a ramp agent who fell asleep in the plane's cargo hold .<n>The pilot radioed air traffic control and said he would make an emergency landing .

Predicted summary with T5-base: a ramp agent fell asleep in the plane's cargo hold . the caller is an employee of a contractor for Alaska airlines . he'll no longer have the option of dozing aboard one of the airline's planes .

Predicted summary with facebook/bart-large-cnn: A ramp agent fell asleep in the cargo hold of an Alaska Airlines flight. The crew and passengers reported unusual banging from the belly of the Boeing 737. The pilot radioed air traffic control and said he would make an emergency landing.

Fig. 1 Original news article, human-written summary, and the summaries produced by the three pre-trained models using CNN-dailymail dataset

Original dialogue: Shelly This year I'm volunteering at the food shelter! Tracy Good 4 u! Jody Gr8! Shelly How about u? Any volunteer work? Tracy Nah. Not into that. Jody Sure! Every year I do some charity 4 Xmas.
Human-written summary: Shelly is voluntering at a food shelter and asks if others do some volunteer work. Tracy is not into that but Jody always does some charity for Christmas.
Predicted summary with google/pegasus-cnn-dailymail: Tracy Good 4 u! Jody Gr8!
Predicted summary with T5-base: Tracy Nah. Not into that. I do some charity 4 Xmas work .
Predicted summary with facebook/bart-large-cnn: Tracy and Jody are doing some charity work for Christmas. Jody is volunteering at a food shelter. Tracy is not into that.

Fig. 2 Original dialogue, human-written summary, and the summaries produced by the three pre-trained models using SAMSung dataset

Bill text (truncated): SECTION 1. SHORT TITLE. This Act may be cited as "Small Business Tax Modernization Act of 2004". SEC. 2. UNIFIED PASS THRU ENTITY REGIME. a Termination of S Corporation Status. 1 No new s corporation elections. Subsection a of section 1362 of the Internal Revenue Code of 1986 is amended by adding at the end the following new paragraph " 3 Termination of authority to make election. No election may be made under paragraph 1 for any taxable year beginning after December 31 2004.". 2 Termination of status. Subsection d of section 1362 of such Code relating to termination is amended by adding at the end the following new paragraph " 4 Treatment as partnership after 2014. An election under subsection a " A shall not be effective for any taxable year beginning after December 31 2014 and " B shall be treated as an election under section 7701 a 2 B for taxable years beginning after such date.". 3 Effective date. The amendments made .....
Human-written summary: Small Business Modernization Act of 2004 Amends the Internal Revenue Code to 1 terminate subchapter S corporation elections after 2004 and subchapter S status after 2014 and to allow privately held domestic corporations in lieu of electing subchapter S treatment to elect to be treated as partnerships for tax purposes 2 set forth rules for the tax treatment of former subchapter S corporations electing partnership status and 3 exclude from net earnings from self employment partnership income attributable to capital.
Predicted summary with google/pegasus-cnn-dailymail: No election may be made under paragraph 1 for any taxable year beginning after December 31 2004 .<n>An eligible corporation may elect to be treated as a partnership for purposes of this title .<n>No gain or loss shall be recognized to the corporation or the shareholders by reason of an election under clause i and II except in the case of an election made by a S corporation after the end of the recognition period.
Predicted summary with T5-base: No election may be made under paragraph 1 for any taxable year beginning after December 31 2004". 2 Termination of status. Subsection d of section 1362 of such Code relating to termination is amended by adding at the end the following new paragraph 4 Treatment as partnership after 2014 . b Election by certain corporations to be taxed as partnership .
Predicted summary with facebook/bart-large-cnn: This Act may be cited as the "Small Business Tax Modernization Act of 2004" No new s corporation elections are allowed. An eligible corporation may elect to be treated as a partnership for purposes of this title.

Fig. 3 Original bill text, human-written summary, and the summaries produced by the three pre-trained models using BillSum dataset

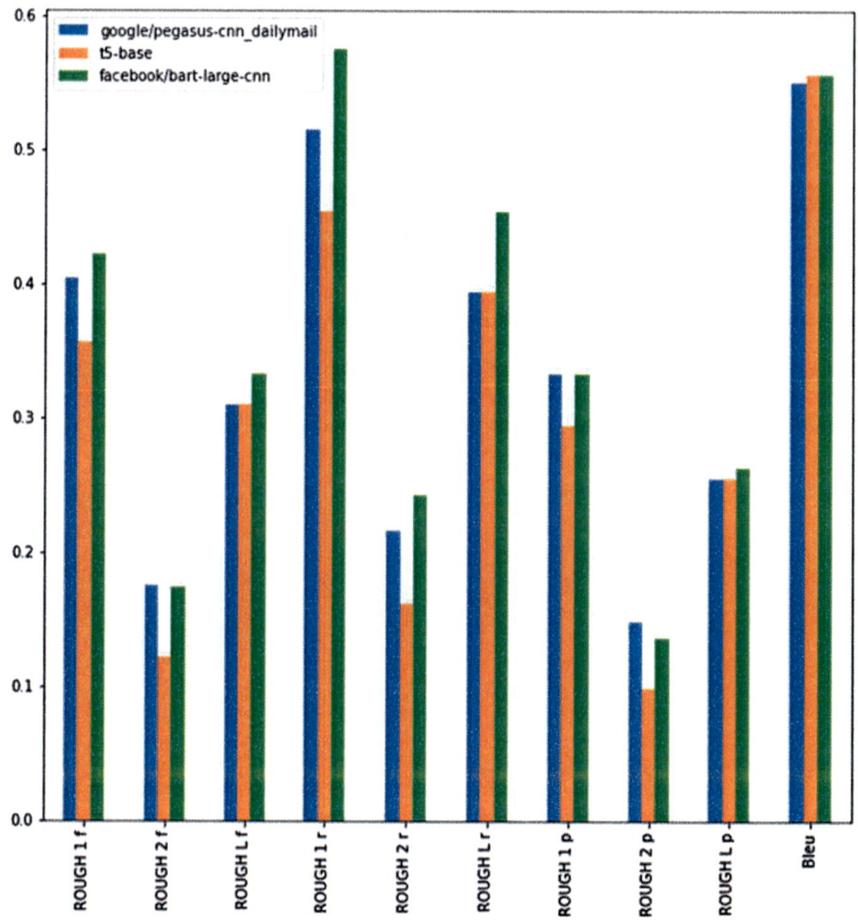

Fig. 4 Performance of three pre-trained models using CNN-dailymail dataset

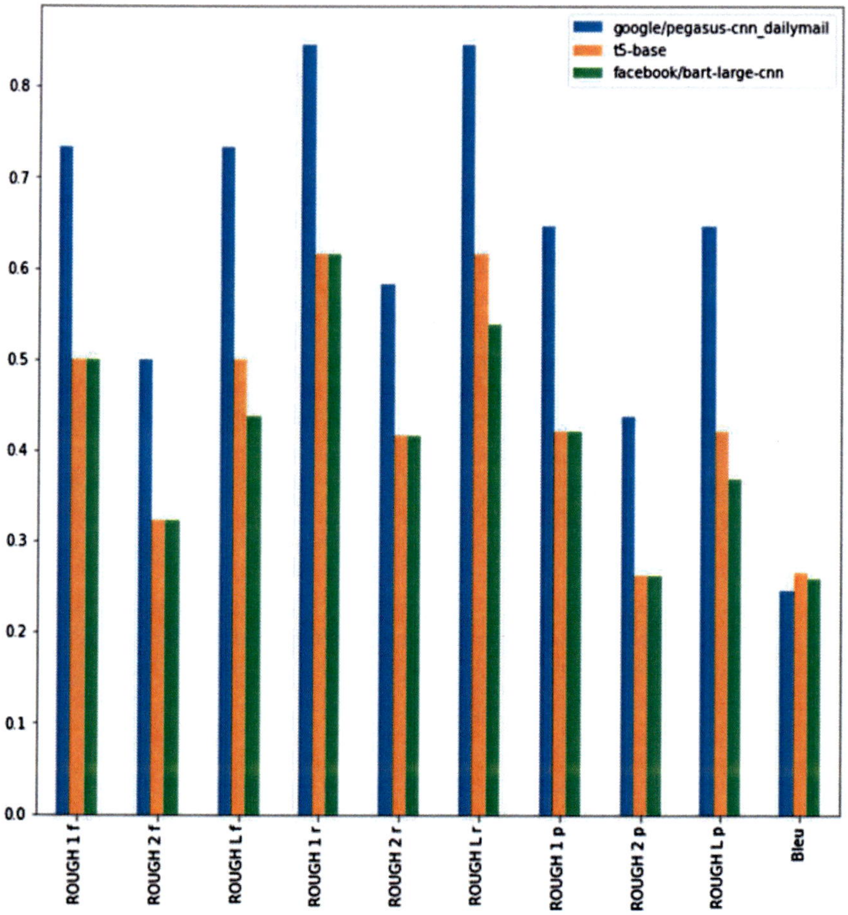

Fig. 5 Performance of three pre-trained models using SAMSum dataset

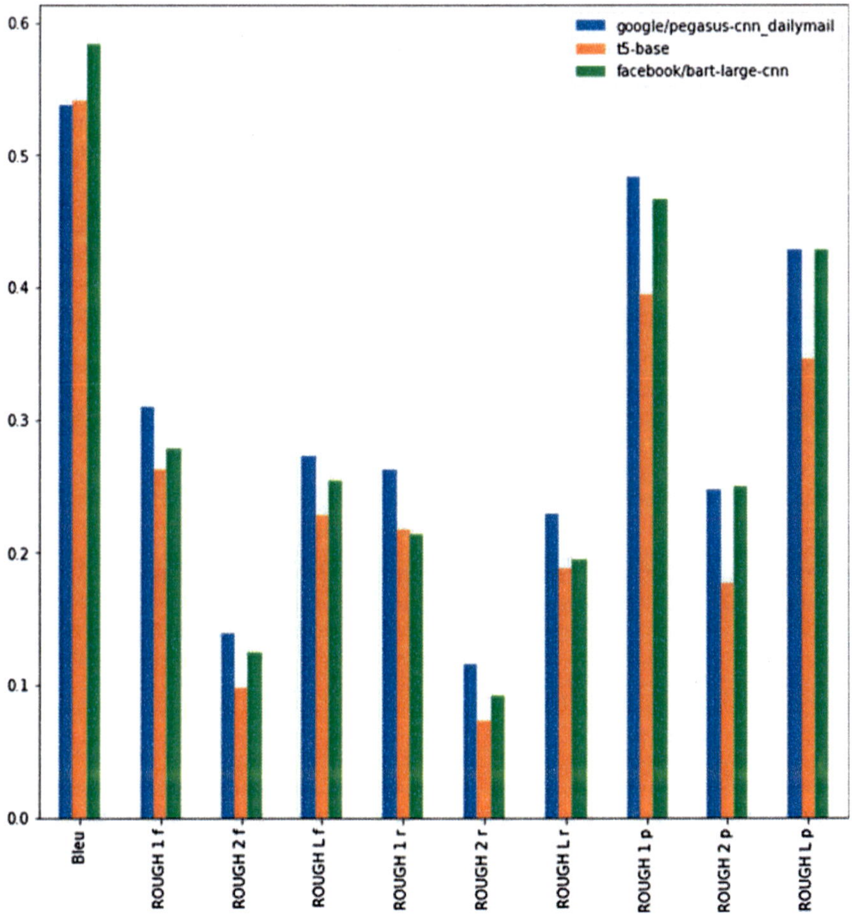

Fig. 6 Performance of three pre-trained models using BillSum dataset

References

1. Luhn HP (1958) The automatic creation of literature abstracts. IBM J Res Dev 2(2):159–165
2. Radev DR, Hovy E, McKeown K (2002) Introduction to the special issue on summarization. Comput Linguist 28(4):399–408
3. Knight K, Marcu D (2002) Summarization beyond sentence extraction: a probabilistic approach to sentence compression. Artif Intell 139(1):91–107
4. Jing H, McKeown K (2000) Cut and paste based text summarization. In: 1st Meeting of the North American chapter of the association for computational linguistics
5. Genest P-E, Lapalme G (2012) Fully abstractive approach to guided summarization. In: Proceedings of the 50th annual meeting of the association for computational linguistics (vol 2: short papers), pp 354–358
6. Hermann KM, Kocisky T, Grefenstette E, Espeholt L, Kay W, Suleyman M, Blunsom P (2015) Teaching machines to read and comprehend. In: Advances in neural information processing systems, vol 28, pp 1693–1701

7. Nallapati R, Zhou B, Gulcehre C, Xiang B et al (2016) Abstractive text summarization using sequence-to-sequence RNNs and beyond. arXiv:1602.06023
8. Gliwa B, Mochol I, Biesek M, Wawer A (2019) Samsum corpus: a human-annotated dialogue dataset for abstractive summarization. arXiv:1911.12237
9. Kornilova A, Eidelman V (2019) Billsum: a corpus for automatic summarization of us legislation. arXiv:1910.00523
10. Lin C-Y (2004) ROUGE: a package for automatic evaluation of summaries. In: Text summarization branches out pp 74–81
11. Papineni K, Roukos S, Ward T, Zhu W-J (2002) BLEU: a method for automatic evaluation of machine translation, In: Proceedings of the 40th annual meeting of the association for computational linguistics, pp 311–318
12. Bahdanau D, Cho K, Bengio Y (2014) Neural machine translation by jointly learning to align and translate. arXiv:1409.0473
13. Luong M-T, Pham H, Manning CD (2015) Effective approaches to attention-based neural machine translation. arXiv:1508.04025
14. See A, Liu PJ, Manning CD (2017) Get to the point: summarization with pointer-generator networks. arXiv:1704.04368
15. Sutskever I, Vinyals O, Le QV (2014) Sequence to sequence learning with neural networks. In: Advances in neural information processing systems, pp 3104–3112
16. Lin J, Sun X, Ma S, Su Q (2018) Global encoding for abstractive summarization. arXiv:1805.03989
17. Shi T, Keneshloo Y, Ramakrishnan N, Reddy CK (2021) Neural abstractive text summarization with sequence-to-sequence models. ACM Trans Data Sci 2(1):1–37
18. Aksenov D, Moreno-Schneider J, Bourgonje P, Schwarzenberg R, Hennig L, Rehm G (2020) Abstractive text summarization based on language model conditioning and locality modeling. arXiv:2003.13027
19. Zhang J, Zhao Y, Saleh M, Liu P (2020) PEGASUS: pre-training with extracted gap-sentences for abstractive summarization. In: International conference on machine learning, PMLR, 2020, pp 11328–11339
20. Raffel C, Shazeer N, Roberts A, Lee K, Narang S, Matena M, Zhou Y, Li W, Liu PJ, Exploring the limits of transfer learning with a unified text-to-text transformer. arXiv:1910.10683
21. Lewis M, Liu Y, Goyal N, Ghazvininejad M, Mohamed A, Levy O, Stoyanov V, Zettlemoyer L Bart: Denoising sequence-to-sequence pre-training for natural language generation, translation, and comprehension. arXiv:1910.13461
22. Rehman T, Sanyal DK, Chattopadhyay S, Bhowmick PK, Das PP (2021) Automatic generation of research highlights from scientific. In: 2nd Workshop on extraction and evaluation of knowledge entities from scientific documents (EEKE'21), collocated with JCDL'21

Chapter 22
Real-Time Left- and Right-Hand Detection for Sign Language Recognition

Archana Ghotkar, Atharva Gokhale, and Durvesh Malpure

1 Introduction

Computer vision-based applications have enabled the existing computers to perform multiple vision-based tasks by making use of relevant techniques and hence obviating human efforts. These techniques can make use of one or more deep learning models that have been trained on suitable data as per the requirements and the purpose of the application. Object detection is one important problem for which many computer vision-based techniques have provided solutions. Object detection is a task which deals with detecting the presence of instances of particular objects in a digital image which is under consideration [1]. In this, every kind of object to be detected in the images is associated with a unique class for identification. Many novel models have been introduced recently which aim towards achieving detections with greater precision and efficiently while reducing the training time on the other hand [1]. However, the selection of the right model depends mainly on their performances on the same data. Based on this analysis of the performance parameters, a particular model can be judged as suitable or otherwise for a particular kind of dataset [1].

Sign language recognition is one of the most important and sought-after tasks in the study of computational linguistics [2]. Sign languages mainly involve the use of hands to make certain gestures having some meaning associated with them. For this, identifying the presence of either of or both the hands and capturing their precise location in a given digital image or a video frame under consideration is vital. Hence,

Atharva Gokhale and Durvesh Malpure are equally contributed for this chapter.

A. Ghotkar · A. Gokhale (✉) · D. Malpure
SCTR's Pune Institute of Computer Technology, Pune, India
e-mail: atharva_gokhale@yahoo.com

A. Ghotkar
e-mail: aaghotkar@pict.edu

L. Mandal et al. (eds.), *Proceedings of International Conference on Computational Intelligence, Data Science and Cloud Computing*, Algorithms for Intelligent Systems,
https://doi.org/10.1007/978-981-19-1657-1_22

many object detection-based techniques which are trained using suitable datasets for the mentioned use case scenarios are implemented for effective detection of the hands [3]. In many sign languages, there are some signs which make use of both the hands collectively with each of the two hands performing a different gesture in order to convey the sign [4]. Hence, there is a need for models which can separately detect left and right hands from an image. As a result, the sign language recognition systems would be able to encompass a greater number of signs [5].

There are many object detector models available but the You Only Look Once (YOLO) [6] family of deep learning-based frameworks provides models which are dedicated for object detection and show better efficiency in the above-mentioned task. They are properly documented and anyone can easily use them in their projects related to sign language. Due to these reasons, the YOLO models have been focused on in this study. In this paper, a detailed analysis of the performance aspects of eight object detection models from the YOLO family has been presented which have been trained for the task of detection of hands separately. Four types of models have been trained which include YOLOv4 Tiny, YOLOv4, YOLOv5 medium and YOLOv5 small. The eight models include each of above models for resolutions 416 and 320.

All of these models were trained on a common dataset that consisted of annotated video frames obtained from sign language videos that consisted of signs being conveyed using either a single or both the hands, with the hands appearing in a variety of hand shapes and orientations. After training, the values of the important performance parameters of the models including the average frames per second (FPS) and mean average precision (mAP) among others were obtained. The mAP metric is a standard metric used in object detection where a mean of average precisions of all the classes is taken to measure the performance of the model. A detailed comparison between all these performance parameters has provided insights necessary for the judgement of the models based on their efficiency for the mentioned task.

The main contributions of this paper are:

- This paper introduces a dataset for left- and right-hand detections which includes annotated frames from Indian Sign Language videos.
- This paper compares and analyses models from YOLO family that are trained to detect left and right hands based on various performance and efficiency metrics such as mAP and FPS.

The rest of the paper is structured as follows. Section 2 surveys this field and discusses the previous work done. Section 3 explains the dataset in detail that has been used for training of all the eight models and discusses the methodology followed for training all the models and the setup steps taken. Section 4 presents the results obtained from training all the models, and performances of all the models have been discussed. The last section (Sect. 5) concludes this paper with some remarks for future work.

2 Related Work

Previous work has been done on Indian Sign Language recognition but several techniques needed expensive sensors such as Microsoft Kinect that worked as motion sensors [7]. There is a need to make the methods based solely on 2D cameras as it is a much cheaper option. Deep learning-based computer vision has made many 2D camera-based approaches possible. An attempt to leverage some of these has been made in this study.

Sridhar et al. have created a similar dataset named 'INCLUDE' consisting of annotated video frames taken from Indian Sign Language videos as a part of their study of sign language recognition which is another publicly available dataset [8]. However, there are differences between the existing datasets (including the one mentioned above) and the one created as a part of this study. The INCLUDE dataset consists of annotated frames pertaining to specific signs of the Indian Sign Language [8]. Thus, as a result, any hand detection model trained on this dataset will be able to detect hands only when the hands represent specific hand shapes belonging to these signs. Hence, there is a need for a dataset which encompasses a wide variety of hand shapes in order to make the hand detection systems more robust.

The goal of object detection is to detect locations of various objects in an image. The most accurate algorithms were found when convolutional neural networks (CNNs) were used [1]. There are two types of object detector models that are based on CNNs. Two-stage detectors use region proposal methods and then detect the objects based on the regions. In contrast to this, single stage detectors directly predict the bounding box of the object and do not rely on region annotations. Different examples of the types of models have been explained below.

2.1 *Two-Stage Detectors*

The R-CNN model [9] was the first in the region-based convolutional neural network (R-CNN) family to show how CNNs may be used to improve object detection. They use a regional proposal model to pose this as a classification and localization problem. Non-maximum suppression and Intersection over Union (IoU) loss were also used in this approach. The faster R-CNN [10] used anchor boxes unlike image pyramids from the models introduced before it which improved the efficiency by an order of magnitude. A feature pyramid network (FPN) [11] was also introduced which helped to increase detection of small objects. This network is used in a number of detection models that are used today. The improvement on faster R-CNN was mask R-CNN [12] which added another branch in parallel for pixel-level object instance segmentation. This not only detects the bounding box of an object, but also the exact pixels which contain that object.

2.2 *Single Stage Detectors*

The most popular single stage detectors are single shot detector (SSD) [13], EfficientDet [14] and YOLO. The Single Shot MultiBox Detector (SSD) was the first single-stage detector to equal the mAP of current two-stage detectors such as the faster R-CNN while preserving real-time speed. EfficientDet uses model scaling and bidirectional feature pyramid network (BiFPN) [14]. The first YOLO was the earliest most successful model in object detection task which gave out the bounding box coordinates and the confidence after a single pass. Building upon that, YOLOv2 [15] and YOLOv3 [16] replaced GoogleNet with DarkNet-19 and DarkNet-53, respectively, as a backbone. All these models were built in the darknet [6] framework. YOLOv4 [17] continued improving in the same framework and is currently the best model in the darknet framework. This model used many improvements such as a bag of freebies and a bag of specials as mentioned in their paper [17]. The YOLOv5 [18] framework developed in the Pytorch library proves to be much more efficient as compared to YOLOv4. It can be easily embedded onto websites as well due to the use of Pytorch library. Although there are many object detectors, this paper focuses the study on YOLO family of models due to them being a popular choice and having a user-friendly documentation.

3 Methodology

3.1 *Dataset*

As the main goal of this study was to detect hands from videos which would be useful for applications aimed at sign language recognition for Indian Sign Language (ISL), Indian Sign Language videos from ISLRTC (New Delhi) YouTube channel were used [19]. A total of 4092 frames were extracted from a diverse set of videos. These images majorly contained either a right hand, a left hand or both in the same image and the person whose hands were shown. As the downloaded videos were of 720p quality, the images obtained had a resolution of 1280 * 720. These videos were taken keeping in mind that the detection performed would be from a frontal perspective. The videos were picked in such a way that there would be a good variety of hand shapes and hand orientations. For all the eight models mentioned, a total of 3692 images out of 4092 were used for training, whereas the remaining 400 images were used for validation. The images containing different combinations of the presence of either a right hand or a left hand or both were annotated accordingly by their bounding boxes. Dataset is provided at: https://gist.github.com/durvesh8/6689bd723fe0511b48dd9230a55a8aec [20].

3.2 Training

Figure 1 depicts the procedure followed for feeding the data to the models and obtaining the expected detections for hands along with the values of the performance parameters.

For the YOLOv4 and YOLOv4 Tiny models, the darknet environment was compiled and then both the models were trained with two different resolutions ultimately leading to four models in total. The pre-trained weights of these models trained on the COCO [21] dataset were used as a part of transfer learning. Most configurations were set to the default settings but some changes were required for this use case. The number of classes were needed to be set to 2 for the YOLO configuration (left hand and right hand) as the original YOLOv4 is trained upon 80 general classes altogether such as car and pedestrian. While configuring, the models were configured for the resolutions: 320 * 320 and 416 * 416. The batch size was set to 64 with subdivisions equal to 16. The YOLOv4 documentation states that a model should be trained for either at least 6000 iterations or 2000 times the number of classes (whichever is the highest). For this study, a total of 6000 iterations were selected accordingly for the YOLOv4 and YOLOv4 Tiny models. The number of filters used were 21.

One of our important modifications which could hamper the model mAP significantly was to turn off flip augmentation by setting the *flip* flag to 0 in the configuration file. This is important for distinguishing between left and right hands. If one were to flip an image having a right hand, it would be identified as a left-hand image. The modifications are the same in the case of YOLOv4 Tiny as well.

The weights pre-trained on the COCO dataset were used for the YOLOv5 models as well. This transfer learning helped in making the training process faster. The same two resolutions were used as they were used in case of YOLOv4 models. Other than this, the number of classes were changed to two for training these models. The YOLOv5 models were trained for 50 epochs. Epochs in YOLOv5 are different

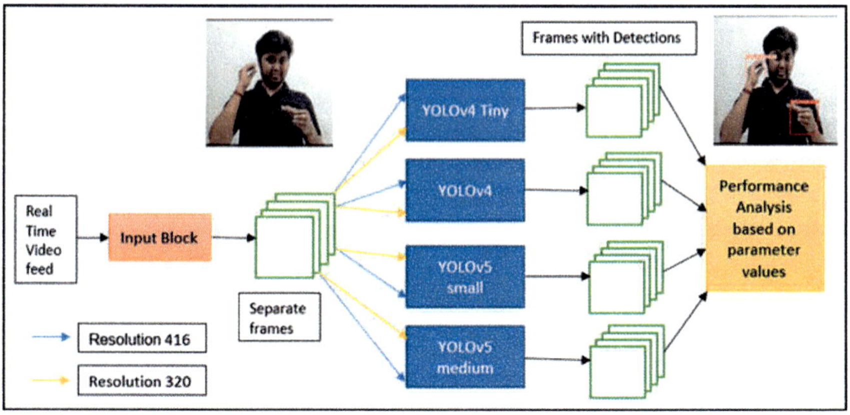

Fig. 1 Procedural block diagram

from iterations in YOLOv4 as epochs mean a pass through the whole data, whereas iterations mean an iteration on a batch of data which is only a fraction of the whole data. Two models YOLOv5m and YOLOv5s were used at two different resolutions each, making the total model count of YOLOv5 models equal to four.

4 Results

The performance metrics considered for result analysis are—mean average precision (mAP), mean average precision for individual classes: left hand (mAP-L), right hand (mAP-R), average frames per second (FPS) and number of billions of floating-point operations per second (in Gflops). Following are the values of the above-mentioned performance parameters for the models:

The values of mAP have been observed and recorded over the training period of the models. On the other hand, values of certain parameters like average FPS have been obtained while testing the models on unseen data which included recorded videos of sign languages. All the results have been obtained by training and testing the models on a Tesla T4 GPU which has a memory size of 16 GB.

4.1 *YOLOv4 Tiny Models*

It can be observed from Tables 1 and 2 that in case of YOLOv4 tiny models for both the resolutions, the total training time for both the models is lowest compared to all other models considered in this study. In addition to this, the mean average precision values have been observed to be slightly lower than the YOLOv4 models. Due to these characteristics, the YOLOv4 tiny model can be used in use-case scenarios, where a model has to be trained on a large dataset in a relatively smaller amount of time. The main reason which causes the YOLOv4 tiny model to have a lower mAP than the YOLOv4 model is the use of a residual block (ResBlock-D) instead of a conventional cross-stage partial block (CSPBlock) [22]. But on the other hand, the ResBlock-D allows tiny models to perform faster object detection [22]. It can also be observed from Table 1 that these models have the lowest number of parameters resulting in a reduced computation complexity which make them suitable for being used in embedded systems and mobile systems [22]. Hence, it can be inferred that YOLOv4 tiny models for real-time hand detection can be used as a part of applications making use of embedded systems for sign language recognition. Figure 2 represents two video frames from an ISLRTC's sign language video [19] in which the left and right hands have been detected using the YOLOv4 tiny models with resolution of 416 and 320, respectively.

Table 1 Values of performance parameters of all the models for resolution: 416 * 416

S. No.	Model	mAP (%)	mAP-L (%)	maP-R (%)	Avg FPS	Train time	GFlops
1.	YOLOv4	**96.74**	96.74	96.92	55.5	14 h 53 m 55 s	59.570
2.	YOLOv5 med	96.6	97.7	95.6	90.91	4 h 0 m 42 s	50.3
3.	YOLOv5 small	96.21	97.3	94.9	125	4 h 14 m 13 s	16.8
4.	YOLOv4 Tiny	95.49	95.12	95.85	116.1	2 h 49 m 27 s	6.789

Bold values indicate highest mAP values

Table 2 Values of performance parameters of all the models for resolution: 320 * 320

S. No.	Model	mAP (%)	mAP-L (%)	maP-R (%)	Avg FPS	Train time	GFlops
1.	YOLOv4	**97.33**	97.80	96.85	73.9	12 h 0 m 0 s	34.249
2.	YOLOv4 Tiny	95.92	96.84	95.01	113.3	2 h 46 m 42 s	4.017
3.	YOLOv5 med	95.02	96.3	93.8	83.33	3 h 45 m 3 s	50.3
4.	YOLOv5 small	93.27	94.5	92.1	125	3 h 41 m 8 s	16.8

Bold values indicate highest mAP values

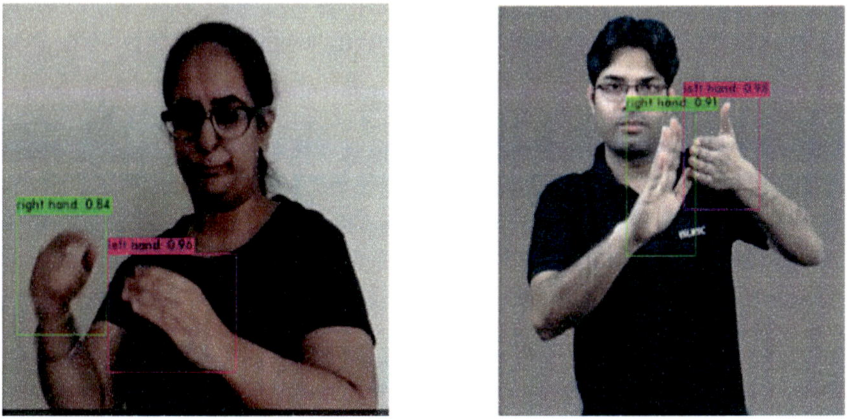

Fig. 2 YOLOv4 Tiny detection for 416 resolution (left) and 320 resolution (right)

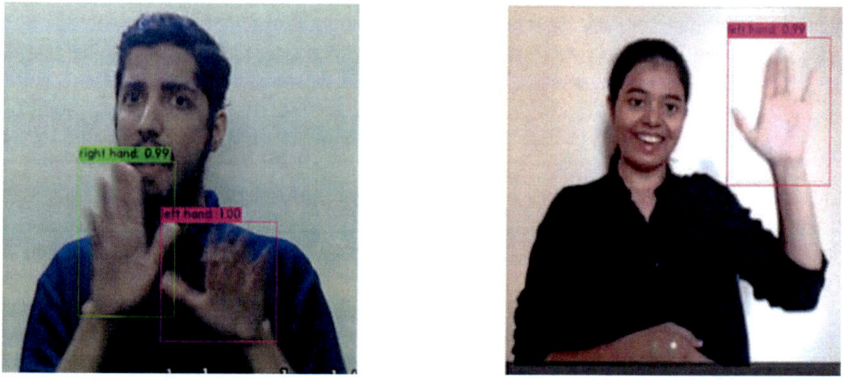

Fig. 3 YOLOv4 detection for 416 resolution (left) and 320 resolution (right)

4.2 YOLOv4 Models

From Table 1, it can be observed that the YOLOv4 model trained with a resolution of 416, took the highest time for training among all the models. Also, the YOLOv4 models have shown the highest mAP values among all the models considered and for both the resolutions considered in this study. The YOLOv4 model trained with a resolution of 320 has shown a mAP value of 97.33% which is the highest among all the models. This model has also shown the highest mAP values in case of detection for individual classes (left hand: 97.80% and right hand: 96.85%).

The complex network structures of the YOLOv4 models which are mainly influenced by the included CSPBlocks and their corresponding layers are the chief governors of the high mAP values [22]. On the contrary, it can be noted from Tables 1 and 2 that the YOLOv4 models show a comparatively poor performance in terms of the average FPS values. As a result, these models, due to their complexity and poor performance in terms of FPS, are difficult to be used in embedded systems [22]. Figure 3 represents two video frames from an ISLRTC's sign language video [19] in which the left and right hands have been detected using the YOLOv4 models with resolution of 416 and 320, respectively.

4.3 YOLOv5 Small Models

The YOLOv5 small models are the least complex in the series of YOLOv5 models which can be inferred from the fact that they have the least number of layers and parameters as compared to other v5 models. Figure 4 represents two video frames from an ISLRTC's sign language video [19] with detections obtained using the YOLOv5 small models with resolution of 416 and 320, respectively.

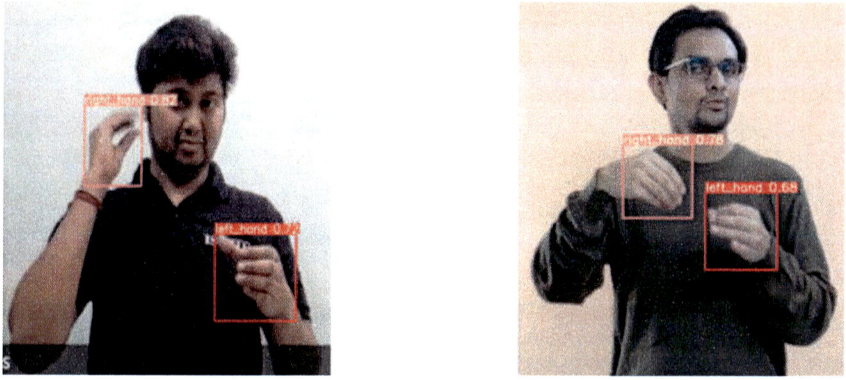

Fig. 4 YOLOv5 small detection for 416 resolution (left) and 320 resolution (right)

From Tables 1 and 2, it can be observed that, at a resolution of 416 the YOLOv5 small model has shown higher mAP value overall as well as for both the classes individually than at a resolution of 320. In addition to this, the average FPS value obtained for these models is the highest among all the models being considered and is 125. From this, it can be inferred that these models can perform the hand detections on video feeds at a better rate as compared with the other models. The YOLOv5 small models can be effectively used for those sign language recognition applications which work on real time or recorded video data fed as input, since the average FPS value has not changed significantly with resolution. It can be observed from Figure 6, that the mAP values have not deflected much and are constantly above 90% with minor variations from the 35th and the 42nd epoch onwards for resolutions of 416 and 320, respectively.

4.4 YOLOv5 Medium Models

The YOLOv5 medium model is the most complex model among the models considered in this study having 308 layers in total. In spite of this, the training time required for this model was lower than the YOLOv4 models for both the resolutions which can be noted from Tables 1 and 2. The average FPS for this model has not been as high as the YOLOv5 small model, but it is still better than that of YOLOv4 model suggesting a more efficient computational capability. Due to this, similar to the YOLOv5 small model, the YOLOv5 medium model can also be used in sign language recognition applications which require video data as input and where continuous recognition is expected. It can be observed from the graph of mAP in Fig. 6, that the mAP values have not deflected much and are constantly above 90% with minor variations after the 35th epoch for both the resolutions. Figure 5 represents two video frames from an ISLRTC's sign language video [19] in which the left and

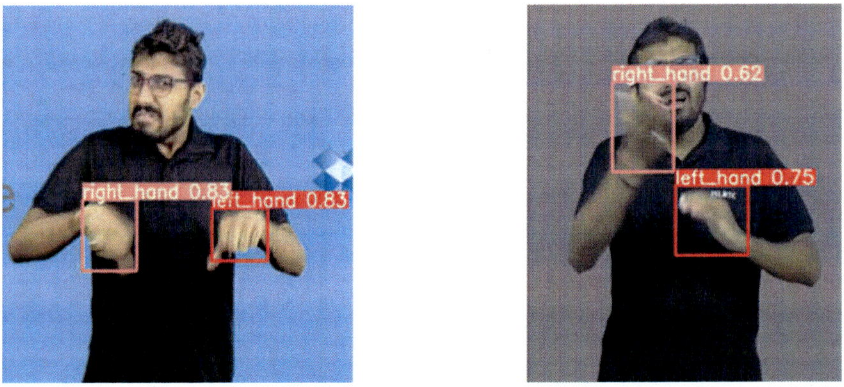

Fig. 5 YOLOv5 medium detection for 416 resolution (left) and 320 resolution (right)

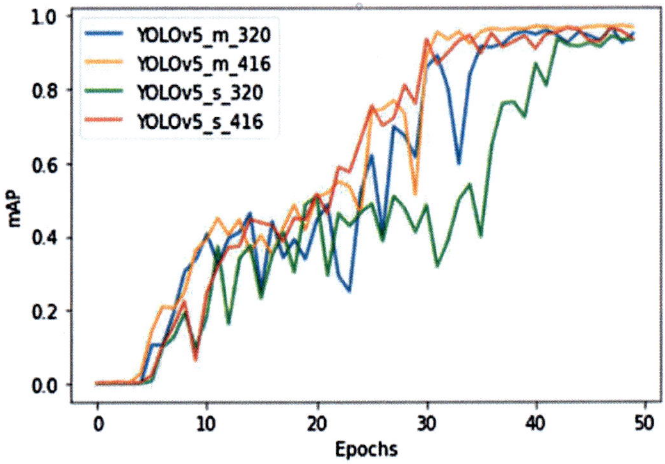

Fig. 6 Variations in mAP values with epochs

right hands have been detected using the YOLOv5 medium models with resolution of 416 and 320, respectively.

The variations in the mAP values with epochs over the training period for the YOLOv5 models can be visualized in the chart displayed in Fig. 6.

It can be noted from Table 1 that for a resolution of 416, the YOLOv4 model has shown an average mAP of 96.74%, whereas the YOLOv5 medium model has shown an average mAP of 96.6% with the difference between the two being 0.14%. This difference can be significant for an application where the model is expected to perform detection with higher accuracy. On the other hand, the YOLOv5 medium model has shown a higher FPS rate than the YOLOv4 model for the same resolution. Thus, the YOLOv5 medium model would be regarded as more suitable in applications where a higher FPS rate has to be achieved and a slight compromise on the mAP would be acceptable. In such scenarios, the difference of 0.14% in the mAP values would not be regarded as significant.

5 Conclusion

In sign language interpretation, hand segmentation and detection, determination of hand location, hand shape, hand orientation and movements are key steps. Most of the research has explored hand detection for various human–computer interaction (HCI) applications. However, accurate detection of left and right hands is one of the essential steps towards sign language interpretation. Due to the availability of multiple models for object detection, the selection of the right model is important. Multiple performance parameters give detailed insights regarding the models under consideration. Such an analysis also helps in determining the trade-off to be considered during model selection by considering the performance factors for which a particular model shows best results with a compromise on some other factors. In addition to generalized hand detection, if a model is more specific in detecting the left and the right hands successfully, then the sign language recognition systems are enabled to work for a higher number of signs, mainly by including the ones involving the use of both the hands. From the analysis presented in this paper, details with respect to important performance parameters for a series of models from the YOLO family for left- and right-hand detections have been put forth. It can be inferred from this study, that the YOLOv4 model has shown better results than the others in terms of mAP. Along with the detection, the localization of the hands with respect to the body, if achieved, can provide information regarding location of the hands with respect to other body parts in the image. This along with the usual detection can provide increased benefits for sign language recognition applications.

References

1. Zou Z et al (2019) Object detection in 20 years: a survey. ArXiv:1905.05055 [Cs]
2. Sinha GR (2017) Indian sign language (ISL) biometrics for hearing and speech impaired persons: review and recommendation. Int J Inf Technol 9(4):425–430
3. Francke H, Ruiz-del-Solar J. Verschae R (2007) Real-time hand gesture detection and recognition using boosted classifiers and active learning. In: Pacific-Rim symposium on image and video technology. Springer, Berlin, Heidelberg, pp 533–547
4. Grif MG, Prikhodko AL (2018) Approach to the Sign language gesture recognition framework based on HamNoSys analysis. In: 2018 XIV International scientific-technical conference on actual problems of electronics instrument engineering (APEIE). IEEE, pp 426–429
5. Bandyopadhyay S (2020) A study on Indian sign language recognition using deep learning approach. Int J Sci Res Dev 8(4):582–585

6. Redmon J, Divvala S, Girshick R, Farhadi A (2016) You only look once: unified, real-time object detection. In: Presented at the proceedings of the IEEE conference on computer vision and pattern recognition
7. Ghotkar A, Kharate G (2015) Dynamic hand gesture recognition for sign words and novel sentence interpretation algorithm for Indian sign language using microsoft kinect sensor. JPRR 10:24–38. https://doi.org/10.13176/11.626
8. Sridhar A, Ganesan RG, Kumar P, Khapra M (2020) Include: a large scale dataset for indian sign language recognition. In: Proceedings of the 28th ACM international conference on multimedia, pp 1366–1375
9. Girshick R, Donahue J, Darrell T, Malik J (2014) Rich feature hierarchies for accurate object detection and semantic segmentation. In: Presented at the proceedings of the IEEE conference on computer vision and pattern recognition
10. Ren S, He K, Girshick R, Sun J (2015) Faster R-CNN: towards real-time object detection with region proposal networks. In: Advanced in neural information processing systems, vol 28
11. Lin T-Y, Dollar P, Girshick R, He K, Hariharan B, Belongie S (2017) Feature pyramid networks for object detection. In: Presented at the proceedings of the IEEE conference on computer vision and pattern recognition
12. He K, Gkioxari G, Dollar P, Girshick R (2017) Mask R-CNN. In: Presented at the proceedings of the IEEE international conference on computer vision
13. Liu W, Anguelov D, Erhan D, Szegedy C, Reed S, Fu C-Y, Berg AC (2016) SSD: single shot multiBox detector. In: Leibe B, Matas J, Sebe N, Welling M (eds) Computer vision—ECCV 2016. Springer, Cham, pp 21–37. https://doi.org/10.1007/978-3-319-46448-0_2
14. Tan M, Pang R, Le QV (2020) EfficientDet: scalable and efficient object detection. In: Presented at the proceedings of the IEEE/CVF conference on computer vision and pattern recognition
15. Redmon J, Farhadi A (2017) YOLO9000: better, faster, stronger. In: Presented at the proceedings of the IEEE conference on computer vision and pattern recognition
16. Redmon J, Farhadi A (2018) YOLOv3: an incremental improvement. arXiv:1804.02767 [cs]
17. Bochkovskiy A, Wang C-Y, Liao H-YM (2020) YOLOv4: optimal speed and accuracy of object detection. arXiv:2004.10934 [cs, eess]
18. Jocher G, Nishimura K, Mineeva T, Vilariño, R (2020) YOLOv5 . https://github.com/ultralytics/yolov5. Accessed 15 Aug 2021
19. Indian Sign Language Research and Training Center (ISLRTC), New Delhi, India. YouTube channel. https://www.youtube.com/channel/UC3AcGIlqVI4nJWCwHgHFXtg. Last accessed 17 Aug 2021
20. Left and right hand detection imagery dataset. https://gist.github.com/durvesh8/6689bd723fe0511b48dd9230a55a8aec. Last accessed 20 Aug 2021
21. Lin T-Y, Maire M, Belongie S, Bourdev L, Girshick R, Hays J, Perona P, Ramanan D, Zitnick CL, Dollár P (2015) Microsoft COCO: common objects in context. arXiv:1405.0312 [cs]
22. Jiang Z et al (2020) Real-time object detection method based on improved YOLOv4-tiny. ArXiv:2011.04244 [Cs]

Chapter 23
Analytical Study on Application of Deep Learning for Sentiment Analysis

Neha Sharma, S. Veenadhari, and Rachna Kulhare

1 Introduction

People in this world have their own opinion about each and every thing around them, and they used to express their opinion in their own way. The common ways of expressing the opinion are commenting on social media, tweets, blogging, etc. Many people use to record their expressions in the form of videos (YouTube) and audios (podcasts), another way to express the thinking in the form of images (Picasa). These thoughts impact others may be positively, negatively or some time it is neutral. The process of classifying these thoughts is named as sentiment analysis. This is basically the process which deals with the analysis of public opinions, sentiments, psychology, thoughts which are social, political, economic, industrial, and educational and many more fields [1].

The research on big data sentiment analysis is becoming the center of attraction as curiosity to sense what the thinking of different personalities is. It is necessary to know because these thoughts further create actions depending on their classification. If few people are sharing their views than it is easy to understand by the people but if millions of people share their point of views on the particular topic, then computer-based analysis plays significant role. Thus, sentiment analysis is done by using machine learning or deep learning methods. Several fields have the involvement of sentiment analysis and it is used in multiple applications also. Some of them are trade and commerce, survey of websites, exploring consumer opinion or response, in

N. Sharma (✉) · S. Veenadhari · R. Kulhare
Department of CSE, RNTU, Bhopal, M.P., India
e-mail: sharmaneha94@gmail.com

S. Veenadhari
e-mail: veenadhari1@gmail.com

R. Kulhare
e-mail: rachna12kulhare@gmail.com

L. Mandal et al. (eds.), *Proceedings of International Conference on Computational Intelligence, Data Science and Cloud Computing*, Algorithms for Intelligent Systems,
https://doi.org/10.1007/978-981-19-1657-1_23

observation of social networking websites, and much more [2]. Sentiment analysis is usually of two types, first one is identifying the text subjectively or objectively and the other one is extraction of feature or aspect. Multiple models are given by researchers for the analysis of sentiments-like traditional machine learning techniques and deep learning techniques. As the Internet users are increasing day-by-day and creation of data is increasing on various platforms. It is important to observe and understand the sentiments of the people by security providers and also government organizations. According to the point of view of the people on the particular topics', actions are taken in short time period. For example, if something against the security of nation or feeling of unity breaking inside the nation is going on viral then, it can easily be trapped by the sentiment analysis and control is taken on such actions and decision is taken in short time period. Machine learning and deep learning techniques are becoming more popular in the recent areas and in natural language processing they are becoming eye-catching. These techniques include support vector machine (SVM), long–short term model (LSTM), Naïve Bayes, convolution neural network (CNN), logistic regression, and many more. These techniques are required because now people do not use only texts for expressing their opinion but also the Hindi–English mixed language is used, emoticons are used, images are also utilized, people share code mixed text and further different sentiments [3–5]. Thus, the sentiment analysis in this twenty-first century is becoming much more useful and significant.

The paper is organized as: Sect. 2 represents the literature review of sentiment analysis work on multiple domains with application of machine learning. Section 3 introduces the brief overview of applications of sentiment analysis. Section 4 is dedicated to give an overview of implementation of deep learning algorithms in different sentiment analysis domain. Section 5 describes the results and discussion of some deep learning approaches. Finally, in Sect. 6, the paper represents the conclusion of this analytical study and gives an insight for future research work.

2 Related Work

To categorize tweets into distinct classes, an optimization-based machine learning approach is investigated by Naresh and Venkata Krishna [1]. In three phases, the suggested model is assessed. The information is initially preprocessed to eliminate any noisy data. The characteristics are retrieved in the second phase using an optimization algorithm, and the revised training dataset is categorized into three classifications: favorable, unfavorable, and balanced in the third phase. In comparison to existing machine learning algorithms, the suggested technique has a highest accuracy of 89.47%. The method is speedier and cuts down on the amount of time it takes to process information. This works well with huge datasets. To increase performance over time, an effective optimization process can be used. Through a series of online and offline investigations, Arulmurugan et al. [2] developed a method for adding emotions into created artificial intelligence and investigated its relevance and effectiveness. A methodology and basic architecture were created to incorporate the proposed technique, and affection processor architectures frequently use it. By combining SVM, Naive Bayes, and neural network algorithms with a combined segmentation and classification model, the suggested approach improves sentence level sentiment recognition accuracy. Birjali et al. [3] put themselves in a tough situation by focusing on the feelings that may lead to suicidal ideation. They suggest, in particularly, to tackle the scarcity of suicide-related terminological resources by developing a suicide-related lexicon. The findings of our experiments show that utilizing Twitter data, we can derive forecasts of suicide attempts using an approach based on machine learning algorithms and semantic sentiment analysis. Furthermore, this investigation verifies the efficacy of performance in terms of exactness on semantic sentiment analysis that could lead to suicidal ideation. Agarwal and Mittal [4] looked at the performance of numerous prominent extracting features strategies, including as unigrams, bigrams, and dependency features, in this chapter. In addition, researchers introduced bi-tagged features for supervised sentiment analysis that include sentiment-rich data while containing less noise. In this paragraph, they also offer a unique feature extraction approach for sentiment analysis based on semantic grouping of characteristics. To overcome the entirely absent issue that machine learning algorithms confront, the suggested technique combines phrases into clusters. Shamantha Rai et al. [5] In this study, tweet is retrieved using a specific word search and then submitted to sentiment classification employing RF, SVM, and NB classifiers to categorize them into favorable, moderate, and unfavorable categories. The machine learning classifiers RF and SVM are included as part of the analysis, and their accuracy is measured using three characteristics and increasing the amount of tweets. Increase for tweets also enhances the accuracy of RF, SVM, and NB. The sentiment analysis of Twitter data utilizing various machine learning techniques is covered in this study conducted by Vanaja [6]. Sentiment classification is a one of a kind approach for resolving issues on social media and identifying people's sentiment in order to examine the emotion rate. As a result, it presents the system for good, bad, and neutral feelings on Twitter's user-generated content. Furthermore, the gathered results will aid in the development of the most robust model feasible for analyzing data effectiveness, attractiveness, and improved performance. The findings of the experiments suggest that machine learning is a good option for forecasting public opinion from social media sites. Sharma et al. [7] used aspect-level sentiment analysis in this research. The efficiency of the process of two machine learning algorithms, the SVM classification techniques, and the NB prediction model is evaluated using accuracy, retention, and F1 measures. The Naive Bayes algorithm outperformed the support vector machine approach in terms of performance. Abd El Jawad et al. [8] evaluated the performance of various machines learning and deep learning methods, as well as providing a new hybrid system for sentiment categorization that combines text analytics and neural networks. The dataset used here contains over one million tweets from five different domains. About 75% of the dataset was used to train the network, while the remaining 25% was used to test it. The study revealed a maximal accuracy rate of 83.7%, demonstrating the effectiveness of the system's hybrid learning technique over traditional supervised methods.

In the realm of machine learning, D'souza and Sonawane et al. [9] applied the dual sentiment analysis (DSA) appears to be highly appealing and intriguing, as it helps to overcome issues including the BoW model, polarization shift classification issues, and so on. Experiment using Amazon product information, especially mobile phone ratings, to see if the research framework can overcome the limits of the current system and improve correctness. Based on the findings and analysis, they can infer that DSA aided in the final decision-making process by allowing customers to acquire accurate insights into the evaluations and, as an outcome, suggest the right products. Gamal et al. [10] focused on the use of SA in Arabic literary content. A dataset of diverse dialects was gathered and automatically classified, containing around 151,500 Twitter messages. To identify and determine the orientation of a particular comment, the NB, LR, PA, RR, SVM, MNB, AdaBoost BNB, SGD, and ME classifications models were employed. To partition the data into a distinct training and testing set, a tenfold cross-validation was used. PA and RR ensure the highest assessment metric values of 99.96% when employing unigram, bigram, or trigram.

3 Application of Sentiment Analysis

The computerized method of evaluating words to identify overall opinion communicated is known as sentiment analysis. Big data analytics, customer care administration, and consumer satisfaction analysis are all common sentiment analysis applications. Here, we are going to discuss some of the applications of sentiment analysis briefly:

Social Media Audit: Sentiment research on social networking will monitor client perception 24 h a day, 7 days a week, in actual environments. When anything unpleasant starts to circulate, you will be able to rapidly reply, and you will be able to bolster your reputation if you get encouraging responses. You will also obtain consistent, reliable information regarding your users, so you will need to track your achievements from period to period. Thousands of comments on Facebook, Instagram, millions of tweets on Twitter are sent every minute and mostly describes the products and services. However, using machine learning tools, you can sift through that kind of information quickly, assessing personal opinion as well as aggregate public opinion across all social media networks. The process of sentiment analysis using social media is presented in Fig. 1.

Customer Service Management and Customer Feedback: The increasing complexity of inquiries, varying themes, and many departments inside an enterprise, not to consider the seriousness of each specific client, make client service administration a difficult task. Sentiment analysis with deep learning natural language processing (NLP) scans standard form of language for intention, attitude, accent, and far more, much like an individual will, to analyze client demands. You can conventionally route client enquiries, chat forums, telephone conversations, and mails to the technical solution based on emotion, which can sometimes represent immediacy. By identifying terms/expressions that suggest negative, sentiment classification can

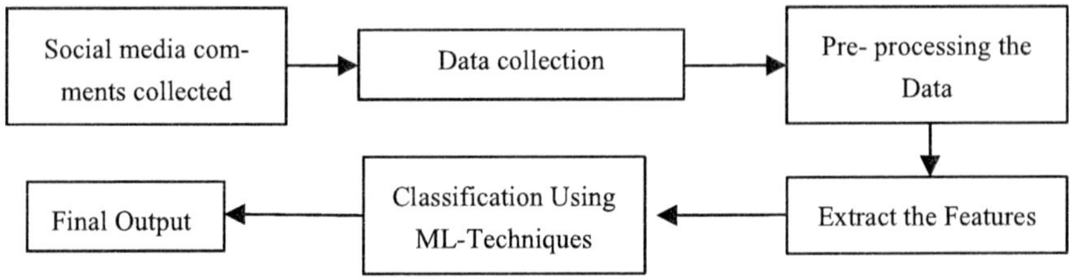

Fig. 1 Social media sentiment analysis

systematically classify millions of client service communications. Text analytics can also be utilized to extract information through vast amounts of client comments. Opinion analysis helps to determine humor, decipher familiar conversation terminologies, and rectify up frequent errors such as incorrect and abused terms. General categorization of customer service sentiment analysis is presented in Fig. 2.

Marketing Strategy and Popularity Tracking: Among the most common uses of sentiment classification in the corporate environment is branding management. Negative comments can quickly accumulate on the Internet, and the more you wait to respond, regrettable the condition will get. Brand image remarks will really be promptly notified using sentiment analysis technologies. You can also track the reputation and credibility of your brand over time or at any specific point in time, ensuring smooth success. Even if you are seeking information about your company in media articles, journals, discussion boards, or online networks, you can turn that information into insights statistical information.

Voice of Customers (VoC): Vast majority of your consumer input, including websites, consumer surveys, conversations, service companies, and mails, should

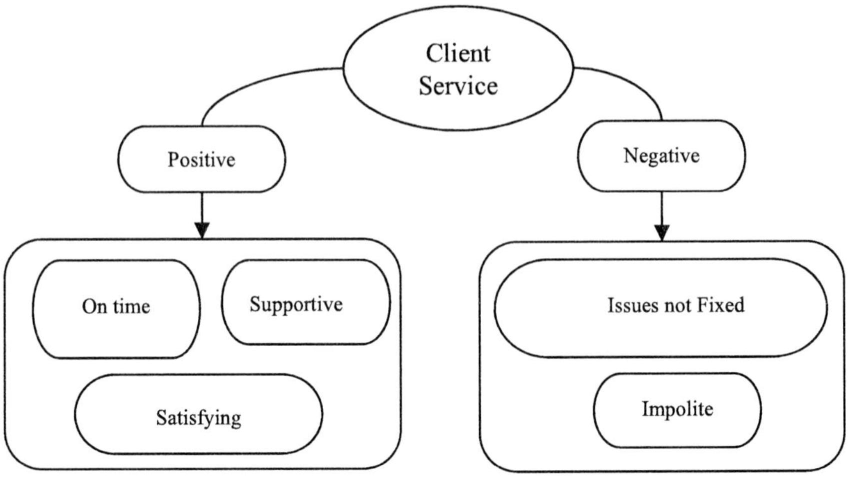

Fig. 2 Customer service sentiment analysis

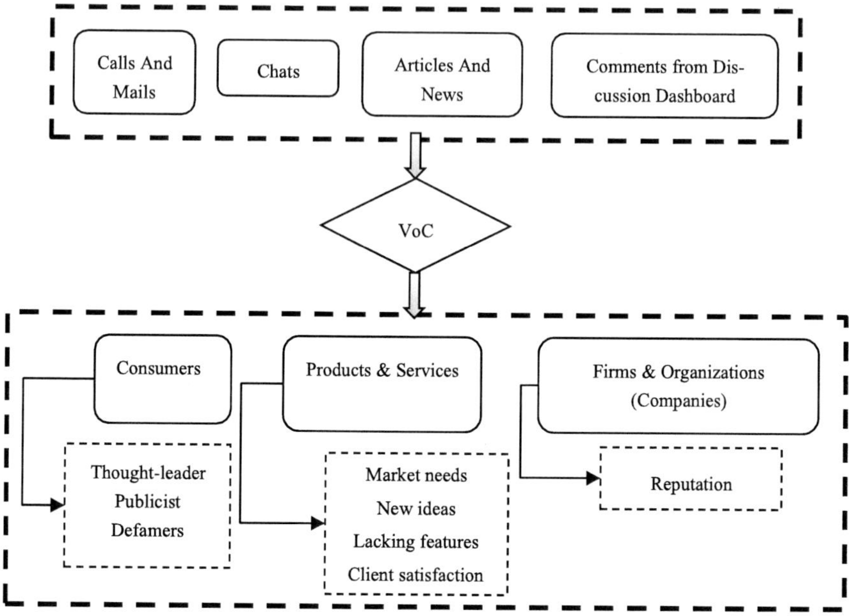

Fig. 3 Voice of customers-based sentiment analysis

be combined and evaluated. Sentiment approach enables administrators to classify as well as arrange the content in order to detect trends and recurrent issues and problems. People can segment their client base to contact clients specifically after learning about them as a whole. For illustration, company may want to offer special offers or incentives to lure unsatisfied clients based on research from a consumer inspection. Alternatively, award people who are exceptionally pleased with your organization, rewarding them to tell each other about the item or brand. Flowchart of VoC is presented in Fig. 3.

Product Reviews—Understand what customers are thinking about a current design soon after it is released, or look through periods of comments you had not seen before. Since artificial intelligence sometimes does not manipulate its outcomes which are not biased, text mining produces superior outcomes over individuals. People can use words to obtain information about a specific product category and build text analytics algorithms to locate exactly what exactly want then.

4 Deep Learning and Sentiment Analysis

This part covers the broad framework for big data sentiment analysis that has been provided in previous research publications. It consists of three key steps: data gathering, feature creation, data extraction, and classification of information, as illustrated in Fig. 4.

Data Acquisition: The database is created in this part of the study. For this, data is first acquired from Internet sources or databases, and then the dataset is preprocessed to make it relevant for further steps.

Preprocessing: Preprocessing is required since the dataset must be cleaned. Data preprocessing is the name for this procedure. This step is required to eliminate any superfluous terminology used in reviews by an individual, such as commas, full stops, colons, or any other special symbols or characters. There are no sentiment values associated with these words. As a result, such phrases must be eliminated in order to decrease further redundancy.

Feature Extraction and Classification: Feature extraction is the process of extracting some useful data from these reviews. This data is a set of characteristics that classifiers may utilize to make decisions. To group any kind of information values into distinct classes, a classification approach is utilized. Natural language

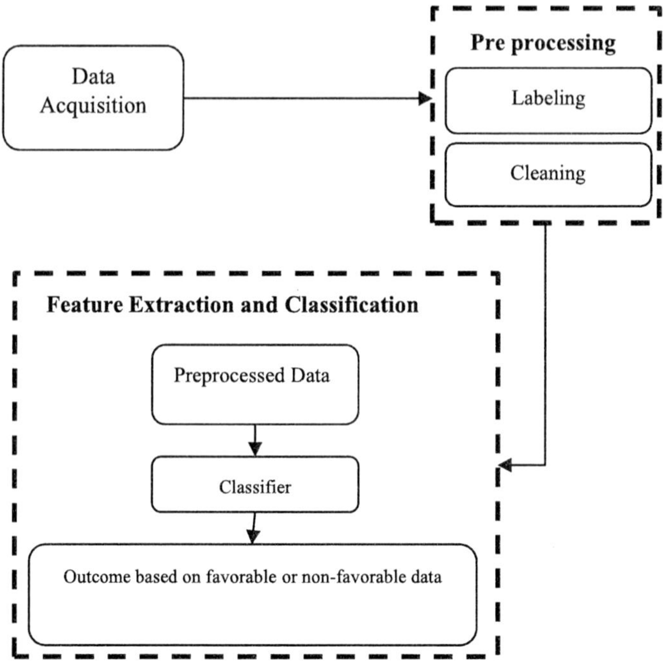

Fig. 4 Framework for sentiment analysis in multiple domains

processing is utilized in this study to categorize data into positive and negative categories. Deep learning has recently gained popularity in NLP. In the section below, we will look at some deep learning approaches for big data sentiment analysis. For example, the word. For example, "the battery is small but the quality is bad." Here it includes both positive and negative sentiment. By using SA, we can use sentiment based on multiple aspect.

4.1 Convolutional Neural Networks (CNNs)

CNNs are a type of neural network that has had a lot of success and invention in the field of digital image processing. In the realm of sentiment analysis, increasingly academics are employing CNNs. Kim [11] developed the most common neural network model for sentence level sentiment categorization. The author experimented with CNN, which was constructed on top of pre-trained word. The results of the experiments demonstrate that pre-trained vectors may be a good feature extractor for NLP applications utilizing neural networks. Encouraged by any of these findings, Zhang and Wallace [12] proposed a one-layer CNN architecture for phrase categorization. Apart from these, how the performance parameter like hyperparameters, filter size, etc., can be varied is also discussed.

4.2 Word Embedding (WE)

Many academics have lately proposed novel techniques to word representation for sentiment analysis. Typical word embedding approaches develop word distributions that are unrelated to the job at hand. This may be overcome in sentiment analysis by leveraging prior information in the form of sentiment labels or opinionated terms from textual data. Hao et al. [13] developed a unique technique for cross-domain sentiment classification using probabilistic embeddings, which retains resemblance structures in embedding space. Yu et al. [14] presented a strategy that uses sentiment intensity ratings from textual data to learn sentiment embeddings.

4.3 Recurrent Neural Networks (RNNs)

The sequential data is modeled using RNN. Sequential data is used in a wide range of situations. For example, a sequence of sentences is converted from one language to another in language translation, an audio clip (sequence that plays over time) is plotted into a text script (sequence of words) in speech recognition, and a series of image sequences is converted into text that explains the exercise shown in video in video activity recognition. By repeatedly processing each word in a phrase, RNNs store the internal states of the inputs. As a result, in order to predict the next word in a phrase, RNN will keep track of all the preceding words and their relationships.

LSTM is one of the most widely used RNN variations, with the capacity to address the degradation problem in conventional RNN and the ability to detect long-term relationships. They become stronger and more adaptable as a result of this. For sentiment categorization, researchers are also integrating several deep learning approaches. Huang et al. [15] presented an architecture that combines two main deep learning networks, LSTM, and CNN and consists of a layer of CNN and two layers of LSTM layered atop CNN. On the other hand, Hassan and Mahmood [16] developed the ConvLstm architecture, which integrated CNN and LSTM for categorizing short texts on top of word2vecYoo et al. [17] presented Polaris, a system that can assess the sentiment trajectory of any activity. Trajectory analysis may be done by categorizing the contents of online posts about a certain occurrence according to the location where another events took place.

4.4 Deep Belief Networks (DBNs)

Unmonitored pre-trained net, which include autoencoders and recursive neural networks, isthe source of DBNs. Layers and layers of uncontrolled networks, such as restricted Boltzmann machines (RBMs) or autoencoders, make up DBNs. For sentence categorization, Ruangkanokmas et al. [18] utilized DBNs with selecting features (DBNFS). Feature extraction begins with pre-processing to delete grammar and stop words, and then the entire manuscript is transformed into feature vectors.

5 Results and Discussion

Deep learning methods have been increasingly popular in the field of NLP in recent years. Convolutional neural networks with several hidden layers between the input and output layers are known as deep neural networks (DNNs). This section contains a thorough description of some of the most common deep learning models, in different domain for sentiment analysis (see Table 1).

To build a classifier, five distinct datasets are used, each of which is divided into an unmarked training set, a labelled training set, and a labelled testing set. DBNs are used to classify sentiment, and they exceed baseline semi-supervised techniques, with the greatest accuracy. Accuracy is evaluated as.

Accuracy = Accuracy refers to the condition of efficiency being right or exact in all circumstances.

$$\frac{\mathrm{TP} + \mathrm{TN}}{\mathrm{TP} + \mathrm{TN} + \mathrm{FP} + \mathrm{FN}} \tag{1}$$

Table 1 DL methods for sentiment analysis

Technique	Domain
CNN [15]	Social media
CNN-LSTM [16]	Movie review
WE [13]	Movie review
GRU [19]	Business review
RNN [20]	Product review
LSTM [21]	Text
DBN [22]	Product review
Attention-based network [23]	Text
Autoencoder [24]	Movie review

FN is true positive, true negative, false positive, and false negative instances, respectively.

According to Fig. 5, the performance comparison of the different DL methods for the prediction of the sentiments using SA is given. Different type of domains is used like social media reviews, movie reviews, text, product review, and business reviews based on the research. Attention-based network [23] has maximum accuracy of 93% used text data. Minimum accuracy is 73% for GRU in [19]. CNN [15] used social media reviews and predict the sentiment having accuracy of 87.2% used business review for SA. The accuracy of CNN_LSTM, WE, RNN, LSTM, DBN, and Autoencoder are 87.2%, 88.3%, 84%, 88%, 91%, 75%, 87%, respectively. Therefore, from this analysis it can be observed that attention-based CNN models or LSTM networks achieves best result while processing textual contents of any domain.

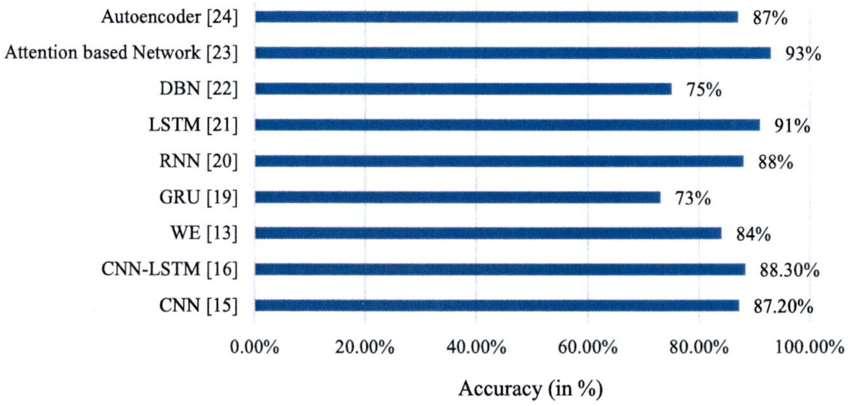

Fig. 5 Performance comparison of DL methods for sentiment analysis

6 Conclusion

This paper is dedicated to explore the machine learning application for big data sentiment analysis in multiple domains. The paper identified that deep learning has showed its efficacy over conventional natural language processing (NLP) methods. First of all, in this paper, an overview is presented on areas of sentiment analysis and application of deep learning in all areas. From result analysis, it can be seen that highest accuracy was achieved by deep attention network and deep recurrent neural network. So, for future research work it can be suggested that attention deep attention network or deep recurrent neural network can be adopted for big data sentiment analysis. As the sentiment analysis field can be effective in many domains review analysis, mental health diagnosis, product analysis, social trend analysis, etc.

References

1. Naresh A, Venkata Krishna P (2021) An efficient approach for sentiment analysis using machine learning algorithm. Evol Intel 14(2):725–731. https://doi.org/10.1007/s12065-020-00429-1
2. Arulmurugan R, Sabarmathi KR, Anandakumar H (2019) Classification of sentence level sentiment analysis using cloud machine learning techniques. Clust Comput 22:1199–1209. https://doi.org/10.1007/s10586-017-1200-1
3. Birjali M, Beni-Hssane A, Erritali M (2017) Machine learning and semantic sentiment analysis based algorithms for suicide sentiment prediction in social networks. Procedia Comput Sci 113:65–72. https://doi.org/10.1016/j.procs.2017.08.290
4. Agarwal B, Mittal N (2016) Machine learning approach for sentiment analysis. 21–45. https://doi.org/10.1007/978-3-319-25343-5_3
5. Shamantha Rai B, Shetty SM, Rai P (2019) Sentiment analysis using Machine learning classifiers: evaluation of performance. In: 2019 IEEE 4th international conference on computer and communication systems, ICCCS 2019, pp 21–25. https://doi.org/10.1109/CCOMS.2019.8821650
6. Vanaja S (2018) Aspect-level sentiment analysis on E-commerce data. In: 2018 International conference on inventive research in computing applications (ICIRCA), Icirca, pp 1275–1279
7. Sharma P, Sharma AK (2020) Experimental investigation of automated system for twitter sentiment analysis to predict the public emotions using machine learning algorithms. Materials Today: Proc xxxx. https://doi.org/10.1016/j.matpr.2020.09.351
8. Abd El-Jawad MH, Hodhod R, Omar YMK (2019) Sentiment analysis of social media networks using machine learning. In: ICENCO 2018—14th International computer engineering conference: secure smart societies, pp 174–176. https://doi.org/10.1109/ICENCO.2018.8636124
9. D'souza SR, Sonawane K (2019) Sentiment analysis based on multiple reviews by using machine learning approaches. In: Proceedings of the 3rd international conference on computing methodologies and communication, ICCMC 2019, ICCMC, pp 188–193. https://doi.org/10.1109/ICCMC.2019.8819813
10. Gamal D, Alfonse M, El-Horbaty ESM, Salem ABM (2018) Implementation of machine learning algorithms in arabic sentiment analysis using N-gram features. Procedia Comput Sci 154:332–340. https://doi.org/10.1016/j.procs.2019.06.048
11. Kim Y (2014) Convolutional neural networks for sentence classification. In: EMNLP 2014—Proceedings of the 2014 conference on empirical methods in natural language processing, pp 1746–1751. https://doi.org/10.3115/V1/D14-1181

12. Zhang Y, Wallace B (2015) A sensitivity analysis of (and practitioners' guide to) convolutional neural networks for sentence classification
13. Hao Y, Mu T, Hong R et al (2020) Cross-domain sentiment encoding through stochastic word embedding. IEEE Trans Knowl Data Eng 32:1909–1922. https://doi.org/10.1109/TKDE.2019.2913379
14. Yu LC, Wang J, Robert Lai K, Zhang X (2018) Refining word embeddings using intensity scores for sentiment analysis. IEEE/ACM Trans Audio Speech Lang Process 26:671–681. https://doi.org/10.1109/TASLP.2017.2788182
15. Huang Q, Chen R, Zheng X, Dong Z (2017) Deep sentiment representation based on CNN and LSTM. In: Proceedings—2017 International conference on green informatics, ICGI 2017, pp 30–33. https://doi.org/10.1109/ICGI.2017.45
16. Hassan A, Mahmood A (2017) Deep learning approach for sentiment analysis of short texts. In: 2017 3rd International conference on control, automation and robotics ICCAR 2017, pp 705–710. https://doi.org/10.1109/ICCAR.2017.7942788
17. Yoo SY, Song JI, Jeong OR (2018) Social media contents based sentiment analysis and prediction system. Expert Syst Appl 105:102–111. https://doi.org/10.1016/J.ESWA.2018.03.055
18. Ruangkanokmas P, Achalakul T, Akkarajitsakul K (2016) Deep belief networks with feature selection for sentiment classification. In: Proceedings—International conference on intelligent systems, modelling and simulation, ISMS, vol 0, pp 9–14. https://doi.org/10.1109/ISMS.2016.9
19. Li B, Cheng Z, Xu Z, et al (2019) Long text analysis using sliced recurrent neural networks with breaking point information enrichment. In: ICASSP, IEEE international conference on acoustics, speech, and signal processing—Proceedings May 2019, pp 7550–7554. https://doi.org/10.1109/ICASSP.2019.8683812
20. Yuan Z, Wu S, Wu F, Liu J, Huang Y (2018) Domain attention model for multi-domain sentiment classification. Knowl Based Syst 155:1–10
21. Shi S, Zhao M, Guan J, et al (2018) A hierarchical LSTM model with multiple features for sentiment analysis of sina weibo texts. In: Proceedings of 2017 international conference on asian language processing IALP 2017, Jan 2018, pp 379–382. https://doi.org/10.1109/IALP.2017.8300622
22. Wang M, Ning ZH, Xiao C, Li T (2018) Sentiment classification based on information geometry and deep belief networks. IEEE Access 6:35206–35213. https://doi.org/10.1109/ACCESS.2018.2848298
23. Yang L, Li Y, Wang J, Sherratt RS (2020) Sentiment analysis for e-commerce product reviews in Chinese based on sentiment Lexicon and deep learning. IEEE Access 8:23522–23530. https://doi.org/10.1109/ACCESS.2020.2969854
24. RongWenge NieYifan, OuyangYuanxin, et al (2014) Auto-encoder based bagging architecture for sentiment analysis. J Vis Lang Comput 25:840–849. https://doi.org/10.1016/J.JVLC.2014.09.005

Chapter 24
Class Biased Sarcasm Detection Using Variational LSTM Autoencoder

Saranya Bhattacharjee, Soumyajit Maity, Rahul Sen, and Sankhadeep Chatterjee

1 Introduction

In the twenty-first century, social media has surfaced as one of the most prominent and instrumental communication mediums. Over the years, it has been a predominant source for propagating information. In the past few years, researchers have demonstrated significant interest in studying textual data from microblogging sites like Twitter, Tumbler, and Reddit to draw insights that help in analyzing the mental status of the public and assist in predicting different human sentiments [16, 17]. In NLP-based research, sarcasm detection has garnered considerable regard and recognition [4, 7]. However, it is experimentally proven that for many socially sensitive issues, non-sarcastic posts outnumber sarcastic posts. This essentially results in a biased prediction with inclination toward the majority class. The literature has revealed that traditional approaches fail to capture semantic information and due to the presence of imbalanced data, experimental predictions are biased. Further investigation demonstrates that no concrete research has been conducted to address this issue in satirical text detection systems.

In this paper, we have worked on LSTM Autoencoder-based Sarcasm Detection while dealing with class imbalance. The compressed latent vectors containing semantic information are obtained and experimental analysis has been conducted.

S. Bhattacharjee · S. Maity · R. Sen · S. Chatterjee (✉)
Department of Computer Science & Engineering, University of Engineering & Management, Kolkata, India
e-mail: chatterjeesankhadeep.cu@gmail.com

L. Mandal et al. (eds.), *Proceedings of International Conference on Computational Intelligence, Data Science and Cloud Computing*, Algorithms for Intelligent Systems,
https://doi.org/10.1007/978-981-19-1657-1_24

2 Literature Survey

In recent times, there has been quite a lot of discussion in the topic of sarcasm detection. In 2018, [15] proposed a technique involving two feature extraction methods viz. interjection and punctuation, in detecting sarcasm in Indonesian Twitter feeds. In 2020, [10] documented an approach to deal with Over-Fitting by means of modifying the Hybrid Approach. The authors presented a thorough and remarkable technique to studying figurative language in Potamias et al. [13], including a hybrid neural network architecture that achieved state-of-the-art performance across all test datasets. Another method for detecting sarcasm using a Contextual Network model (C-Net) was investigated [8]. The authors [9] illustrated an Adversarial and Auxiliary Features-Aware BERT (AAFAB) network which recorded the results that outperformed baseline traditional models. Chunyan Yin et al. [20] proposed a novel multi-task deep neural model (MT_SS) for detecting collective sarcasm detection and sentiment analysis. Han et al. [6] developed four different types of memory network models for text sentiment classification using CNN and BLSTM. A sophisticated algorithm was put forward by the authors [19] by introducing a novel method to study cross-lingual sentiment, incorporating dynamic weighting method. To study the effects of Class Imbalance on streaming data, the authors [14] created a concept drift detection model using Adadelta optimizer-based deep neural network. The authors [18] worked on a thorough and comprehensive study on detection of sarcasm. [1] studied contextualized data and presented a remarkable analysis. Banerjee et al. [2] demonstrated a thorough and elaborate analysis in social media sarcasm detection with respect to imbalanced dataset.

3 Methodology

3.1 Corpus Creation

The initial steps consist of extracting English tweets pertaining to serious topics like the military, Donald Trump, jobs, and election. The selected topics are found to generate primarily serious posts, resulting in less sarcastic posts [11]. A second segregation process is performed to effectively sort out the extracted sarcastic tweets. To locate the sarcastic tweets, we utilized hashtags such as #sarcasm, #sarcastic, #lol, and #not. The tweets were retrieved by using the Twitter API. The training dataset which we have developed comprises a total of 55,000 tweets, including 42,000 non-sarcastic and 13,000 sarcastic tweets. For our dataset, the imbalance ratio is 3.23. It has been noticed that in case of extreme imbalance, the folds in Cross-Validation hardly capture any data from the positive class (sarcastic tweets in our case). Hence, we have worked on a Balanced Testing regimen. The Balanced Testing method employs distinct testing set with an equal number of samples from each class. This results in a more accurate testing distribution for evaluating the models. We constructed a balanced set of 3000 sarcastic and non-sarcastic tweets to test our approaches.

3.2 Deep Neural Network-Based Approach

For our research, we have implemented a deep learning-based approach involving a Bidirectional LSTM Variational Autoencoder to obtain the text embeddings and attain syntactic representation of the textual data. According to recent studies [2], the amount of sarcastic tweets in a real corpus taken from Twitter is significantly lower than the number of non-sarcastic tweets. Previous research sought to minimize the class imbalance issue by using data augmentation methods and utilizing NLP approaches, but was unable to effectively represent text in vector form. The proposed architecture is shown in Fig. 1. Bidirectional LSTMs have been employed in our experiments because it has the strength to grasp the context of a particular word for prediction. Also, Variational autoencoders employ a probabilistic approach. It helps in assessing and obtaining smooth latent representations. The sequences in a batch have all been padded and are all the same length, 52. The Embedding layer accepts as input a 2-D tensor of shape (samples, sequence length). The layer gives back a 3-D floating-point tensor (samples, sequence length, embedding dimensionality). The dimension for the Embedding layer has been selected to be 150. A memory unit of size 128 has been chosen for the initial bi-LSTM layer, and hence the number of dimensions of each hidden state (for each word) is 2 * 128 = 256. The second bi-LSTM layer has a unit of size 64, however, the forward and backward pass gets concatenated and gives an output shape of (None, 128). The experiment has shown that latent vector of size 128 is most suitable for capturing the compressed latent vector representation. The bi-LSTM Variational Autoencoder has been trained and the Encoder is later used to obtain latent vectors. During the unsupervised training, we have taken sparse categorical crossentropy as our loss function and Nesterov-accelerated Adaptive Moment Estimation, or Nadam, has been set as the optimization algorithm [5]. A dropout value of 0.25 and a learning rate of 0.001 have been selected.

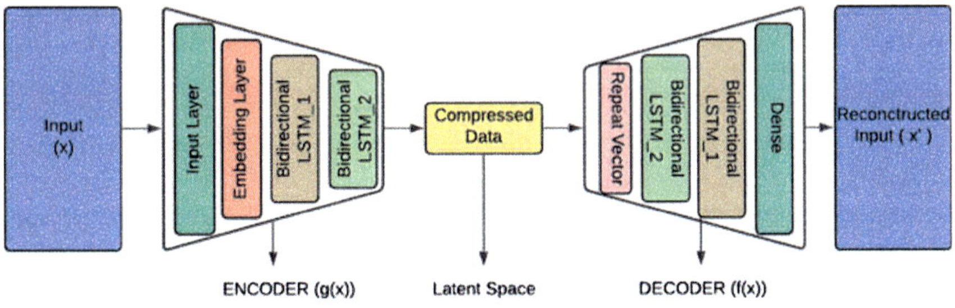

Fig. 1 Proposed bidirectional LSTM variational autoencoder architecture

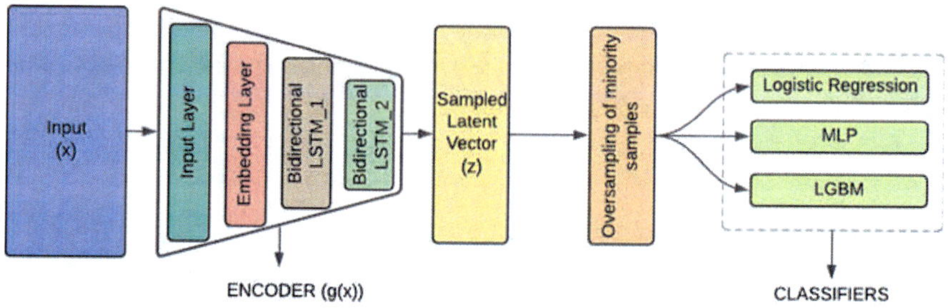

Fig. 2 Methodology for generating latent vectors from the encoder, then oversampling the minority samples and applying classification algorithms

ReLU was set to be the activation function for the training operation. However, after obtaining the structured latent representation, the data remains to be imbalanced and hence the oversampling techniques are employed on the training samples and classification algorithms have been administered (as shown in Fig. 2).

4 Results and Discussion

In this section, we have thoroughly analyzed and recorded the performance of each of the Minority Oversampling techniques and the different Classifiers to reflect the efficiency of our proposed experimental set-up. Our simulations have been conducted on an Intel Core i5-1035G1 processor, Intel UHD Graphics 620, 8 GB RAM, Windows 10 Home 21H1, and TensorFlow 2.5.0.

4.1 Experimental Setup

The Bi-LSTM Variational Autoencoder has been trained on the merged dataset, composed of the training and test dataset. Once the text data is structured, the training dataset is 100 % oversampled with SMOTE and its varients. The oversampling techniques which we have used in our experiment are-(I) SMOTE, (II) ADASYN, (III) Boderline-SMOTE, (IV) SVM-SMOTE, and (V) KMeans-SMOTE. To study the performance of our proposed architecture, we have evaluated on three different classification algorithms, viz. Logistic Regression (LR), Light Gradient Boosted Machine (LGBM) and Multi-layer Perceptron (MLP). We have implemented Grid Search [3] to determine the hyperparameters of our model. And the classifiers are implemented using *scikit-learn* [12] package.

Logistic Regression classifier has been applied with 'saga' solver,the maximum iteration is set to be 100,000 and L1 penalty term has been added. The default parameters have been utilized for LGBM classifier and in case of MLP classifier, learning rate at a constant value of 0.001 and adam solver have been set.

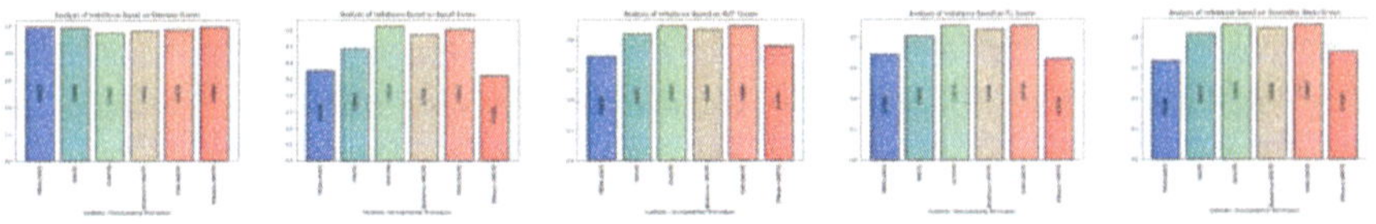

(a) (b) (c) (d) (e)

Fig. 3 Analysis of minority oversampling techniques based on performance metrices: **a** Precision **b** recall **c** AUC **d** F1-score **e** geometric mean

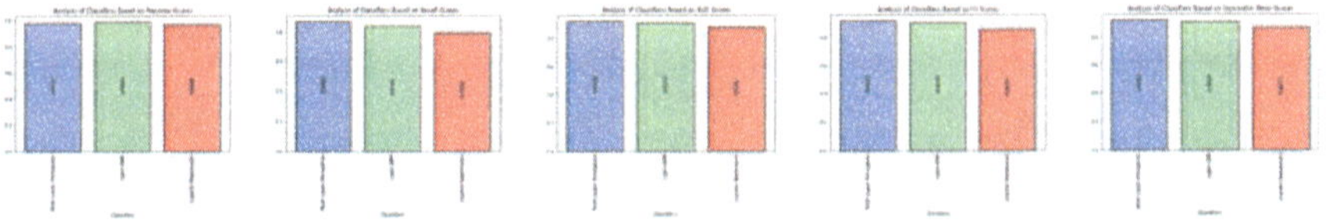

(a) (b) (c) (d) (e)

Fig. 4 Analysis of classifiers based on performance metrices: **a** Precision **b** recall **c** AUC **d** F1-score **e** geometric mean

4.2 Performance Analysis

Figure 3 illustrates how each of the Minority Oversampling techniques have performed in relation to the Imbalanced Dataset. The comparative analysis has demonstrated that SVM-SMOTE and ADASYN have performed the best, while on the other hand Kmeans-SMOTE has led to a low recall score. SVM-SMOTE has documented the greatest geometric mean score (Fig. 3) of 0.8816, over a score of 0.6449 for the unbalanced data. The AUC scores (Fig. 3) documented by SVM-SMOTE, ADASYN, Borderline-SMOTE, SMOTE, and Kmeans-SMOTE are 0.8887, 0.8867, 0.8685, 0.8377, and 0.7584, respectively. All of the data collected for each performance metric calculation has shown that the synthetic data performed significantly better than the imbalance data and exhibited a major improvement.

To study the efficiency of our proposed architecture, the classification algorithms we have employed are thoroughly evaluated on the basis of precision, recall, AUC score, F1-score, and Geometric mean score. From Fig. 4, it is noticed that all the three classifiers have captured the semantic representation. MLP classifier has recorded the highest geometric mean score (Fig. 4) of 0.9195, while LGBM and LR documented an individual geometric mean score of 0.9064 and 0.8677, respectively. The AUC scores (Fig. 4) recorded by MLP, LGBM and LR are 0.9218, 0.9104, and 0.8792, respectively, which imply that the classifiers have successfully distinguished between the classes. The F1-scores (Fig. 4), being the harmonic mean of precision and recall, noted down by the classifiers, viz. MLP, LGBM, and LR are 0.9163, 0.9018, and 0.8559, respectively.

Table 1 has documented an extensive review of the various classification models incorporated with the minority oversampling techniques. The geometric mean has presented a deep insight into the study and helped us analyze the performances. The

Table 1 A comparative study of the models

Model	AUC	Precision	Recall	F1-Score	Geometric mean
LR SVMSMOTE	0.9553	0.9630	0.9470	0.9549	0.9552
MLP ADASYN	0.9448	0.9603	0.9280	0.9438	0.9446
MLP SVMSMOTE	0.9406	0.9681	0.9113	0.9388	0.9402
LGBM ADASYN	0.9335	0.9723	0.8923	0.9306	0.9325
LGBM SVMSMOTE	0.9288	0.9812	0.8743	0.9247	0.9272
LGBM BorderlineS-MOTE	0.9283	0.9755	0.8786	0.9245	0.9270
LR ADASYN	0.9240	0.9320	0.9146	0.9232	0.9239
MLP Border-lineSMOTE	0.9223	0.9699	0.8716	0.9181	0.9209
MLP SMOTE	0.9101	0.9892	0.8293	0.9022	0.9065
LR Borderli-neSMOTE	0.8956	0.9396	0.8456	0.8901	0.8942
LGBM SMOTE	0.8943	0.9945	0.7930	0.8824	0.8885
MLP Imbalance	0.8933	0.9970	0.7890	0.8809	0.8872
MLP KmeansS-MOTE	0.8911	0.9957	0.7856	0.8783	0.8849
LR SMOTE	0.8885	0.9928	0.7826	0.8753	0.8821
LGBM imbalance	0.8710	0.9964	0.7446	0.8523	0.8617
LGBM KmeansS-MOTE	0.8668	0.9986	0.7346	0.8465	0.8566
LR imbalance	0.7701	0.9963	0.5423	0.7023	0.7356
LR KmeansS-MOTE	0.7326	0.9964	0.4670	0.6359	0.6828

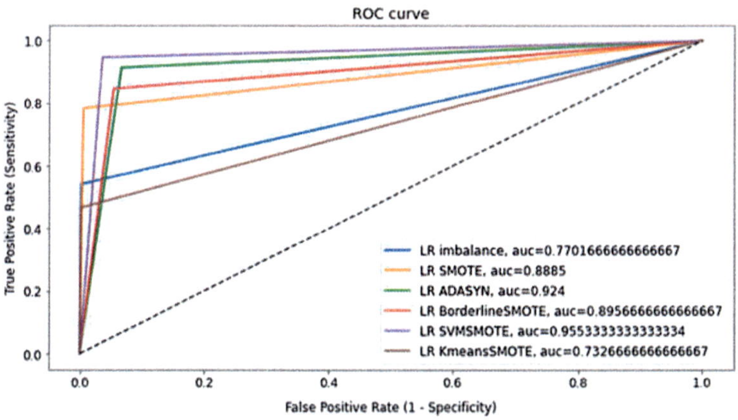

Fig. 5 Analysis of ROC curves on LR classifier

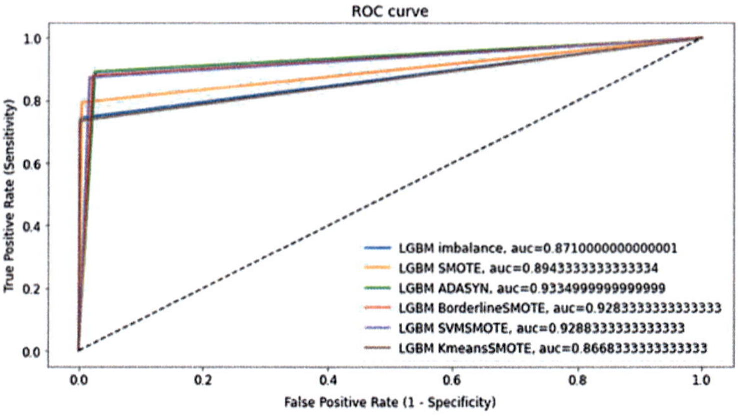

Fig. 6 Analysis of ROC curves on LGBM classifier

precision scores achieved by MLP Kmeans-SMOTE and LR Kmeans-SMOTE are 0.9957 and 0.9964, respectively. LR SVM-SMOTE, MLP ADASYN, and LGBM ADASYN have shown notable improvement on the recall score of 0.9470, 0.9280, and 0.8923 over their respective imbalance classification, viz LR imbalance, MLP imbalance, and LGBM imbalance, given by 0.5423, 0.7890, and 0.7446, respectively. The geometric mean recorded by LR SVM-SMOTE, MLP ADASYN, MLP SVM-SMOTE, and LGBM ADASYN are 0.9552, 0.9446, 0.9402, and 0.9325. LR Kmeans-SMOTE has reported the lowest geometric mean score of 0.6828. The F1-scores portrays a better understanding of the incorrectly classified cases. A high F1-score denotes a high precision and recall score, observed in case of LR SVM-SMOTE and MLP ADASYN.

The ROC curves have been plotted and the AUC score for the respective classification algorithms have been calculated. It is very much evident from the plot that

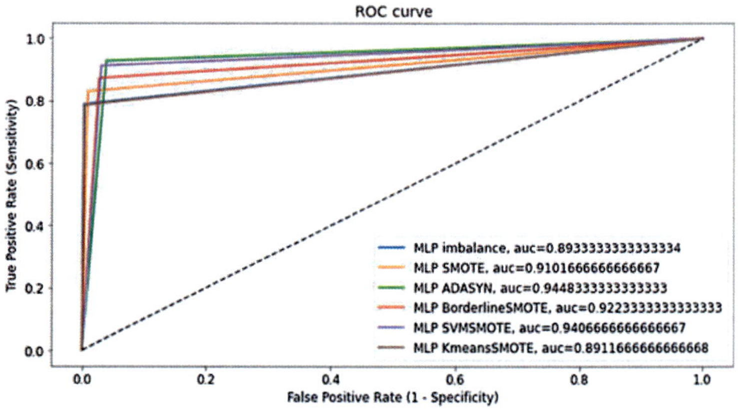

Fig. 7 Analysis of ROC curves on MLP classifier

LR SVM-SMOTE (Fig. 5) has shown the highest performance and recorded an AUC score of 0.9553. It has exhibited a prominent improvement from the AUC score of 0.7701 shown by LR imbalance. However, the AUC score for LR Kmeans-SMOTE fails to display much improvement. At the same time, the overall performance of MLP and LGBM classifiers have been quite promising. ADASYN has displayed the highest AUC scores of 0.9448 and 0.9335 for MLP (Fig. 7) and LGBM (Fig. 6), respectively. LGBM Borderline-SMOTE and LGBM SVM-SMOTE have recorded an AUC score of 0.9283 and 0.9288. Meanwhile, MLP SVM-SMOTE has documented a noticeable AUC score of 0.9406.

5 Conclusion

The current study proposed a bidirectional LSTM Variational autoencoder to extract the most important features in form of latent vectors. Firstly, an unsupervised training phase has been employed to learn the latent vectors. Then, imbalanced latent vectors have been resampled using resampling techniques viz., SMOTE, ADASYN, Borderline-SMOTE, SVM-SMOTE, and KMeans-SMOTE. After oversampling, the balanced dataset of tweets in latent vector form has been used to train three well known classifiers. Experimental results have indicated that the proposed LSTM autoencoder is capable of producing highly accurate latent representation of tweets and the same can be used to balance the dataset as well. Performance comparison has revealed that SVM-SMOTE oversampling technique is most successful in mitigating the class imbalance problem in sarcasm detection using tweets.

References

1. Bamman D, Smith NA (2015) Contextualized sarcasm detection on twitter. In Ninth international AAAI conference on web and social media
2. Banerjee A, Bhattacharjee M, Ghosh Kushankur, Chatterjee S (2020) Synthetic minority oversampling in addressing imbalanced sarcasm detection in social media. Multimedia Tools Appl 79(47):35995–36031
3. Bergstra J, Bengio Y (2012) Random search for hyper-parameter optimization. J Mach Learn Res 13(2)
4. Cai Y, Cai H, Wan X (2019) Multi-modal sarcasm detection in twitter with hierarchical fusion model. In: Proceedings of the 57th annual meeting of the association for computational linguistics, pp 2506–2515
5. Dozat T (2016) Incorporating nesterov momentum into adam
6. Han H, Liu J, Liu G (2018) Attention-based memory network for text sentiment classification. IEEE Access 6:68302–68310
7. Hernández Farías DI, Prati R, Herrera F, Rosso P (2020) Irony detection in twitter with imbalanced class distributions. J Intell Fuzzy Syst 39(2):2147–2163
8. Jena AK, Sinha A, Agarwal R (2020) C-net: contextual network for sarcasm detection. In: Proceedings of the second workshop on figurative language processing, pp 61–66
9. Kumar A, Narapareddy VT, Gupta P, Srikanth VA, Neti LBM, Malapati A (2021) Adversarial and auxiliary features-aware bert for sarcasm detection. In: 8th ACM IKDD CODS and 26th COMAD
10. Lestari S, Rahmadsyah A, Lubis RMF, Gunawan M et al (2020) HAR-MI with COSTE in handling multi-class imbalance. In: 2020 8th international conference on cyber and IT service management (CITSM). IEEE, pp 1–4
11. Pandey R, Kumar A, Singh JP, Tripathi S (2021) Hybrid attention-based long short-term memory network for sarcasm identification. Appl Soft Comput 106:107348
12. Pedregosa F, Varoquaux G, Gramfort A, Michel V, Thirion B, Grisel O, Blondel M, Prettenhofer P, Weiss R, Dubourg V et al (2011) Scikit-learn: machine learning in python. J Mach Learn Res 12:2825–2830
13. Potamias RA, Siolas G, Stafylopatis AG (2020) A transformer-based approach to irony and sarcasm detection. Neural Comput Appl 32(23):17309–17320
14. Priya S, Uthra RA (2021) Deep learning framework for handling concept drift and class imbalanced complex decision-making on streaming data. Complex Intell Syst 1–17
15. Rahayu DAP, Kuntur S, Hayatin N (2018) Sarcasm detection on Indonesian twitter feeds. In: 2018 5th international conference on electrical engineering, computer science and informatics (EECSI). IEEE, pp 137–141
16. Ren L, Lin H, Xu B, Yang L, Zhang D (2021) Learning to capture contrast in sarcasm with contextual dual-view attention network. Int J Mach Learn Cybern 1–9
17. Xu N, Wang J, Tian Y, Zhang R, Mao W (2021) Ananet: modeling association and alignment for cross-modal correlation classification. *arXiv preprint* arXiv:2109.00693
18. Yaghoobian H, Arabnia HR, Rasheed K (2021) Sarcasm detection: a comparative study. *arXiv preprint* arXiv:2107.02276, 2021
19. Yilmaz SF, Kaynak EB, Koç A, Dibeklioğlu H, Kozat SS (2021) Multi-label sentiment analysis on 100 languages with dynamic weighting for label imbalance. IEEE Trans Neural Netw Learn Syst
20. Yin C, Chen Y, Zuo W (2021) Multi-task deep neural networks for joint sarcasm detection and sentiment analysis. Pattern Recogn Image Anal 31(1):103–108

Chapter 25
A Systematic Survey of the Chatbot Evolution

Srestha Sadhu, Ayusha Burman, and Lopa Mandal

1 Introduction

A chatbot is a computer application that is used to do conversations with users using a human's natural language. It is designed in such a way so that humans feel that they are conversing with another human while having a conversation with the chatbot. In the year 1996, the first chatbot was created by Weizenbaum named ELIZA, which was able to make a simple conversation; its goal was to behave as a psychologist. The first system which passed through some sort of Turing Test was PARRY developed in the year 1975. Sonia et al. [1] proposed a chatbot whose key objective was to serve as an additional school tool and as a mediator between teachers and the students. With the advancement of technology and the advent of natural language processing, artificial intelligence, machine learning, etc., chatbots are being applied in various fields such as business, healthcare, administration, automated telephone answering system, education.

Chatbots are often referred to as the most feasible and advanced form of interaction between machines and humans. Researchers are working and implementing various technologies to make the conversation more human-like. The two different tasks of the chatbot are:

1. User request analysis
2. Returning the response

S. Sadhu (✉) · A. Burman
Department of Computer Science and Engineering, University of Engineering and Management, Kolkata, India
e-mail: sresthasadhu0110@gmail.com

L. Mandal
Department of Information Technology, Institute of Engineering and Management, Kolkata, India

L. Mandal et al. (eds.), *Proceedings of International Conference on Computational Intelligence, Data Science and Cloud Computing*, Algorithms for Intelligent Systems,
https://doi.org/10.1007/978-981-19-1657-1_25

User Request Analysis
This is the chatbot's most important task. Here, it analyzes the user's request and identifies the user's intent to extract the relevant entities to provide the right answers. This is the primary task and the most important step that a chatbot performs. If it fails to read the user's request, then it will fail to deliver relevant answers to the user's request.

Returning the Response
After analyzing the user's request and extracting relevant entities, now it is the task to provide pertinent responses. The response could be:

- A predefined and common text.
- A text recovered from a knowledge database containing different answers.
- A contextualize information that is user data-oriented.
- The information is stored in enterprise systems.
- The consequence of an action that the chatbot has executed by interacting with backend applications.

Chatbots are being developed using various tools and technologies. To get the desired outcomes, a combination of several AI techniques such as machine learning and natural language processing is used.

The current work aims to present the systematic survey of different types of available chatbots and different methodologies applied to develop those chatbots. It also highlights the drawbacks of the existing system and focuses on research gaps in the given field. The organization of the paper is as follows. Section 2 describes different types of chatbots, while Sect. 3 describes various methodologies. The work is concluded in Sect. 4 with highlighting research gaps in existing work and future scope of research.

2 Related Works

Chatbots are found to be used for different purposes. Different works on chatbots worth mentioning are:

Endurance: Alzheimer's patients suffer from short-time memory loss. The chatbot aims to acknowledge divergences in conversational branches that may designate a problem with instant remembrance. Moreover, since the chatbot is a cloud-based solution, physicians and family members can view the log taken from the chatbot to review the condition or the mental health of the patient.

U-Report: UNICEF is using chatbots named U-Report for people living in developing nations to speak out about their most urgent needs. The chatbot focuses on large-scale data collection via polls. This chatbot is not for conversation. It sends out prepared polls, and users can respond with their input. UNICEF then uses this feedback to help people resolve their issues.

Casper: This chatbot takes on the role of a conversational representative, focusing on providing insomniacs with someone to talk to while the rest of the world sleeps.

Bahja et al. [2] in a paper discussed the various points that need to be considered while designing a chatbot. The points are as follows:

- Planning
- Pedagogy
- Policies and regulations
- Ethics
- Artificial intelligence
- Privacy and security
- Personalization
- Performance
- Evaluation
- Standards
- Institutional
- Integration.

The educational system is changing dramatically as artificial intelligence and chatbots have become more prevalent. In this COVID-19 pandemic, it has been observed that massive online open courses (MOOC), e-learning, and m-learning have taken the limelight. Though everything is being held in the online mode, the education system did not take a backseat. A lot of online education emerging platforms are providing various courses according to user's recommendations. It also analyzes the interest of the user and suggests courses accordingly.

In the year 2015, a survey on chatbot design techniques in speech conversation systems has been made by Abdul-Kader et al. [3]. They surveyed selected studies on chatbot designs and had made comparisons between different chatbot designing techniques. In their paper, it has been observed that there is no common approach for chatbot designing. Chatbots are being developed using various approaches. Setiaji and Wibowo [4] published a paper in 2016 in which a user interface has been created to transmit and receive messages from the user using natural language processing (NLP) in teaching computer languages such as Java, Pascal, PHP, and SQL.

In normal classroom education, teaching methods are determined by the teachers. The systems like e-learning, m-learning, distance learning, etc., are being adopted nowadays to help learners to learn at their own pace. It has been observed that some kids may not be able to learn at the same rate as others. Every learner possesses unique capabilities and aptitude; hence, self-learning gives them the liberty to learn their subject at their own pace. Therefore, scientists came up with the idea of e-learning, which is a type of learning that makes use of electronic technologies to provide access to educational content.

Various advantages have been laid out by e-learning, such as:

- Consistency
- Cost-effectiveness
- Flexibility

- Remote operability
- Simplicity.

There are various types of e-learning, namely:

- Web-based learning: In this type of learning, it can be used as learning materials that are available through a web browser such as materials that are packaged in the form of CD-ROMs.
- Online learning: In this type of learning, courses are available online, and one needs to browse through the Internet to find the appropriate course.
- Distance learning: Distance learning is concerned with the interaction between the learner and the instructor who are at a distance from each other. It does not only mean uploading videos and sending the educational material to the student. The educator must maintain contact with the student. Fabio et al. [5] are working to expand the use of the proposed approach in a variety of scenarios to improve e-learning systems.

Francesco et al. [6] gave a detailed study of chatbot in the field of e-learning. After performing various assessments, the results obtained by the experimental campaign were satisfying and showed a good perspective.

Souali et al. [7] have introduced traceability, which is an ability to trace and track an item's movement from one position to another by keeping a complete record of all activities, changes, and relations. An accessible e-learning approach has been proposed that combines both traceability and recommendation for students through chatbots, which provides the result as input and indicators for the traceability recommendation engine, and then finally, it suggests the resulted resources as a form of recommendation.

Another type of learning which is commonly used nowadays is m-learning. It is a short form of mobile learning. Mobile learning is the education or tutoring controlled through portable computing gadgets. As mobile technology is becoming more and more popular, it is making our life easy. From a calculator to a calendar to a music player to a clock, everything is there on our mobile. So, why not make education such portable as well. With advancing of mobile technology, developers made mobile learning possible. Many schools and colleges are supporting the concept of e-learning and m-learning. If used in the right way, mobile learning could be very beneficial for learners.

Gaglo et al. [8] suggested a conversational chatbot for academic remediation in the framework of COVID-19, where moodle learning environment has been employed, and the solution alters students' perceptions and encourages them to participate in this new method of online learning. This technique allows professors to identify students' weaknesses and also helps students to improve without fear of being judged or humiliated.

Pham et al. [9] gave an explicit illustration of how mobile learning can be used to learn the language English using a chatbot. The research determined how online users utilize English learning apps on their mobile devices. In nearly two months, the number of users that downloaded and installed the app was about 26,000, of which about 14,000 installed the chatbot integrated version. However, many users did not use all functions of the system. Specifically, the number of people who have used chatbots at least once was 3649.

Davies et al. [10] proposed a personalization of the e-learning process using AI-powered chatbot integration. The outcomes have been primarily divided into two categories (1) the use of chatbot and (2) without the use of a chatbot. They took a few parameters like the number of the tasks finished effectively (60–80% of correctness) where 87% of the usage of chatbot and 64% without the use of a chatbot. In addition, the second parameter turned into the wide variety of tasks completed efficiently (>80% of correctness) where 51% through the use of chatbot whereas 22% without using chatbot and so forth.

Agrebi et al. [11] proposed deep reinforcement learning for personalized recommendation of distance learning, where the study aimed to suggest the most appropriate courses for learners based on their profile, needs, competence, etc.

In 2019, Shukla et al. [12] proposed an enhancing LMS experience through AIML-based and retrieval-based chatbot. In the retrieval-based database chatbot mechanism, the system can generate responses based on specific words. If a query does not have the words, then a response will not be generated. The chatbot cannot generate new responses. To make a more productive rule-based chatbot, keywords from the inquiry will be taken out with the assistance of R-language pattern matching.

In 2019, Rooein [13] proposed data-driven Edu chatbots using iChat technology. The goal of this study is to turn possibility into reality in the context of Edu chatbots. The research could lead to the implementation of a chatbot that can readily adapt to the current subject or course, allowing content managers to do so without the need for development skills.

In the 2019, Rahmaoui et al. [14] proposed recommending moodle resources using chatbot. For recommendation, a content-based and collaborative approach has been used. The first recommendation phase is content-based, and it is done by computing feature similarities between available learning materials in moodle (courses, tips, books, etc.). The trace-based recommendation checks the historical records. It can be tried either at the beginning to look for alike queries or at the end to check if one or more of these proposed resources were already disregarded or rejected.

In 2020, Rajkumar and Ganapathy [15] worked on NLP to implement bio-inspiring learning style chatbot inventory using brain computing interface, where there were several experiments performed. In experiment 1, a chatbot has been implemented to classify individuals into extroverts and introverts. In experiment 2, machine learning algorithms and clustering processes have been applied, which are performed using the WEKA tool. In the future, the plan is to conduct the study with a greater number of learners. The current experiment does not discuss another human trait called ambiverts. In the future, this trait will take into account as well.

In 2020, Wu et al. [16] proposed a chatbot that is based on two different models: One is a retrieval-based model, and another one is a QANet model. The applied QANet model gave back a result of EM/F1 score of 70.8/80.1 at best with SQuAD v1.1 datasets. It was contrasted the instructing time of the Chinese datasets to the trained QANet model and fundamental RNN-based seq2seq model. The results show that QANet is five times faster in training than the RNN-based seq2seq model. Participants were asked to distinguish between responses generated by their chatbot and those generated by a human in a response context design test. They have collected questionnaires from 53 participants. The average chosen ratio between human-generated and chatbot-generated replies is 52% for human-generated answers and 48% for chatbot-generated answers across all queries.

1. The first category was feelings of isolation and detachment, which means how many of their participants had feelings of isolation and detachment in three different situations.
    (a) E-learning
    (b) E-learning with teacher services
    (c) E-learning with the chatbot
2. The second was the comparison between their chatbot's E-learning course-related QA performances to that of the teacher counseling service.
3. The third category was the average user experience score of their chatbot.

In 2020, Prof. Modak et al. [17] used machine learning and NLP to show the machine learning disability detection using LMS. This study highlights an e-learning system created using moodle, which is an open-source learning environment between the tutors and the students. Based on the work done, they conclude that the students in between the age group (11–13) having a learning disability can be solved. The drawback of the system is that the speech-to-text conversion is not up to the mark, hence, leading to certain incorrect conversions. The system can be further expanded to capture the written responses of the LD students through optical character recognition.

In 2021, Ashok et al. [18] have done a systematic survey of cognitive chatbots in personalized learning framework, where they found the usage of chatbots in various contexts, and the main purpose of the chatbots is to provide guidance and maintain the relationship with the users.

3 Methodologies Used

Many papers have been surveyed from 2015 to 2021, where many methodologies have been used. Some researchers in the year 2017 used ANN to evaluate the strategies and experience in the tutoring model of an intelligent tutoring system [19]. In the year 2019, algorithms like C4.5, ID3, and NBTree have been used for multiple intelligence for encouraging critical thinking as the basis for developing e-learning activities [20]. Except for all these different classification algorithms like naïve Bayes, maximum entropy, and SVM, which have been used for the admission system of an educational institute and prediction of user interest while taking admission [21, 22]. SVM is also used for classifying learners' emotions in the e-learning environment [23]. In the year of 2016, Nayden et al. [24] used artificial intelligence technology shortly known as AI for a personal learning management system. It has been observed that the researchers proposed one architecture of machine translation which will translate Bengali to English and English to Bengali [25] to support multilingual conversation using chatbots.

3.1 Natural Language Processing (NLP)

NLP or natural language processing is a branch of AI which cares about giving computers the power to know, interpret, and manipulate human language. NLP employs a wide range of techniques to explain and handle human language, ranging from statistical and machine learning methods to rule-based and algorithmic approaches.

Natural language processing is not a recent science. It has been continuously evolving and advancing. The increased interest in a human-to-machine conversation, the availability of big data, and enhanced algorithms have contributed more to the advancement of this technology. NLP is used by many researchers in their work [26–28].

The most commonly researched tasks in NLP are as follows:

Text and speech processing, morphological analysis, syntactic analysis, lexical semantics, and relational semantics.

Advantages of NLP:

Various advantages of NLP are, it assists users to ask different types of questions and get relevant responses. Implementation of NLP needs couple of steps like parsing, tokenization, stemming, language detection, part of speech tagging, etc. It does not provide any unwanted and unnecessary information; rather, it helps computers to communicate to humans in any language because it is very time-efficient as well as it improves the accuracy of documentation.

3.2 Recurrent Neural Network (RNN)

Recurrent neural networks (RNNs) are a type of neural network. It is powerful and robust. It belongs to the most favorable algorithms in use because it is the only one with internal memory. Since it has its internal memory, RNNs can remember the things they received. This enables them to make extremely accurate predictions about what will happen next.

The most relevant question to come up is what sequential data is? To put it another way, sequential data is simply ordered data in which related items appear one after the other.

Types of RNNs:

- One to one

- One to many
- Many to one
- Many to many.

Advantages:

It can operate on the input of any length. RNN has its internal memory so that it can recall each detail throughout the time, which is very obliging in any time series predictor. Another important advantage of RNN is that as the input size is large, the model size does not increase.

3.3 Deep Learning

Deep learning is a crucial part of data science. It is a sort of machine learning and AI that copies the way humans acquire specific sorts of knowledge. It includes statistics and predictive modeling. Deep learning, to put it another way, can be viewed as a means to automate predictive analytics. If the comparison has been done between the traditional machine learning and deep learning algorithms, then traditional machine learning algorithms are in a way linear, and deep learning algorithms are piled during a way of accelerating complexity and abstraction.

The learning process is superintended in traditional machine learning. When informing the computer what kinds of items it should look for, the programmer must be very specific. This strenuous process is called feature extraction, and the computer's success rate completely depends upon the programmer's ability to accurately define the features. The advantage of deep learning is that the program builds the characteristic set by itself without supervision. Unsupervised is not only faster, but it is also usually more accurate. Initially, the computer program might be supplied with training data. The program uses the knowledge it receives from the training data to make a characteristic set for the object given. It will look for patterns of pixels in the digital data. In each iteration, the estimated model becomes more compound and more accurate.

It has been observed that an ubiquitous learning environment was also used to explore the effects of web-mediated computational thinking [29].

Using machine learning (regression analysis) and emotional state mapping theory, an affective system has been developed which can judge the emotional state of the learner and bring that person back to a more balanced state of mind. The survey results show that the system is found to be workable in 75% of the cases [30].

4 Conclusions

Chatbot technology is a continuously evolving technology. It has been seen from the present study that how it is used in different works of life including education.

Chatbots have become our household necessity with the advent of Amazon's Alexa and Apple's Siri. Chatbots are being used in business and administration, in the medical field, in sports, and in every aspect of our life.

The present study also found out some challenges in the present chatbot systems.

Currently, the main challenge that the existing chatbots have is to understand the human inputs and responses and maintaining the context. Most of the existing bots work just on the pattern matching of inputs and then try to find a scripted response which matches with the input. This method does not lead to a fully satisfying talk or a conversation with a clear goal. So, work needs to be done toward overall improvement in more satisfying conversation.

Existing e-learning systems are not commonly found to have voice-based chatbots incorporated with them. Future visions of e-learning systems can be made much more effective with voice-based chatbot assistant to make more satisfying interactions with learners.

It has been observed that various methodologies are used by many researchers. It is seen that if hybrid algorithms like retrieval-based model and a QANet models are used, then it yield better results.

References

1. Mendoza S et al (2020) Supporting student-teacher interaction through a chatbot. In: International conference on human-computer interaction. Springer, Cham
2. Bahja M, Hammad R, Hassouna M (2019) Talk2Learn: a framework for chatbot learning. In: European conference on technology enhanced learning. Springer, Cham
3. Abdul-Kader SA, Woods JC (2015) Survey on chatbot design techniques in speech conversation systems. Int J Adv Comput Sci Appl 6(7)
4. Setiaji B, Wibowo FW (2016) Chatbot using knowledge in database: human-to-machine conversation modeling. In: 2016 7th international conference on intelligent systems, modeling and simulation (ISMS). IEEE
5. Clarizia F et al (2018) Chatbot: an education support system for student. In: International symposium on cyber space safety and security. Springer, Cham
6. Kita T et al (2018) A discussion on effective implementation and prototyping of voice user interfaces for learning activities on moodle. In: CSEDU, no 1
7. Souali K, Rahmaoui O, Ouzzif M (2018) Introducing traceability based recommendation approach using chatbot for e-learning platforms. In: International conference on advanced intelligent systems for sustainable development. Springer, Cham
8. Gaglo K et al (2021) Proposal of conversational chatbots for educational remediation in thecontextofcovid-19. 2012. In: 3rd International conference on advanced communication technology (ICACT). IEEE
9. Pham XL et al (2018) Chatbotas an intelligent personal assistant for mobile language learning. In: Proceedings of the 2018 2nd international conference on education and E-learning 2018
10. Davies JN et al (2020) Personalization of e-learning process using AI-powered chatbot integration. In: International scientific-practical conference. Springer, Cham
11. Agrebi M, Sendi M, Abed M (2019) Deep reinforcement learning for personalized recommendation of distance learning. In: World conference on information systems and technologies. Springer, Cham

12. Shukla VK, Verma A (2019) Enhancing LMS experience through AIML base and retrieval base chatbot using R language. In: 2019 International conference on automation, computational and technology management (ICACTM). IEEE
13. Rooein D (2019) Data-driven EDU chatbots. In: Companion proceedings of the 2019 world wide web conference
14. Souali K et al (2019) Recommending moodle resources using chatbots. In: 201915th International conference on signal-image technology &internet-based systems (SITIS). IEEE
15. Rajkumar R, Ganapathy V (2020) Bio-inspiring learning style chatbot inventory using brain computing interface to increase the efficiency of e- learning. IEEE Access 8:67377–67395
16. Wu EH-K et al (2020) Advantages and constraints of a hybrid model K-12 e-learning assistant chatbot. IEEE Access 8:77788–77801
17. Modak M et al (2020) Machine learning based learning disability detection using LMS. In: 2020 IEEE 5th international conference on computing communication and automation (ICCCA). IEEE
18. Ashok M et al (2021) A systematic survey of cognitive chatbots in personalized learning framework. In: 2021 Sixth international conference on wireless communications, signal processing and networking (WiSPNET). IEEE
19. Mandal L, Bhattacharya S, Basu PN. An approach to incorporate teachers' individual teaching and evaluation strategies and experience in tutoring model of an Intelligent Tutoring System. Int J Hybrid
20. Jommanop T, Mekruksavanich S (2019) E-learning recommendation model based on multiple intelligence. In: 201914th International joint symposium on artificial intelligence and natural language processing (iSAI-NLP). IEEE
21. Bal S et al (2021) An intelligent chatbot for admission system of an educational institute and prediction of user interest in taking admission. In: Proceedings on 2nd global conference on artificial intelligence and applications. CRC Press, Taylor & Francis (InPress)
22. Baletal S. Sentiment analysis of online reviews for educational institutions
23. Mandal L et al (2017) Intellimote: a hybrid classifier for classifying learners' emotion in a distributed e-learning environment. Turkish J Electr Eng Comput Sci 25(3):2084–2095
24. Nenkov N et al (2016) Artificial intelligence technologies for personnel learning management systems. In: 2016 IEEE 8th international conference on intelligent systems (IS). IEEE
25. Bal S, Mahanta S, Mandal L (2021) Bilingual machine translation: Bengali to English. In: Proceedings of international conference on computational intelligence, data science and cloud computing: IEM-ICDC2020, vol 62. Springer
26. Clarizia F et al (2021) E-learning and industry 4.0: A chatbot for training employees. In: Proceedings of fifth international congress on information and communication technology, Springer, Singapore
27. Mandal L, Dutta S (2021) Chatbot for query processing in E-learning system. In: Proceedings of international conference on computational intelligence, data science and cloud computing: IEM-ICDC, vol 62. Springer
28. Dutta D (2017) Developing an intelligent chat-bot tool to assist high school student's for learning general knowledge subjects. Georgia Institute of Technology
29. Tsai C-W et al (2017) Exploring the effects of web-mediated computational thinking on developing students' computing skills in a ubiquitous learning environment. Interactive Learn Environ 25(6):762–777
30. Mandal L et al. An affective and adaptive E-learning system: a machine learning based approach.

Part III
Cyber Security and Blockchain

Chapter 26
Secured Network Using Penetration Methodologies

Aswathy Mohan, G. Aravind Swaminathan, and N. Jeenath Shafana

1 Introduction

Nowadays, cybercriminals are targeting systems with several new complex vulnerabilities. The purpose of this study is to conduct vulnerability testing using penetration testing tools and analyze the vulnerabilities found on the website [3]. Vulnerability refers to the security weakness of the target system which allows the hacker to update or make changes to the applications and obtain root user privileges through privilege escalation methodology. These vulnerabilities are used by hackers to attack the target system. Penetration testing refers to ethical hacking, which involves the pen testers attacking the target system like hackers but with the legal consent of the organization. This is done to identify the set of vulnerabilities present in the system. After successful pen testing, the pen tester will generate a report which informs the owner regarding the types of vulnerabilities present. So, by utilizing this report, they could overcome the attacks from hackers.

Here in this research paper, a demonstration is done on the topic of how a target system can be accessed by a pen tester using automated tools and the latest methodologies. The pen testers use a virtual pen testing laboratory to perform this operation. This is done in VMware or a virtual box environment. Penetration testing is the massive field in security systems. It deals with most of the common things that usually

A. Mohan
Computer Science and Engineering, Anna University, Tirunelveli, Tamil Nadu, India

G. Aravind Swaminathan (✉)
Computer Science and Engineering, Francis Xavier Engineering College, Tirunelveli, Tamil Nadu, India
e-mail: aravindswaminathan.g@francisxavier.ac.in

N. Jeenath Shafana (✉)
Computer Science and Engineering, SRM IST, Ramapuram, Chennai, India
e-mail: jeenathn1@srmist.edu.in

L. Mandal et al. (eds.), *Proceedings of International Conference on Computational Intelligence, Data Science and Cloud Computing*, Algorithms for Intelligent Systems,
https://doi.org/10.1007/978-981-19-1657-1_27

a developer forgets to cover during the development process [4]. Here, I have used a Linux system as a target machine and a host system as Kali Linux. Because, Kali Linux consists of all the automated tools available for ethical hacking techniques. So, no need to install any of the tools needed. For learning purposes and for practicing by security researchers, various learning platforms for ethical hacking are available. These platforms like VulnHub and TryHackMe contain various target systems with complex levels of vulnerabilities. From this space, we could download the specific target system and import it into our VMware environment. So, for performing the pen test, we must download and import the target machine and host machine, which is Kali Linux into VMware or virtual box.

Through this VAPT analysis process, various methodologies, automated tools, and different phases in ethical hacking are analyzed. After successful completion of penetration testing and vulnerability assessment, the pen tester generates a report which contains the list of vulnerabilities present in the system. So, by implementing the solutions given by the pen tester, the organization could eliminate the threats from cybercriminals or hackers.

2 VAPT Technique and Phases

VAPT technique refers to the combination of vulnerability assessment and penetration testing. Vulnerability assessment involves the identification of vulnerabilities in the target machine. Penetration testing involves ethical hacking in which the pen tester will be attacking the target system like a hacker but with the legal authority from the organization. Finally, the pen tester will provide a generated report which contains the solutions or guidelines to protect the target system from the attack of hackers. In penetration testing process, the process can be divided into seven phases, i.e., information gathering, automated scanning, discovered vulnerability, attack generation, response analysis, report generation, and pen tester database [5]. The phases include:

1. Planning
2. Reconnaissance/information gathering
3. Vulnerability assessment
4. Pen testing
5. Login in the target system
6. Privilege as a normal user
7. Privilege as a root user
8. Report.

Here, initially, the pen tester decides the target system and the type of testing. The next phase is information gathering. Here, the information regarding the target system is collected. Vulnerability assessment (VA) or vulnerability analysis (VA) or vulnerabilities is the process of scanning the system or software or a network to find out the flaws and weakness in that [13]. The next stage is vulnerability assessment, which identifies the set of vulnerabilities available in the system through various methodologies and automated tools available. By exploiting the identified vulnerabilities, the pen tester will access the target system as a normal user. Then, by using privilege escalation methodology, he could enter as a root user who has admin privileges to access the confidential data. Finally, before exiting, the system is converted to the settings as before pen testing and all changes are cleared. Finally, all these processes are reported. Using this report, the organization's owner could eliminate the security flaws to overcome the attack from hackers properly.

3 Proposed System

In this paper, the proposed system contains the target system and host machine downloaded and imported in VMware or virtual box environment. Here, I have selected the target system from the VulnHub learning platform for owasp students. Vulnerability arises when an application directly indicates any object without any encoding or authentication [8]. The target system is a Linux machine, and its name is a djinn. The host machine is always Kali Linux since it contains almost all automated tools available.

Network security is one of the major concerns of world [6]. We all know that the systems on the Internet are increasing day by day and so the vulnerabilities. These vulnerabilities must be found before the attacker. [6] This can be done with the help of penetration testing. Penetration testing is used to check or evaluate the security posture of an organization or network [6]. Web applications vulnerabilities allow attackers to perform malicious actions that range from gaining unauthorized account access to obtaining sensitive data. The number of reported web application vulnerabilities [7] (Fig. 1).

Algorithm of Proposed System.

Step 1: Start.

Step 2: Finding IP of Target System.

Step 3: Scanning the target system IP using scanning tools like Nmap to know the information regarding open ports and services.

Step 4: Access open port and service.

Step 5: Directory Bruteforcing to get information regarding directories or files in the target system.

Step 6: Accessing Suspicious directories through the web browser.

Step 7: If it contains suspicious links for passwords then we will be using password cracking tools and encryption-decryption techniques to find the password.

Step 8: SQL Injections to implement malicious SQL commands to run a shell in the target system to log in.

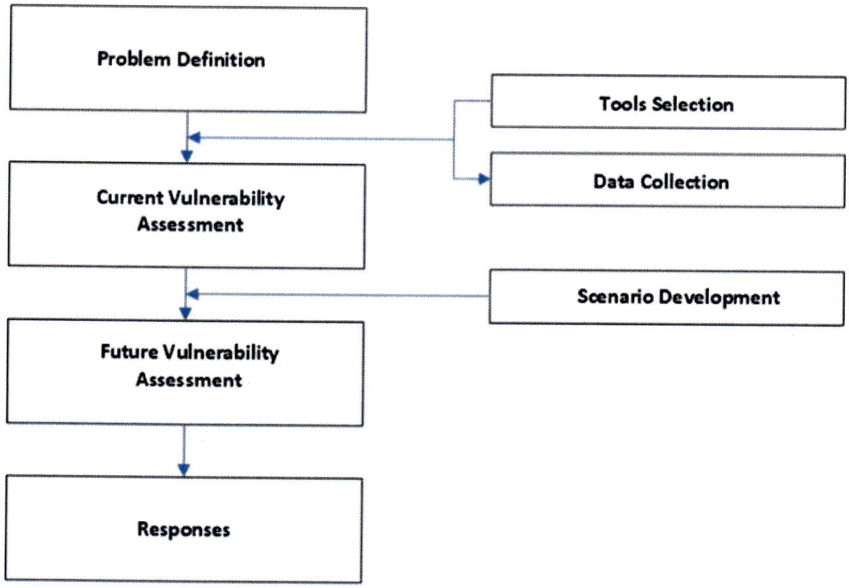

Fig. 1 Workflow of the proposed system

Step 9: Burp suite tool for setting up proxy IP so that the pentesters original IP won't be visible and can attack the target system through the uploaded payload using wordlist available.

Step 10: Uploading PHP Reverse shell in the target system to get a shell for login.

Step 11: Use of linpeas to transfer files from host to target when network is not available.

Step 12: Use malicious JavaScript to attack the system.

Step 13: Use of Metasploitable framework for target system login.

Step 14: SUID ID for acting as root user in target system so that the pen tester will be having full privileges.

Step 15: Initially login as a normal user.

Step 16: Using modular hacking techniques the user becomes the root user to get full privileges using Privilege Escalation methods.

Step 17: Access files from the home directory of the root user.

Step 18: Clearing and Exiting.

Step 19: Stop.

First, let us start with the scanning and enumeration to find the IP of our target machine (djinn), and then we will try to gain as much knowledge as we can about our box, respectively.

Step 1: To find the IP address of the target machine.

We can use different commands to achieve this like arp-scan, Nmap, or netdiscover. Here, we use arp-scan in this case,

$ sudo arp-scan -l.

Now, we have got the IP of the target machine, i.e., 192.168.0.115, we can start to enumerate the target to get.

Step 2: Let us scan our target to see which ports are opened and what services they are running.

$ nmap -sV 192.168.0.115 -o firstscan.

Discovered that ports 21 (FTP), 1337 (waste), 3771 (HTTP) are open on the target machine. We can also see the version of the services running. Also, FTP has anonymous login enabled by default, and it shows three file listings.

Step 3: Let us connect to the FTP port and transfer those files to our attacker machine (Fig. 2).

As we can see, there is some information, we found out that a "game" is running on port 1337, and there are some creds and a username nitish81299 in the other two files.

Step 4: Let us check port 1337.

When we tried to visit the page on the browser it says, "connection was reset". Then, I tried to connect to it using Netcat.

$ nc 192.168.0.115 1337.

And it worked. Found that a game where we must answer mathematical problems, and it says if answer them correctly 1000 times, we will get a reward. That sounds time-consuming. Let us move on to the next level.

Step 5: Let us check HTTP port 7331.

It simply loads a webpage with no functionality to exploit. Also, the source code had nothing interesting to look at.

Step 6: Let us do directory and files enumeration using gobuster on port 3771.

$ gobuster dir-u http://192.168.0.115:7331-wusr/share/dirb/wordlists/comman.txt.

That did not have any results. After that, I decided to try a different wordlist.

$gobuster dir-u http://192.168.0.115:7331-w/usr/share/dirb/wordlists/big.txt (Fig. 3).

And it worked. We can see two listings as /genie and /wish. After visiting /wish, we can see that it has an input box, and we can input commands to run (Fig. 4).

First, the id command was tried, and it worked properly. Then, I searched for Netcat binary with which command, and it was present. The next step is trying to get a reverse shell with nc, but that did not work. I tried different reverse shells, but nothing seems to work. Whenever tried it said, "wrong choice of words". Maybe it's doing filtering for certain words or symbols. I decided to go back to our game and make a Python script for it. I looked up online and found a few articles with the same "waste" service, and they were using pwn library to make the script, so I decided to

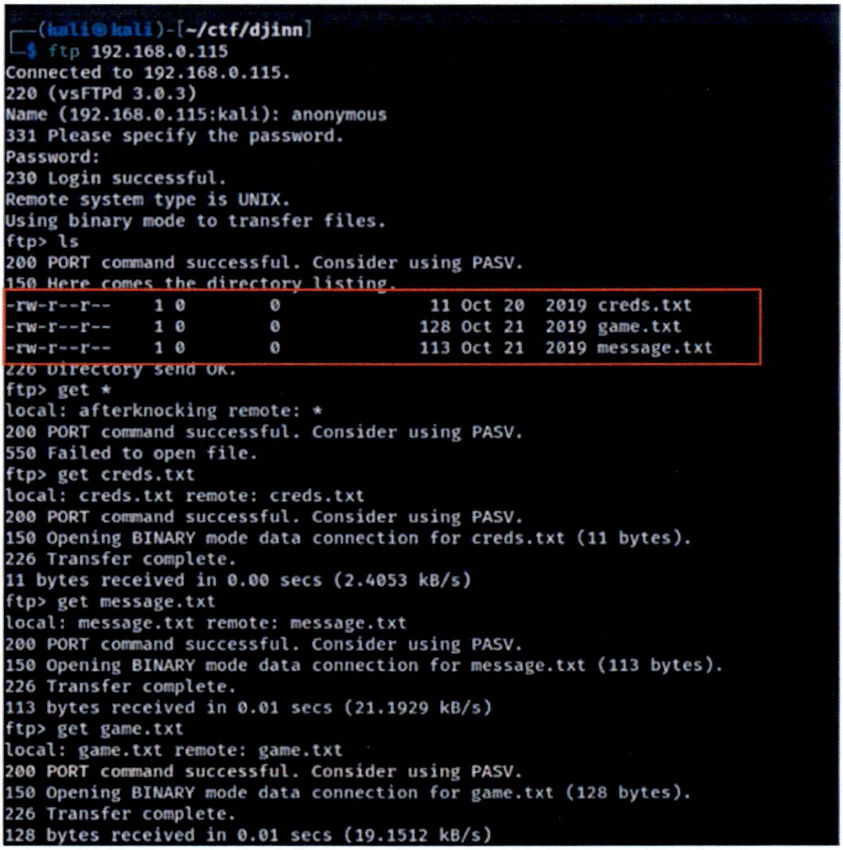

Fig. 2 File transfer to target machine

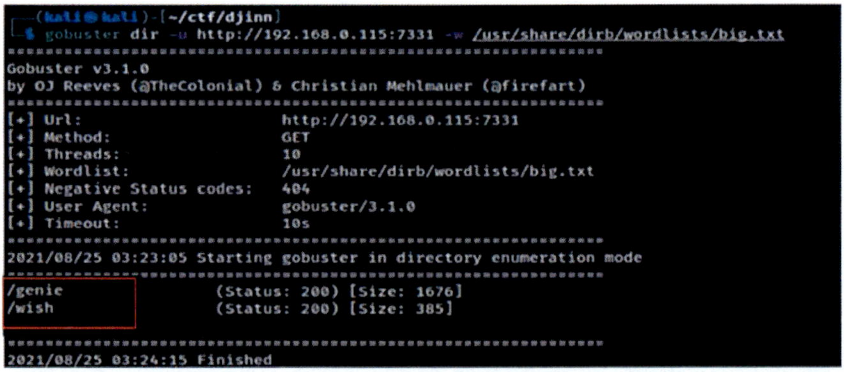

Fig. 3 Wordlist comparison

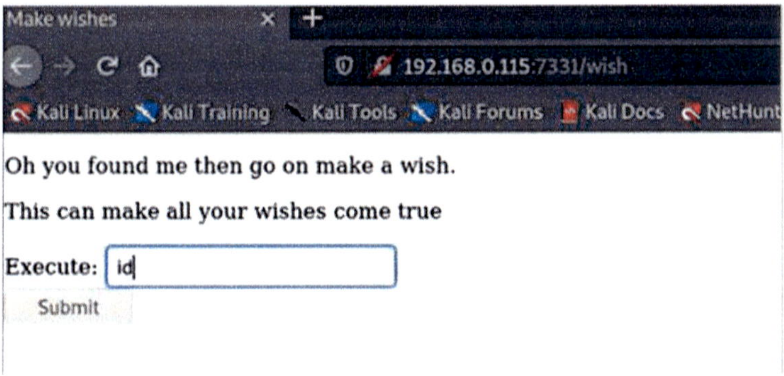

Fig. 4 Input box

use that too for my script. After many tries, I finally made it work and it gave the results with some hint as port numbers.

After that, searched for port knocking on google and found this article about port knocking on Linux and followed it. Manual penetration testing needs lot of expertise in playing with HTTP requests and response. An expert penetration tester would know the possible attacks that can be performed on a particular end point by fuzzing the HTTP requests [8].

$ knock 192.168.0.115 1356 6784 3409.

It opened port 22(SSH) on the target machine. Then tried to log in with the creds we got in the creds.txt file through FTP port but that did not work. All this for nothing it seems. Then decided to go back to /wish and try more commands. Again, searched online for bypassing symbol filtering to get a reverse shell and found an article.

Step 7: Let us try the method specified in the article. It is using base64 to encode the command, so it does not have symbols in it, and then while executing, we are just decoding the base64 string and passing it to bash command to run it (Fig. 5).

$ echo "YmFzaCAtaSA + JiAvZGV2L3RjcC8xOTIuMTY4LjAuMTIxLzY4OD YgMD4mMQo = "| base64 - d|bash (Figs. 6 and 7).

And we got a shell.

Step 8: Let us check the home directory to find any other users. We found Nitish and Sam. Inside Nitish, there was a hidden directory called/. dev with some creds.txt files with possible credentials for Nitish users. Vulnerability assessment and penetration testing are a special approach to eliminate various security threats from the web

Fig. 5 Encrypting command using base64

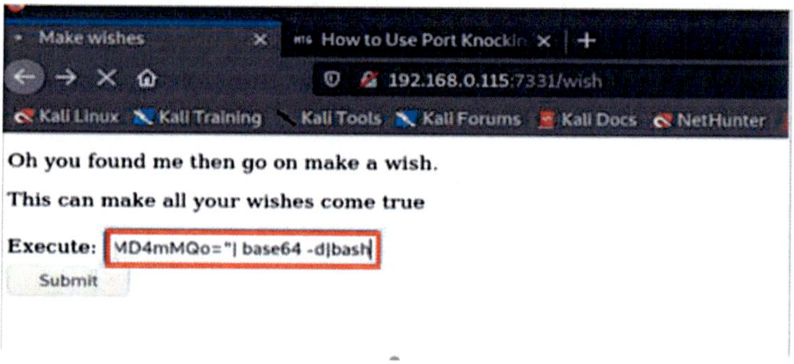

Fig. 6 Command execution

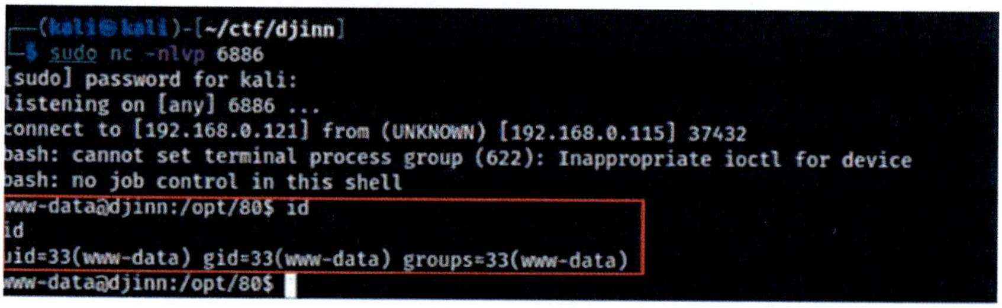

Fig. 7 Reverse shell

application. By focusing on high-risk vulnerabilities such as SQL injection, cross site scripting, local file inclusion and remote file inclusion [9].

Step 9: Let us try to login into Nitish with the credentials.

$ su nitish (Fig. 8).

Step 10: Let us try to run sudo-l to see what commands we can run on behalf of other users.

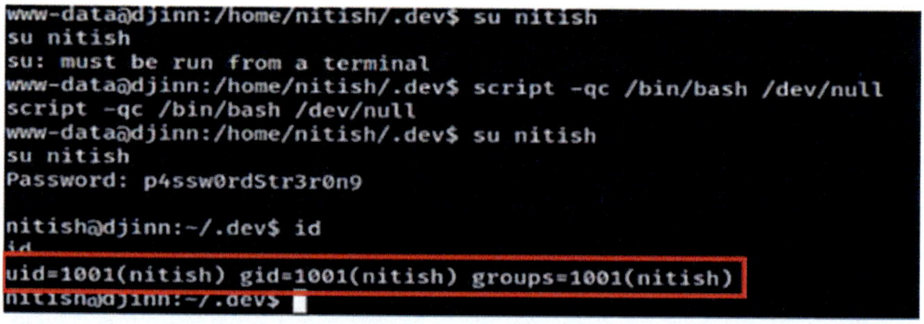

Fig. 8 Login to user Nitish with credentials

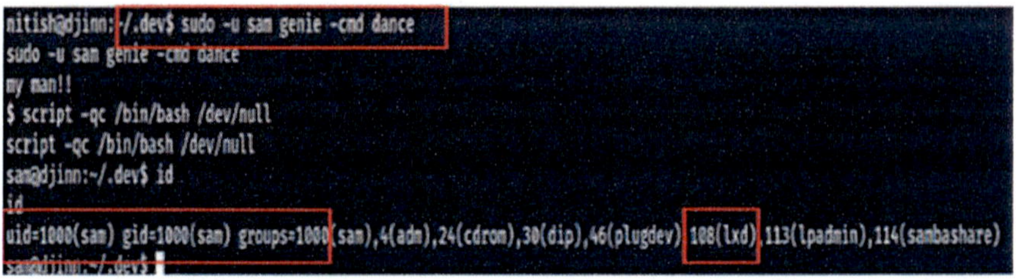

Fig. 9 Shell as Sam

We can run some binary called genie on behalf of Sam. After fiddling around a little bit with it, found an option called -cmd in the man page of genie.

$ sudo -u sam genie -cmd dance (Fig. 9).

And we got a shell as Sam.

We can also see from here that Sam is a member of the lxd group.

Step 11: Let us try privilege escalation using the lxd group. First, we need to build an Alpine image on our attacker system. After that, we will transfer the image to the attacker machine in the/tmp folder. Then, we will run lxd init to set up the basic configuration on the victim machine. Then, we will build a container with that image using the lxc command (Fig. 10).

Finally, I got root shell.

Step 12: Let us get the root flag. It is in /mnt/root/root directory.

Here, through VAPT methodologies and tools, I have accessed the target machine as the root user to access the confidential data. Next stage refers to reporting the whole process. By using the report guidelines, the organization owners could eliminate the security flaws and improve the security of the system. Penetration testing is different from security functional testing. The latter demonstrates the correct behavior of the system's security controls, while penetration testing determines the difficulty for someone to penetrate an organization's security controls against unauthorized access to its information and information systems [10].

4 Future Work

Modern web applications resulted in complex vulnerabilities, which led to severe breaches that resulted in the stealing of data. VAPT technique resulted in the discovery of the latest complex vulnerabilities which are exploited by cybercriminals or hackers. This paper discussed the methodologies and tools through which vulnerabilities are identified and eliminated. In the future, I will be analyzing virtual machines with the most modern vulnerabilities using the latest methodologies and all automated tools. Vulnerabilities are big flaws in system security and information assurance. The main

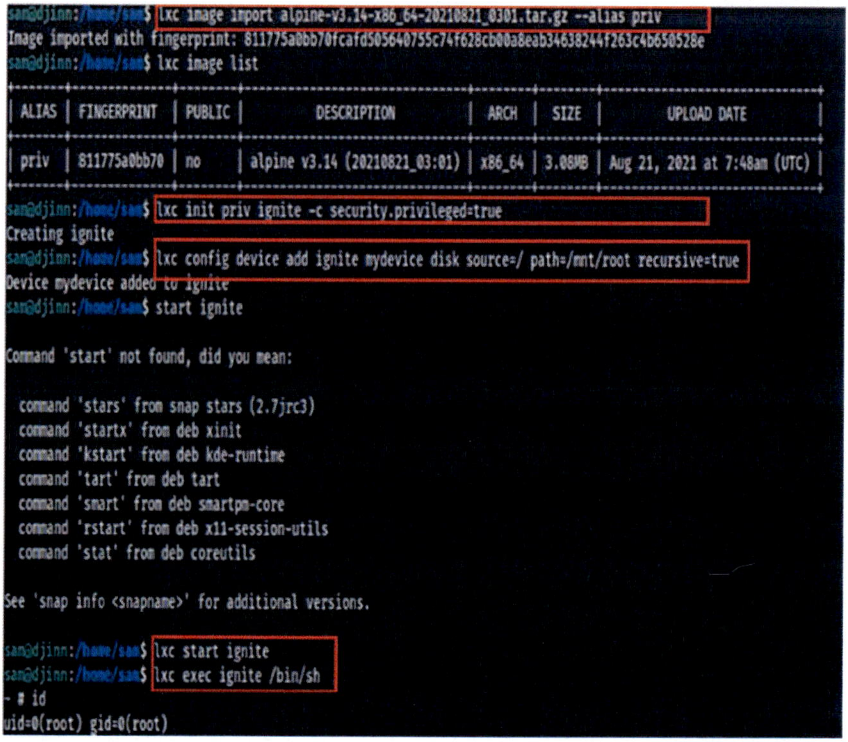

Fig. 10 Root shell

intent of this paper is to recognize openness and flaws in networks and web applications using penetration testing to protect the institutions from cyberthreats. There are many scanning methods suggested by many authors to identify the weakness [11].

5 Conclusion

This paper deals with the phases in vulnerability assessment and penetration testing using Kali Linux and the automated tools available. After successful testing, the pen tester generates a report which is utilized by owners of the organization to eliminate the security flaws in the network. In this proposed work, we can see how the vulnerabilities are identified and exploited by the pen tester to access the target machine with admin privilege. So, this technique is used in cyberdefense to protect organizations from cybercrimes. Security on networks is of the utmost priority, and methodologies like penetration testing help us achieve frequent quality checks in place. With an ever-growing population and their heavy dependence on technologies, there is an ever-increasing security and data threat which immediately calls for an unprecedented and highly secure set of methods, which are practical in nature, have minimal footprint, and can be employed across platforms and throughout the interdisciplinary industries for one common goal to "achieving security on networks and mitigating and identifying threats" [12]. Vulnerabilities assessment is defined as automatically identify weakness of the system via a software rather manually, whereas penetration testing mainly refers to a form of stress testing which exposes weakness in a vulnerable networks and sets standards to cover these vulnerabilities in the network [13].

References

1. Jaiswal A, Raj G, Singh D (2014) Security testing of web applications: issues and challenges. Int J Comput Appl 88(3): 0975–8887
2. Goel JN, Mehtre BM (2015) Vulnerability assessment & penetration testing as a cyber defence technology. Elsevier Procedia Comput Sci 57(2015):710–715
3. Mahtuf FR, Hatta P, Wihidiyat ES, Laboratorium P (2019) Virtual untuk Simulasi Uji Penetrasi Sistem Keamanan Jaringan. J Inf Technol Comput Sci 4(1)
4. Mundalik SS. Penetration testing: an art of securing the system (Using Kali Linux). Int J Adv Res Comput Sci Softw Eng Rese Pap
5. Upadhyaya J, Panda N, Acharya AA. Attack generation and vulnerability discovery in penetration testing using SQL injection. Int J Comput Sci Eng 2(3). E-ISSN 2347-2693
6. Singh H, Dr. Jangra S, Dr. Verma PK (2016) Penetration testing: analyzing the security of the network by hacker's mind. IJLTEMAS V(V). ISSN 2278-2540
7. Đurić Z (2014) WAPTT—Web application penetration testing tool 1582-7445 2014 AECE. Adv Electr Comput Eng 14(1)
8. Nagendran K, Adithyan A, Chethana R, Camillus P, Bala Sri Varshini K (2019) Web application penetration testing. Int J Innov Technol Explor Eng 8(10). ISSN 2278-3075
9. Hasana A, Dr. Mevab D. Web application safety by penetration testing. Int J Adv Stud Sci Res (IJASSR). ISSN 2460 4010, Indexed in Elsevier-SSRN
10. Bacudio AG, Yuan1 X, Bill Chu2 B-T, Jones1 M (2011) An overview of penetration testing, Int J Netw Secur Appl (IJNSA) 3(6)
11. Sri Devi R, Mohan Kumar M. Testing for security weakness of web applications using ethical hacking. IEEE Xplore. ISBN 978-1-7281-5518-0
12. Pandey R, Jyothindar V, Chopra UK. Vulnerability assessment and penetration testing: a portable solution implementation
13. Yaqoob I, Hussain SA, Mamoon S, Naseer N, Akram J, ur Rehman A (2017) Penetration testing and vulnerability assessment. J Netw Commun Emerging Technol 7(8). ISSN: 2395-5317

Chapter 27
An Artificial Intelligence Assisted Defensive Framework for Securing Cyberspace

Keshav Kaushik and Susheela Dahiya

1 Rise of Cyber Threats in Industry 4.0

The first generation of industries was based on steam, responsible for mechanization, waterpower system, and steam system. Then, with the advent of electricity, the second generation of industry came into the picture, accountable for assembly lines and mass production of various items. The third generation of industries came up with computers and automation, responsible for the connected world. Today, we are changing according to Industry 4.0. PCs and motorization will get together through and through new ways, with mechanical independence related remotely to PC systems equipped with AI counts that can learn and control put forth concentrated effort adequacy with near no responsibility from human managers. Industry 4.0 [1] introduced the "splendid creation line," in which advanced physical systems screen the material strategies of the mechanical office and choose decentralized decisions. The physical structures become the Internet of Things, passing on and planning both with each other and with individuals continuously by methods for the remote Web. Now adopting Industry 4.0 [2] will also arise in cybersecurity threats, which can be prevented with the help of artificial intelligence. The popular cyber-attacks, which would have been prevented if a proper intelligent defensive system was placed, are as follows.

K. Kaushik · S. Dahiya (✉)
School of Computer Science, University of Petroleum and Energy Studies, Dehradun, India
e-mail: susheela.iitr@gmail.com

L. Mandal et al. (eds.), *Proceedings of International Conference on Computational Intelligence, Data Science and Cloud Computing*, Algorithms for Intelligent Systems,
https://doi.org/10.1007/978-981-19-1657-1_28

1.1 JP Morgan Chase

In 2014, the assault on the American banking giant prompted the information of 83 million air conditioning tally holders and 7,600,000 independent companies were undermined—influencing 66% of the family units in the US. The source of the breach was traced to only one server in the bank's system, in which the security team had failed to activate two-factor verification, a fundamental security feature that limits framework access. A significant update that programmers need a lucky tiny opening to dispense genuine harm [3].

1.2 TalkTalk

In 2015, TalkTalk got interchangeable with digital security disappointment, when assaulters had the option to get to the individual information of 156,959 UK clients, including the financial balance subtleties and sort codes of around 15,000. The hack costs the business £60 m and prompted the loss of approximately 100,000 clients. At the same time, an inside and out investigation by the Information Commissioner's Office has quite recently fined the organization a record £400,000 for security failings. The Information Commissioner said, "TalkTalk ought to and could have accomplished more to protect its client data. It didn't, and we have made a move." [4]

1.3 Sony Pictures

In November 2014, a gathering calling themselves 'Guardian of Peace' accessed Sony's frameworks and continued to release classified data, including messages, representative subtleties, and even duplicates of unreleased movies. Unimaginably, the hack ended up being supported by North Korea in a fight at Sony Pictures' forthcoming film, The Interview, a parody about a plot to kill its pioneer Kim Jong-un. What's more, while Sony was unmistakably the casualty of a coercion plot, numerous specialists accept the harm could have been limited if the messages and data taken had been scrambled. Sony's security activity additionally neglected to recognize the hack until it got open, giving the assailants a lot of time to the exact extreme [5].

2 Solving Cybersecurity Problems with AI

The disturbing truth about continuously standard advanced attacks is that most organizations and the computerized security industry are not prepared to secure the data. Past the nonappearance of preparation on the business level, the automated security workforce itself is moreover having an incomprehensibly difficult stretch remaining mindful of intrigue. By 2021 [5], there are evaluated to be a dumbfounding 3.5 million unfilled advanced security positions the world over. That suggests advanced security specialists should compensate for those unfilled spaces, either by working or by working longer hours. A depleted progressive security breach infers it will in general, be hard to respond to risks reasonably and effectively. The rapidly expanding threat levels make it irksome (if unquestionably possible) [6] to manage all pieces of security without using an especially immense gathering of specialists. However, the risk loosens up to associations taking everything into account, and data volume and speed are developing as digitization proceeds. In [7], the unexplainable and incomprehensibility are defined as complementary results. The Chen addresses the challenge of high-level tactics by proposing a cyber defense framework that integrates strategic, operational, and tactical levels. Several counterattacks related to AI are analyzed in [8], and various AI models face cyber threats that may disturb the decisions, learning, and sampling.

Even though there has been a lot of research into fingerprint recognition technologies, most of it has concentrated on using advanced equipment. Since such instruments are too expensive, some researchers have turned to digital cameras as an external option for fingerprint photographs. Such outlets, on the other hand, present new problems in terms of picture accuracy. Most optical cameras compress shot photographs before saving them, potentially resulting in material loss. In [9], the authors address the need to decide the best fingerprint image compression ratio to ensure high precision in fingerprint recognition systems. This research is based on a huge dataset of raw images produced in-house. As a result, all fingerprint information is saved in order to calculate the compression ratio correctly.

The paper focuses on a solution using artificial intelligence, which will help the cyberspace implement security in a much dynamic and advance way. In the wake of thinking about the present status of digital protection, the execution of AI frameworks with digital security can fill in as a genuine defining moment on the Internet. These frameworks accompany various generous advantages that will help digital security expert professionals to get ready for taking on digital assaults and shielding the endeavor. New AI calculations [10] use machine learning (ML) to adjust after some time. New ages of malware and digital assaults can be hard to recognize with ordinary digital security conventions. They advance after some time, so progressively powerful moves toward it are essential. Digital security arrangements that depend on ML use information from earlier digital assaults to react to newer assault, however, comparable hazard. Every attack has a signature; signatures can be attacked exploit, payloads, or may be changed in the current behavior of the victim system if we notice deeply some vulnerabilities like SQL Injection (SQLi), memory corruption,

XML External Entity (XXE), or even the remote access malware and other attacks that are exploited by acting on the system explicitly by the evil hackers. If we talk about SQLi, an attacker to get information about the methods stored in the database can change the queries used by default. Moreover, the SQLi vulnerability can be exploited to escalate privilege, which can be even more dangerous.

Memory level corruption, stacks are the first and foremost memory space when one talk about memory corruption, stacks are used to store local variables and function arguments that the program itself can use later. For an attack to be successful, attacker need to inject an exploit to a place where the system will process it in a way as desired by the attacker also known as injection points. Injection points are the places/fields in which attacker can inject/deliver its exploit; it can be a form field for a Web application, and it can be a user field in the File Transfer Protocol (FTP) server running on the victim machine or various other parameters accepting inputs directly from the user.

3 Buffer Overflow and Placing AI in Solution

In the case of buffer overflow (BOF), we try to overwrite critical memory sections, which can then overwrite the registers like Extended Stack Pointer, Extended Base Pointer, and Extended Instruction Pointer which are affected by buffer overflow. The attacker's fuzzing string can fuzz any injection point, and if it crashed the application, there is a very high chance for the application to be vulnerable to buffer overflow. The overflow will typically look like the image provided below as Fig. 1. This vulnerability can be exploited to gain remote access to the victim's machine or perform denial of service making that service go down.

The attack can be prevented with certain mechanisms like compiler protection, ASLR, REBASE, DEP, SafeSEH, and few others. This mechanism of buffer overflow is explained below:

Fig. 1 Memory visualization

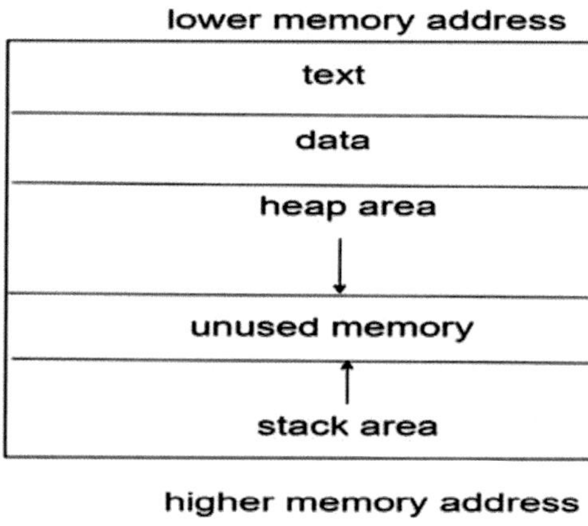

Compiler Protection When we compile the program, compilers can warn us for the usage of vulnerable functions used while coding.

ASLR It stands for Address Space Layout Randomization, which randomizes the address layout of memory areas with a particular sled.

Rebase Changes address every time it is rebooting.

DEP Data Execution Protection is the area of memory that cannot be used to execute instructions.

SafeSEH Structure Exception Handler (SEH) can be used to perform buffer overflow in the program itself at the time of execution nSEH and SEH gets predefined and thus prevent SEH overwrite which can be exploited by POP, POP, RET instruction set.

All those protections explained above can be bypassed. Any protection mechanism for BOF is bypassable not because of its limitation but also due to human errors. This can be dangerous when done in a corporate environment and need AI to get us a solution. As explained above, every attack has a signature linked to it. We have discussed two vulnerabilities that can cause any organization a huge loss: SQLi and buffer overflow. About the SQLi and its signature, when we perform a penetration testing to any Web application vulnerable to SQLi, we try to get back more data than desired and required. However, due to lack of intelligence, SQL database does what its application layer implementation instructs it to do. AI provides solution to such types of problem by training the database to react cognitively on that situation and sending no, limited, or full data even if the query is made by privileged account. It is based on various signatures like time at which the query is made, what is the IP, and is it linked without network or is known in some way of the many other criteria. By going through that signature, we can determine if any malicious activity is going on. However, most people think that this will violate the basic working of a database, and it may be true but for good. The data can still be available to the admins who require it, but they need to handle intelligence that guards that data.

Such solutions are not permanent, but we can go for a much intelligent solution by implementing AI that can judge whether the execution of a payload is essential or not, by sensing previously collected analytics of the system environment. If necessary, that execution can be performed with least privilege. The least privileged execution depends on the systems' analytics of its execution environment along with the signature of that execution. For example, is it directly forced to be executed by CPU, backtracking the software, and understanding reasoning like "why will an FTP server execute a reverse shell connection?" and many more signatures. Using that we can get to know which processes should run and which should not. Artificial intelligence relies upon savvy robotization to trigger hazard warnings that people might not have the opportunity and assets to scan for. Simulated intelligence frameworks that legitimately handle dangers all alone do as such as per a normalized method or playbook. As opposed to the inconstancy (error) that accompanies a human touch, AI frameworks do not commit errors in playing out their capacity. Thusly, every danger will be reacted to in the best and appropriate manner. Human–machine groups can be the way to understanding progressively complex cybersecurity challenges. More bosses are relying upon robotization to understand their most squeezing difficulties. Cybersecurity experts are among the workers utilizing computerization to use their time for adequate and reinforce work execution.

4 Intelligent Defensive Framework Using AI

This paper proposes an intelligent defensive framework based on cognitive working with reliable and substantial data sets. In the proposed framework, the data sources are used to generate traffic for analytics with less error rate. Active Directories, VPNs, and domain controllers are used for resource information for system analytics and attack-based signatures. Models that take proper decisions under a cyber-attack are discussed below.

4.1 Cyber Analytics Based on Available Data Sources

This model is preferred to be used with the network devices. This analytics can help us to determine if anything abnormal is going on in the network, like Lateral Movement etc. As shown in Fig. 2, this model is based on data sources that majorly comprise of user data, endpoint data, network data, and application data.

4.1.1 User Data

User data is used in cyber analytics for collecting and analyzing user accounts and activities. The user data plays a crucial role in analytics and is used for tracking the suspicious activities done by any user that can help predict the cyber-attacks.

4.1.2 End Point Data

End point data is basically used to install agents used to analyze the system's internal components. The data collected from end point data source is essential for the proper functioning of endpoint solutions.

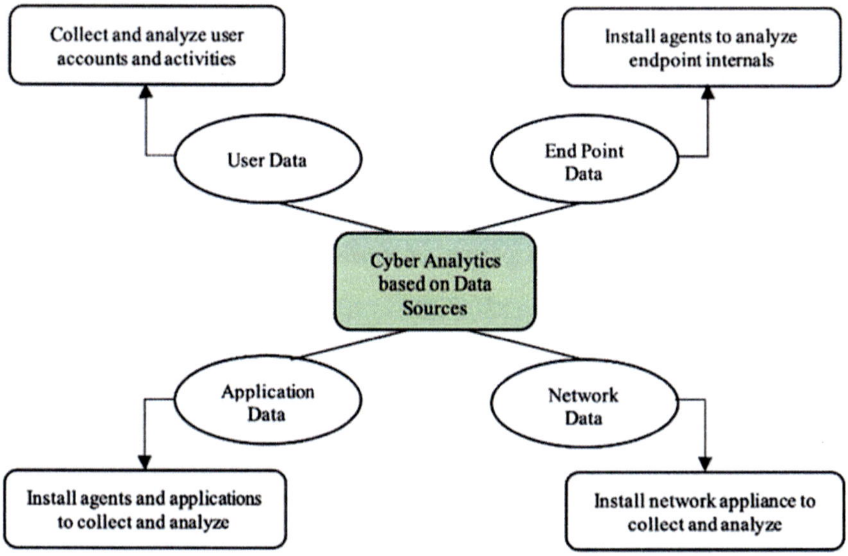

Fig. 2 Cyber analytics based on data sources

4.1.3 Network Data

The network sources are used to install network appliances to collect and analyze data from them which is further utilized in cyber analytics and can give critical information about network-related attacks.

4.1.4 Application Data

The application data collected from various installed agents is used for cyber analytics that is very crucial and is helpful in the defensive framework.

4.2 Cyber Analytics Based on Learning Methods

Since self-learning is an integral part of this model, the analytics will get more accurate and efficient over time. This model can detect some zero-days and most of the malware and misconfiguration-based exploitations.

As shown in Fig. 3, this framework categorizes cyber analytics based on various learning methods like supervised, unsupervised, reinforcement, and semi-supervised learning methods.

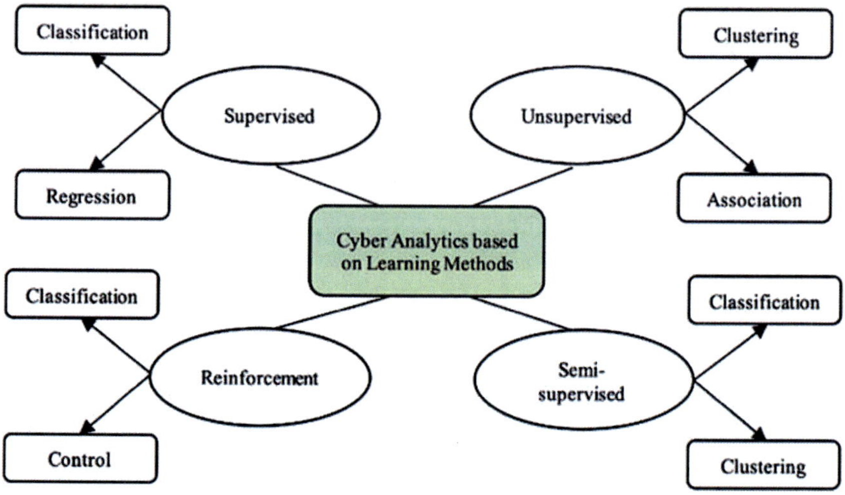

Fig. 3 Cyber analytics based on learning methods

Supervised learning is further subdivided into classification and regression, whereas unsupervised is classified into clustering and association. The semi-supervised learning method is based on clustering techniques and classification methods. Moreover, reinforcement is also subdivided into classification and control methods.

4.3 Cyber Analytics Based on Models and Algorithms

This model is suggested to be used for the endpoint protections, mainly the hosts. This can be an intelligent and effective replacement for a host-based intrusion detection system (HIDS). This model will improve over time, and due to its learning capabilities, it can make most of the attacks worthless even if a vulnerable service is deployed and running. Cyber analytics is based on various model algorithms like clustering, association rules, deep learning, regression, pattern matching, and classification.

All these techniques have several subtypes that help them in further categorization and is displayed in Fig. 4. This framework is based on popular algorithms and models of artificial intelligence that is being used in various practices.

4.4 Cyber Analytics Based on End Objectives

This model is based on a relative outcome to meet the end business objectives. This can be incomplete without using the above models and their analytics. Sometimes,

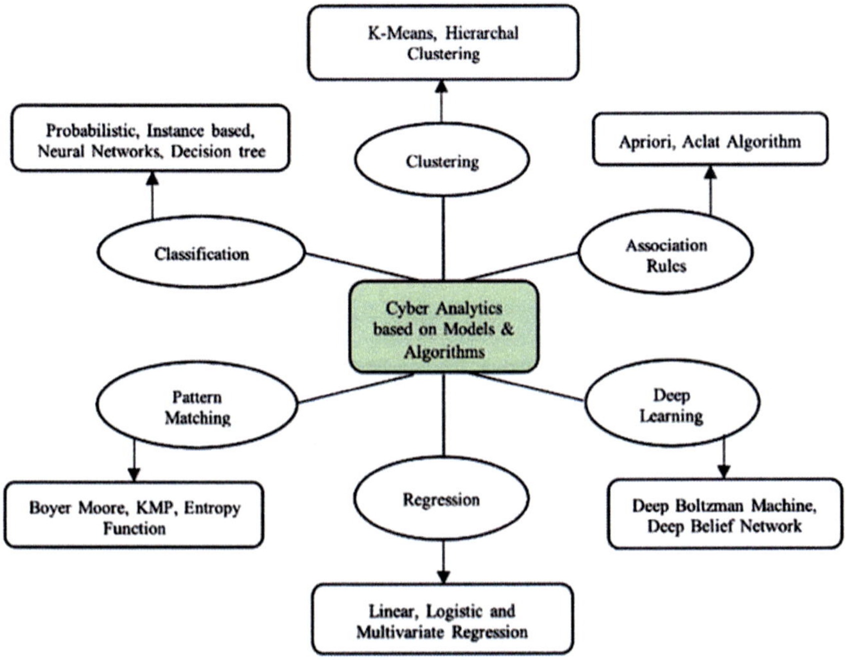

Fig. 4 Cyber analytics based on various AI algorithms in computer science

you need to compromise performance a little bit to ensure "usable security." As shown in Fig. 5, this framework categorizes cyber analytics based on end objectives and each analytics is responsible for performing some analytics based on its type and characteristics. Prescriptive analytics is responsible for information about best course of action for analysis and response. Descriptive analytics contains the information about the current security status and the performance of the entire system. Another analytics is detective analytics that is responsible for the information that is hidden, bypassed, or unknown threats. This type is crucial in this framework and is responsible for early detection and treatment. Predictive analytics is for the prediction of hazards that are likely to happen next based on the predictive algorithms and pattern matching. Diagnostic analytics is responsible for mainly troubleshooting purpose and taking care about the how and why of any incident. This analytics finds the root cause of the incident and gives the main reason behind it.

5 Characteristics of Proposed Intelligent Defensive Framework

Each of the cyber analytics method discussed in the above section is unique, can work independently, and can also work in combination with other analytics method

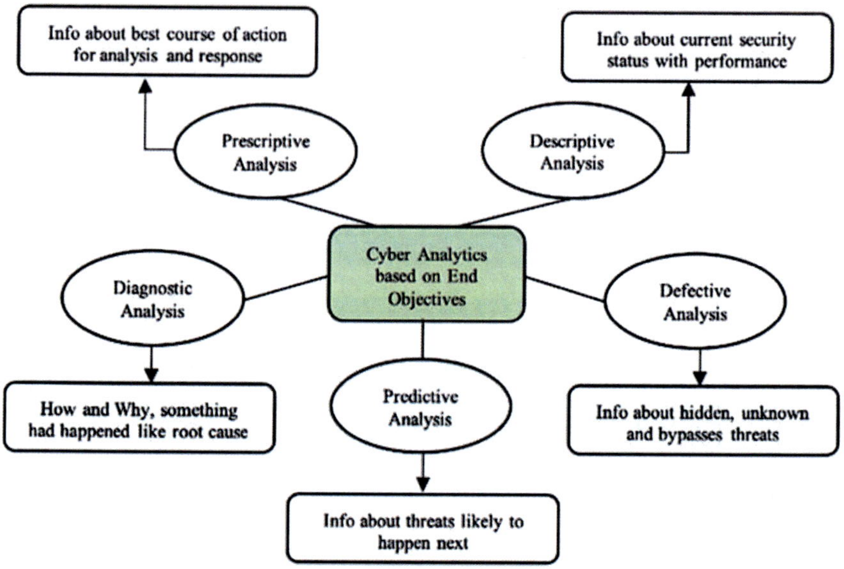

Fig. 5 Cyber analytics based on end objectives

based on the event, response time, attack type, scale of the attack, and risk associated. The proposed framework will be capable of giving better results than other defensive frameworks because of the following expected characteristics.

- Looking forward to scanning files and the coding and the DNA of the file (bit level) and predicting its nature once executed.
- Aims at detecting the malicious files around 70 ms before their actual execution and hence take actions as per user settings.
- Do not run a constant scanning mechanism rather their program stays dormant throughout. It only escalates its privileges once a file is to be executed, caned, and again become dormant.
- The primary focus is on preventing the endpoints and collecting data from the sources just in case forensic analysis or reporting is required.
- These models can be used individually to predict cyber-attacks and can also be used collaboratively, thus increasing the defensive level.
- Cyber analytics based on end analytics aims for descriptive and predictive analysis, and thus, it is a beneficial defensive framework.

6 Conclusion

Due to the increase in cyberspace, there is a promising future of artificial intelligence in securing it. AI is helpful yet just when taken care of huge amounts of applicable information. The enterprise gets to organize, incorporating live bundles of the wired system, remote measurements from WLAN controllers, SYSLOG information from various system servers, and other system information sources. AI is utilized to rapidly dissect all this different information, corresponding it across various system layers. This is something that is not essentially conceivable with individuals attempting to associate it physically. Solo learning is preparing man-made reasoning (AI) calculations utilizing data that is neither grouped nor named and permitting the calculation to follow up on that data without direction. The framework proposed in this paper is a reactive approach where we are taking into account that the system is being trained with the help of artificial intelligence and learning to increase the strength the security applied in the cyber world and help us to fight against any threat or identify a new type of threat. Also, the proposed framework will be active only when it detects some malicious file/s, take action before their execution, and becomes dormant again. The framework mentioned in this paper needs to be implemented on a large scale to get satisfying results.

References

1. Liao Y, Deschamps F, de Loures EFR, Ramos LFP (2017) Past, present and future of Industry 4.0—A systematic literature review and research agenda proposal. Int J Prod Res 55(12):3609–3629. https://doi.org/10.1080/00207543.2017.1308576 (Taylor and Francis Ltd.)
2. Lu Y (2017) Industry 4.0: a survey on technologies, applications and open research issues. J Ind Inf Integr 6:1–10. https://doi.org/10.1016/j.jii.2017.04.005 (Elsevier B.V)
3. Larry M (2014) JP Morgan Chase warns customers about massive data breach. https://www.forbes.com/sites/larrymagid/2014/10/02/jp-morgan-chase-warns-customers-about-massive-data-breach/?sh=a8e0a631e8ab. Accessed 08 Apr 2021
4. TalkTalk data breach customer details found online. BBC News. https://www.bbc.com/news/business-48351900. Accessed 08 Apr 2021
5. Lewis D (2014) Sony pictures: the data breach and how the criminals won. https://www.forbes.com/sites/davelewis/2014/12/17/sony-pictures-how-the-criminal-hackers-won/?sh=743b7abb5e02. Accessed 08 Apr 2021
6. Lezzi M, Lazoi M, Corallo A (2018) Cybersecurity for Industry 4.0 in the current literature: a reference framework. In: Computers in industry, vol 103. Elsevier B.V., pp. 97–110. https://doi.org/10.1016/j.compind.2018.09.004
7. Yampolskiy RV (2020) Unexplainability and Incomprehensibility of AI. J. Artif. Intell. Conscious 07(02):277–291. https://doi.org/10.1142/s2705078520500150
8. Li J (2018) Cyber security meets artificial intelligence: a survey. In: Frontiers of information technology and electronic engineering, vol 19, no 12. Zhejiang University, pp 1462–1474. https://doi.org/10.1631/FITEE.1800573
9. Alsmirat MA, Al-Alem F, Al-Ayyoub M, Jararweh Y, Gupta B (2019) Impact of digital fingerprint image quality on the fingerprint recognition accuracy. Multimed. Tools Appl. 78(3):3649–3688. https://doi.org/10.1007/s11042-017-5537-5
10. Zubritskaya IA (2019) Industry 4.0: digital transformation of manufacturing industry of the Republic of Belarus. Digit Transform 0(3):23–38. https://doi.org/10.38086/2522-9613-2019-3-23-38

Chapter 28
Controlled Chaos Position Transformation and Noise Generation-Based Keyless Speech Encryption Technique

Tanul Kumar Srivastava, Aniruddha Ghosh, Himadri Mandal, Amit Phadikar, and Pritom Adhikary

1 Introduction

Digital data are prone to unauthorized access. Thus, maintaining the confidentiality of the digital data with new and efficient encryption techniques has become a concerning issue as crypto-attackers are now aided with numerous attacking techniques and machine learning [1] algorithms, because of this there will always be a need of more efficient and secure encryption technique.

Cryptography [2–9] is a science of protecting actual information from inappropriate use to maintain the confidentiality of original information. Cryptography, based on key generation is classified as symmetric-key and asymmetric-key. Symmetric-key cryptography, utilizes a single key for both the encryption and decryption processes. This whole technique requires mutual agreement for key selection between the sender and the intended recipient before they transfer digital data securely. At the same time asymmetric-key cryptography employs separate keys for the encryption and decryption processes.

To maintain the confidentiality and liability of speech data several researchers had developed different techniques for speech encryption. Teerakanok et al. [9] proposed a speech encryption technique by implementing Parallel Cryptographic Algorithm (PCA), to prevent speech signals from Brute force attack and to improve the efficiency and the execution time of encryption and decryption. Sathiyamurthi et al.

T. K. Srivastava · A. Ghosh · H. Mandal (✉) · P. Adhikary
Department of Electronics and Communication Engineering, Calcutta Institute of Technology, Ulubaria, Howrah, West Bengal, India
e-mail: himadrimandal2007@gmai.com

A. Phadikar
Department of Information Technology, MCKV Institute of Engineering, Liluah, Howrah, West Bengal, India

L. Mandal et al. (eds.), *Proceedings of International Conference on Computational Intelligence, Data Science and Cloud Computing*, Algorithms for Intelligent Systems,
https://doi.org/10.1007/978-981-19-1657-1_29

[10] developed a Fast Fourier Transform (FFT) and 3-D-Lorenz-Logistic Chaotic map-based speech encryption technique to improve several issues like residual intelligibility in encoded signal, inferior quality in decrypted signal, etc. Ramakrishnan et al. [11] established a highly secured technique to transmit speech signal through telephone line by utilizing multi-level chaos-based modulation technique. Mosa et al. [12] developed a speech encryption technique by utilizing segment permutation. Belmeguenai et al. [13] designed a speech encryption technique, in which important segments of the speech signal is encrypted while other remains the same. Tamimi et al. [14] proposed a private key cryptosystem to protect different audio streams. Their experimental analysis shows that their algorithm is more suitable for high quality speech signals rather than low quality speech signals Moreno-Alvarado et al. [15] introduced a Compressive Sensing (CS) technique to encrypt the incoming audio signals. Etem et al. [16] proposed a data encryption scheme based on Self-Generation of True Random Numbers. Farsana et al. [31] developed as speech encryption technique based on nonorthogonal quantum state with hyper chaotic key stream.

Different research had worked on multi-level encryption techniques to provide better security like Kaur et al. [17] developed a four level encryption system for speech signals. Belazi et al. [18] worked on single/multiple round of permutation, diffusion, and different DNA encoding techniques. All the aforesaid implementations are good and sustainable for speech encryption and communication but regardless of all the efforts, the said field is still open for improvement in regards of robustness, security, fidelity, and computational effort.

A two stage keyless speech encryption technique to improve the security and robustness against crypto-attacks is developed in this paper. The developed algorithm encrypts the speech in two stages. First stage of encryption is based on controlled chaos position transformation, and second stage is based on controlled noise generation. In the proposed algorithm, speech sample positions are randomized with the help of controlled chaos position transformation system and then noise is generated in the randomized speech samples with the help of controlled noise generation system. Secret noise generation level of the noise generation system increases the security and robustness of the developed algorithm and it is attached in the end of encrypted speech with the help of combinational masking scheme, which makes the algorithm keyless.

The major contributions of this work are as follow:

- Masking of noise generation level makes the technique keyless.
- It has considerably less number of encryption stages than other multi-level encryption techniques.
- Decryption does not damages the original speech samples.
- Controlled chaos position transformation system increases the robustness of the encryption technique.
- Controlled noise generation system increases the security of the encryption technique.

The rest of the article contains encryption and decryption algorithm in Sect. 2 and 3, respectively, experimental analysis in Sect. 4, overall performance analysis in Sect. 4, and conclusion in Sect. 5.

2 Encryption Technique

Figure 1 depictes the block diagram of encryption technique. In first block, i.e., Speech sample preprocessing, speech sample array is created and its length is modified by adding 0's in the end of it until its length becomes a perfect square. In second block, i.e., Controlled position transformation system, length extended array is transformed into a square matrix then position transformation of samples is done by Eq. (1). After changing the position of every sample this block will again transform the matrix into an array.

$$D(i, j) = X\left(j, \left((r-1)(i-1) + j - \left\lfloor \frac{(r-1)(i-1) + j - 1}{r} \right\rfloor r\right)\right) \tag{1}$$

where $D(i, j)$, is the transformed speech sample matrix.

$X(i, j)$, is the original speech sample matrix.

r, is the length of row and column.

$i \wedge j \in [1, r]$.

Formula used in this system only works for square matrix, i.e., in 2-Dimensions but after transforming that matrix into an array, i.e., in 1-Dimension the new positions of sample does not make any kind of pattern or formula, it behave like a chaotic distribution. In third block, i.e., Controlled noise generation system, noise is introduced in the randomized speech sample array with the help of Eq. (2).

$$H(i) = \begin{cases} Q(i) - Q(i-1);\, i \in [1, k] \text{ and } Q(0) = 0 \\ Q(i) - Q(k);\, i \in [k+1, L] \end{cases} \tag{2}$$

where $H(i)$, is the noisy speech sample array.

$Q(i)$, is the randomized speech sample array.

L, stands for length of randomized sample of speech array.

k, is the noise generation level.

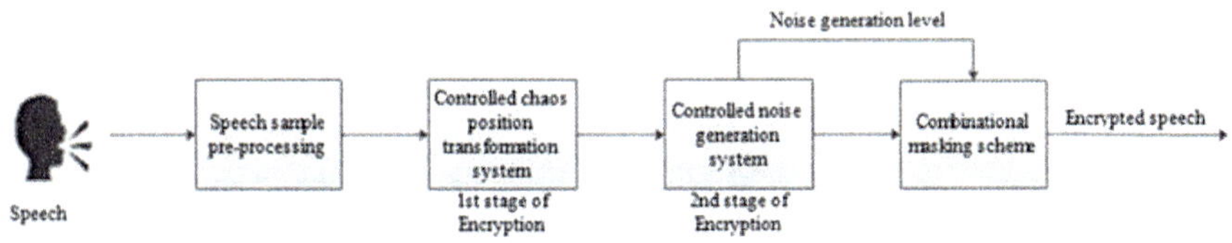

Fig. 1 Encryption block diagram

$i \in [1, L]$.

In fourth or last block, i.e., Combinational masking scheme, noise generation level 'k' is converted to its equivalent binary sequence of length 'n' then with the help of n bit hider combinational logic, and hider binary sequence of that noise generation level is calculated. Figure 2 explains the generation of 10 bit hider binary sequence using 10 bit hider combinational logic and Table 1 explains the truth table of hider combinational logic. In the end hider binary sequence is attached to the noisy speech sample array as depicted in Fig. 3. To generate encrypted speech sample. This combinational masking scheme makes this encryption technique keyless.

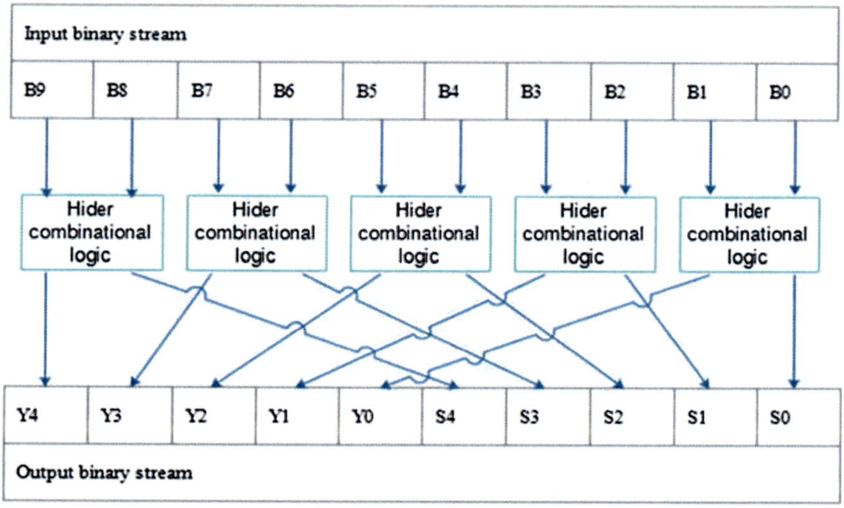

Fig. 2 10 bit hider combinational logic

Table 1 Hider combination logic truth table

A	B	Y	S
0	0	1	1
0	1	0	0
1	0	1	0
1	1	0	1

Noisy speech sample array	Hider binary sequence of 'k'

Fig. 3 Encrypted speech format

2.1 Encryption Algorithm

Step 1: Create an array using original speech samples.
Step 2: Extend the length of speech sample array by adding 0's in the end, until its length becomes a perfect square number.
Step 3: Using controlled chaos position transformation system transform the length extended speech sample array into a square matrix and change the position of every samples and again transform that randomized speech sample matrix into an array.
Step 4: Using controlled noise generation system generate noise in the randomized speech sample array of noise generation level '*k*'.
Step 5: Using combinational masking scheme convert the noise generation level into its unique hider binary sequence and attach it in the end of noisy randomized speech sample array.

3 Decryption Technique

Block diagram of decryption technique is depicted in Fig. 4. In first block, i.e., Speech sample preprocessing, encrypted speech sample array is created. In second block, i.e., Combinational unmasking scheme, last part of the sample array, i.e., hider binary sequence part is separated from the encrypted speech sample array and noise generation level '*k*' is extracted from it with the help of *n* bit exposer combinational logic. Figure 5 explains the generation of exposer binary sequence using 10 bit hider combinational logic of a 10 bit binary sequence and Table 2 explains the truth table of exposer combinational logic.

In third block, i.e., Controlled noise removal system, noise is removed in the noisy randomized speech sample array with the help of Eq. (3). With the help of extracted noise generation level.

$$R(i) = \left\{ \sum_{m=1}^{i} W(m); i \in [1, k] \sum_{m=1}^{p} W(m) + W(i) \right\}; i \in [k+1, l] \tag{3}$$

where $R(i)$, is the noise removed randomized speech sample array.

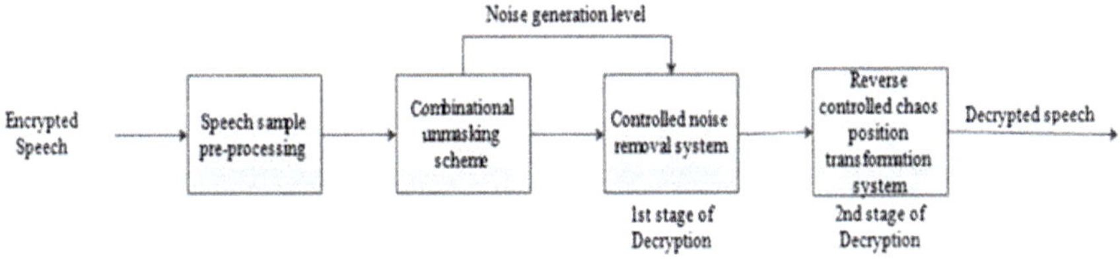

Fig. 4 Decryption block diagram

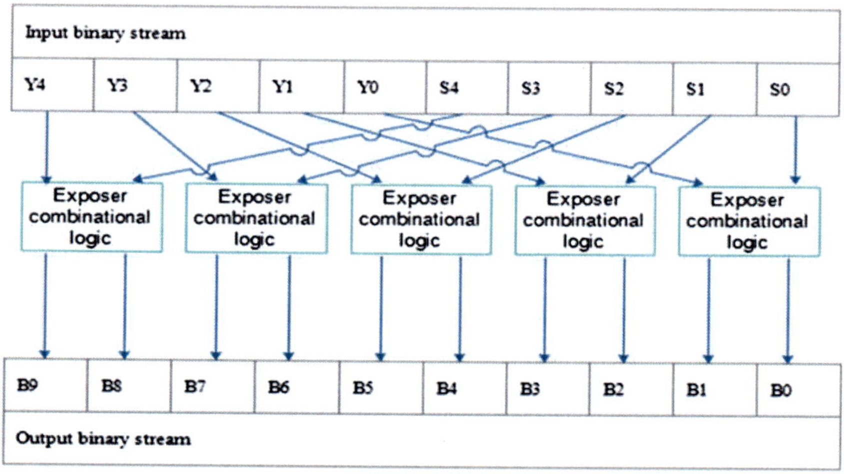

Fig. 5 10 bit exposer combinational logic

Table 2 Exposer combinational logic truth table

Y	S	A	B
1	1	0	0
0	0	0	1
1	0	1	0
0	1	1	1

$W(i)$, is the noisy randomized speech sample array.
l, stands for length of noisy randomized sample array of speech.
k, is the noise generation level.
$i \in [1, l]$.

In fourth block, i.e., Reverse controlled chaos position transformation system, Noise removed speech sample array is transformed into a square matrix then the speech samples are placed its original position with the help of Eq. (4). After placing samples its original position speech sample matrix is again converted into an array.

$$B\left(j, \left((r-1)(i-1)+j-\left\lfloor \frac{(r-1)(i-1)+j-1}{r} \right\rfloor r\right)\right) = Y(i, j) \quad (4)$$

where $B(i, j)$, is the decrypted speech sample matrix.
$Y(i, j)$, is the noise removed randomized speech sample matrix.
r, is the length of row and column.
$i \wedge j \in [1, r]$.

3.1 Decryption Algorithm

Step 1: Create an array using encrypted speech samples.
Step 2: Using combinational unmasking scheme separate the hider binary sequence of noise generation and extract noise generation level.
Step 3: Using controlled noise removal system generated noise is removed from the encrypted speech sample array.
Step 4: Using reverse controlled chaos position transformation system transform the noise removed randomized speech sample array into a square matrix and change the position of every samples to its original position and again transform that decrypted speech sample matrix into an array.

4 Experimental Analysis

The proposed technique is simulated in octave-6.1.0 (GUI) [21]. The test was conducted on 5 recorded speech files with WAV extension, having duration of 30 s with the sample rate of 40 kHz. The visualization of the actual speech, encipher speech, and decipher speech are depicted in Figs. 6, 7 and 8.

Different performance parameter like Peak Signal to Noise Ratio (PSNR) [16], Signal to Noise Ratio (SNR) [17], Correlation, Number of Sample Change Rate (NSCR) [31], Unified Average Changing Intensity (UACI) [31], and Position Residual Deviation (PRD) [31] are used to analyze the developed encryption technique. The obtained values of all performance parameter are summarized in Table 3.

The clarity between the recorded speech and the encrypted speech is tested with the help of SNR evaluation. The PSNR investigation is the measurement of the quality of the encrypted audio. Higher PSNR value reflects better encryption quality.

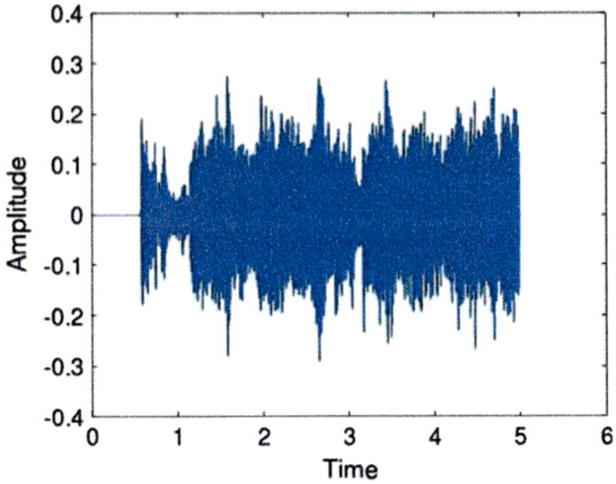

Fig. 6 Original speech

Fig. 7 Encrypted speech

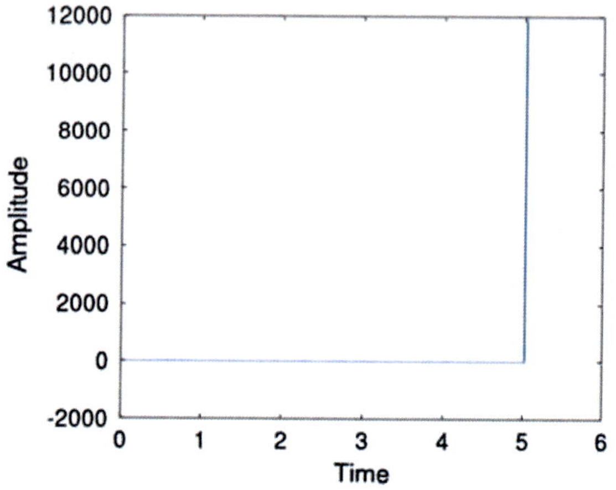

Fig. 8 Decrypted speech

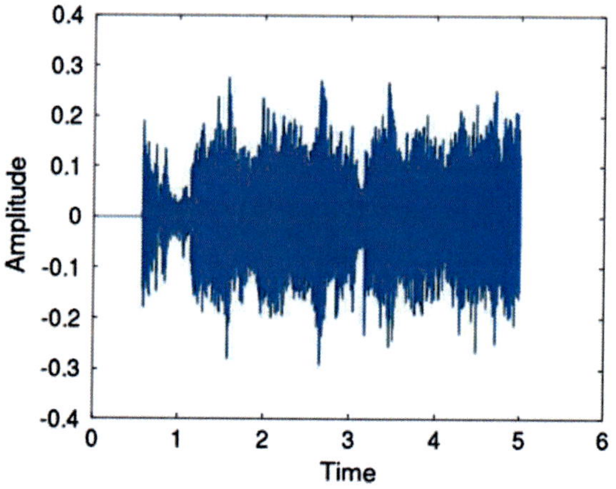

Table 3 Performance analysis

Speech file	SNR (O vs. E)	PSNR (O vs. E)	Correlation (O vs. E)	Correlation (O vs. D)	NSCR	UACI	PRD
T1.wav	−48.58	−26.14	0.0000006	0.965	99.30	43.23	6.34×10^4
T2.wav	−49.06	−25.33	0.0000006	0.992	99.95	43.56	6.89×10^4
T3.wav	−47.85	−26.36	0.000004	0.986	99.65	42.89	7.31×10^4
T4.wav	−49.10	−26.45	0.000005	0.972	99.12	43.02	6.05×10^4
T5.wav	−49.22	−26.02	0.000003	0.974	99.95	43.98	6.27×10^4
Avg	−48.762	−26.06	0.00000264	0.9778	99.594	43.336	6.57×10^4

O = Original speech signal, E = Encrypted speech signal, and D = Decrypted speech signal

$$\mathrm{SNR} = 10\left[\frac{\sum_{i=1}^{n} X(i)^2}{\sum_{i=1}^{n}(X(i) - Y(i))^2}\right] \tag{5}$$

$$\mathrm{PSNR} = 10\left[\frac{\mathrm{Max}(X)^2}{\mathrm{MSE}}\right] \tag{6}$$

Correlation expresses the relation between two signals. If there is no relation relationship between them, the correlation result will be close to 0 but if they are related the value of their correlation will be close to 1.

$$\mathrm{Correlation} = \frac{\sum_{i=1}^{n}\left(X(i) - \underline{X}\right)\left(Y(i) - \underline{Y}\right)}{\sqrt{\sum_{i=1}^{n}\left(X(i) - \underline{X}\right)^2 \sum_{i=1}^{n}\left(Y(i) - \underline{Y}\right)^2}} \tag{7}$$

where X is the original audio specimen, $\underline{X}$ is the mean of original audio sample, Y is the encrypted speech sample, $\underline{Y}$ is the mean of encrypted speech sample, and MSE is the Mean Square Error between X and$Y$.

A good encryption technique stand against differential attacks. To check its strength against differential attacks, NSCR and UACI are used.

$$\mathrm{NSCR} = \sum_{i=1}^{n}\frac{d(i)}{n} \times 100\% \tag{8}$$

$$d(i) = \{1; Y(i) \neq E(i); Y(i) = E(i) \tag{9}$$

$$\mathrm{UACI} = \left[\frac{\sum_{i=1}^{n} Y(i) - E(i)}{n \times M}\right] \times 100\% \tag{10}$$

where Y and E are the encrypted and original speech samples, respectively, n is the length of Y, M is the maximum sample value of Y. The base value of NSCR and UACI is prescribed as 100% and 30%, respectively.

The other way to analysze the variation between the coded and original audio signal is PRD. High value of PRD indicated the high deviation in encrypted and original speech.

$$\mathrm{PRD} = 100 \times \sqrt{\frac{\sum_{i=1}^{n}(X_i - Y_i)^2}{\sum_{i=1}^{n} X_i^2}} \tag{11}$$

where X is the original speech sample, and Y is the encrypted speech sample.

The proposed encryption technique is compared with existing algorithms. Table 4 presents various quality matrices such as SNR, correlation coefficient, NPCR, UACI among acual, and coded signals.

It is observed that the proposed technique is more superior to other existed works in terms of SNR and security. The performance analysis of the technique shows

Table 4 Overall performance analysis

Quality matrices	[11]	[13]	[18]	[19]	[31]	Developed algorithm
SNR	−33.746	−10.63	−44.8	−23.89	−12.45	−48.762
Correlation	0.0233	0.0011	0.0491	0.002	0.00669	0.00000264
NSCR (%)	99.9982	99.63	99.6	99.87	99.6	99.596
UACI (%)	33.1197	33.8	33.1	33.21	33.6823	43.336

49.29% improvement in SNR and 17.40% improvement in UACI. Speech encryption algorithm employing Fast Furior Transform and 3-D-Lorenz–logistic chaotic map [11], chaotic encryption of speech signals [13], A four level speech signal encryption algorithm [18], Well organized cryptosystem methods: S-boxes and permutation–substitution-based encryption [19] and Speech Encryption Algorithm Based on Nonorthogonal Quantum State with Hyperchaotic Keystreams [31] are taken for comparison.

5 Conclusion

This article presents a multi-level, keyless audio encryption/decryption algorithm presumed on controlled chaos position transformation and controlled noise generation combinational masking scheme. Controlled chaos position transformation system increases the robustness of the encryption technique, Controlled noise generation system increases the security of the encryption technique, and Combinational masking scheme makes the encryption algorithm keyless. The analytical results shows the improvement in correlation coefficient, SNR (by 41.29%), and UACI (by 17.4%). The profound result analysis concluded that the proposed algorithm found to be superior to prevent different statistical attacks with a good fidelity at the output over the other algorithms and also admirable choice for the cryptologists to develop practical applications.

The present work can be further improved and elaborated to develop a complete trans-receiver hardware module in reconfigurable platform to support voice-based access control system.

References

1. Chang S, Cohen T, Ostdiek B (2018) What is machine learning? Phys Rev D 97:056009
2. Kahate A (2013) Cryptography and network security. Tata McGraw-Hill Education
3. Satti MVK (2007) Quasi group based crypto-system
4. Yoon JW, Kim H (2010) An image encryption scheme with a pseudorandom permutation based on chaotic maps. Commun Nonlinear Sci Numer Simul 15(12):3998–4006

5. Tamimi AA, Abdalla AM (2012) A double-shuffle image-encryption algorithm. In: Proceedings of the International conference on image processing, computer vision, and pattern recognition (IPCV). The Steering Committee of The World Congress in Computer Science, Computer Engineering and Applied Computing (WorldComp), p 1
6. Yahya AA, Abdalla AM (2008) A shuffle image-encryption algorithm. J Comput Sci 4(12):999
7. Kaur M, Kaur S (2014) Survey of various encryption techniques for audio data. Int J Adv Res Comput Sci Softw Eng 4(5):1314–1317
8. Al-kateeb ZN, Mohammed SJ (2020) A novel approach for audio file encryption using hand geometry. Multimedia Tools Appl 79(27):19615–19628
9. Teerakanok T, Kamolphiwong S (2009) Accelerating asymmetric-key cryptography using parallel-key cryptographic algorithm (PCA). In: 2009 6th International conference on electrical engineering/electronics, computer, telecommunications and information technology, vol. 2. IEEE, pp 812–815
10. Sathiyamurthi P, Ramakrishnan S (2020) Speech encryption algorithm using FFT and 3D-Lorenz–logistic chaotic map. Multimedia Tools Appl 1–19
11. Sathiyamurthi P, Ramakrishnan S (2017) Speech encryption using chaotic shift keying for secured speech communication. EURASIP J Audio Speech Music Process 2017(1):1–11
12. Mosa E, Messiha NW, Zahran O, Abd El-Samie FE (2011) Chaotic encryption of speech signals. Int J Speech Technol 14(4):285
13. Belmeguenai A, Ahmida Z, Ouchtati S, Djemii R. A novel approach based on stream cipher for selective speech encryption
14. Tamimi AA, Abdalla AM (2014) An audio shuffle-encryption algorithm. In: The world congress on engineering and computer science
15. Moreno-Alvarado R, Rivera-Jaramillo E, Nakano M, Perez-Meana H (2020) Simultaneous audio encryption and compression using compressive sensing techniques. Electronics 9(5):863
16. Etem T, Kaya T (2020) Self-generated encryption model of acoustics. Appl Acoust 170:107481
17. Kaur H, Sekhon GS (2012) A four level speech signal encryption algorithm. Int J Comput Sci Commun 3(2):151–153
18. Belazi A, Khan M, Abd El-Latif AA, Belghith S (2017) Efficient cryptosystem approaches: S-boxes and permutation–substitution-based encryption. Nonlinear Dyn 87(1):337–361
19. Farsana FJ, Gopakumar K (2016) A novel approach for speech encryption: Zaslavsky map as pseudo random number generator. Procedia Comput Sci 93:816–823
20. Eaton JW, Bateman D, Hauberg S, Wehbring R (2014) GNU Octave version 6.1.0. Available from https://www.gnu.org/software/octave/download.html
21. Nagakrishnan R, Revathi A (2020) A robust cryptosystem to enhance the security in speech based person authentication. Multimedia Tools Appl 79(29):20795–20819
22. Fridrich J (1998) Symmetric ciphers based on two-dimensional chaotic maps. Int J Bifurcation Chaos 8(6):1259–1284
23. Lin Q-H, Yin F-L, Mie T-M, Liang H-L (2004) A speech encryption algorithm based on blind source separation. In: 2004 International conference on communications, circuits and systems (IEEE Cat. No. 04EX914). IEEE, vol 2, pp 1013–1017
24. Lin Q-H, Yin F-L, Mei T-M, Liang H (2006) A blind source separation based method for speech encryption. IEEE Trans Circuits Syst I Regul Pap 53(6):1320–1328
25. Yilmaz O, Rickard S (2004) Blind separation of speech mixtures via time-frequency masking. IEEE Trans Signal Process 52(7):1830–1847
26. Lima JB, da Silva Neto Eronides F (2016) Audio encryption based on the cosine number transform. Multimedia Tools Appl 75(14):8403–8418
27. Davis GM (ed) (2018) Noise reduction in speech applications. CRC
28. Thirupalu U, Kesavulu Reddy E (2019) A new cryptosystem for ciphers using transposition techniques. Int J Eng Res Technol (IJERT) 8(4):402–406
29. Farsana FJ, Gopakumar K (2020) Speech encryption algorithm based on nonorthogonal quantum state with hyperchaotic keystreams. Adv Math Phys 2020:12. Article ID 8050934

Chapter 29
The Study of Major Web Application Vulnerabilities and Their Implementation for Learning Purpose

Mukund Kr Kedia, Md Faisal Farooquee, and Baisakhi Das

1 Introduction

Recent years have seen a high rise in the number of cyber-attacks on private and government organizations, leading to huge loss of assets and sometimes even life. Many cases of ransomware have come forward where the entire system's data gets encrypted by attackers. Then the victim has to pay a huge ransom in order to get their data back safely [1]. Many things these days are connected to the Internet, so hackers capture the big industry machinery and make them behave abnormally which even leads to blasts. Attackers use various techniques to break the traditional and modern security and breach the system or network. The most common form of attack is on web applications [2]. All the organizations have their web application today, which is used for carrying out the different functions. It contains the company's confidential data, customers' private information which is to be kept safe and is the responsibility of the organization. Therefore, web application security has become a crucial issue in today's world and every company is trying to strengthen its security framework. For this purpose, lots of security engineers and researchers are being hired and constant steps are being taken to secure the application. The web is the most commonplace for attacks and there are numerous vulnerabilities present in it. For that security, testing is to be done regularly of such applications to find out the flaws that could attack the surface and fix it before it catches the notice of hackers.

This paper focuses on the study of vulnerable web applications. Most of which is in PHP. Some of them are set up for the localhost, while some are hosted on servers and available to us as a website. Both of those things have been combined and tried to create a website that has a section for individual practice of security researchers and students.

M. K. Kedia · M. F. Farooquee · B. Das (✉)
Institute of Engineering & Management, Kolkata, India
e-mail: baisakhi.das@iemcal.com

L. Mandal et al. (eds.), *Proceedings of International Conference on Computational Intelligence, Data Science and Cloud Computing*, Algorithms for Intelligent Systems,
https://doi.org/10.1007/978-981-19-1657-1_30

The paper is organized as follows. Literature review in this field of web application vulnerability which has inspired the thought of the approach for this paper is explained in Sect. 2. Section 3 shows the web application vulnerabilities. The methodologies used are discussed in Sect. 4, with the results in Sect. 5. Section 6 concludes the paper.

2 Literature Review

There have been many vulnerable web applications made before, with easy to advance vulnerabilities. DVWA, bWAPP, OWASP juice shop are some of the common vulnerable web apps present. They are all based on PHP, which is the backend coding that is done using PHP. They are to run using XAMPP. XAMPP provides the support to create a local server and also provides support for databases, generally MySQL.

DVWA contains very few vulnerabilities at different difficulty levels. It is very nice for someone who has gotten into the web security field to get an idea of different vulnerabilities [3].

Damn Vulnerable Web App (DVWA) is a PHP/MySQL web application that is damn vulnerable. It is to be an aid for security professionals to test their skills and tools usage, help web developers better understand the processes of securing web applications and help teachers/students to teach/learn web application security in a class environment.

bWAPP contains a lot of vulnerability options with many difficulty levels. It is generally the second application that a newbie tries to get a hint of web application security testing. Buggy web application (bWAPP) is a free and open-source deliberately insecure web application. bWAPP helps security enthusiasts, developers, and students to identify web vulnerabilities and think for their remediation. bWAPP prepares one to conduct successful penetration testing and ethical hacking assessments. It makes uses MySQL database. It can be hosted on Linux/Windows with Apache and MySQL. It is supported on XAMPP [4].

OWASP juice shop too contains a wide range of vulnerabilities for hands-on practice and is also quite preferred among all the people in this field [5].

Also, other than these there are some open-source projects for such web applications. There is one such application 'DVNA' on GitHub. This was now implemented in node.js with all the backend code for request and response in a single file. The app has a good interface and automatically loads the database just like the previous PHP applications discussed [1].

There is a website called 'Portswigger Web Security Academy'. This contains a lot of vulnerabilities and a full explanation for each of them. This is not only used by newbies in this field but also by serious professionals due to great documentation. My paper is also intended to provide documentation for all the vulnerabilities shown so that it can give proper understanding to the user. Portswigger academy is completely online, and we cannot see its database and backend code, while the previous apps discussed were open-source and could be downloaded completely to run on a local server. Portswigger academy runs on its own website [2].

3 Web Application Vulnerabilities

The vulnerabilities discussed in this paper are taken from the OWASP top 10. OWASP top 10 is a list of the most severe vulnerabilities that are found throughout the year based on their impact. The list contains vulnerabilities like SQL Injection, Cross-site scripting, Broken authentication, Broken access control, XML external entity, Misconfiguration, Improper serialization, and deserialization. The paper contains and takes inspiration from these vulnerabilities.

SQL injection is where the attacker can exploit the user database and get sensitive information not meant for that user by inputting and sending a malicious payload. There are a lot of ways where this vulnerability could be exploited. Many organizations have shifted from SQL to NoSQL databases, but still, those using RDBMS are prone to it most of the time [6].

Cross-site scripting (XSS) is another vulnerability that may and may not have a very serious impact. Here the attacker is allowed to inject JavaScript code into the website's input box, due to improper checks and input sanitization. Attackers take advantage of this vulnerability to put malicious code into the browser which could give out sensitive information not intended to be shown [7].

XML external entity (XXE) is also on a rise and can cause large impacts if not checked. XML is a format for data in a webpage that contains different entities filled with data. If not properly checked, attackers are able to add an external entity to the XML data and put the malicious payload in that entity, which gets executed to in turn execute the unintended operations [8].

File upload vulnerability is when the application does not correctly check the file upload type or extension. Suppose there is an option to upload the profile image somewhere, and the user uploads a PHP reverse shellcode in place of it, so now it can gain a shell on the server which is critical [9].

Broken authentication is a vulnerability where there is a loss of a proper authentication mechanism. Maybe proper password complexity is not set while registering a user, maybe there are no checks on repeated login attempts by a user. This can enable an attacker to successfully execute brute force attacks on the login page [10].

Broken access control is a vulnerability where an attacker can register into a system as a simple or regular user and exploit some flaws to get access to administrators or even different user's accounts which they are not authorized to. This can happen in case improper checks on cookies and headers on the request data.

Sensitive information disclosure is a vulnerability where the error handling mechanisms are not properly configured and they give out some server data in case of any anomaly. Developers often output server messages on the frontend for testing the application but forget to replace the messages. This information can be a source of critical information for the hackers who can take advantage of it to find vulnerabilities and exploit them.

4 Methodology

The work done is total with Node.js for the backend and the basic HTML, CSS, and JavaScript for the frontend. The backend part is put inside the controller and model files for different vulnerabilities, where the controller files manage the request, processing, and response part, and model files manage interaction with the database as per the requirement of the controller. There is a resources folder, which contains all the frontend files for HTML, CSS, and JavaScript. The node modules folder contains the file for modules that we use in the program. All of them need to be installed. The package.json file contains the information about the application like the list of dependencies, starting command, author, etc. The work done for the research was on the Ubuntu (Linux distro) platform, and it should work as well on other platforms like Windows too.

There is a list of different vulnerabilities inspired from the OWASP Top 10 list, and a try to demonstrate most of them. Users will get a nice experience on how these vulnerabilities work along with the code to even see what code made these work and what code will disallow them.

Further, there is an option to visit a site that is based on an e-commerce website. The website is completely vulnerable to different kinds of vulnerabilities that the user could find in it. The documentation for the whole implementation part is also to be included in the course of time which gives a more detailed explanation of the research and paper.

5 Results

Explanation is provided about how an application can be made to solve some of the issues prevalent in the digital world focusing on web security. Such an application would be started using a terminal or command prompt and then with the opening of the URL at a specific address (Fig. 1). The researcher can visit any specific section and understand the working of that vulnerability as shown in Fig. 2. As the backend code will be open-source, the researchers could very well go and visit the code to understand it, and also modify it to research more. Different npm modules could be used and their working could be observed for different payloads. Different other tools can be integrated with the paper implementation for ease of research, like OWASP ZAP, Burp suite, Postman, etc. as in Figs. 3, 4. A playground platform is also integrated where the researchers could go to practice on the learning based on

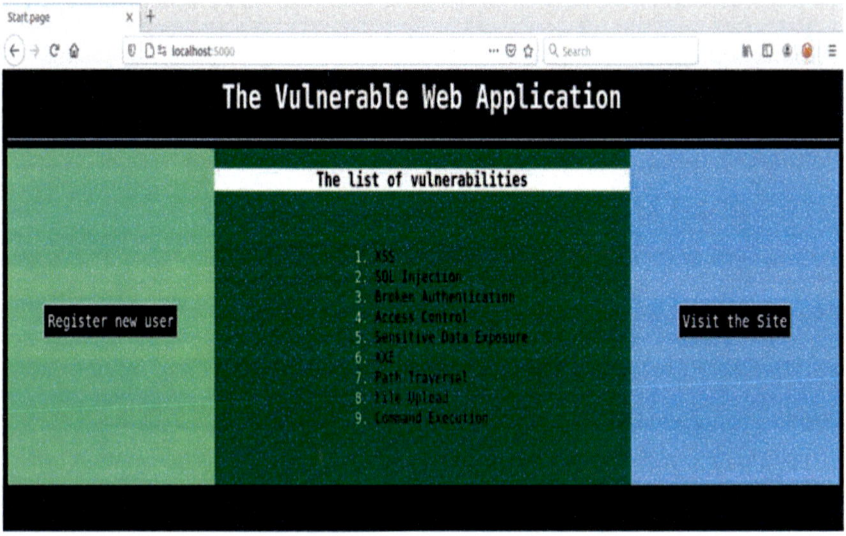

Fig. 1 The homepage of website (demonstrating different vulnerabilities)

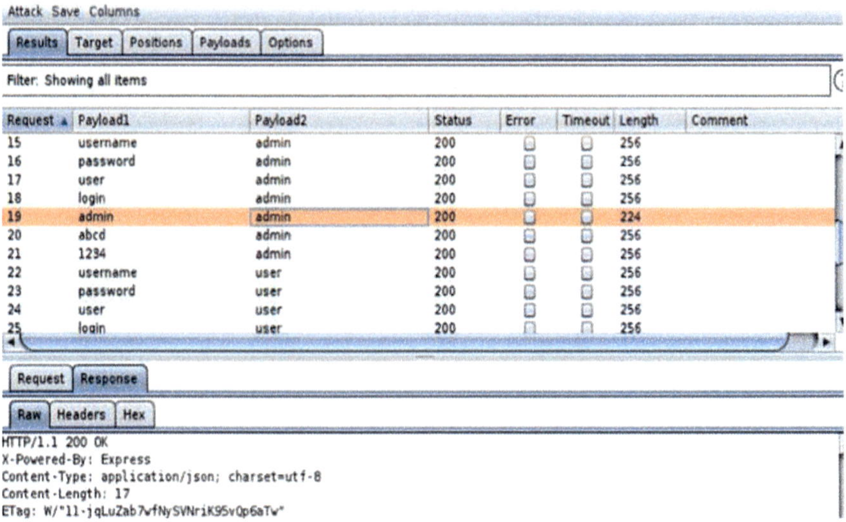

Fig. 2 Brute force attack successful (using Burp Intruder)

individual vulnerabilities. This would be good for beginners and the source code provided in the back end will be helpful for advanced researchers (Fig. 5).

Fig. 3 Cookie is admin set to true (Burp Repeat request)

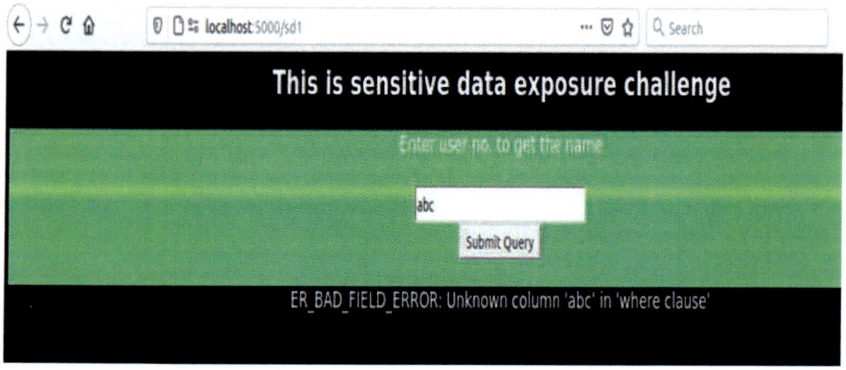

Fig. 4 Sensitive data exposed on frontend (error handling not done properly)

6 Conclusion

The major web application vulnerabilities are discussed in the paper. Most of the prior works were done in PHP, however this paper focuses on the current Node.js technology. Node.js has been used in this technology because most of the web applications that are being made today use Node.js for the backend development. PHP was the most used language for this purpose for a long time, and we can still say that the applications using PHP are still in abundance, but all the modern applications are using Node.js and its variants like Angular, React, and Vue. So, researching for the OWASP vulnerabilities in Node.js will be more advantageous for the user.

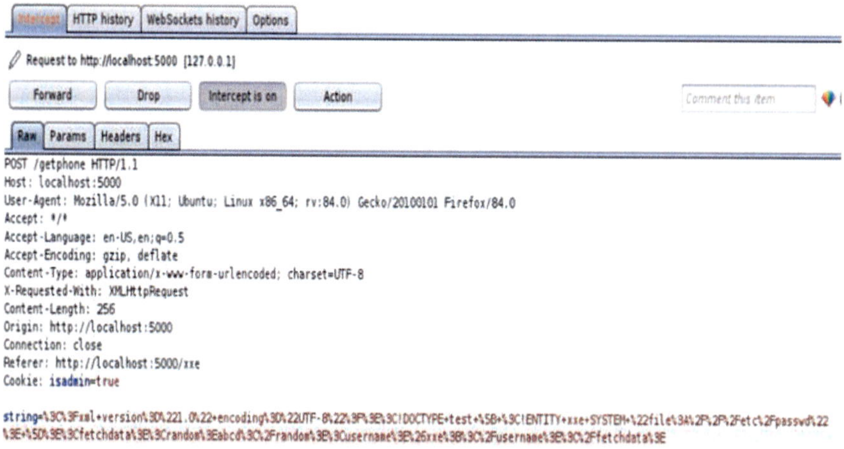

Fig. 5 Modifying the XML request (to fetch /etc./passwd system file)

References

1. Jaiswal A, Raj G, Singh D (2014) Security testing of web applications: issues and challenges. Int J Comput Appl
2. Bhatt D (2018) Cyber security risks for modern web applications: case study paper for developers and security testers. Int J Sci Technol Res 7(5)
3. Wang B, Liu L, Li F (2019) Research on web application security vulnerability scanning technology. In: 2019 IEEE 4th advanced information technology, electronic and automation control conference (IAEAC)
4. Wu Q, Liu X (2014) Research and design on web application vulnerability scanning service. In: 2014 IEEE 5th International conference on software engineering and service science
5. Atashzar H, Torkaman A, Bahrololum M, Hesam Tadayon M (2011) A survey on web application vulnerabilities and countermeasures. In: 6th International conference on computer sciences and convergence information technology (ICCIT)
6. Wood R. DVWA—Damn vulnerable web application. https://github.com/digininja/DVWA
7. Mesellem M. bWAPP an extremely buggy web app. http://itsecgames.com/
8. The OWASP Foundation, OWASP Juice Shop. https://owasp.org/www-project-juice-shop/
9. Appsecco, Damn Vulnerable Node.js Application. https://github.com/appsecco/dvna
10. Portswigger, Web Security Academy. https://portswigger.net/web-security/dashboard

Chapter 30
The Study of the Issues Related to Orphan Blocks

Bhargavi Jha and Baisakhi Das

1 Introduction

Bitcoin is a computerized payment system that follows the idea of decentralized network to validate transactions. In the network, the validation is carried out by the network's peers and no central authority is needed. Bitcoin users conduct transactions by sending digital messages that identify who should be debited, credited, and where the exchange (if any) should be put [1]. A single block of the blockchain contains the current block header that is, the hash value of the current block, hash value to the previous block to connect the blocks in the mainchain, timestamp, Nonce, and the transactions present in the block, storing them in the form of a Merkle tree, as shown in Fig. 1 [1, 2]. The Bitcoin system verifies transactions in a coded shape in a data structure referred to as the blockchain, which is maintained via a neighborhood of participants, recognized as miners. At the coronary heart of the Bitcoin system, is the computational technique referred to as mining, which includes the answer of a computationally difficult cryptographic problem.

Bitcoin miners acquire copies of all transactions as they are generated. They take a look at the blockchain to check out the records of the Bitcoins concerned in every transaction. If the proposed transaction has ample Bitcoin credit, then it is normal for incorporation into the block that the miner is presently working on [1, 3].

A double SHA-256 hash is used to identify each transaction. Miners pool transactions and utilize their hashes, along with the hash that is currently at the top of the blockchain, as inputs to the cryptographic problem in order to solve it. These problems are solved by the miners in the network. The process is called as blockchain mining, and this contains all the records and transactions which were part of the calculation process. Every transaction in the block earns the miner a reward [2].

B. Jha · B. Das (✉)
Institute of Engineering and Management, Kolkata, India
e-mail: baisakhi.das@iemcal.com

L. Mandal et al. (eds.), *Proceedings of International Conference on Computational Intelligence, Data Science and Cloud Computing*, Algorithms for Intelligent Systems,
https://doi.org/10.1007/978-981-19-1657-1_31

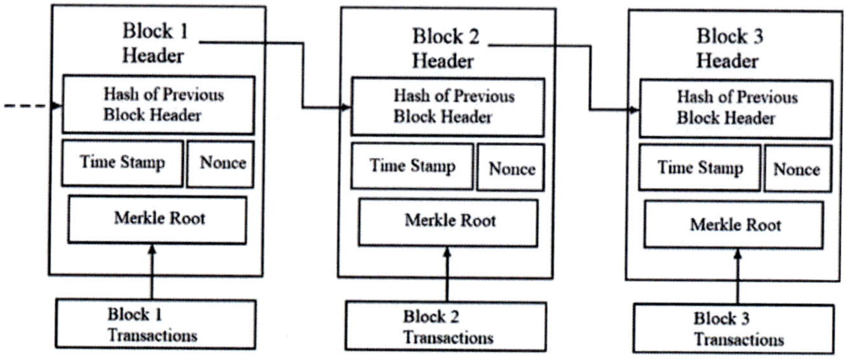

Fig. 1 Representation of blockchain

Further in this paper, we discuss the various methods to reduce the occurrence of orphan blocks. Section 2 describes the blockchain forks, Sect. 3 describes two methods to prevent the orphan risk. Section 4 describes different methods to manage them if they get formed somehow. Section 5 gives the conclusion of the paper.

2 Blockchain Forks

Blockchain forks, which take place when two miners independently locate and post a new block referencing the identical preceding block, happen commonly in permissionless blockchains such as Bitcoin. In the case of a fork of the chain, Bitcoin's rule for attaining consensus is choosing the longest chain and discarding the different chain as an orphan or stale. In this case, the miner who mined these blocks are paid nothing and the transactions existing in these blocks are rolled back. Because the maximum size of an orphan chain is six, an individual must wait for at least six confirmations to know that their block is finally a part of the main chain. This additionally makes the Bitcoin transactions slow. [3, 4] On average the range of blocks introduced to the chain per hour is six, i.e., a new block every ten minutes. So the consumer has to wait for an hour to verify that his transaction has been established or rolled back [3].

It has been observed that the consensus rule for Bitcoin is vulnerable to selfish mining. It involves a group of selfish miners controlling the production of blocks by withholding the specific block units. The working of selfish mine is as follows. When a pool miner mines a block, it informs a specific pool of dishonest miners, not all of the miners present in the network.

This process creates a secret extension in the blockchain and it continues to work further using the same procedure. The honest miners are oblivious of the blocks in the hidden extension and continue to mine and submit their mined blocks and options according to the preferred rules or protocols.

The computational energy handy to the trustworthy miners is higher than that on hand to the mining pool. So, with likelihood one, the public branch will sooner or later turn out to be as long as the pool's secret extension. However, if we observe the same for a short span of time, the possibility that secret extension will last longer than the public branch is very high. The mining pool is giving up the nearly assured revenue it would receive if it revealed its recently mined block in return for a chance that its secret department would live long enough for it to obtain non-permanent control of the mining process [5].

If the lead grows to two or more units, the pool can publish a single block each time the honest community mines a block, and two blocks when the lead shrinks to one unit. In this approach, the pool may work on its own blockchain model while allowing the honest community to waste time looking for blocks that have little possibility of becoming part of the long-term blockchain.

3 Methods to Reduce the Orphan Risk

To begin, we use the sub chain technique to lower the formation of orphan blocks, for minimizing the orphan block production rates. Then we learn about Eyal and Sirer's parameter, which is the percentage of a dishonest community that mines a previous block launched by the pool in reaction to a block mined by the honest community.

3.1 Sub Chains to Reduce Orphan Risks

One of the strategies to minimize the orphan chance is the use of sub chains can minimize the chance of creation of orphan blocks. The chance of this event relies upon on how shortly information of a new block spreads throughout the network. [6, 7] Let the time taken by the block to propagate be τ, then the probability of finding any other block at some point of the propagation interval $0 < t < \tau$ by the community is given as

$$P_{\text{orphan}} = \int_0^{\tau} \frac{1}{T} e^{-\frac{t}{T}} = 1 - e^{-\frac{\tau}{T}}, \tag{1}$$

The probability distribution for the arrival time of legitimate proof of work is given as (1/t)e-t/T. This is because only the most recently completed transactions are propagated, sub chains reduce block propagation time. This reduces the likelihood of orphaning.

Assuming that miners produce equal-sized Δ-blocks with an common duration of ΔT, every Δ block is scaled down via the sub chain factor, $T/\Delta T$, such that $\Delta Q = (\Delta T/T)Q$. The propagation time (cf. Asm. 1) is thus $\tau = z\Delta Q + \tau 0$, from which it follows that

$$P_{\text{orphan}} = 1 - e^{-\frac{\tau_0}{T}} e^{-\frac{zQ\Delta T}{T^2}}. \quad (2)$$

A sub chain with $T/\Delta T = X$ would allow approximately X times greater transactions per second at the identical level of orphaning risk as without the sub chain. However, the minimum beneficial sub chain verification time is limited since the community cannot reach consensus on the sub chain faster than the network's latency [8].

3.2 Eyal and Sirer's Parameter

Another approach to minimize the orphan danger can be Eyal and Sirer's parameter γ, which is the percentage of the truthful community that mine on a previously secret block launched by using the selfish mine pool in response to the truthful community mining a block [2, 9]. When there is no variability in the propagation delay, the triangle equality dictates that $y = 0$ (at least in the Poisson community model we considered). However, when the variety in propagation delay grows, the value of might rise remarkably quickly. There is a very high possibility that one of the pools beats the direct discussion if the secret block is launched by all the pools as soon as they are informed about the discovery of public block. However, if all serious miners transmitted the true miner's block in the same way, a counter-balancing effect may emerge. As a result, it is a good topic to check into [2, 3, 10].

4 Methods to Manage Orphan Blocks

Two methods are used to minimize the orphan risks. But even after minimizing, the orphan blocks are most likely to occur. So, following are the methods to manage them once they are created. GHOST protocol that is earlier being used in uncle blocks, which are comparable to orphan blocks in Ethereum, which can be utilized to orphan blocks as nicely with some modifications, and MWPoW is a decentralized blockchain consensus protocol that works like a mining pool and may be used to validate numerous proofs of work at the same time.

4.1 GHOST Protocol

A stale block is any block that is not part of the dominant chain. Orphaned or stale blocks are not rewarded, and their payload is ignored. If a block is not a descendant of the major chain, it can still be referred to in terms of uncle block. They can be categorized as nephew blocks, i.e., when the stale block is the direct descendant of

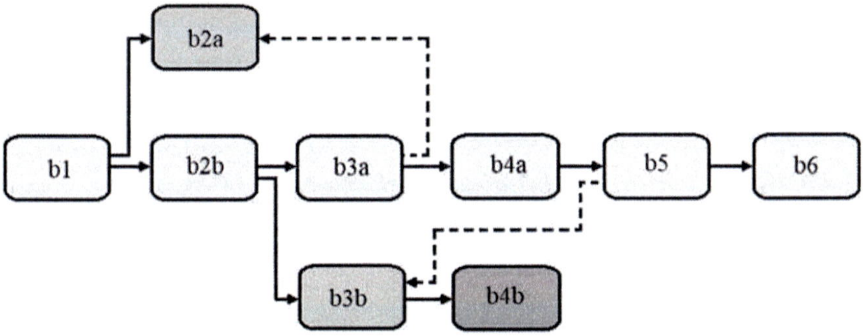

Fig. 2 Sample of Ethereum blockchain with GHOST protocol. Due to the inclusion of block b3a with distance one, block b2a is a first-generation uncle block in this diagram. Block b3b is an uncle block from the second generation. Because its father, block b3b, is not part of the main chain, block b4b is a stale block that cannot be used as an uncle block

any of the uncle blocks and is at a maximum distance of 6 blocks after the fork [11, 12].

As a result, uncle blocks have a valid header. Their payload, however, is ignored. One normal block can also reference up to two uncle blocks. The miner of the referring block is rewarded for incorporating it, which is known as a block reward (Fig. 2).

The payout of an uncle block miner depends on how many blocks have been passed after the uncle block was referred. The range varies from adding 7–8 blocks immediately to adding 2–8 blocks after 6 blocks [13].

As a result, the inclusion of an uncle block by a nephew block occurs two blocks later on average. If a fork is longer than one block, any subsequent blocks may not be referred to as uncle blocks.

The Ethereum community strives to maintain the average block time stable. This is accomplished by dynamically adjusting the proof of work complexity in response to the rate at which blocks are generated. Previously, the network's block creation rate was determined entirely by normal blocks. As a result, an exploit known as uncle mining became conceivable. Uncle blocks are now included in the calculation of the block generation rate as well [14].

The GHOST protocol is the name for this method. The Ghost protocol in Ethereum (Greedy Heaviest Observed Subtree) is a solution to mitigate the stale block problem that plagues blockchains with short block times. The Ethereum version of Ghost only travels down seven levels—or returns seven levels in the block chain's peak [6, 12].

- An uncle who is included in a block must be a direct descendant of the new block and not more than seven blocks below it in height.
- Its parents should be listed, as well as the number of uncle blocks.
- The miner gets an extra 3.125% for each uncle in the block, and the uncle's miner gets 93.75% of the preferred block reward.

- An uncle must have a valid block header, and he must be distinct from all other uncles in previous blocks and the block in which he is forming.
- It cannot be a direct ancestor of the block in the making.

4.2 *Multiple Winners Proof of Work Protocol (MWPoW)*

The Multiple Winners Proof of Work protocol (MWPoW) is a consensus protocol that aims to boost blockchain scalability while maintaining decentralization. We inspire users to use and engage in blockchain technology, minimizing the reward expectation gap that exists between the mining that is done on the blockchain and in a mining pool. A block is not placed by a single miner, but by a team of miners. If they jointly mined a block, miners receive remuneration depending entirely on their input. Finally, we increase blockchain scalability by lowering the time it takes to accept a block: any node may calculate the Support Rate of a block without having to wait a lengthy time. Any node can calculate a particular Support Rate of a block without waiting for later block confirmations [15, 16]. To fit into the data flow, either the block measurement or the block interval can be safely and reasonably dynamically changed to create a correct proportion for time allocation in block intervals [17, 18].

The following are some definitions for the term [5]:

Calculation Power Claim—Calculation Power Claim is the hash difficulty that a miner seeks for in each round of the mining game.

Valid Block—If all of the transactions, New Joins, and Shares in a block are valid, the miner deems it legitimate; the Shares and New Joins must have been known to the miner for more than 90% of the time.

New Join—A participant's Calculation Power Claim as well as his or her wallet address are kept in the New Join data set (the wallet address is used for receiving remuneration). In New Join, there is a Nonce field that is used to alter New Join's hash. The hash of a New Join must meet at least the Calculation Power Claim of this New Join in order for it to be valid.

Try Range (TR) is a numeric interval in the block header for the Nonce.

Branch Choosing Restrictions—The miner is not allowed to shift branches if he or she is given two shares of a block. This is the rule for this round of game. If a miner broadcasted two or more lawful Shares for blocks of different branches at the same block height, he would be expelled from the game, and his contribution would not be counted toward the Support Rate.

Countable Share—When at least two genuine shares of blocks are sent by a miner, the block's support rate includes the difficulties of their shares, and the reward for announcing the block is received by the miner if the game is won later.

Acceptance Difficulty—In each round of mining, the first block that reaches the Acceptance Difficulty shall be added to the mainchain. The Acceptance Difficulty is changed according to how much time the winner block takes to achieve the Acceptance Difficulty.

Entrance Difficulty—When a block reaches Entrance Difficulty, it is broadcast to the network. The difficulty of a new round is changed based on how many blocks attained the previous round's Entrance Difficulty.

Block Size—The total number of transaction permits in a single block.

Share—When transmitting, Share is a container for Nonce. A miner's Nonce inside a Share must satisfy at least 25% of the miner's Calculation Power Claim by making the block's hash match it.

Support Rate—It is defined as the ratio of the sum of the problems of the Countable Shares for the branches that originate from this block to the sum of the difficulties of all Countable Shares of all the branches in the blockchain since the block height of this block.

Following is the procedure for MWPoV:

1. Mining: Miners try to produce a block and find a Nonce that meets Entrance Difficulty in their Try Ranges; if a miner's block meets Entrance Difficulty and is approved as a Valid Block by other miners, they will search their Try Range for a Nonce that meets Acceptance Difficulty.
2. Get a Try Range: Miners with New Joins that are embedded in a block will be assigned.
3. Ranges Register power: A new miner creates a New Join and submits it to the system.
4. Rearrange the Try Range and start over: After each round of the game, the invalid users will be expelled without exception. That means the try range of the miners will be canceled, who fail to ship shares that stand for at least 50 percent of the power they claimed. Then, new miners are added, and all of the genuine miners' Try Ranges are reconfigured. A fresh round of the game begins after that. Miners who previously submitted a New Join and were not expelled do not need to re-register power to participate in the new round of the game [5].
5. Reward: If the miner submitted a sufficient number of genuine Shares for the winner block, the reward amount will be distributed at the next block height.

5 Conclusion

The benefits of blockchain present the necessity of blockchain in more than a few fields. However, there are countless risks to blockchain. One of the issues that greatly affect the value and utilization of blockchain is the formation of orphan blocks. Although the value of blockchain might also be degraded by the orphan block problem, previous research did not pay a lot attention to lookup on orphan block. Therefore, the motive of this study attempts to decrease the prevalence of the orphan blocks and if they get formed, they should prepare for the orphan block trouble of blockchain. This study offers a number of techniques to do so. In this paper, we discuss how the problem can be minimized by using techniques like sub chains and Eyal and Sirer's parameter and various other methods to control them if they are still formed.

Along with that, the GHOST protocol used in Ethereum can be applied to Bitcoin as well with some adjustments to make it extra dependable and convenient. The end result of this study contributes to limit the incidence of orphan block and prepare for the orphan block problem of blockchain. Along with this, we introduced a model of MWPoW with distinctive definitions of the Support Rate, branch choosing, and the unique characteristic of block interval/block size adjustment. This allows a couple of miners to work at the same time, reducing the number of orphan blocks created.

References

1. Nakamoto S. Bitcoin: a peer-to-peer electronic cash system. Whitepaper, 2008. [Online]. Available: https://bitcoin.org/en/bitcoin-paper
2. Moustapha BA (2020) The effect of propagation delay on the dynamic evolution of the Bitcoin blockchain. Digital Commun Netw 6. Available: https://doi.org/10.1016/j.dcan.2019.01.006
3. Ghosh E, Das B. A study on the issue of blockchain's energy consumption. Available [online]: https://link.springer.com/chapter/https://doi.org/10.1007/978-981-15-0361-0_5
4. Gobel J, Gobel J, Keeler HP, Krzesinski AE, Taylor PG Bitcoin blockchain dynamics: the selfish- mine strategy in the presence of propagation delay. [Online]. Available: https://arxiv.org/abs/1505.05343v2
5. Xu Y, Huang Y MWPoW: Multiple winners proof of work protocol, a decentralisation strengthened fast-confirm blockchain protocol. Available [online]: (hindawi.com)
6. Goldenfein J, Hunter D (2017) Blockchains, orphan works, and the public domain" 41 COLUM. J.L. & ARTS 1 [Online]. Available: https://www.cryptocompare.com/coins/guides/what-is-the-ghost-protocol-for-ethereum/
7. Neudecker T, Hartenstein H Short paper: an empirical analysis of blockchain forks in bitcoin. Available: https://link.springer.com/chapter/https://doi.org/10.1007/978-3030321017_6#:~:text=We%20provided%20an%20empirical%20analysis,observed%2050%25%20block%20propagation%20percentile
8. Rizun PR. Subchains: a technique to scale bitcoin and improve the user experience. Available: http://ledger.pitt.edu/ojs/ledger/article/view/40
9. Sompolinsky Y, Zohar A. Secure high-rate transaction processing in bitcoin (full version). [Online]. Available: https://www.ifca.ai/pub/fc15/89750476.pdf
10. Li Q-L, Chang Y-X A new theoretical framework of pyramid Markov processes for blockchain selfish mining [Online]
11. Chang S-Y, Park Y, Wuthier S, Chen C-W. Uncle-block attack: blockchain mining threat beyond block withholding for rational and uncooperative miners. Available: https://link.springer.com/chapter/https://doi.org/10.1007/978-3-030-21568-2_12
12. Wood G (2017) Ethereum: a secure decentralised generalised transaction ledger. Yellowpaper, vol. EIP-150 Revision (759dccd—2017-08-07). [Online]. Available: https://ethereum.github.io/yellowpaper/paper.pdf
13. Carlsten M, Kalodner H, Matthew Weinberg S, Narayanan A. On the instability of bitcoin without the block reward. [Online]. Available: https://www.cs.princeton.edu/~arvindn/publications/mining_CCS.pdf
14. Lerner SD (2016) Uncle mining, an ethereum consensus protocol flaw. Wordpress Blog. [Online]. Available: https://bitslog.wordpress.com/2016/04/28/uncle-mining-an-ethereum-consensus-protocol-flaw/
15. Decker C, Wattenhofer R. Information propagation in the bitcoin network. [Online] Available: https://ieeexplore.ieee.org/document/6688704
16. Baccelli F, Mathieu F, Norros I, Varloot R (2013) Can p2p networks be super-scalable? In: INFOCOM. IEEE, pp 1753–1761

17. Perez-Sola C, Donet JAD, Herrera-Joancomart J (2018) The bitcoin p2p network. In: Proceedings of the International conference on financial cryptography and data security. Springer, Berlin, Germany, pp 87–102
18. Baccelli F, Norros I, Mathieu F. Performance of p2p networks with spatial interactions of peers. [Online]. Available: http://hal.inria.fr/inria-00615523v2

Chapter 31
A Comprehensive Study on Blockchain-Based Data Security in IoT to Build Reputation for Intelligent Transport Systems

Amrita Halder and Maumita Chakraborty

1 Introduction

Mobility has become a key concern in our everyday life, irrespective of any geographic location. It may be going to schools, colleges, offices, hospitals, travelling or may be for any other purpose. Intelligent Transport System (ITS) aims to transform our existing transportation system into a digitally smart system, making life easier for the commuters. If commuters have prior knowledge about traffic, it would help them in several ways like, avoiding traffic congestion, knowing the real time availability of seats in buses and other vehicles. This would reduce the travel time and also provide comfort and safety. ITS is thus an integrated system for people, roads and vehicles, to focus on the issues like road safety, efficiency, comfort, traffic congestion, maintenance cost, etc.

ITS has already been accepted by a lot of countries due to its endless possibilities. One such great example of ITS could be seen in the city of Glasgow. ITS is used in transportation to provide regular information about public buses, their arrival and departure timings, the availability of seats, the current location of the vehicles, the time taken to reach a particular destination, next location of the vehicle and also the density of passengers inside the vehicle. All these information are very useful for the daily commuters. The vehicles are embedded with sensors that allow commuters to get the real time information.

A. Halder (✉)
Department of Computer Science and Business Studies, Institute of Engineering and Management, Y-12, Block-EP, Sector V, Salt Lake, Kolkata 700091, India
e-mail: amrita.halder@gmail.com

M. Chakraborty
Department of Computer Science and Technology, University of Engineering and Management Kolkata, University Area, Plot No. III – B/5, New Town, Action Area – III, Kolkata 700156, India

L. Mandal et al. (eds.), *Proceedings of International Conference on Computational Intelligence, Data Science and Cloud Computing*, Algorithms for Intelligent Systems,
https://doi.org/10.1007/978-981-19-1657-1_32

Another example of such ITS application could be seen in Kolkata (India) Public Transport—"Pathadisha". It is a unique bus status app which displays live bus status and up-to-date schedules, destination alarms, etc.

ITS uses combined communications and data processing technologies for:

- Providing users with prior information about traffic, real time information of vehicles, seat availability, local conveniences, etc.
- Improving the flexibility of people and goods.
- Amplifying safety, reducing traffic blockage and managing incidents efficiently.
- Meeting transport demands.

Figure 1 shows what an intelligent transport does, starting from improving the mobility of people and goods, responding to traffic related queries of people, increasing safety, reducing traffic congestion, managing incidents effectively, keeping records of all transit surveillance, etc. ITS mainly focuses on the following systems:

- Traffic Management System
- Traveller Information System
- Vehicle Control System
- Public Transportation System
- Rural Transportation System
- Commercial Vehicles Operations System.

Fig. 1 What intelligent transport does [15]

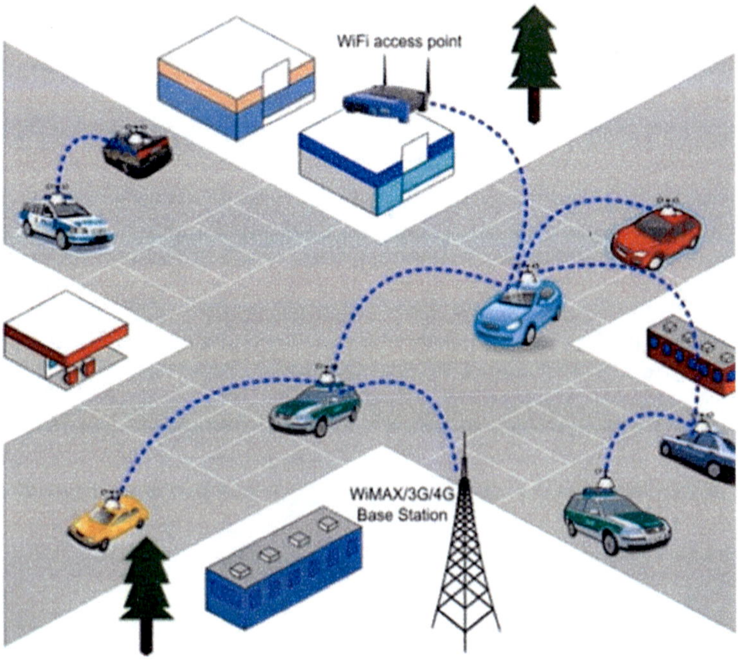

Fig. 2 Connected vehicles in ITS [12]

Figure 2 shows how VANET provides Internet connectivity to all the vehicular nodes and how computer technology and wireless communication equipped vehicles can communicate within themselves to avoid collisions, focus on road safety, dynamically schedule routes and monitor real time traffic conditions.

Within a vehicular network, sharing the traffic information proves to be very helpful in the implementation of an intelligent traffic management, few of the examples are car accident warnings, road construction notices, driver route changes to reduce traffic congestion, etc.

In the next couple of years, autonomous driving will take over the transportation system, where traffic information will be exchanged more frequently and more immediately. However, if the revealed traffic occurrence is not correct, the driving path will be deceptive, and the driving reciprocation may be in danger.

With the increase in population, the demand of transportation has also increased, this in turn has led to increased number of vehicles along with the amount of data available from the vehicles. Security has become the primary concern for vehicular networks. It is very important to secure the information of both commercial users and the public and hence it is necessary to build a reliable system where the unreliable nodes are eliminated. Data are often stolen, and information are breached. There have been numerous hacking and data leak incidents in the past that have shaken the trust of people to keep their data and personal information with the transportation companies.

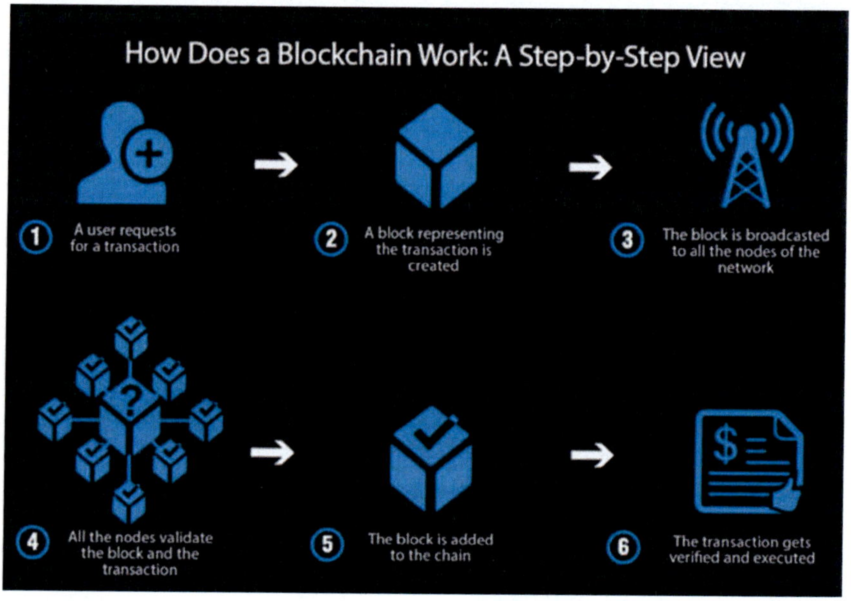

Fig. 3 Step by step approach of how Blockchain works [7]

Blockchain technology has helped to secure the data and information, leaving very less scope for data leaking and hacking, steps being explained in Fig. 3. Blockchain provides a very high level of security making it almost impossible to violate. It provides high transparency during transactions, so any kind of fraud happening in the open-source ledger of Blockchain cannot stay hidden and can be easily identified. This makes the transportation system protected from fraud and paves a way for secure and safe online transactions.

Blockchain has already been used commercially in the financial sectors such as cryptocurrencies like Bitcoin [11] (in Fig. 4), Ethereum [9] (in Fig. 5), Ripple [1] (in Fig. 6), but its decentralized nature (peer-to-peer communication) and immutability has made it very popular in recent times. Blockchain is being introduced to several other domains, few of them include: Food tracking and tracing [5], Open-Source Software Management [3], Law [10] etc.

This paper is segmented into four sections. Section 1 gives us an overview of the problem domain (exploring how Blockchain is used to create a tamper-proof network). Section 2 gives us an overview of the related works in the same domain. Section 3 shows the comparative analysis of similar articles and Sect. 4 includes the conclusion.

Fig. 4 Bitcoin [17]

Fig. 5 Ethereum [16]

Fig. 6 Ripple [2]

2 Related Work

In this section we will discuss the related works done so far in terms of how Blockchain technology can be used to make a tamper-proof network, focussing on data privacy and security in the ITS. Managing reputation within a distributed system is one of the most important aspects of Blockchain applications. The term reputation is basically the belief or reliability that is created inside a network. Blockchain does the computation and stores this belief in some quantitative form, which is relevant for building trust relations amongst the parties. Reputation in blockchain ensures a peer-to-peer human and machine ecosystem that can survive strategic bad actors sustainably.

In Ref. [20], authors Zhao et al. has created a Blockchain-based fair payment system with reputation, using a new architecture called secure pub-sub (SPS). In SPS, publishers can publish their topic on the Blockchain and the subscribers can show interest message by payment driven subscription for the topic, keeping the confidentiality and reliability of data and secures transaction of currency. The author has not used any third-party pub-sub services and the validity of the proposed SPS is also analysed. The malicious subscribers will not be able to takeback the deposit and the malicious publisher will be given 0 score for his/her reputation value.

In Ref. [4], authors Dennis et al. proposed a Blockchain technology where he considered E-commerce platform to resolve the reputation issue in the system. The author created a new Blockchain to minimize the inflation of Bitcoin Blockchain. The network focuses on two main aspects, (1) To combat past documented attacks on reputation systems and (2) To establish a new flexible reputation system which can be implemented into any network. The author has proposed to link the users with their IP address or email to avoid the creation of multiple identities of a user. Every individual user has the option to mark the transactions as positive denoted as 1 or negative denoted as 2, based on the file received. For every positive transaction the user receives a file from the sender with the sender's signature and private key. The positive transaction has a reputation score, timestamp and hash of the received file. Each user gets a reputation score from the average of all his/her reputation scores. Based on the limitations faced before, an analysis was conducted to come up with a best fit solution.

2.1 Related Works in Vehicular System:

In Ref. [19], author Zhang has proposed the first ever survey on trust management for vehicular ad-hoc network (VANET). Here the author has focussed on effective trust management for (VANET). The article discusses about the hurdles caused due to the important characteristics of VANET environments. A survey on the existing trust model was carried out in multi agent systems, mobile ad-hoc networks and vehicular ad-hoc networks, to identify the key issues. None of the trust models, so far, has archived all the desired properties or paid attention towards the robustness. Based on these studies, the author suggested the desired properties for an effective trust management in VANET. This research catered improvisation on managing trust effectively to develop a safer vehicular system. Both governments and business organizations focussed on road safety which reduced the number of car accidents and traffic congestions.

In Ref. [6], author Hîrtan et al. has proposed to establish reputation in Intelligent Transport System where the main focus is to preserve privacy of users. The author has proposed a consensus (an algorithm which is used by all the peers within the network to reach a common agreement about the current state of the distributed ledger) within the decentralized network which might consist of unreliable nodes. The primary actors being vehicles and a central server (for storage of data and computing trust), the interactions between the vehicles and the central authority are encrypted. A consensus algorithm is built, where the vehicles are clustered based on their geographical locations. The resultant of the algorithm stores the data in a secure manner using Blockchain, which is permanent in nature and cannot be modified. The research also considers the ageing of the road data after a certain threshold. In the final analysis the system provides the users with the best possible travel route based on the reliable data whilst conserving the confidentiality.

In Ref. [13] authors Singh et al. have stated how IoT is exposed to vulnerabilities and are prone to cyber-attacks. The evolution of Blockchain has come up with the solution to such issues. In this article, the author has presented a model on how the security of IoT works using Blockchain. The author discussed the 4 pillars of Blockchain (Consensus, Ledger, Cryptography and Smart contract). Since IoT requires expensive hardware, the author derived that IoT cannot be the only participant of the Blockchain network. This article has introduced reward system to build up trust within vehicles. Just like other reward systems, Blockchain also has its own way of rewarding its users, wherein users receive and redeem loyalty tokens that are compatible across several programmes. These tokens never expire or lose value. The reputation is calculated by other participants in the Blockchain based on few constrains, and it does not degrade if the vehicle provides wrong information. The vehicles communicate directly with other vehicles, which belong to the same geographical area. The authors have created consensus based upon Proof of Driving where the transmitted message is validated by the vehicles only and only if 50% or more cars in the network agrees to it.

In Ref. [14] authors Singh et al. have focussed on self-driven car which would allow the vehicle to communicate with other vehicles within the communication network. Due to intrusions in the network, the author has proposed a Local Dynamic Blockchain (LDB) technology, LDB branching, branching and unbranching algorithms for automating the branching process, which use a unique crypto ID known as Intelligent Vehicle Trust Point (IVTP) for a secure and trustworthy network. IVTP makes sure a trusted environment for communication is created which would not share user's personal information. The Blockchain model suggested in this article has the ability to perform in real time and it is able to tackle the enormous amount of message traffic, avoiding the congestion. Performance is evaluated based on real time traffic scenarios.

In Ref. [8] authors Lu et al. have mentioned that in any vehicular ad-hoc networks, the public key infrastructure alone is not enough to preserve the data integrity. The main challenge is to prevent vehicles from forging messages and preventing them from cyber-attacks. The author has suggested a Blockchain-based Anonymous Reputation System [BARS] to maintain the trust communication and privacy in VANET by breaking the connection between real identities and public keys. With the idea of using two Blockchains named Blockchain for certificates (CerBC) and Blockchain for revoked public keys (RevBC), a trust model was built. Trust in Blockchain technology increases confidence in the operations of a computational system. To create trust, one needs to verify relations between few objects like unique tokens, each of which contains the history of each previous owner.

The model highlights on two actors, Certificate Authority (CA) and the Law Enforcement Authority (LEA). The connection between the public key and the actual identity of the vehicle is eradicated with the CA, which handles the issuing and cancellation of the certificates. Whilst the LEA manages the BARS by keeping records of the public key and the real identity, in case there are any disputes. The LEA is aware of the correlation between the public key and the actual identity of a user. All the measures are stored in the Blockchain for messages (MesBC) and the communications between the vehicles are encrypted using the public–private keys. This model resulted in improvement of trustworthiness and highly efficient vehicle privacy protection.

In Ref. [18] authors Yang et al. have stated how an exposed traffic incident may be deceived, putting the vehicular information in danger. To resolve such issues, author proposed a Proof of Event consensus. The relevant data are collected from the Roadside Units (RSU) and are transmitted to the vehicles passing by. The vehicles will be asked to verify the accuracy. If the users validate the data within a particular time, the RSU will add the event in the local chain. The author has introduced a two-phase transaction on Blockchain which would send warning messages in suitable regions and time periods. The proposed system is difficult to implement since the installation and the interconnection of RSUs over a large geographical location would cost a lot. The proposed system is also unable to build reputation and hence for a certain geographic location, where fake traffic is expected from few malicious vehicles and is not supported by a trustworthy vehicle, the event will be verified by the system.

3 Comparative Study

This section shows the comparison between various works, discussed in the previous section, Sect. 2. A Table has been formed with the objective to analyse the various aspects of Blockchain and how each of the related works are implementing it in their work. Although some of the research papers aim to reach a common goal, they have diversities in their approach, methodologies and behaviours. The results obtained are also different for each of the work being considered.

Table 1 compares each of the work based on their objectives, the methods they have applied to reach the objective. We have also checked whether the confidentiality is being maintained in their work or not, reputation matrix is used or not, whether it is feasible to implement the work they have demonstrated, etc. The related works are also compared based on the factors like data integrity (the accuracy and consistency of data stored) and robustness (ability of the system to resist change). Finally, the results obtained have been compared and the unachieved areas or future scopes have been noted.

4 Conclusion

In this paper, we have discussed about different existing works related to usage of Blockchain in Intelligent Transport System (ITS). The papers mainly talk about how Blockchain technology can be used to make a tamper-proof network, focusing on data privacy and security in the ITS. Some of them do talk about how to manage reputation within a distributed system, which is one of the most important aspects of Blockchain applications. In the comparative study sections, we have discussed and compared different research works on vehicular ad-hoc network, focusing on the above-mentioned aspects, proposed by different academicians. Each of them is unique and has a specific purpose. The models put forward by authors and their methodologies to achieve the goal were different from each other. Although the goal was common, i.e., to create a secure and trustable network, the ways of achieving it were different. Blockchain has certainly proved itself to provide a tamper-proof network.

To take a step further, we are aiming to achieve results by conducting experiments which would reach consensus within a decentralized network, comprising of unreliable nodes.

Blockchain technology has an expansive application across several industries. It has the capability of changing almost every industry. It is already being used to accelerate identity managements, supply chain analysis, smart contracts and many more. We have discovered the potentials of Blockchain and probably a lot more remains to be discovered in the near future.

Table 1 Comparative analysis of related works on vehicular network using Blockchain technology

	Hîrtan et al. [6]	Lu et al. [8]	Singh et al. [13]	Singh et al. [14]	Yang et al. [18]	Zhang [19]
Objective	To build reputation system to preserve privacy	To achieve anonymous authentication and trust communication for VANETs	To secure data transmission between the devices, connected through Internet	Proposes a self-driven car which would allow vehicles to communicate with other vehicles within the communication network	To validate traffic events and trust in VANET	To discuss the challenges for establishing trust management and point out key issues
Methodology Used	Sim2Car—a microscopic traffic simulator	Blockchain-based anonymous reputation system (BARS)	Does not discuss anything about any methodology. Has discussions only, there is no practical work done	LDB branching and unbranching algorithms, using a unique crypto ID known as Intelligent Vehicle Trust Point (IVTP)	BTEV framework which introduces PoE consensus mechanism	Theoretical results were obtained by conducting survey
Guaranteed Confidentiality	Yes	No	No	No	Yes	No
Use of reputation matrix	Yes	Yes	No	No	No	No
Implementation feasibility	Feasible	Not feasible	Feasible	Feasible	Not feasible	Not feasible
Data integrity	Achieved	Not achieved	Proposed	Achieved	Not achieved	Not achieved
Robustness	Robust	Robust	The author has proposed for robustness, but there is no proof that the outcome was robust	Robust	The author has proposed for robustness, but there is no proof that the outcome was robust	Not robust

(continued)

Table 1 (continued)

	Hîrtan et al. [6]	Lu et al. [8]	Singh et al. [13]	Singh et al. [14]	Yang et al. [18]	Zhang [19]
Result achieved	The model provided its users an ideal travel route based on reliable data, whilst maintaining confidentiality	BARS effectively improved the trustworthiness of broadcasted messages and protects vehicle privacy with high efficiency	IoT was benefited from the functionalities introduced by Blockchain technology	IVTP ensured a trusted environment for communication, having the ability to perform in real time and tackle enormous amount of traffic, avoiding congestion	Vehicles could access to traffic information efficiently and it saved a lot of power consumption	Proposed a list of important properties like Decentralize Trust Establishment, Coping with Scarcity, Event/Task and Location/Time Specific, Scalable, Integrated Confidence Measure, System Level Security, Sensitive to Privacy Concerns and Robustness
Unachieved Points	Accurately identifying the GPS position of users	No practical implementation	IoT was not a full member of a Blockchain network due to its high-end hardware requirements	IVTP could have been used as a limited cryptocurrency that can be exchanged in return for services like traffic information	The proposed system is unable to build reputation and difficult to implement	The proposed model was not implemented in practical

References

1. Armknecht F, Karame G, Mandal A, Youssef F, Zenner E. Ripple: overview and outlook. Available online: https://www.researchgate.net/publication/281631024_Ripple_Overview_and_Outlook. Published on August 2015
2. CMC Markets. What is Ripple? https://www.cmcmarkets.com/en/learn-cryptocurrencies/what-is-ripple. Last accessed on 22 Sept 2021
3. D'mello G, González-Vélez H (2019) Distributed software dependency management using blockchain. In: Proceedings of the 27th Euromicro International conference on parallel, distributed and network-based processing, PDP 2019, 13–15 February 2019. Pavia, Italy, pp 132–139
4. Dennis R, Owen G (2015) Rep on the block: a next generation reputation system based on the blockchain. In: Proceedings of the 2015 10th International conference for internet technology and secured transactions (ICITST), 14–16 December 2015. London, UK, pp 131–138
5. Galvez JF, Mejuto JC, Simal-Gandara J (2018) Future challenges on the use of blockchain for food traceability analysis. TrAC Trends Anal Chem 107:222–232
6. Hîrtan LA, Dobre C, González-Vélez H. Blockchain-based reputation for intelligent transportation systems. Available online: https://doi.org/10.3390/s20030791. Published on 31 Jan 2020
7. Inlea. Intelligent Transport System (ITS). https://inlea.com/but-how-does-blockchain-really-works/. Last accessed on 29 Sept 2021
8. Lu Z, Wang Q, Qu G, Liu Z (2018) BARS: A blockchain-based anonymous reputation system for trust management in VANETs. In: Proceedings of the 2018 17th IEEE International conference on trust, security and privacy in computing and communications/12th IEEE International conference on big data science and engineering (TrustCom/BigDataSE), 1–3 August 2018. New York, NY, USA, pp 98–103
9. Mascarenhas JZG, Ziviani A, Wehmuth K, Vieira AB. On the transaction dynamics of the Ethereum-based crypto currency. Available online: https://academic.oup.com/comnet/article-abstract/8/4/cnaa042/6056228?redirectedFrom=fulltext. Published on 31 Dec 2020
10. Millard C (2018) Blockchain and law: incompatible codes? Comput Law Secur Rev 34:843–846
11. Nakamoto S (2008) Bitcoin: a peer-to-peer electronic cash system. Available online: https://bitcoin.org/bitcoin.pdf. Accessed 31 Jan 2020
12. ScienceDirect. Vehicular Ad-hoc Network System (VANET). https://www.sciencedirect.com/topics/computer-science/vehicular-ad-hoc-network/. Last accessed 29 Sept 2021
13. Singh M, Singh A, Kim S (2018) Blockchain: a game changer for securing IoT data. In Proceedings of the 4th IEEE world forum on internet of things, WF-IoT 2018, 5–8 February 2018. Singapore, pp 51–55
14. Singh M, Kim S (2018) Branch based blockchain technology in intelligent vehicle. Comput Netw 145:219–231
15. TechVlogz. Intelligent Transport System (ITS). https://techvlogz.com/learning/next-generation-technology/what-is-intelligent-transport-system-and-how-it-works/. Last accessed 29 Sept 2021
16. The Economic Times. Indians follow Ethereum to Stratosphere. https://economictimes.indiatimes.com/tech/technology/indians-follow-ethereum-to-stratosphere/articleshow/86025555.cms. Last accessed 29 Sept 2021
17. US NEWS. The history of bitcoin. https://money.usnews.com/investing/articles/the-history-of-bitcoin. Last Accessed 29 Sept 2021
18. Yang Y, Chou L, Tseng C, Tseng F, Liu C (2019) Blockchain-based traffic event validation and trust verification for VANETs. IEEE Access 7:30868–30877
19. Zhang, J (2011) A survey on trust management for VANETs. In: Proceedings of the 2011 IEEE International conference on advanced information networking and applications, 22–25 March 2011. Singapore, pp 105–112
20. Zhao Y, Li Y, Mu Q, Yang B, Yu Y (2018) Secure pub-sub: blockchain-based fair payment with reputation for reliable cyber physical systems. IEEE Access 6:12295–12303

Chapter 32
A Survey on IoT Protocol and Threat Matrix

Kajari Sur and Indraneel Mukhopadhyay

1 Introduction

The IoT [1–4] refers to a connection of computational physical nodes (things) which have the processing power to communicate amongst themselves and with external entities via the use of sensors and actuators. These nodes or things use simple sensors like temperature sensors, humidity sensors and pressure sensors to complex sensors like optical and infrared sensors, giving it a wide range of computing and sensorial capabilities, which helps the IoT application to interact more efficiently with the real-life environment. IoT sensor market fell 10–15% due to COVID-19 as no new industrial projects, temporary shutdowns of production facilities. Current IoT device market share is USD 8.4 billion which is projected to grow more than threefold to USD 29.6 billion by 2026 [5]. With the arrival of IoT, many real-time applications have been developed which was never thought in various domains, such as defence, health care, logistic management and so on. These specific applications have specific needs and constraints (scalability, coverage and energy) [6] that encourage a global interoperability in IoT. There have been many protocols that have been implemented to facilitate communication in different types of IoT applications. But each IoT manufacturer in order to maximize their own market share wants to stick with their own communication protocol, and therefore, the goal of creating a protocol for the global interoperability remains a distant dream. The allied market force manufacturers to produce their IoT devices as quickly as possible. But the manufactures forget two things interoperability with different two IoT devices and security of the devices. They only look towards their profit margins. To make different IoT device communicate with each other, a common protocol needs to agree upon. Once the common

K. Sur (✉) · I. Mukhopadhyay
Institute of Engineering and Management, Kolkata, West Bengal, India
e-mail: reenew@gmail.com

L. Mandal et al. (eds.), *Proceedings of International Conference on Computational Intelligence, Data Science and Cloud Computing*, Algorithms for Intelligent Systems,
https://doi.org/10.1007/978-981-19-1657-1_33

protocol is there it will result in improved efficiency, accuracy and finally will lead to economic benefit for all both for manufactures and users of the IoT devices.

It is found that IoT, as of now, is highly vulnerable:

1. Out of every 10 IoT devices 6 devices provide user interfaces which are vulnerable to XSS attack and have weak credentials [7]
2. If we take into account all the IoT devices present at least 90% of such devices collect sensitive personal information which when stored in the cloud or mobile application is open for cyber-attacks [7]
3. 70% of IoT devices use unencrypted network service for communication [7]
4. 70% of the IoT device along with their cloud or mobile application is easy target for hackers [7]
5. 80% of the IoT device along with their cloud or mobile application is not protected properly via password of sufficient complexity and length [7].

Here in the first section, we discussed on different layers of IoT system. Then IoT eco-system and different threats at different layers are discussed. Then the comparison between different protocols used in IoT system has been pointed out [7].

2 IoT Layers

The different layers in an IoT system have been depicted in Fig. 1.

Each of these layers is important towards the implementation of the IoT device eco-system:

> The application layer is the layer with which bears the responsibility for IoT-based services, which its been intended to do so. Say for example taps of a button in the mobile application it turns on a coffee maker.
> Middleware layer is where the infrastructure needs are taken into account to be able to support the architecture.
> The network layer is responsible of transmission and communication of the data that has been collected by the sensors. It plays an important role towards global interoperability amongst other smart nodes, servers and network devices.
> IoT devices have a sensor layer where we have the sensors and actuators of the connected devices. This particular layer is responsible for collecting the data for the real-time environment it has been deployed in. The devices in this layer interact with the real-time environment.

3 IoT Applications Eco-system and Threats

IoT device eco-system is connected via the Internet. All nodes can either use the cloud or mobile application and each of the IoT devices eco-system there will be

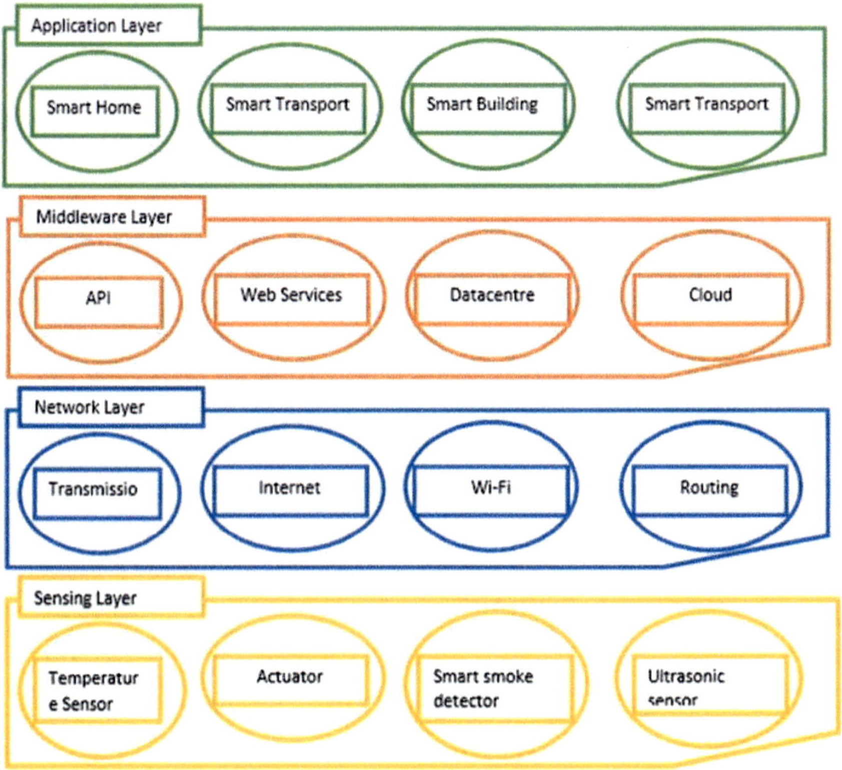

Fig. 1 Layer in IoT system

several IoT protocols and applications that exist with specific requirements. Let's take an example for the implementation of few protocols like ZigBee, WirelessHart. They are all IoT short-range protocols which can be effectively used when we try to implement a smart implementation in an office or factory. Interconnection can be also done via private network. These subsystems integrate actuator–sensor and monitoring applications as per the need of the factory. But for bigger implementation, i.e. few square kilometre long-range protocols, such as LoRaWan, SigFox or NB-IoT. The use of cloud services will help in implementing the solution. Instead of private network strip antenna can be used as gateway to communicate between the devices [8].

Digitalization has made our life super easy but security is the biggest challenge. Cyber-attacks are no more a fun. It is very organized business. Every system can be compromised but your matured action will help to protect your resources and enable to operate security in the digital world. Security hygiene is essential to ensure data security, and IoT eco-system is no different. We were able to go over many related work where we were able to find security loophole which made it easy to take control of IoT devices and its allied networks. A technophobic, i.e. hacker who can take the full control of the IoT environment [9–12]. Once the hacker has control he can control all the nodes of the network. In research papers [9, 10], it clearly stated that the hacker after gaining access to the IoT network very easily control the door locks and the alarm and other connected devices in a smart home implementation, whereas in few IoT application it was found that an hacker very easily could take control of any Bluetooth-enabled device in the allied network of the eco-system. Once the hacker has control of the device, the hacker may salvage sensitive data stored in the device, use it to spy the user, sell it in the dark web or use it as a relay to spread the malware attack (Mirai malware [13] or its variants). The issue becomes even more dangerous in terms of safety critical systems used in the healthcare industry [14]. Once the IoT device has been compromised, the patient is in the harm's way, and it becomes very hard to trace the intruder [15]. Every 39 s devices are getting compromised and as the number of IoT devices is growing at an exponential rate it and its eco-system are also prone to be exploited to retrieve the valuable information [16]. We can think of solutions towards prevention of security loopholes via Blockchain implementation but global interoperability will still remain the issue [17].

In an IoT environment threats can be classified as per different layers is shown in Fig. 2.

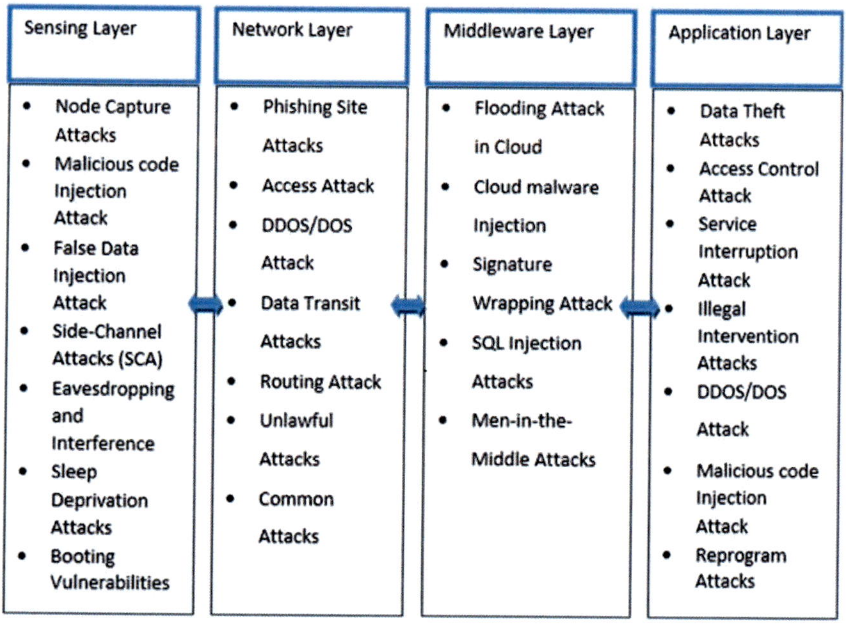

Fig. 2 Different attacks in IoT devices

4 Comparison of Various IoT Protocol:

In order to go towards global interoperability between all IoT devices we have to keep in mind the following two points:

- We have to identify a generic approach of IoT protocol so that the comparison of IoT protocol stacks and its documentations and specifications have to be agreed upon by the IoT manufactures.
- We have to make all attacks towards IoT devices a generic approach. So in the case of an attack on IoT eco-system we can depend on another protocol for safely rebooting the IoT system without much data being compromised.

We have reviewed around 14 different protocol and security specific to either specific protocol, Open Systems Interconnection model (OSI) layer or on the entire eco-system. The classification is as follows:

Focuses on the security on protocol
Focuses on the security on OSI layer
Focuses on IoT systems.

The first group [18–22] focuses on the security of a protocol stack, such as Z-Wave, ZigBee, BLE, WirelessHart or IPv6 Low Power Wireless Personal Area Networks (6LoWPAN). After analysis of the security vulnerabilities, we were able to compare the same with the help of Table 1.

The second group [23–27] tackled the problem of comparing protocols, but focuses on OSI layer, such as data link, network, transport or session layer. In the second group analyse the network keys which are provisioned over and done with the application layer and many other strong interactions throughout the stack. Analysis led to the full-stack understanding of IoT protocols by describing and analysing all layers.

Finally, the third group [28–31] provided a global view of security in IoT eco-system. When we are talking of an IoT eco-system, we have to keep in mind IoT model is so abstract that it is inflexible to fit the security complications of IoT protocols into it.

5 Conclusion

In this paper, we were able to compare various protocols which are commonly used in IoT eco-systems. Whenever we talk about IoT devices, we are always open to vulnerabilities. Our goal will be to propose a comprehensive interoperability protocol for IoT eco-system and also to have tried to bring down the threat matrix down via the use blockchain technology [17].

Table 1 Comparison of different IoT protocols

		[18]	[19]	[20]	[21]	[22]	[23]	[24]	[25]	[26]	[27]	[28]	[29]	[30]	[31]
Protocol stack(1)	OS4I					√	√	√	√	√		√			√
	BLE		√								√	√			√
	ZigBee				√			√			√	√			√
	Z-Wave	√												√	√
	WirelessHart			√										√	
	LoRaWAN													√	
	SigFox														
Layer(2)	APP	√	√	√	√	√			√	√					
	TRANS	√	√	√	√	√	√		√	√					
	NWK	√	√	√	√	√	√				√				
	DL	√	√	√	√	√		√							
	PHY	√	√	√	√	√		√							
System(3)	3-Layers											√	√	√	√
	Protocol-level	√	√	√	√	√	√								

References

1. Al-Fuqaha A, Guizani M, Mohammadi M, Aledhari M, Ayyash M (2015) Internet of things: a survey on enabling technologies, protocols, and applications. IEEE Commun Surv Tutorials 17(4):2347–2376
2. Da Xu L, He W, Li S (2014) Internet of things in industries: a survey. IEEE Trans Ind Inf 10(4):2233–2243
3. Gubbi J, Buyya R, Marusic S, Palaniswami M (2013) Internet of things (IoT): a vision, architectural elements, and future directions. Future Gener Comp Syst 29(7):1645–1660
4. Sicari S, Rizzardi A, Grieco LA, Coen-Porisini A (2015) Security, privacy and trust in internet of things: the road ahead. Comput Netw 76:146–164
5. IoT sensors market with COVID-19 impact by sensor type, network technology, vertical, application, and geography (North America, Europe, APAC, RoW)—Global Forecast to 2026. https://www.marketsandmarkets.com/Market-Reports/sensors-iot-market-26520972.html. March 2021
6. Lin J, Yu W, Zhang N, Yang X, Zhang H, Zhao W (2017) A survey on internet of things: architecture, enabling technologies, security and privacy, and applications. IEEE Internet Things J 4(5):1125–1142
7. El Jaouhari S, Bouabdallah A, Bonnin J-M (2017) Chapter 14 Security issues of the web of things. In: Sheng QZ, Qin Y, Yao L, Benatallah B (eds) Managing the web of things. Morgan Kaufmann, pp 389–424. ISBN 9780128097649
8. Kaur MJ, Maheshwari P (2016) Building smart cities applications using IoT and cloud-based architectures. In: International conference on industrial informatics and computer systems (CIICS). IEEE, Sharjah, pp 1–5
9. TH News (2019) How drones can find and hack internet-of-things devices from the sky, https://thehackernews.com/2015/08/hacking-internet-ofthings-drone.html. Accessed 21 Aug 2019
10. TH News (2019). Z-wave downgrade attack left over 100 million IoT devices open to hackers. https://thehackernews.com/2018/05/z-wavewireless-hacking.html. Accessed 21 Aug 2019
11. TH News (2019) Blueborne: critical bluetooth attack puts billions of devices at risk of hacking. https://thehackernews.com/2017/09/blueborne-bluetoothhacking.html. Accessed 21 Aug 2019
12. TH News (2019) Hackers could turn LG smart appliances into remote-controlled spy robot. https://thehackernews.com/2017/10/smart-iot-devicehacking.html. Accessed on 21 Aug 2019
13. Hallman R, Bryan J, Palavicini G, Di Vita J, Romero-Mariona J (2017) IoDDoS—the internet of distributed denial of service attacks—a case study of the Mirai malware and IoT-based botnets. In: IoTBDS 2017.
14. TH News (2019) Hackers can remotely access syringe infusion pumps to deliver fatal overdoses. https://thehackernews.com/2017/09/hackinginfusion-pumps.html. Accessed 21 Aug 2019
15. Radcliffe J (2011) Hacking medical devices for fun and insulin: breaking the human SCADA system. Black Hat
16. TH News (2019). Casino gets hacked through its internet-connected fish tank thermometer. https://thehackernews.com/2018/04/iot-hackingthermometer.html. Online. Accessed 21 Aug 2019
17. Mukhopadhyay I (2021) A comprehensive study of pros and cons on implementation of blockchain for IoT device security. In: Chakraborty M, Singh M, Balas VE, Mukhopadhyay I (eds) The "essence" of network security: an end-to-end panorama. Lecture notes in networks and systems, vol 163. Springer, Singapore. https://doi.org/10.1007/978-981-15-9317-8_4
18. Fouladi B, Ghanoun S (2013) Security evaluation of the z-wave wireless protocol. Black Hat, USA, 24
19. Ryan M (2013) Bluetooth: with low energy comes low security. In: 7th USENIX workshop on offensive technologies. WOOT, USENIX Association
20. Raza S, Slabbert A, Voigt T, Landernäs K (2009) Security considerations for the wireless HART protocol. In: International conference on emerging technologies and factory automation (ETFA). IEEE, pp 1–8

21. Baronti P, Pillai P, Chook VWC, Chessa S, Gotta A, Hu Y (2007) Wireless sensor networks: a survey on the state of the art and the 802.15.4 and ZigBee standards. Comput Commun 30(7):1655–1695
22. Sain M, Kang YJ, Lee HJ (2017) Survey on security in internet of things: state of the art and challenges. In: International conference on advanced communication technology (ICACT). IEEE, pp 699–704
23. Nguyen KT, Laurent M, Oualha N (2015) Survey on secure communication protocols for the internet of things. Ad Hoc Netw 32:17–31
24. Reziouk A, Laurent E, Demay J-C (2016) Practical security overview of IEEE 802.15.4. In: International conference on engineering & MIS (ICEMIS). IEEE, pp 1–9
25. Brachmann M, Garcia-Morchon O, Kirsche M (2011) Security for practical CoAP applications: issues and solution approaches. In: 10th GI/ITG KuVS Fachgespraech Sensornetze (FGSN11), pp 15–16
26. Karagiannis V, Chatzimisios P, Vazquez-Gallego F, Alonso-Zarate J (2015) A survey on application layer protocols for the internet of things. Trans IoT Cloud Comput 3(1):11–17
27. Lounis K, Zulkernine M (2020) Attacks and defenses in short-range wireless technologies for IoT. IEEE Access 8:88892–88932
28. Mendez DM, Papapanagiotou I, Yang B (2017), Internet of things: survey on security and privacy. CoRR abs/1707.01879. arXiv:1707:01879. http://arxiv.org/abs/1707:01879
29. Oracevic A, Dilek S, Özdemir S (2017) Security in internet of things: a survey. In: 2017 International symposium on networks, computers and communications (ISNCC), pp 1–6
30. Alaba FA, Othman M, Hashem IAT, Alotaibi F (2017) Internet of things security: a survey. J Netw Comput Appl 88:10–28
31. Ammar M, Russello G, Crispo B (2018) Internet of things: a survey on the security of IoT frameworks. J Inf Sec Appl 38:8–27

Part IV
Cloud and ICT

Chapter 33
A Study of Different Encryption Techniques to Generate Keys in Cloud Computing Security

P. S. S. Akilashri and A. Ahadha Parveen

1 Introduction

Like many other technical services, cloud computing brought several advantages. It enabled a huge number of data and services to be stored. In addition, this platform resolved the problem of limited resources by sharing valuable resources with various users and reducing the cost of services. In the cloud environment, there were three major entities: (i) Cloud Service Provider (CSP), (ii) User, and (iii) Data Owner(DO). Numerous users save their private data here on the cloud server and CSP [1] is used by DOs and users. Current cloud environments data management is a crucial expected feature. The platform needs to be resistant against security risks for dependability and performance of resources [2]. Cloud computing has become one of the most substantial subjects in security investigation in recent years. This includes data storage security, security of the network, and software.

Cryptography is one of the leading large data security solutions. The basic purpose of cryptography is to transport data or messages via an untrusted medium between the sender and recipient so that the original data is not read by an attacker or a malicious user. The cryptosystem can be split into two different methods, the use of encryption key and decryption key. The coding and decryption processes are carried out using the same key [3] with symmetric key encryption. Diverse keys or values for encrypting and decrypting data or communication are used in asymmetric key encryption. A key is private and the public key is delivered to each side here. The private key remains confidential and the public key can be public with everyone. In recent years, numerous researchers have suggested several measures aimed at enhancing data security and improving cloud presentation [4, 5].

P. S. S. Akilashri · A. Ahadha Parveen (✉)
Department of Computer Science, National College, Affiliated to Bharathidasan University, Trichy620001, India
e-mail: ahadha.parveen@gmail.com; parveena@jmc.edu

L. Mandal et al. (eds.), *Proceedings of International Conference on Computational Intelligence, Data Science and Cloud Computing*, Algorithms for Intelligent Systems,
https://doi.org/10.1007/978-981-19-1657-1_34

The usage of encryption, which is dependent upon many of the security features in the encryption and decoding process, is a vital way of creation. The usage of encryption is a powerful instrument for privacy, integrity, and data security. In this symmetry encoding, one secrete key is used for encoding and decrypting, and two keys in the asymmetrical one, one public and one private, for encoding the two forms of public encoding, which has an ancient history. The key can be resisted when the key is robust, even the attacker who discovers all systems info suffers concerning the generation of the encryption key.

2 Literature Review

Shabir et al. [6] in this paper, an analysis is done on the traditional encryption techniques involved in securing data that is stored in the cloud environment. The basic requirements that are to be satisfied by the encryption techniques are categorized here. Those basic requirements are confidentiality, access control, reliability, scalability, accountability, fault tolerance, revocation, and collision resistance. This work is mainly focused on analyzing various Attribute-Based Encryption (ABE). The variations of ABE are Key Policy ABE (KP-ABE), Expressive Key Policy ABE (EKP-ABE), Cipher text policy ABE (CP-ABE), Key policy weighted ABE (KP-WABE Hierarchical Attribute-Based Encryption (HABE), Cipher text policy Attribute set-based encryption (CP-ASBE), Hierarchical Identity-based Encryption (HIBE), Cipher text policy-weighted ABE (CP-WABE).

Soofi et al. [7] in this paper, a survey is done on the data confidentiality algorithms used in cloud computing for information security. The algorithms that were employed for enforcing data confidentiality in the cloud environment are Hierarchical ABE (HABE), Proxy re-encryption (PRE), Geo encryption methods, AES, RC5, Role-based Encryption (RBE), Rivest Shamir Adleman (RSA), DES, Simplified DES (SDES), Secure Socket Layer (SSL), and mixed encryption procedures. RSA contributes to 33% of the cloud applications, SSL and AES contribute to 13% each, Geo, RBE, PRE, HABE, and diverse algorithms contribute to 7% each, DES and SDES contribute to 6%. All these approaches were found to be validated using methods like Comparative analysis, statistical analysis, and experiments.

Manjusha and Ramachandran [8] in this article, various attribute-based encryption techniques are analyzed and compared. The variations of ABE discussed in this paper are key policy ABE, cipher text policy ABE. The rejection rate of key policy ABE and cipher text policy ABE are more than the multi authority ABE. Also, multi authority-based ABE takes minimal time for execution when compared to the other two. These are the observations through the analysis performed by the authors. The future extensional work of this study is to enhance the multi authority-based ABE such that it supports homomorphism.

Shynu and Singh [9] the contribution in this paper includes an access control background called Privacy Preserving SE (PPSE). This protocol is mainly for multi expert schemes in cloud computing. The authors claim this system to be efficient, fine-grained, and flexible in cloud storage service. A very smaller sum of attributes is used for key extraction when compared to the existing key extraction methods. Cryptanalysis and the resistivity of the proposed work toward that cryptanalysis are explained in the context of collision attack, insider attack, and compromised key attack. The major entities of this protocol are central authority data owners, data consumers, and cloud servers.

Kaur et al. [10] through enhancing the level of confidentiality, privacy, integrity, authentication, the proposed work in this paper aims to give high-level security to data that is stored in the cloud. This method utilizes standard encryption techniques to protect data against attacks. There are two phases in this model Data storage is the first phase and retrieval is the second phase. All the security measures are taken through in the retrieval process. The user can choose the storage type and it could be either private storage or public storage or hybrid storage. Encryption techniques like AES, Secure Hash Algorithm (SHA) are employed in all three types of storage options. In private storage, to enhance the authentication service, a distinct token generation method is used. A two-tier security architecture is implemented in hybrid storage and public storage supports high speed data encryption and decryption.

The [11] have examined to which degree this skepticism legitimized, by introducing to the Cipher Cloud. The Cipher Cloud is a system that lets clients keep their information secretly on open cloud structures. To accomplish this, the Cipher Cloud utilizes a two-advance encryption procedure; in which, all the information sent from a customer to a cloud worker or the other way around is kept completely encoded and private. The most exhaustive security controls expected to ensure the most delicate information may not be ensured out in the open CC models, while they can be acknowledged in private CC structures. As the most encouraging CC method, this paper proposes encryption methods, which nearly provide the information classification simply like private cloud prototypes.

Hemalatha et al. [12] an analysis of security issues and the security measures taken in the cloud environment is done in this article. Also, a discussion on cloud computing techniques, classifications, characteristics, and encryption methods is done. The characteristic set of cloud computing involves elasticity, network access, resource pooling, and on request self-service. The types of cloud environment are public, private, hybrid, and community cloud. The challenges in cloud security are integrity, data location, confidentiality, availability, privacy, authentication, storage, backup, and recovery.

Mehmood et al. [13] an attempt for analyzing lightweight cryptography procedures in the cloud and Internet of Things (IoT) is done in this paper. It is observed that the combination of cloud and IoT minimizes the computational cost due to their coordinated services. There are some commonalities between cloud and IoT and they include applications over the Internet, services, storage features, efficiency, energy, and computational ability. There is also a comparison done on the various cryptographic methods that are used in cloud and IoT. Also, the authors have done a comparison of different algorithms based on their computational complexity. To achieve confidentiality, authenticity, and integrity, cryptographic techniques are used. The authors have claimed that cloud storage is best suitable for IoT devices. CC is a recently developing field that offers numerous open doors for clients to information preparing and information stockpiling out the gadgets. Nonetheless, IoT gadgets produce broad information so the security of the information is a difficult errand. CC is considered the most ideal alternative for information stockpiling of IoT gadgets. The extraordinary sorts of cryptography can be utilized to beat this test. This investigation summarizes the writing to portray lightweight cryptography procedures to determine security issues.

Suthar et al. [14] a study of symmetric encryption methods in CC environment is done in this article. Present day cryptographic procedures, for example, encryption and encoding are broadly utilized in CC to ensure secrecy and respectability of the clients information put away on the cloud. These calculations affect computational overhead which influences the exhibition of the cloud. An examination is done on AES, DES, and 3DES symmetric encryption strategies and MD5 and SHA-256 encoding methods. AES is one of the great possibilities for symmetric key encryption, whereas DES and 3DES create all the more overhead. If there should be an occurrence of encoding, MD5 is quicker yet SHA is by all accounts safer. Rather than giving level choices to encryption and encoding, the authors suggest taking a choice from the cloud information proprietor about the affectability of the information and execution required for the equivalent. This client information can be utilized to choose mixes of these cryptographic natives.

Liu et al. [15] The contributions include a multi owner data allocation system. It allows multiple users to share data in an untrusted environment. This method supports efficient dynamic group management. Newly added authenticated users can decrypt data without the support of the data owners. A novel revocation list that does not require secret key updating is utilized for user revocation. The amount of revoked users does not affect the computational cost and time of encryption. Also, secrecy and privacy-preserving access control mechanism are used that assures members of a group to namelessly using the resources in the cloud environment. In case of any dispute, the group director can reveal the true identity of the data owners. A security examination is also performed to describe the functioning ability of the proposed work.

Song et al. [16] an investigation is done on encryption techniques, key management methods in this paper. An evaluation is done to find the attributes that affect the system's presentation and efficiency. Finally, a comparison is performed on the encryption methodologies to suggest good mechanisms for future work that can yield high performance and good scalability. The study is made with AES, Public key cryptography, searchable symmetric encryption (SSE), Public key encryption with keyword search, deterministic encryption methods, order preserving encryption methodologies, fully encryption over integers, homomorphic evaluation of AES, and key management. The traditional key management scheme includes a server, a key provider, and a third party.

Goyal and Khurana [17] the cryptographic techniques are intended to improve security in actual life track and speed up. This is somewhat a Feistel figure which gives genuine use for the procedures from grouped (blended/symmetrical) mathematical assortments. AES is joined with Segmentation and validation calculation to recover the execution of the security. Additionally, Key extension is finished to make the AES safer. The procedures are pipelined to haste up AES. Our techniques are mixture strategy which functions as changing the first documents (information) into encoded/scrambled sort utilizing AES and Small Encryption Process. The mixture procedure is deliberated for effectiveness and better execution. In the encryption design, information is encoded utilizing a small AES-128 encryption procedure and confirmation SHA that modifications it into an unintelligible code text. In that encodes the data which think about an irregular book through the possibility of cryptography and afterward information/text for the client. Our projected method gets the limited computational time and memory usage which is re-enacted in MATLAB.

Ghosh and Karar [18] distributed computing assumes a significant job in sharing information and assets to different gadgets through information redistributing. During sharing assets, it is a move undertaking to give and get to control and secure compose activities. The principle issue is to give secure peruse and compose tasks cooperatively also, to lessen computational over-burden by viable key administration. In this paper, a safe and productive information joint effort plot BH-WABE for secure information composing and capable access control has been proposed. Here, weight is doled out to each trait dependent on its significance, and information is scrambled utilizing access control approaches. The cloud specialist co-op stores the re-appropriated information and quality power renounce or refreshes the qualities by allotting various characteristics depending on the weight. The beneficiary can get the information document compared to its weight to lessen the computational over-burden. The proposed BH-WABE gives plot obstruction, multi authority security, and fine-grained get to control in terms of security, unwavering quality, and effectiveness. The presentation is contrasted and the conventional hybrid attribute-based encryption (HABE) plot as far as information privacy, adaptable access control, information joint effort, full designation, fractional decoding, and check are halfway marking.

Jeyabalu and Krishnamoorthy [19] distributed computing, a sort of relational figuring offers administrations on request. There emerge various weaknesses and dangers that endanger the confirmation of the cloud administrations. Security administrations stay predominant despite standard strategies offered by specialist organizations, particularly on an open cloud. An encryption administration is one of the regular methods to ensure the plain information, by changing over it into an incoherent structure. This paper brings an intriguing application administration that gives security as a support of the clients. This creative help works dependent on asymmetric key encryption conspire. Encryption keys in the cryptographic procedure uncover the quality of encryption. Different keys are engaged with this encryption procedure, among them, one intriguing key age is from hybridization of improved code square fastening encryption activity and another from environment motivated hereditary calculation. Inspiration for planning this half and half calculation is to limit the execution time and extra room limit. The exploratory examination is achieved for sight and sound documents, including plaincontent, pictures. Execution of the proposed encryption calculation is investigated utilizing different measurements that uncover the quality and quality of the calculation regardless of different assaults. Assessment reports express that this application administration relic's potential for verifying sight and sound documents with better fulfillment while outside as requested in distributed computing conditions.

Shehzad et al. [20] cloud computing and enormous information have given an answer for putting away and preparing huge measures of complex information. Notwithstanding the way that they are extremely helpful, the danger to information security in the cloud has gotten a matter of incredible concern. The security shortcoming in Hadoop, which is an open source system for huge information and CC, has mishaps since it is sending in numerous operational zones. Diverse symmetric, hilter kilter, and half-breed encryption plans have been functional on Hadoop for accomplishing a reasonable degree of information security. In this paper, a fresh mixture encryption plot, which joins symmetric key calculation utilizing pictures as mystery keys and uneven information key encryption utilizing RSA, is proposed. The recommended plot diminished the overhead of the mystery key calculation cycles when contrasted with the other existing encryption plans. Along these lines, it is protected to guarantee that the proposed plot holds a sufficient security level and makes information encryption more productive.

Sangeetha and Vaidehi [21] a cloud-based framework is proposed in this work. The objective of this work is to split the public health records into two sections called Personal Domain (PSD) and Public domain (PUD). The personal domain is for the patients and their relatives or friends. The public domain is for the health care firms, doctors, medical researchers, nurses, etc. Different procedures for data handling and key generation are implemented for PSD and PUD. This method is claimed to give good access control. Privacy preserving decentralized collusion resistant attribute-based encryption (PPDCRABE) is used for implementing PUDs. This method is said to be free against collusion and time efficient.

Kumar et al. [22] this work has a review of particular image encryption calculations and presumes that the disordered methodology displays extraordinary vulnerability and gives a high-level of security. In addition, the study reveals that Number of Pixels Change Rate (NPCR) does not rely upon the key affectability; along these lines, to accomplish palatable NPCR in the block-based encryption strategies, the estimations of a scrambled block ought to be subject to the next code blocks. Moreover, results show that scrambling alone is not adequate to offer amazing security; there ought to be a replacement alongside rearranging. Subsequently, this audit derives that a superior image encoding method has the accompanying qualities to give exceptional insurance; (1) Deliver huge keyspace, (2) High key sensitivity (3) Generate a uniform histogram, (4) Content to Shannon's disarray and dispersion property, (5) Decrease connection adequately between two nearby pixels, (6) Provide vulnerability in the framework, and (7) High NPCR esteem (close to 100%) and appropriate UACI rate (close to 33%). In addition, other than these given traits, an encryption strategy ought to be sufficiently quick to encipher an image. Finally, an end is that all analyzed encryption calculations in this paper, are successful and have their own benefits and bad marks in regard to speed and security compromise.

Brandao [23] the authors have discussed the significance of the authentication process and the importance of encrypting data in a cloud framework to safeguard security. As per the observation of the authors of this article, the main issues in cloud security are authentication, access control, policy integration, services monitoring, data security, regulatory complaints, privileged user access, location of data, issues in trust, and data recovery. The technical problems in the cloud are operating systems, transaction management, memory management, databases, virtualization, concurrency control, networking, load balancing, and scheduling the resources. The cloud computing environment issues include data integrity, data segregation, data transfer, data security, data location, data availability, server access and applications, network security, virtual machine security, compliance policy, and patches management.

Radhadevi and Kalpana [24] a modified Elgamal cryptosystem is proposed in this work. It is a combination of the Elgamal cryptosystem and Rabin carp system. Here, the algorithm is split into three sections, namely, key generation, encryption, and decryption. The strength of the Elgamal cryptosystem is based on Discrete Logarithm Problem (DLP). DLP states that it is easy to compute exponentiation but hard to perform the inverse. Due to improvements in technology, this DLP could be easily resolved. Thus, this has led to the alteration in this work. The authors have claimed that this method is resistive to mathematical attacks, brute force attacks, and known plaintext attacks. To this modified Elgamal cryptosystem, another strength is added, and it is the integer factorization problem. Also, a comparison of the traditional Elgamal cryptosystem and the modified Elgamal cryptosystem is done in this article.

Saha et al. [25] in this article, an effort is made to encrypt and decrypt images using AES. AES is a block cipher that is known to be secure against most cryptanalytic attacks. Since AES processes data in matrix format, it is very suitable for encrypting images since the image when represented in pixels could be handled in matrix form. Cryptanalysis on AES encryption and decryption is discussed along with the justification for why AES is suitable for image encryption. A symmetrical random function builder was introduced in this paper [26] (SRFG). In this area, the technique of creating keys is a new one. The results were shown to be strong in the case of resistance against AES attacks, depending on the AES algorithm. In this paper, [27] powerful encryption was provided using a DNA-based method for symmetrical encryption applications. To construct a key for encryption, an N-size permutation is created using four vectors formed by private key input. The rules produced with the same flipping bus are restored and reused for identifying new cryptographic keys, as much as new cryptographic keys are necessary.

Gayathri et al. [28] in order to measure the effectiveness of these methods, a number of similar simulated tests were undertaken and the best criteria were: FRR = 0.061, FAR = 0.023 for fuzzy extracts with a key 192-bit long length. The presentation of the AES and the ECC algorithms in this paper [29] produces the private key based on the two methods. Generally, AES has a 128-bit key length with 10 redundancies, to avoid effective security for consumers, as 196-bit bits are utilized with a 12-fold redundancy on the key. The activities of users may be very secure and their data can be kept confidential with this improvement.

To provide security for electronic health record attribute-based cryptosystem and blockchain concept is employed in this work. Confidentiality of the data is ensured using ABE and Identity-Based Signature (IBS) is generated using the Digital Signature (DS) method. A new cryptographic primitive is introduced here. It is the

Table 1 Table type styles

Algorithms	Fine-grained access control	Computation overhead efficiency	User revo cation	Scalability and efficiency	Collision resistance	Association of attributes	Association of access policy
ABE	Low	AVG	AVG	AVG	Below AVG	With cipher	With key
KP-ABE	Low if re-encryption than high	Most of computation overhead	Low	AVG high if broadcast type	AVG	With cipher	With key
EKP-ABE	Better than KP-ABE	Reduced computation overhead	AVG	Higher than KP-ABE	Above AVG	With cipher	With key
CP-ABE	AVG realization of complex access structure	AVG	Low	AVG is not efficient in modern enter prise environment	Good	With key	With cipher
CP-ASBE	Higher than CP-ABE	Lower than CP-ABE	Above AVG CP-ABE	Better than	Good	With key	With cipher
HIBE	Comparative low	Most of computation overhead	–	Better: lower when compare with ABE	Good	–	–
HABE	High	Some of overhead	AVG	Above AVG	Good	With key	With cipher
HASBE	High	Less than all of above	Above AVG	High	Good	With key	With cipher
CP-WABE	Very high	High	Above AVG	High	Good	With cipher	With cipher
KP-WABE	Very high	Low	Above AVG	High	Good	With cipher	With key
MA-WABE	Very high	Low	Very High	Very high	Very good	With cipher	With cipher

combination of an attribute, identity-based encryption with signature. To trace the medical data and ensure its integrity, blockchain concept is employed. Also, the authors have discussed medical insurance. All these implemented technologies are cloud-based. A blockchain utilizes a bounded information structure to store information, utilizes the agreement convention between conveyed hubs to create and refresh information and uses the cryptography strategy to guarantee security, and utilizes mechanized scripting code to run astute agreements.

In the research study conducted, various cloud-based security mechanisms for data safety have been studied. The study is focused on encryption techniques in the cloud and new encryption schemes introduced to enhance data security in cloud situations and authentication management schemes. Different works by various authors were analyzed along with their corresponding strengths and future scope of extension. Numerous mechanisms including security mechanisms, authentication protocols, and cloud security challenges are discussed and an effort has been made in this research to address some of the major drawbacks that occur during data management in the cloud computing environment (Fig. 1).

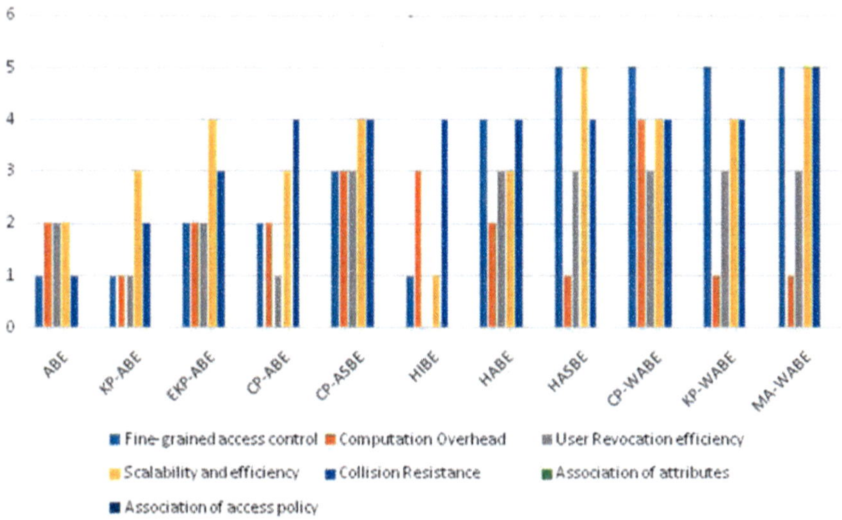

Fig. 1 Comparison of different algorithms under different parameters

Table 1 illustrates a comparison between different algorithms under the following parameters including fine-grained access control, computation overhead, user revocation efficiency, scalability and efficiency, collision resistance, an association of attributes, and association of access policy. These are measured in low, average, above average, high, and very high. The DES algorithm has 192bits of key size and it is very effective against brute force attack. AES has various key sizes that include 128, 192, and 256 bits, as well as RSA, has various key sizes that include 128, 256, 512, 1024, and 2048 bits.

3 Summary

Cloud services are now a fundamental component of our business life and provide a significant potential to speed up business by rapidly developing its resources and creating new prospects for cooperation. Cloud service provides us with agile resources. Indeed, the cloud provides enterprises, organizations, and even countries with various benefits. The cloud is nevertheless exposed to several security concerns, even with several advantages. Therefore, safety is the biggest hurdle in cloud adoption. The customer and suppliers are aware of the dangers to security. This research examines several key generation algorithms and their advantage with the disadvantage of the security and privacy of sensitive user data stored in the cloud environment of cloud. We realized the efficacy and the safety of algorithms utilized in all forms of algorithms when analyzing the proposed research into key creation in numerous types. The study offers a brief introduction and study of algorithms and offers some help to improve encryption schemes so that the network's security is guaranteed.

References

1. Chang V, Kuo YH, Ramachandran M (2016) Cloud computing adoption framework: a security framework for business clouds. Future Gener Comput Syst 57:24–41
2. Subramanian N (2018) Jeyaraj recent: a security challenges in cloud computing. Comput Electr Eng 71:28–42
3. Kumar S, Wollinger T (2006) Fundamentals of symmetric cryptography. In: Lemke K, Paar C, Wolf M (eds) Embedded security in cars. Springer, Berlin, Heidelberg, pp 125–143
4. Namasudra S (2019) An improved attribute-based encryption technique towards the data security in cloud computing. Concurrency Comput Pract Exper 31:e4364. https://doi.org/10.1002/cpe.4364
5. Namasudra S, Roy P (2016) Secure and efficient data access control in cloud computing environment: a survey. Multi-agent Grid Syst Int J 12 (2):69–90
6. Shabir MY, Iqbal A, Mahmood Z, Ghafoor AU (2016) Analysis of classical encryption techniques in cloud computing. Tsinghua Sci Technol 21(1):102–113
7. Soofi SS, Khan MI, Amin F (2014) Encryption techniques for cloud data confidentiality. Int J Grid Distrib Comput 17(4):11–20
8. Manjusha R, Ramachandran R (2014) Comparative study of attribute based encryption techniques in cloud computing, In: 2014 international conference on embedded systems (ICES), 2014, pp 116–120. https://doi.org/10.1109/EmbeddedSys.2014.6953101
9. Shynu PG, Singh KJ (2018) Privacy preserving secret key extraction protocol for multi-authority attribute-based encryption techniques in cloud computing. Int J Embedded Syst 10(4):287–300
10. Kaur R, Singh RP (2014) Enhanced cloud computing security and integrity verification via novel encryption techniques. In: International conference on advances in computing, communications and informatics (ICACCI), IEEE, pp 1227–1233
11. Kaur M, Singh R (2013) Implementing encryption algorithms to enhance data security of cloud in cloud computing. Int J Comput Appl 70(18):16–21
12. Hemalatha N, Jenis A, Donald AC, Arockiam L (2014) A comparative analysis of encryption techniques and data security issues in cloud computing. Int J Comput Appl 96(16)
13. Mehmood MS, Shahid MR, Jamil A, Ashraf R, Mahmood T, Mehmood A (2019) A comprehensive literature review of data encryption techniques in cloud computing and IoT environment. In: 8th international conference on information and communication technologies (ICICT), pp 54–59
14. Suthar K, Kumar P, Gupta H, Patel H (2012) Analytical comparison of symmetric encryption and encoding techniques for cloud environment. Int J Comput Appl 60(19):16–19
15. Liu X, Zhang Y, Wang B, Yan J (2013) Mona: secure multi-owner data sharing for dynamic groups in the cloud. IEEE Trans Parallel Distrib Syst 24(6):1182–1191
16. Song C, Park Y, Gao J, Nanduri SK, Zegers W (2015) Favored encryption techniques for cloud storage. In: IEEE first international conference on big data computing service and applications 267:274
17. Goyal R, Khurana M (2017) New design of tiny-block hybridization in AES. Int J Comput Netw Inf Secur 9:46–53
18. Ghosh S, Karar V (2018) Blowfish hybridized weighted attribute-based encryption for secure and efficient data collaboration in cloud computing. Appl Sci 8(7):1119. https://doi.org/10.3390/app8071119
19. Jeyabalu M, Krishnamoorthy K (2018) Hybridization of ICBC and genetic algorithm for optimizing encryption process in cloud computing application service. Fundamenta Informaticae 157:79–109

20. Shehzad D, Khan Z, Da H, Bozku Z (2016) A novel hybrid encryption scheme to ensure hadoop based cloud data security. Int J Comput Sci Inf Secur (IJCSIS) 14(4)
21. Sangeetha D, Vaidehi V (2016) A secure cloud based personal health record framework for a multi owner environment. Ann Telecomm 72:95–104
22. Kumar Mohit, Aggarwal Akshat, Garg Ankit (2014) A review on various digital image encryption techniques and security criteria. Int J Comput Appl 96(13):19–26
23. Brandao PR (2011) The importance of authentication and encryption in cloud computing framework security. Int J Data Sci Technol 4(1):1–5 (2018). **24** Sharma P, Sharma S, Dhakar RS (2011) Modified elgamal cryptosystem. In: 2nd international conference on computer and communication technology, pp 439–443
24. Radhadevi P, Kalpana P (2012) Secure Image encryption using AES. Int J Res Eng Technol 1(2):115–117
25. Saha R, Geetha G, Kumar G, Kim T (2018) RK-AES: an improved version of AES using a new key generation process with random keys. Hindawi Secur Commun Netw. Article ID 9802475,11
26. Hussain Sm, Al-Bahadili H (2016) A DNA-based cryptographic key generation algorithm. In: International conference on security and management, SAM'16, pp. 338
27. Lozhnikov P, Buraya E, Sulavko S, Eremenko A (2016) Methods of generating key sequences based on keystroke dynamics, dynamics of systems, mechanisms and machines (dynamics). In: Date of conference, pp 15–17
28. Gayathri P, Umar S, Sridevi G, Bashwanth N, Srikanth R (2017) Hybrid cryptography for random-key generation based on ECC algorithm. Int J Electr Comput Eng (IJECE) 7(3):1293–1298
29. Wang H (2018) Wang H, Song Y (2018) Secure cloud-based EHR system using attribute-based cryptosystem and blockchain. J Med Syst 42:152. https://doi.org/10.1007/s10916-018-0994-6

Chapter 34
Diverse Contemporary Algorithms to Resolve Load Balancing Issues in Cloud Computing—A Comparative Study

Lopa Mandal and Jayjeet Dhar

1 Introduction

Cloud computing has gained much importance in the present scenario [1]. According to NIST, cloud computing is defined as "a model for enabling convenient, on-demand network access to a shared pool of configurable computing resources (e.g. networks, servers, storage, applications and services) that can be rapidly provisioned and released with minimal management effort or service provider interaction" [2].

Cloud computing offers three main service models: Software as a Service (SaaS), Platform as a Service (PaaS) and Infrastructure as a Service (IaaS). There are three deployment models in cloud computing, viz. public cloud, private cloud and community cloud [3–7]. Resource utilization is a very important task in this context.

In cloud environment, load balancing is a technique to efficiently distribute tasks to available cloud resources, to ensure their complete utilization. It improves the overall performance of the system and reduces cost and the user's response time [8].

Load balancing strategies for cloud environment can be classified broadly into the following two categories: static and dynamic approaches.

Static algorithms: static load balancing algorithms use the prior knowledge about the nodes and assign the jobs at the beginning of implementation [9]. In this approach, there is no scope of reconfiguration with the changing scenario. Mostly, these are suitable for homogeneous and non-dynamic stable environments. Round Robin is a very famous static load balancing algorithm [10].

Dynamic algorithms: dynamic load balancing algorithms take into consideration the different attributes of the nodes' capabilities both prior to and during runtime and network bandwidth. These algorithms continuously track the situation for proper implementation [9].

L. Mandal (✉) · J. Dhar
Institute of Engineering & Management, Kolkata, India
e-mail: mandal.lopa@gmail.com

L. Mandal et al. (eds.), *Proceedings of International Conference on Computational Intelligence, Data Science and Cloud Computing*, Algorithms for Intelligent Systems,
https://doi.org/10.1007/978-981-19-1657-1_35

In this paper, an organized review of the current load balancing algorithms which are aimed at resolving load balancing issue in cloud computing is presented. Different journal articles and conference papers have been collected for the present study to analyse the data and generate insightful result. Google Scholar, Scopus and Web of Science were adapted as our data source. Consequently, the search process covered articles available in reliable databases that are technically and scientifically peer-reviewed: IEEE Xplore, Springer Link, Elsevier and ACM Digital Library among others.

The rest of this paper is organized as follows: in Sect. 2, the related works are reviewed, and algorithms already proposed by researchers to solve the relevant problems are discussed. Section 3 describes the problem statement, and in Sect. 4, the experimental setup is explained. In Sect. 5, experiment and results are stated, and the performance of relevant approaches is compared based on load balancing metrics. In Sect. 6, the paper concludes with the discussion of the possible areas of enhancement and potential scope of improvement of load balancing algorithms in cloud environment.

2 Related Works

Load balancing is cloud environment which is a complex task. Different researcher proposed different optimization techniques to achieve this. Researchers gave emphasis on the significance of choosing the appropriate node in the cloud for executing jobs in an efficient manner.

Round Robin algorithm is a well-known cloud load balancing algorithms. It allocates each incoming request to the next available server in the queue [11, 12]. The algorithm does not have any scope to record the status information. Without knowing the status, if new job is allocated to a node, there is a possibility that the node will be overloaded. To handle this problem, an improved Round Robin algorithm is proposed known as "Round Robin based on the load degree evaluation".

Round Robin algorithm firstly selects the initial node arbitrarily and then assigns tasks to multiple nodes in Round Robin (RR) directive [13]. The work load is equally distributed to the processors though the job processing time for distinct procedures is dissimilar. The real-world example of this algorithm is web server where http requests are identical and dispensed equally.

It is also stated that the classical Round Robin algorithm is a static algorithm, used to allocate resources according to time slices and execute in first-in-first-out (FIFO) order but is pre-emptive in nature. Hence, within the time quantum, the tasks must complete, otherwise, CPU is pre-empted to next waiting process [14]. As the algorithm does not consider the parameters like processing times for each individual requests, so if the virtual machine is not free, then the new job needs to wait in the queue [15].

Another load balancing algorithm in cloud environment is throttled load balancing algorithm (TLB). In this case, the load balancer maintains an index table of virtual machines (VM) along with their states. The states are busy or available. When new job request comes from the client, the indexed table is scanned and the first available VM that can perform the job is identified to assign the job. The VM id is then sent to the data centre. The data centre then communicates the VM to take that job. The allocation then is communicated to the load balancer, and the index table is updated accordingly.

If no VM is available, then the job requests are queued in the data centre. When any VM becomes free, it is assigned the job from the queue [16–18].

Modified throttled algorithm and priority-based modified throttled algorithm (PMTA) is the modified version; the TLB algorithm aimed to improve average execution time. It can manage the resources more efficiently in comparison to the load balancing algorithm discussed earlier.

Equally spread current execution (ESCE) load balancing algorithm maintains a load balancer to allot incoming new jobs to the free virtual servers. The algorithm is named so as it works on equally spreading the execution load on different virtual machine. Size of the process has been used in some cases to make this a priority-based dynamic load balancing algorithm [19–22]. Since load balancer spreads the load on to different lightly loaded nodes, after checking the size of the incoming process, it is known as spread spectrum technique. Equally spread current execution is called active monitoring load balancing algorithm by some researchers as well [23, 24].

The algorithm has got some similarities with the TLB algorithm. But, there are certain changes in maintaining the VM index table. Here, when a new job request comes, the index table is searched for the VM with least load. This VM id is then returned to the data centre so that the new job can be allocated to that VM. The job allocation count of that VM is increased then. After the job is over, the allocation count of that VM is decreased.

Ant colony optimization (ACO), is another algorithm used in load balancing in cloud environment. This algorithm is based on the concept of the pheromone trail laying and the behaviour of ant colonies [25]. The artificial pheromone information is updated based on the result of ants' search experience and existing heuristic information.

The ant colony algorithm is composed generally by the following steps: at first, the variables are initialized. Then, different iterations are taken into account for building of solution and also the local pheromone is updated. Finally, the updating of the global pheromone is done until end condition [26]. In [26, 27], an improved load balancing algorithm is proposed which aims to allow better response time on the basis of ACO technique.

Another research states that ant colony optimization (ACO), which is a swarm intelligence-based algorithm. It uses the technique of ants to search for food source randomly. It is based on the fact that ants lay pheromones on path, and the path with highest pheromones is considered as the best solution [28].

A load balancing algorithm is also derived based on the honey bee behaviour adopted to find food [29]. When comparing the foraging behaviour of honey bees with load balancing, the tasks are similar to the honey bees and the VMs are similar to the food sources. Finding a suitable VM is similar to a honey bee finding a food source [30]. In this algorithm, if a VM is overloaded, the new jobs will be submitted to an under loaded VM and the current load of that VM is tracked so that the waiting jobs can select an appropriate VM.

Honey bee behaviour-inspired load balancing (HBB-LB) [31] is also based on the foraging behaviour of honey bees. The advantages of the algorithm are low response time and low makespan, but it has low scalability.

Another related algorithm is enhanced bee colony algorithm [32]. It is also based on the foraging behaviour of honey bees. In this algorithm, the advantages are low VM migration and low makespan. But, the algorithm lacks scalability. Honey bee foraging behaviour is a self-organizing and a nature-inspired algorithm where performance is improved by increasing the system size, but overall throughput does not increase with increase in system size.

Particle Swarm Optimization is another algorithm used in this cloud load balancing [33]. This algorithm is based on swarm intelligence, where the behaviour of flocks of fishes and birds are used for optimization. In the context of cloud load balancing, particles are considered as tasks and VMs are the food sources. At first, tasks are assigned to available VMs for execution. As the performance of different VMs is different, some VMs get overloaded over time. Then, the tasks are migrated from overloaded VMs to underloaded VMs. It not only reduces execution time, PSO technique reduces the searching time of suitable VM for migration, and it also makes the system energy efficient [34, 35].

3 Problem Description

It has been established by the researchers that efficient load balancing strategies are required to optimize the cloud resource allocation. It is important to have certain metrics to determine the effectiveness of load balancing techniques in cloud computing [36]. Some of them are:

- **Response time**: It is the total time taken to serve a request. The response time expected to be minimized.
- **Cost**: It involves the migration cost, processing cost and storage cost. An efficient algorithm always tries to minimize this factor.
- **Scalability**: It determines how the system is capable of accommodating and balancing load in cloud environment.
- **Throughput**: This parameter tells us about the number of tasks completed per unit of time. For better performance of the cloud, a high throughput is required.
- **Fault tolerance:** It shows the algorithm's potential to handle the situation of fault and its recovery power from the failure.
- **Resource Utilization**: It is the degree to which the resources of the system are utilized. A desirable load balancing algorithm provides maximum resource utilization.

- **Migration time**: The time required to move a job from one machine to other machine in cloud environment is known as migration time. It is expected to be minimized for overall improvement in performance of the cloud system.
- **Makespan**: It is the maximum time required to complete a given sequence of jobs.

Study of different scheduling algorithms in the cloud computing environment shows that some researchers have given importance on reduction of execution cost, whereas others have focused on, resource utilization, response time, throughput, etc. The present work aims to compare existing cloud load balancing algorithms based on improved resource utilization, optimization criteria, response time and the costs incurred for the implemented algorithms. It also aims to compare the load balancing algorithms on the basis of their pros and cons so that the future researchers can get a useful insight about the discussed algorithms.

4 Experimental Setup

CloudSim toolkit is a very popular simulation platform to run and execute cloud simulations. By using this tool, whole bunch of objects of a standard data centre can be designed and developed in a simulated environment. It is a Java-based GUI environment where different parameters can be configured for the purpose of experiment.

The main components of Cloud Analyst are region, user base, internet, internet cloudlet, data centre controller and VM load balancer. The importance of the components is described below:

Region: The primary interface is divided into six main continents of this world. The data centre and the user base response time are displayed on the main interface refer Fig. 1.

User Base: User base represents a group of users that generates traffic required for the simulation.

Internet: It is used to define the characteristics of the internet applied during the simulation, including the latencies and available bandwidths between regions, the current traffic levels and current performance-level information for the data centres.

Internet Cloudlet: It is the grouping of user requests. It represents different details regarding the group of requests which is used for routing through the internet.

Data Centre Controller: It represents the controls on different data centre activities.

VM Load Balancer: It is used by the data centre controller to identify the VM that can be assigned to the next group of user requests for being processed.

For the present work, five user bases (UB1 to UB5) have been used and are distributed in five regions (0 to 4). Distribution of user bases in different regions is shown in Fig. 1. Number of virtual machines used in each data centre is six and the service broker policy chosen is "Closest Data Centre" in the Main Configuration section. The simulation is repeated for different simulation duration of 1, 2, 3 and

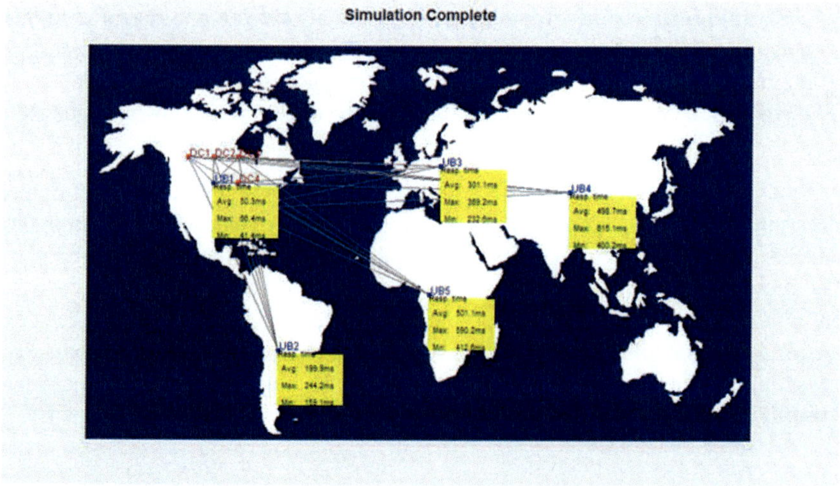

Fig. 1 Distribution of user bases in different regions in Cloud Analyst Tool

4 h. User base configuration and data centre configuration of simulation in Cloud Analyst Tool used for the present work are shown in Fig. 2 (for 3 h) and Fig. 3.

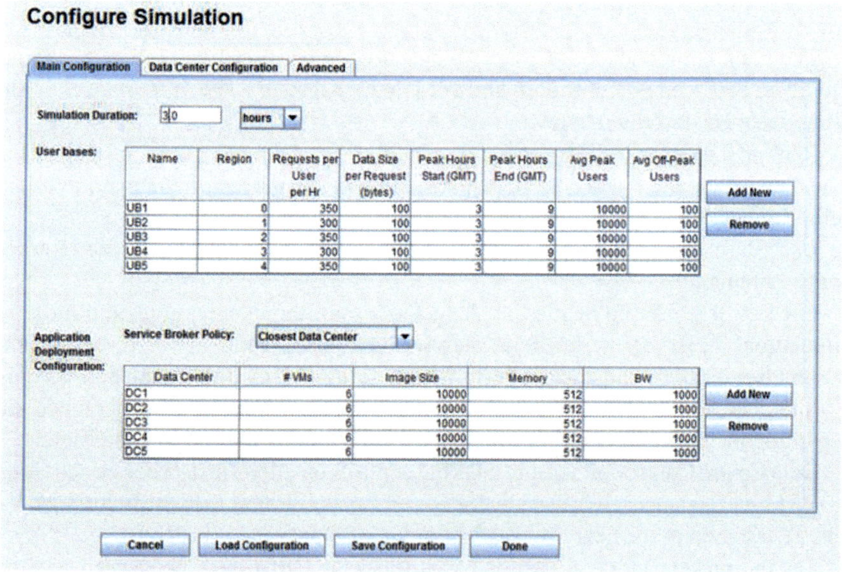

Fig. 2 User base configuration of simulation in Cloud Analyst Tool for 3 h

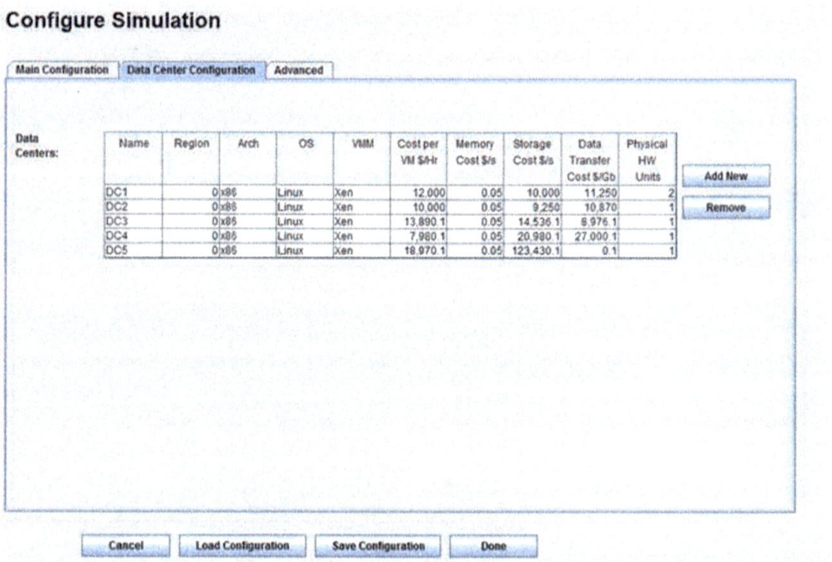

Fig. 3 Data centre configuration in Cloud Analyst Tool

5 Experiment and Results

The present work mainly focuses on six algorithms, viz. Round Robin, throttled, ant colony, honey bee, ESCEL, PSO. Experiments show their relative performances in terms of average response time and data centre processing time. The comparative simulation results after one hour, after two hours and after three hours are shown in Tables 1, 2 and 3, respectively. All the six algorithms are compared in terms of their data centre processing time after one hour, two hours, three hours and four hours of simulation as shown in Fig. 4. It is seen that that honey bee algorithm takes maximum time in each case. Similar comparison has also been done in terms of average response time as shown in Fig. 5. It is found that average response times of all the six algorithms are increased after four hours of simulation. Average response time after one hour of simulation is found maximum in case of particle swarm optimization. Average

Table 1 Simulation results for different cloud scheduling algorithms after 1 h of simulation

Parameters	Round Robin	Throttled	Ant colony	Honey bee	ESCEL	PSO
Average response time (ms)	307.99	308.02	307.99	308.06	307.98	308.65
Data centre processing time (ms)	0.34	0.35	0.36	0.42	0.36	0.35

Table 2 Simulation results for different cloud scheduling algorithms after 2 h of simulation

Parameters	Round Robin	Throttled	Ant colony	Honey bee	ESCEL	PSO
Average response time (ms)	307.84	307.82	307.86	307.92	307.85	307.82
Data centre processing time (ms)	0.34	0.35	0.36	0.44	0.36	0.35

Table 3 Simulation results for different cloud scheduling algorithms after 3 h of simulation

Parameters	Round Robin	Throttled	Ant colony	Honey bee	ESCEL	PSO
Average response time (ms)	308.11	308.12	308.13	308.18	308.12	308.08
Data centre processing time (ms)	0.35	0.35	0.36	0.44	0.36	0.35

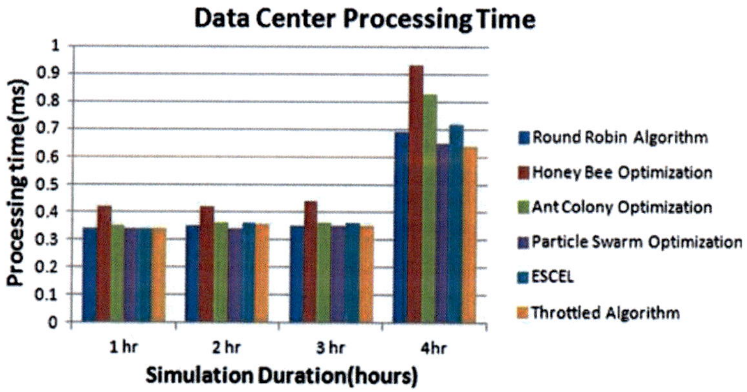

Fig. 4 Data centre processing for 4 h of simulation

response time after four hours of simulation is found maximum in case of honey bee algorithm. Total cost incurred during 4 h of simulation has also been shown, refer Fig. 6 (Tables 4 and 5).

6 Conclusion

The paper widely reviews various techniques in the area of load balancing solutions and presents an organized comparative review of state-of-the-art load balancing

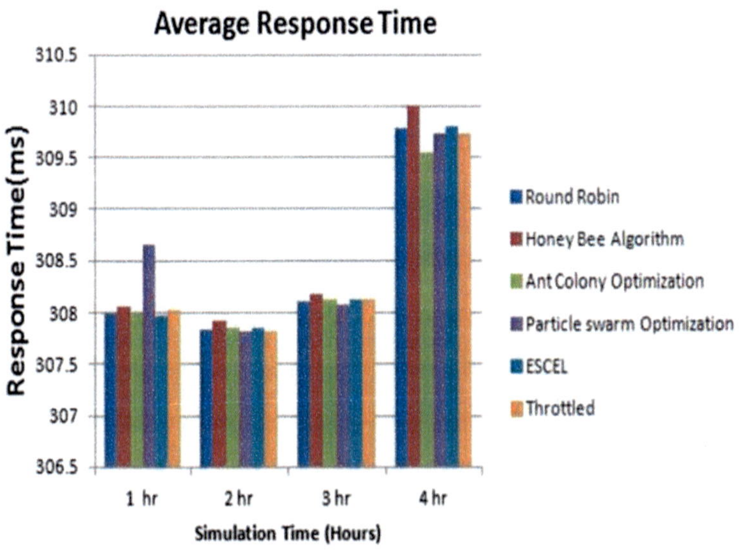

Fig. 5 Average response time for 4 h of simulation

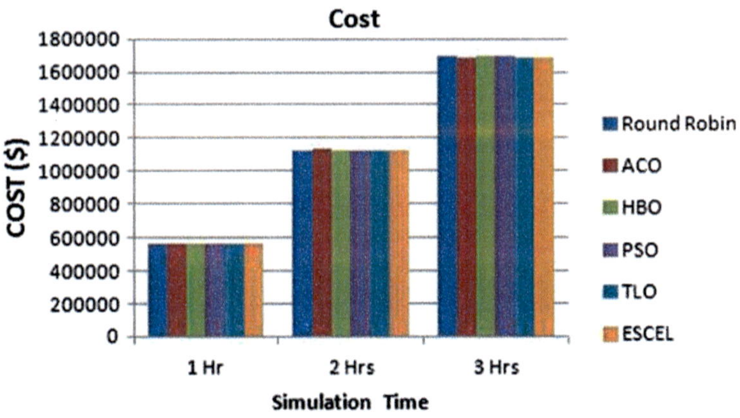

Fig. 6 Total cost incurred during 4 h of simulation

techniques in cloud environment. Based on the available literature, the present study classified the domain of research broadly into subdomains related to concepts of dynamic load balancing and static load balancing. The present work mainly focused on six algorithms, viz. Round Robin, throttled, ant colony, honey bee, ESCEL and PSO. The pros and cons of each load balancing algorithm have also been highlighted.

The present work also compared the existing cloud load balancing algorithms based on improved resource utilization, optimization criteria, response time and the costs incurred for the implemented algorithms.

Table 4 Performance of various load balancing algorithms based on certain metrics

Algorithms	Performance	Response time	Data centre processing time	Fault tolerance	Cost	Scalability	Resource utilization	Makespan	Complexity
Round Robin	Good for short CPU burst	Fast	Fast	No	Costly but cheaper than PSO	No	Low	Moderate	Simplest
Throttled load balancing	Good	Fast	Moderate	Yes	Least	Yes	Moderate	Moderate	Simple
Ant colony optimization (ACO)	Moderate	Moderate	Slow	Yes	Moderate	Yes	High	Fast	Moderate
Honey bee optimization	Moderate	Moderate	Slow	No	Moderate higher than ACO	No	High	Fastest	Moderate
Equally spread current execution load (ESCEL)	Moderate	Slow	Moderate	Yes	Lower than ACO, HB, PSO and RR	Yes	High	Fast	Simple
Particle swarm optimization	Best	Fastest	Moderate but faster than ESCEL	Yes	Costliest	Yes	High	Moderate	Highest complexity

Table 5 Pros and cons of load balancing algorithms based on certain metrics

Algorithms	Pros	Cons
Round Robin	– Easier to implement – Performs a limited amount of tasks in an efficient way – Fixed time quantum – Performs better for short CPU burst	– Does not work well in heterogeneous environments – Some data centres might be overloaded, while others might remain idle at the same time – Not scalable – Larger tasks take long time – More context switches due to short quantum time
Throttled load balancing	– Good performance – Reduces response time – Organized use of table makes processing faster	– If the VM is not ready to execute, then it may suffer starvation
Equally spread current execution load (ESCEL)	– Better utilization of resources – Minimizes makespan	– Costlier – Data centre processing time greater with respect to Throttled, Round Robin, PSO
Ant colony optimization	– The node with least load is found at first in the search – Decentralized, no single point of failure – Ants can collect information faster – Minimizes makespan	– The large number of ants can cause network overhead – Number of ants and points of initialization of ants and are not clear – Only availability of node is being considered; other factors should also be taken into consideration
Honey bee optimization	– Reduces makespan – High resource utilization – Minimizes response time – Increases throughput – Scalable	– Network overheads – Need to explore heterogeneous task – High priority tasks cannot work without VM machine
Particle swam optimization (PSO)	– Good performance – High resource utilization	– Too much complexity

The overall data collected in this study will be helpful to acquaint the researchers with the nitty–gritty of the load balancing issues in cloud computing and its available algorithmic solutions. The drawbacks of the discussed algorithms have also been highlighted. Future research can focus on improving the performance of these algorithms and can also propose new algorithms to make the overall system to be more energy efficient.

References

1. Panwar R, Mallick B (2015) Load balancing in cloud computing using dynamic load management algorithm. In: 2015 International conference on green computing and Internet of Things (ICGCloT)
2. Mell P, Grance T (2009) The NIST definition of cloud computing. National Institute of Standards and Technology, 53
3. Hwang K et al (2016) Cloud performance modeling with benchmark evaluation of elastic scaling strategies. IEEE Trans Parallel Distrib Syst 27(1):130–143
4. Wang L, Tao J, Kunze M, Castellanos AC, Kramer D, Karl W (2008) Scientific cloud computing: early definition and experience. In: IEEE, pp 825–830
5. Mishra SK, Sahoo B, Parida PP (2020) Load balancing in cloud computing: a big picture. J King Saud Univ-Comput Inf Sci 32(2):149–158
6. Arora P, Wadhawan RC, Ahuja ESP (2012) Cloud computing security issues in infrastructure as a service. Int J Adv Res Comput Sci Softw Eng 2(1)
7. Motta G, Sfondrini N, Sacco D (2012) Cloud computing: an architectural and technological overview. In: 2012 International joint conference on service sciences (IJCSS). IEEE
8. Ghumman NS (2016) Cloud computing model and its load balancing algortihms. In: 2016 3rd International conference on computing for sustainable global development (INDIACom). IEEE
9. Al Nuaimi, K et al (2012) A survey of load balancing in cloud computing: challenges and algorithms. In: 2012 Second symposium on network cloud computing and applications (NCCA). IEEE
10. Sotomayor B et al (2009) Virtual infrastructure management in private and hybrid clouds. IEEE Internet Comput 13(5)
11. Ghosh S, Banerjee C (2018) Dynamic time quantum priority based round robin for load balancing in cloud environment. In: 2018 Fourth international conference on research in computational intelligence and communication networks (ICRCICN), IEEE, pp 33–37
12. Enokido T, Aikebaier A, Takizawa M (2010) A model for reducing power consumption in peer-to-peer systems. IEEE Syst J 4(2):221–229
13. Jain S, Saxena AK (2016) A survey of load balancing challenges in cloud environment. In: System modeling & advancement in research trends (SMART), international conference. IEEE
14. Pavithra B, Ranjana R (2016) A comparative study on performance of energy efficient load balancing techniques in cloud. In: International conference on wireless communications, signal processing and networking (WiSPNET). IEEE
15. Domanal SG, Ram Mohana Reddy G (2013) Load balancing in cloud computing using modified throttled algorithm. In 2013 IEEE international conference on cloud computing in emerging markets (CCEM). IEEE
16. Ray S, De Sarkar A (2012) Execution analysis of load balancing algorithms in cloud computing environment. Int J Cloud Comput: Serv Arch (IJCCSA) 2(5):1–13
17. Kaur P, Kaur PD (2015) Efficient and enhanced load balancing algorithms in cloud computing. Int J Grid Distrib Comput 8(2):9–14
18. Ghosh S, Banerjee C (2016) Priority based modified throttled algorithm in cloud computing. In: International conference on inventive computation technologies (ICICT), vol 3. IEEE
19. Bagwaiya V, Raghuwanshi SK (2014) Hybrid approach using Throttled and ESCE load balancing algorithms in cloud computing
20. Nitika, Shaveta, Raj G (2012) Comparative analysis of load balancing algorithms in cloud computing. Int J Adv Res Comput Eng Technol 1(3):120–124
21. Ahmed T, Singh Y (2012) Analytic study of load balancing techniques using tool cloud analyst. Int J Eng Res Appl, pp 1027–1030
22. Suguna S, Barani R (2015) Simulation of dynamic load balancing algorithms. Bonfring International Journal of Software Engineering and Soft Computing 5(1):1. [23] Kushwaha M, Gupta S (2015) Response time reduction and performance analysis of load balancing algorithms at peak hours in cloud computing. Int J Comput Appl

23. Hota A, Mohapatra S, Mohanty S (2019) Survey of different load balancing approach-based algorithms in cloud computing: a comprehensive review. Comput Intell Data Min, pp 99–110
24. Patel S et al (2015) CloudAnalyst: a survey of load balancing policies. Int J Comput Appl 117(21)
25. Liu C, Fengrui Mu, Zhang W (2021) Cloud computing demand elasticity algorithm based on ant colony algorithm. Recent Adv Electr Electron Eng (Formerly Recent Patents on Electrical & Electronic Engineering) 14(1):37–43
26. Ragmani A et al (2016) A performed load balancing algorithm for public Cloud computing using ant colony optimization. In: 2016 2nd international conference on cloud computing technologies and applications (CloudTech), IEEE
27. Liao T et al (2014) Ant colony optimization for mixed-variable optimization problems. IEEE Trans Evol Comput 18(4):503–518
28. Abualigah L, Diabat A (2020) A novel hybrid antlion optimization algorithm for multi-objective task scheduling problems in cloud computing environments. Cluster Comput, pp 1–19
29. Babu KRR, Samuel P (2016) Enhanced bee colony algorithm for efficient load balancing and scheduling in cloud. In: Innovations in bio-inspired computing and applications. Springer International Publishing, pp 67–78
30. Sheeja YS, Jayalekshmi S (2014). Cost effective load balancing based on honey bee behaviour in cloud environment. In: 2014 First international conference on computational systems and communications (ICCSC). IEEE
31. Dhinesh Babu LD, Venkata Krishna P (2013) Honey bee behavior inspired load balancing of tasks in cloud computing environments. Appl Soft Comput 13(5):2292–2303
32. Karaboga D, Basturk B (2008) On the performance of artificial bee colony (ABC) algorithm. Appl Soft Comput 8(1):687–697
33. Ali MF, Batarfi OA, Bashar A (2015) A simulation-based comparative study of Cloud Datacenter scalability, robustness and complexity. In: 2015 IEEE Seventh international conference on intelligent computing and information systems (ICICIS). IEEE
34. Devaraj A, Saviour F, Elhoseny M, Dhanasekaran S, Laxmi Lydia E, Shankar K (2020) Hybridization of firefly and improved multi-objective particle swarm optimization algorithm for energy efficient load balancing in cloud computing environments. J Parallel Distrib Comput 142:36–45
35. Wig A, Khushwah RS et al (2015) An efficient distributed approach for load balancing in cloud computing. In: International conference on computational intelligence and communication networks. IEEE Press
36. Milani AS, Navimipour NJ (2016) Load balancing mechanisms and techniques in the cloud environments: Systematic literature review and future trends. Elsevier J Netw Comput Appl

Chapter 35
Gradient-Based Intra-prediction in H.265/HEVC to Improve the PSNR

Vijayalaxmi S. Patil and Suvarna Nandyal

1 Introduction

Network bitrates have been steadily increasing in remote locations but have been decreasing in broader areas. High-bandwidth connections to the house are standard, and hard drives, flash memory, and optical media have greater storage capacity than at any time previously. When it comes to the rate per sent or stored bit steadily decreasing, it is doubtless no longer. It should be obvious why video compression is significant (and why there is such a serious effort to improve it). Compression (coding) usually has two significant advantages. The primary advantage is that it enables digital video format storing and transmission, with supported uncompressed or raw video.

Even though the frame rate or size is lower, current web performance cannot handle real-time uncompressed video. The secondary advantage is that compression of video may allow for more efficient storage and transmission of data. Over the last two decades, video and image compression have been a hot topic of research and implementation, with numerous groundbreaking compression and decompression methods and algorithms suggested and created.

Creating standard compression encoding and decoding means has been critical to correctly communicating across products from various manufacturers to stimulate competition, interworking, and more options. As a result, several critical global standards for image and video compression have been developed, such as the MPEG, JPEG, H.264, and H.265/HEVC standards. These guidelines focus on two critical points: a decoded syntax for reconstructing visual information and a coded syntax for compactly describing visual data.

V. S. Patil (✉) · S. Nandyal
PDA College of Engineering, Kalaburagi, India
e-mail: vijayalaxmispatil21080@gmail.com

L. Mandal et al. (eds.), *Proceedings of International Conference on Computational Intelligence, Data Science and Cloud Computing*, Algorithms for Intelligent Systems,
https://doi.org/10.1007/978-981-19-1657-1_36

In recent years, digital signal processing and transmission technology have made multimedia software more flexible and firmer. The ISO/IEC MPEG and ITU-T VCEG Joint Video TERM (JVT) work on a new raw video image coding standard (compression). The new standard is H.265/HEVC and MPEG-4 part-10 new video coding and is considered the video compression standard for the future generation. As demonstrated, the whole design of H.265/HEVC is identical to the previous video coding standard, which can be a hybrid coder. H.265 is primarily concerned with the efficient compression of video frames.

These standards' key characteristics include far better compression than any preceding standard. It also has various characteristics that support stable, dependable, and efficient transmission across a variety of networks, with a focus on widely used video compression applications. Nonetheless, compared to prior video criteria with the same quality PSNR, H.265/HEVC can achieve a 50% bit rate reduction using unique techniques and broad rate distortion optimization (RDO) techniques. To achieve the best coding affectivity, RDO techniques in H.265 compare different combinations of nodes, such as specific intra- and inter-prediction modes. Due to this, video coding's difficulty and computational overhead in H.265 increased significantly.

2 Literature Survey

Pastuszak et al. [1] proposed a well-organized H.265/advanced video coding (HEVC) encoder architecture that employs the RDO technique to achieve large throughputs. At simultaneously MB process as the RD mode examination, motion estimation is carried out. To obtain the accurate vector motion prediction, every time, the examination is done by dividing and MB levels. Xu et al. [2] proposed a practical intra-prediction module. The AVS coding standards are presented first, along with an introduction to its principles. High-efficiency architecture is designed using the intra-prediction technique analysis.

Azimi et al. [3] presented an altered H.265/HEVC standard to define HDR content properly. A new Lagrangian multiplier is utilized to save a lesser bit rate and minimize distortion of HDR video. The Lagrange multiplier has been implemented in the H.265/HEVC software. This approach could produce high-quality compressed HDR video. Peng et al. [4] devised a model for intra-prediction that is both efficient and effective. The transform orders affect the coding gain with adequate directional modes, thorough decomposition, and better interpolation. A directional filtering transform is explained for enhanced correlation among H.265 intraframe coding samples. With this approach, both intra-block and inter-block correlations experiment.

Jiao et al. [5] explained a pipelining approach for selecting intra-prediction mode w.r.t 4×4 blocks. The selection of intra-prediction mode for H.265/HEVC 4 with GPU streaming has been created. A specific technique with optimizing instruction and maximizing combined memory was developed to use the GPU's fine-grained parallelism. Kim et al. [6] proposed an efficient approach for MB mode determination and effectively choosing the mode in H.265/HEVC for intra-prediction high profile. They improved mode selection for prediction by using the current MB variance. H.265/HEVC effectively was also studied by Zhang et al. [7], Lee [8], Mansour [9], [3, 10], and others.

Bitstream coder carriers compress H.265/HEVC by performing prediction, transformation, and encoding. All the time, depending on data compression already collected, selected between the current frame (which means intra-prediction) or another frame (which means inter prediction). In one macroblock (MB), the quantity of mode permutations using the RDO process for intra-prediction in H.265/HEVC is $N8 \times (16 \times N4 + N16) = 4 \times (16 \times 9 + 4) = 592$. A 16×16 luma block has N16 modes, an 8×8 chroma block has N8, a 4×4 luma block has N4, and an 8×8 chroma block has N8 modes. To pick the best MB quality, the H.265/HEVC encoder conducts 592 RDO computations. As an outcome, the encoder's difficulty in existing systems rises dramatically. In the proposed work, there are $1 \times (2 \times 16 + 4) = 36$ mode combinations. The paper below has comprehensive explanations.

3 Methodology

Figure 1 depicts the proposed methodology for intra-prediction of H.265/HEVC. The input video is first transferred to the block for frame generation, which generates frames. Pre-processing is applied to each frame based on the input selected. Image scaling and color conversion are examples of this. Then, 'I' had to figure out how to make a frame out of many photographs (GOP). An 'I' frame indicates the beginning of each frame. 'I' frames are always placed after P and B frames. Reconstructing 'I' frames does not require any additional information because they always contain the

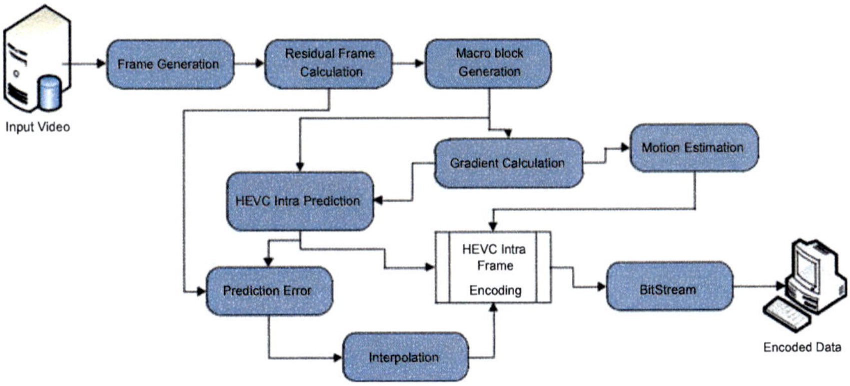

Fig. 1 Architecture of the proposed system

entire image. Furthermore, in the 'I' frames, the texture information for the entire plan generates 16 × 16 and 4 × 4 macroblocks (MB).

Homogeneity determination is performed on the calculated 4 × 4 block. The homogeneity of the block is determined by applying the DHT transformation to it. The measured energy is matched with the value, which assets to be the threshold. A gradient-dependent forecast is made if the energy exceeds the fixed value of the threshold. When calculated energy falls below the threshold, DC mode is used to make the prediction. Because the DC mode is not focused on direction, it usually provides a better prediction than other modes. The DC prediction uses approximately recently decoded pixels in rows and columns to predict corresponding pixels in the 4 × 4 block.

Each pixel is predicted individually during the prediction. The luminance value is stored at the current block, resulting in increased values of PSNR in many circumstances. As a result, they have DC mode a greater chance of succeeding the mode of prediction selection. All four prediction modes are implemented for the 16 × 16 MB. The cheapest model is regarded as the best in this case. The DC mode is used to anticipate the 8 × 8 chroma (U, V) block once more. The anticipated MB obtained at every step is then transferred to the interpolation block, which improves the MB's PSNR. The final smoothed MB with good PSNR comes from the interpolation block. Each block has a thorough explanation below.

3.1 Gradient-Based Prediction

If the 4 × 4 MB's similarity exceeds a threshold, the MB is sent to the gradient prediction block. The spatial features of the textures in the homogenous block are comparable. This MB has a tight relationship with the block next door. As a result, the direction of gradient in this MB tends to point in a single direction. As a result, calculating the angles of these routes yields the proper mode for the determined direction. Figure 2 shows the many modes for various directions. The graphic illustrates that nine modes are defined for various orientations. However, a few in the first and

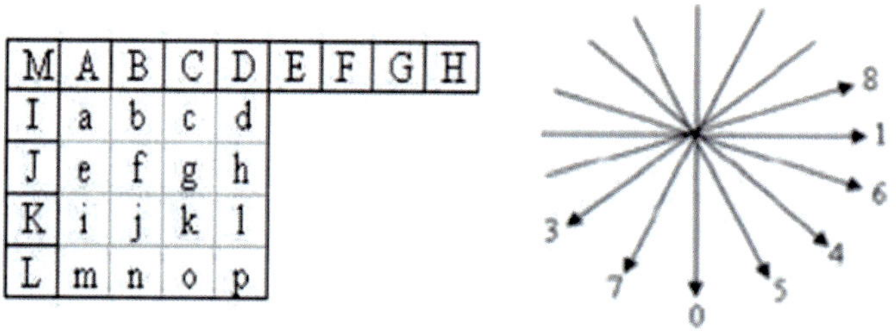

(a) (b)

Fig. 2 Prediction modes for different gradient direction second quadrants are not. As a result, the DC mode is used in these situations [11]. The partial derivative of the image in both x- and y-axes are combined to generate a gradient vector for calculating the gradient at an image location. This is seen in the following equation.

$$\nabla I = \left(\frac{\partial I}{\partial x}, \frac{\partial I}{\partial y}\right) \tag{1}$$

This gradient appears to be a vector that represents the speed with which the image changes while traveling in either x or y the directions. However, this does not apply to calculations in other directions. The arbitrary direction in the location (x + cos, y + sin) is defined as the direction change when we change from location by a tiny amount. Taking into account the intensity I (x, y) at a specific site (x, y),

$$I(x + \nabla \cos\theta, y) = I(x, y) + \nabla \cos\theta \frac{\partial I(x, y)}{\partial y} \tag{2}$$

This denotes movement in the x-direction; the intensity change equals the length traveled multiplied by the intensity derivative in the x-direction. As a result, as we move a little, the derivative provides us with the intensity change. The angle is determined utilizing equation (3). The angular value is future compared to specified thresholds to determine the appropriate mode.

$$\text{orientation} = \frac{\left(\left(a \tan 2\frac{\partial I}{\partial x}, \frac{\partial I}{\partial y}\right) + pi\right) * 180}{pi} \tag{3}$$

3.2 *DC Mode for Prediction*

Suppose the 4 × 4 block MB's homogeneity is lesser than the threshold value. In that case, the MB is forwarded through the prediction block of DC mode. In similar circumstances, DC mode is applied straightly to MB. The significant block with maximum textures is used along DC mode to predict MB because the DC mode does not define any direction. The nearby pixel value is replaced with the value of entire pixels in DC mode prediction. The preferred pixels are a, b, c, d, i, j, k, and l, as seen in Fig. 2 (a). The stages for making a DC prediction are listed below.

- Calculate the anticipated samples if all of these reference pixels are accessible using (a + b + c + d + i + j + k + l + 4) ≫ 3.
- When pixels are missing, 128 is the expected value with whole luma block [8] that are used instead.

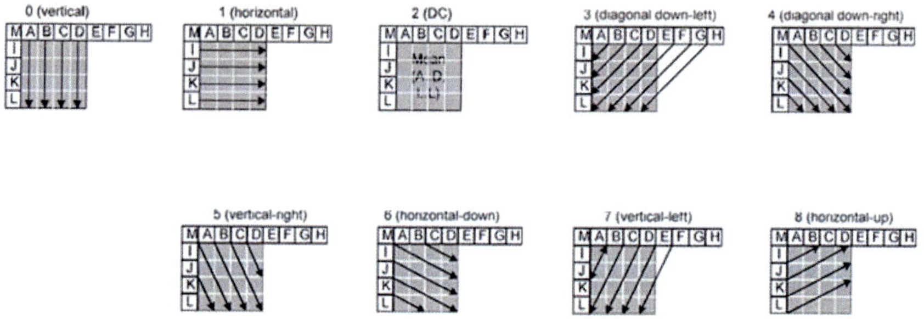

Fig. 3 9 Modes for 4 × 4 luma blocks

3.3 The Mode Prediction for 16 × 16 Luma and 8 × 8 Chroma MB

The 16 × 16 luma prediction is made in a region with a high degree of homogeneity. There are four different prediction modes available for this forecast. Modes of DC, vertical, horizontal, and plane suggesting are the four predicted modes. The plane prediction mode is accepted in these modes, identical with the modes utilized in 4 × 4 intra-prediction. A combination of weighted horizontally and vertically neighbor pixels is utilized in the planar prediction. On this, the whole four planes are MB. The mode having the lowest price is considered the better mode. The same four modes are considered in 8 × 8 chroma. However, the best DC mode is the prediction model (Fig. 3).

3.4 Interpolation

Each projected MB steps of interpolation is to improve the PSNR. The interpolation strategy, as shown in Fig. 4, consists of eight sets of polynomial equations. These eight equations can be obtained by partitioning each block into 4 × 4 blocks and splitting it into eight blocks with a reference frame. The reference frame is created by optimizing several sets of frames while maintaining a GOP size of 15. After that, each block is utilized to obtain the coefficient for the set of equations.

After that, the predicted MB is provided to calculate the interpolation block, which applies all polynomial equations. The error computation block receives the eight sets of 16 values in 1 D. The error computation is done with the original MB to see whether the equation is better than the expected values. The 16 best-interpolated values are chosen as the best-interpolated values. These values are converted from 1D to 2D to obtain the anticipated MB with better PSNR. For adaptability, a few steps in the flow can be changed.

The proposed system's algorithm is listed below.

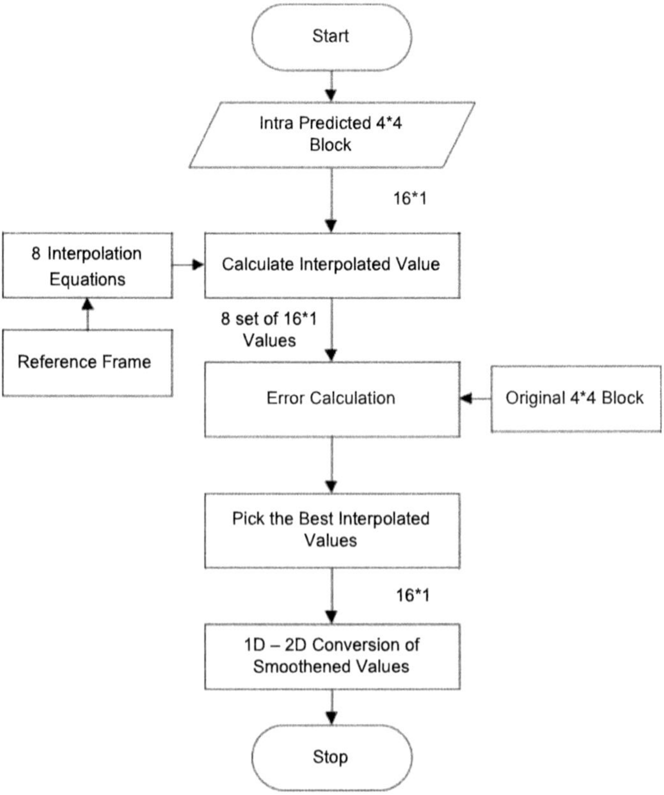

Fig. 4 Flowchart representing interpolation to get smoothened values

---

Input : *Input Frame*

Output : *Interpolated Frame*

Step 1: Create a block for input Frame.

Step 2: if the image is 16 × 16, applied to four modes.

Step 3: if it is 4 × 4, check for homogeneity.

Step 4: if the determined greater homogeneity than the threshold, the gradient is computed.

Step 5: The appropriate orientation is calculated depending on the gradient.

Step 6: Compare orientation values over a range of angles to find the optimal mode.

Step 7: Interpolate the Predicted frame's final frame.
Step 8: Reconstruct the interpolated frame.

4 Experimental Result

The results acquired at each phase of the process are shown in this section. As illustrated in Fig. 5a, the result generated from one RGB first frame for first video, second video, and third video is shown. After preprocessing, the frame is considered an 'I' frame. It is transformed into 16×16 and 4×4 blocks depending on texture data contained in all planes. The homogeneity determination check for the 4×4 block is performed using DHT once more.

Computation of energy is determined by comparison of the threshold value. The plot of gradient found for one macroblock is shown in Fig. 5b (MB). A gradient-dependent forecast is made while exceeding energy with a threshold value. The

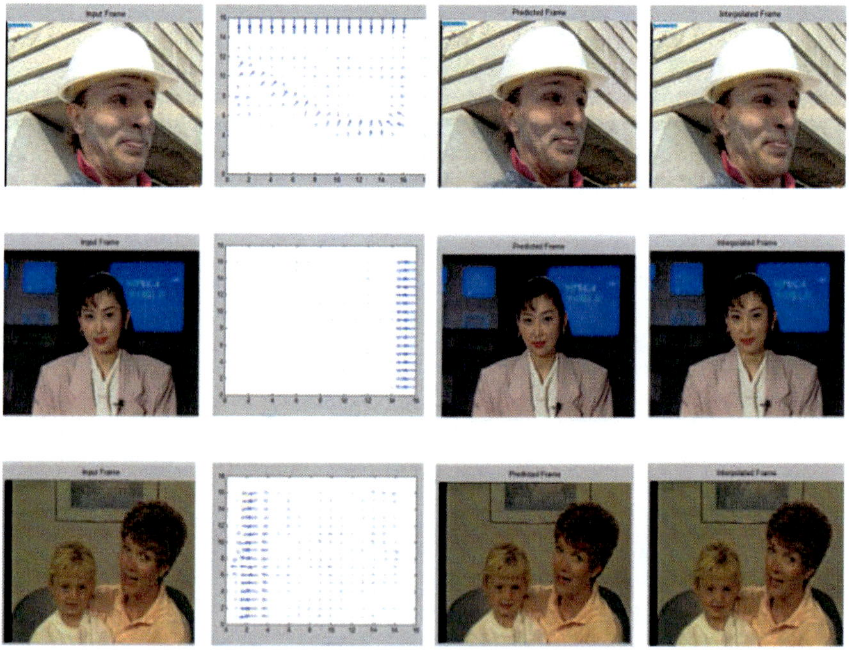

(a) (b) (c) (d)

Fig. 5 **a** Select video as input; **b** plot of calculation of gradient; **c** predicted frame for each select frame; **d** interpolated RGB final image optimal mode for the prediction is derived depending on the calculation of gradient as shown in Fig. 5c for varied quantization parameter = 20, 24, 28, 32 for three movies. The interpolation block receives the predicted MB for each MB. Figures 6 and 7 show the PSNR and time graphs for each of the three videos, based on six frames from each.

The final interpolated RGB frame 1 is shown in Fig. 5d. The value of before PSNR and later the frame for various QP values is determined and presented in Table 2 when interpolation is included to increase the final better quality output. Table 1 shows that, when compared to the present H.265, the suggested approach provides a good PSNR value.

The same procedure is followed for the remaining frames. As a result, our proposed method greatly reduces number of RDO calculations from 1 * (4 * 16 + 4) = 68, 2 * (4 * 16 + 2) = 132, 4 * (9 * 16 + 4) = 592, and 1 * (3 * 16 + 2) = 50 [11] modes

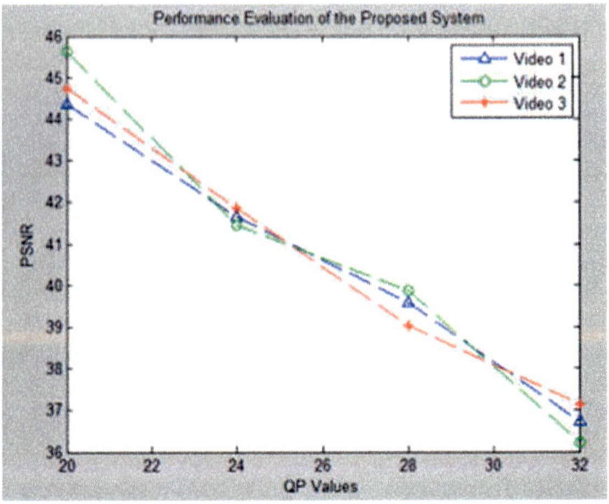

Fig. 6 Average PSNR graph for six frames computed for three videos

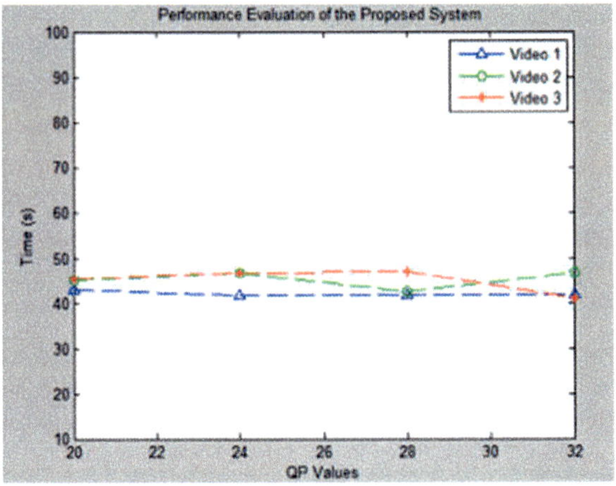

Fig. 7 Average time graph for six frames calculated for all the three videos

Table 1 Table for comparing values of average PSNR found from existing and proposed system

Sequence	QP	Paper	PSNR
	20	Saha et al. [12]	34.885
Foreman		Changsung Kim et al. [13]	38.3075
		Proposed	40.82

Table 2 Table for comparing values of PSNR before and after interpolation

Videos	Quantization parameter	PSNR		
		FRAME 1		
		PSNR before interpolation	PSNR after interpolation	Elapsed time in seconds
Video 1 (Foreman)	20	42.82	44.36	43.082655
	24	40.62	41.65	41.632461
	28	37.19	39.57	46.716965
	32	34.74	36.72	42.405948
	20	43.28	45.62	45.208599
Video 2 (Akiyo)	24	40.52	41.44	46.769455
	28	37.39	39.87	42.592223
	32	34.47	36.22	46.690972
Video 3 (Mother_Daughter)	20	42.34	44.73	45.494613
	24	40.67	41.87	46.695881
	28	37.92	39.03	47.089172
	32	35.44	37.12	41.004173

which are adopted in the existing RDO calculation in H.265/HEVC video coding to $1 \times (2 \times 16 + 4) = 36$.

5 Conclusion

The video sequence is compressed using the H.265/HEVC video coding standard specified as a proposed method in this paper. To obtain identical PSNR as a full search, the suggested method employs a gradient-based technique and an interpolation mechanism. This method could minimize the number of mode options in each MB participating in RDO computation to 36. After examining this codec using foreman video and various publicly available videos, the video quality is proven to be better trustworthy and assured on the premise of the higher performance of transcoding procedures. Furthermore, the technique is simple, with the video data being well-protected. The new strategy yielded higher gains and accuracy than the existing methods.

References

1. Pastuszak G (2015) Architecture design of the H.265/AVC encoder based on rate-distortion optimization. IEEE 25(11):1844–1856
2. Xu S, Xing Y, Wei X (2011) Design of intra prediction module in H.265 and AVS dual modes video decoder chip. IEEE 2:635–638
3. Azimi M, Pourazad MT, Nasiopoulos P (2014) Rate-distortion optimization for high dynamic range video coding. In: IEEE, pp 310–31
4. Peng X, Xu J, Wu F (2010) Directional filtering transform for image/intra-frame compression. IEEE Trans Image Process, 19(11)
5. Jiao L, Zhou J, Chen R (2011) Efficient parallel intra-prediction mode selection scheme for 4 × 4 blocks in H.265. IEEE, pp 527–530
6. Kim T, Hwang U, Jeong J (2011) Efficient block mode decision and prediction mode selection for intra prediction in H.265/AVC high profile. IEEE, pp 645–649
7. Grois D, Peng X, Mulayoff A, Itzhaky B, Marpe D, Hadar O (2013) Performance comparison of H.265/MPEG-HEVC, VP9, and H.265/MPEG-AVC encoders. IEEE, pp 394–397
8. Lee C, Kim C-S (2008) Rate-distortion optimized compression of high dynamic range videos. IEEE, pp 1–8
9. Mansour H, Fallah YP, Nasiopoulos P, Krishnamurthy V (2009) Dynamic resource allocation for MGS H.265/AVC video transmission over link-adaptive networks. IEEE 11(8)
10. UlAlam T, Ikbal J, UlAlam T (2012) Fast DC mode prediction scheme for intra 4 × 4 block In H.265/AVC video coding standard. (IJACSA) Int J Adv Comput Sci Appl 3(9)
11. Elyousfi A, Tamtaoui A, Bouyakhf E (2010) Fast intra prediction algorithm for H.265/AVC based on quadratic and gradient model. Int J Electr Electron Eng
12. Saha A, Mallick K, Mukherjee J (2007) SKIP prediction for fast rate-distortion optimization in H.265. IEEE 53(3):1153–1160
13. Kim C, Jay Kuo C-C (2007) Feature-based intra-InterCoding mode selection for H.265/AVC. IEEE 17(4)

Chapter 36
Semantic Web Architecture-Based Examination System for Smart Learning Environment

Subhabrata Sengupta, Anish Banerjee, Tanmoyee Roy, Sakshi Didwania, Srishti Chakraborti, and Satyajit Chakrabarti

1 Introduction

Smart e-learning is the new educational content that is focused on the learner's use of technology at their fingertips. It depends on how the software and hardware are made into use for quality education/training. The classification model is proposed for ordering the e-learning materials/substances which are accessible uninhibitedly over the Web. The substance is arranged based on different data sources like catchphrases, meta-keywords, and the tally of watchwords on the specific material identified with the subject. Rating, regular visits and how often the student is visiting that substance, time spent on the specific page, and level hardness of the substance are teamed up and recognized to fitting classes of materials as fundamental, moderate, and progressed levels. This system comprises all the data about the learning materials and their example levels. Search inquiries are frequently short, and the basic learner's aim might be questionable. This makes it trying for the framework to anticipate potential plans, just one of which might relate to the current learner. To resolve this issue, the system frequently broadens the outcome rundown and presents records pertinent to various goals of the question. The system initially distinguishes a scientific categorization of explanation for open-space search questions by investigating enormous scope inquiry reformulation information inspected from browser logs. This scientific categorization drives clients to a bunch of inquiry layouts and a basic yet powerful opening filling calculation. The system further uses this model as a wellspring of

S. Sengupta (✉) · A. Banerjee · T. Roy · S. Didwania · S. Chakraborti
Department of Information Technology, Institute of Engineering & Management (IEM), Kolkata 700091, India
e-mail: subhabrata.sengupta@iemcal.com

S. Chakrabarti
Department of Computer Science & Engineering, Institute of Engineering & Management (IEM), Kolkata 700091, India

L. Mandal et al. (eds.), *Proceedings of International Conference on Computational Intelligence, Data Science and Cloud Computing*, Algorithms for Intelligent Systems,
https://doi.org/10.1007/978-981-19-1657-1_37

feeble management to consequently produce explanatory inquiries for preparing. Besides, we propose regulated and support learning models for producing explaining questions gained from frail oversight information. The system likewise researches techniques for producing up-and-comer responses to each explaining question, so clients can choose from a bunch of predefined answers. Performance measurement is racially critical to dissect the advancement and is presently practical because of the gigantic enhancements and plenitude in information stockpiling, access, and preparing. Significant boundaries are gotten through online tests and that boundaries are determined from the essential ones. From these, we have attempted to form the most proficient boundary to decide the presentation list of the student which is determined based on successful time use, inadequate time use, dynamic window, right-wrong endeavors, an absolute dynamic time during the test brought by cursor development, and AI delegating. A prescient investigation is done over these inferred boundaries and the presentation file of the understudies, and accordingly, we utilize the Facebook prophet calculation for this prescient demonstration. In contrast with these forecast results, we examine the genuine outcomes, and hence, near insightful criticism is created and outrageous inconsistencies are followed and featured.

2 Proposed System Characteristic

While nations are at various focuses in their COVID-19 disease rates, overall, there are as of now more than 1.2 billion youngsters in 186 nations influenced by school terminations because of the pandemic. With this unexpected shift away from the homeroom in numerous pieces of the globe, some are puzzling over whether the selection of Internet learning will keep on continuing post-pandemic, and how will such a shift affect the overall education platform. Regardless of whether or not it is language applications, digital mentoring, video conferencing instruments, or net gaining knowledge of programming, there has been a massive flood in use given that COVID-19. The overall Web is a useful asset for investigating yet is bottlenecked by the way that clients need to explore through the worldwide substance where productivity is decreased because of the presence of obscure substances. In the proposed framework, we target furnishing a more adaptable educational program with a customized level of trouble and a further developed presentation investigation apparatus that examines the understudy's exhibition on an assortment of variables other than the quantity of right and mistaken endeavors in an assessment. An ideal learning system must be fast, accurate, and relevant—achieved by a distributed learner-oriented, customizable dynamic (nonlinear) form of learning as provided by the proposed system. Main characteristics presented by the system include:

- Personalized curriculum
- Better adaptability
- Smarter and efficient Web access
- Advance student's performance index analysis

- Distributed content with subdivisions for repeated concept clearing, reference, and a modular approach
- Open system
- Adaptive navigation
- Curriculum sequencing.

3 Comparative Study of the Present Web-Based Learning and the Proposed System

See Table 1.

Table 1 Difference between present e-learning system and the proposed e-learning system

Parameters	Present system	Proposed system
Nature of the content	Static, pre-recorded	Dynamic can be inserted, updated, or deleted
Domain of accessibility	Confined to the content which is uploaded for the particular system	The domain is the entire WWW
Scalability	Knowledge is divided into chunks	Knowledge is knit together in a large bubble
Information retrieval	Keep track of the scores based on the assignments	Keep track of scores based on the screen time, active window time, time span in the questions, time lag between the questions, etc.
Performance analysis	Analyzes the performance of students based on the no. of correct and incorrect attempts in the assessments	Analyzes the performance of the learners based on derived parameters and performance index on an individual basis
Cost	Expensive	Cheaper as compared to the existing system
Time effectiveness	Decreases the speed of learning as the learner has to go through the previously learnt topic over and over again	Increases the efficiency of learning as the learner is able to choose topics and subtopics according to one's need and understanding
Readability	Designed as per the general structure	Personalized for every learner

4 Components of the Proposed System

As discussed earlier, the proposed framework includes three segments to be specific: Semantic Web Architecture, information retrieval, and performance analysis. Semantic Web Architecture is a procedure utilizing which a learner can explore through the World Wide Web to discover customized and significant substance. Information retrieval (IR) is utilized to extricate pertinent data as per a person's data prerequisites. Performance analysis is utilized to look at a student's exhibition on a more profound level other than considering the right and mistaken endeavors. Every one of these areas is talked about comprehensively underneath.

4.1 Semantic Web Architecture

The Semantic Web is a thriving piece of innovation as it gives an intelligent environment between the learner or educator and the machine with which it is working. The advancement of the Semantic Web continues in layers. Each layer that is added to the Semantic Architecture ought to follow two standards:

- Downward Compatibility—Should be capable to interpret statistics at lower layers
- Upward partial understanding—Should have the capability to in part apprehend new layers.

Adding layers in this manner gives the technology interoperability. To accomplish similarity, the Semantic Web utilizes vocabularies, scientific classifications, and ontologies which contain a bunch of characterized terms and each is basic to the capacity to communicate the significance of the information [1]. A blend of a construction language and a cosmology language gives these abilities. We portray every one of these layers beneath.

4.1.1 Unicode Layer and URI

Unicode layer and URI provide a simplified means for the identification of resources like a Web site, an image, a document, or a person: basically, any unit with an identity. Unicode, utilized for PC character portrayal, is a widespread standard encoding framework: It addresses all languages dissimilar to other encoding systems. The motivation behind a URI is to unambiguously indicate an identifier to address an asset uniformly, distinguishing data portrayal develops, including classes, properties, and people. As there is no uncertainty, it becomes conceivable to all information that alludes to a given asset. It is the utilization of URIs that gives the Semantic Web a basic advantage over different advancements [2].

4.1.2 XML Layer

XML layer portrays the document content. XML composition then again gives lexical help for authorized XML documents. The idea is to give organized data that portrays, locates, and clarifies data assets making it easier for resources to be recovered [2].

4.1.3 RDF and Data Interchange Layer

URI is utilized to target Web assets and uses graph models to depict relationships among different resources. The motivation behind RDF is to give a standard system to offering statements about resources and their characteristics, and making assertions about resources is the premise of information representation [2].

4.1.4 SPARQL—Query, Ontology (OWL), RIF, and RDFS Layer

The foundation of Semantic Web, ontology, gives semantic executable by the machine and for a better man to application communication—a shareable domain. OWL has been based upon RDF and RDFS and has the same XML-based syntax. It fulfills the Semantic Web's necessities of giving negligible contribution from people and supporting programming prerequisites for a language with express importance. Adding additional vocabulary to ontologies, expanding RDFs with ontological constructs for depicting object-oriented classes, properties, and individuals. The ontology language utilizes RDF and RDFs, XML Schema data types and OWL namespaces [2].

4.1.5 Unifying Logic Layer

Unifying logic layer passes on as an establishment layer for consolidating the two previously mentioned layers into a whole, to coordinate requests and draw in rules over information addressed in the RDF alongside ontologies related with it and schemata. Different works in this area have centered on joining rules with query-infused facilities, with a blend of rules and grammar [2].

4.1.6 Proof Layer

Proof layer is for validation of particular statements. All the more succinctly, this is the layer of provenance knowledge [2].

4.1.7 Trust Layer

Trust layer depends upon the data source which can deny unfortunate applications or clients admittance to these sources. It is everything except a piece of trust and conviction between information sources and units of users [2].

4.1.8 UI and Application Layer

UI and application layer goes about as an interface the clients associate with so it must satisfy them alongside the applications [2].

4.1.9 Vertical Layers

Vertical layers like crypto are utilized for encryption purposes and for digital signature. Existing in the fifth to sixth layer is used to establish a Web of trust. XML signatures, by applying it to the resource content, can be identified [2] (Fig. 1).

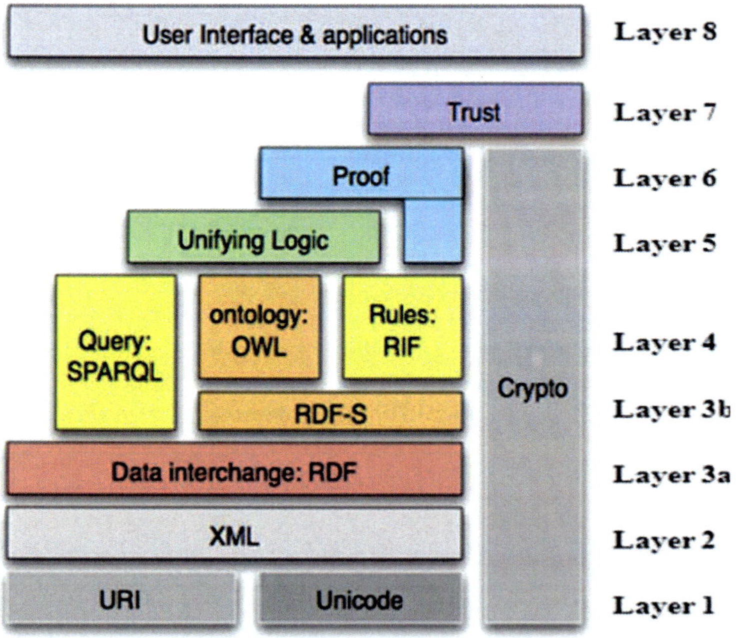

Fig. 1 Diagrammatic representation Semantic Web block architecture [2]

4.2 *Information Retrieval*

Information retrieval can be described as a software program that deals with the organization, storage, retrieval, and contrast of data from file repositories, especially textual information. Information retrieval is the recreation of acquiring fabric that can commonly be documented on an unstructured nature, i.e., generally textual content that satisfies statistics want from inside giant collections stored on computers [3]. Information retrieval comes into motion when the consumer enters a search question into the net shape, the place, the machine extracts, and the key phrases and tries to discover the most applicable content material for the same. The NLP algorithms preprocess the looking out key phrases to locate and map consequences extra efficiently. RAKE algorithm is used to extract the key phrases with growing significance past a threshold post- and pre-processing (elimination of quit words, lemmatization, tokenization, etc.). The key phrases are then mapped with the matters and path gadgets, and from the chosen team, cosine characteristic helps in the righteous effects with the most efficiency (Fig. 2).

4.2.1 Removal of Stop Words

Stop phrases are in many instances used phrases such as a, an, the do not add any unique meaning to a search query. Stop phrases are these phrases in the textual content that no longer add any which means to the sentence and their elimination will no longer have an effect on the processing of textual content for the described purpose. They are eliminated from the vocabulary to minimize noise and to minimize the dimension of the function set (Table 2).

The technique of changing information to something a laptop can apprehend is referred to as pre-processing. One of the main types of pre-processing is to filter out vain data. In herbal language processing, vain phrases (data) are referred to as cease words [4]. Stop phrases are a set of phrases that a search engine has been programmed to ignore, each when indexing entries for looking and when retrieving them as the end result of a search query [4] (Figs. 3, 4, and 5).

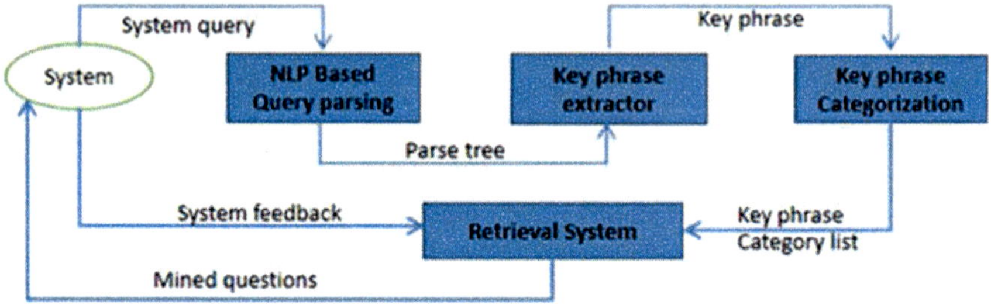

Fig. 2 Proposed mining model for question analysis and information retrieval for smart relevant searches for smart e-learning systems

Table 2 A demonstration of sample text after removal of stop words

Sample text with stop words	Sample text without stop words
Can listening be exhausting?	Listening, Exhausting
I like reading, so I read	Like, reading, read

Fig. 3 RAKE-NLTK for key phrase extraction on a sample sentence that extracts the keywords from the sample input statements

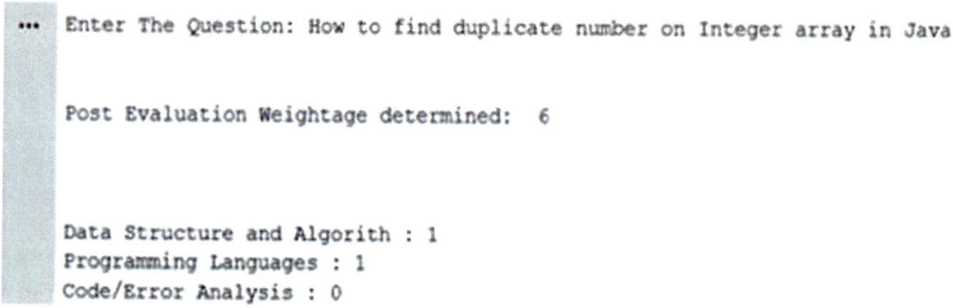

Fig. 4 Class and weightage determination in sentences to understand and classify outputs

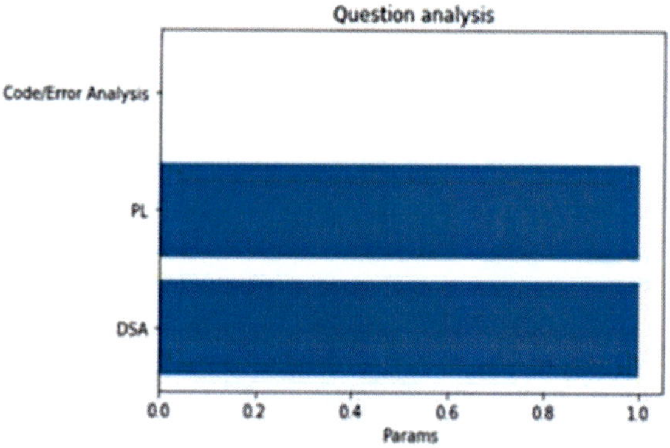

Fig. 5 Key phrase-based domain determination [4]

4.2.2 Tokenization

Tokenization is breaking the raw textual content into small chunks. Tokenization breaks the raw textual content into words, and sentences are referred to as tokens.

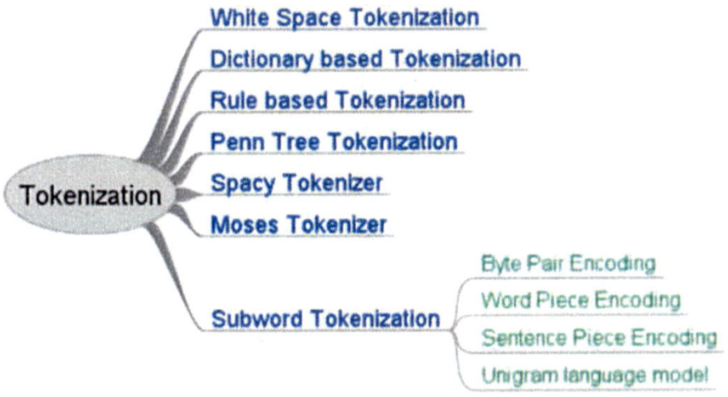

Fig. 6 Types of tokenization

These tokens assist in perceiving the context of growing the mannequin for the NLP. The tokenization helps in decoding the means of the textual content via examining the sequence of the words. Tokenization can be finished into separate phrases or sentences. If the textual content is cut up into phrases, the usage of some separation method is known as phrase tokenization and the identical separation completed for sentences is known as sentence tokenization [5] (Fig. 6).

4.2.3 Lemmatization

Lemmatization is the method of grouping collectively the extraordinary inflected varieties of a phrase so they can be analyzed as a single item. Lemmatization is comparable to stemming; however, it brings context to the words. So, it hyperlinks phrases with comparable meanings to one word. Lemmatization is quite possibly the most incessant printed content pre-handling technique utilized in natural language processing (NLP) [5]. The root phrase is referred to as a stem in the steaming process, and it is referred to as a lemma in the lemmatization process. The apparent gain of lemmatization is that it is extra accurate. So, if we are dealing with an NLP utility such as a chatbot or a digital assistant the place perception of the means of the talk is crucial, and lemmatization would be useful (Fig. 7).

Fig. 7 Lemmatization of texts to find root word

4.2.4 RAKE Algorithm

Rake or Rapid Automatic Keyword Extraction is a keyword extraction algorithm that operates on individual documents and is additionally very useful in dealing with multiple documents, in particular the kind of textual content that follows particular grammar conventions. Rake is primarily based on inspecting the frequency of phrase appearance and its co-occurrence with different phrases in the text. It observes the key phrases that regularly incorporate more than one phrase with minimal lexical meaning or are otherwise acknowledged as stop words, and these stop words are eliminated to structure the significant key phrases. The input parameters for the RAKE algorithm incorporate a listing of stop phrases and phrase delimiters [1]. It uses stop words and phrase delimiters to partition the record into candidate keywords, and these candidate key phrases are typically the phrases that assist a developer in extracting the actual keyword vital to get data from the document.

4.2.5 Long Short-Term Memory (LSTM)

Long Short-Term Memory (LSTM) is a distinctive sort of recurrent neural networks (RNN), successful in mastering long-term dependencies [6]. Remembering facts for longer durations of time is their default behavior. LSTMs additionally have a chain-like structure; however, the repeating module has a distinct structure. Instead of having a single neural community layer, there are 4 interacting layers, speaking in a very different way [7] (Fig. 8).

To enhance Internet search quality and serve a better search experience for users, it is vital to seize semantic data from personal queries that comprise the user's intentions in net search. Long Short-Term Memory (LSTM) has made extremely good achievements in capturing semantic statistics and predicting the semantic relatedness of two sentences [7]. In this study, thinking about the similarity between predicting the relatedness of sentence pair mission and semantic search, we furnish a novel channel to system semantic search task: see semantic search as an odd predicting the relatedness of sentence pair task. Moreover, an LSTM-based neural network model is recommended which is satisfactory for anticipating the semantic relatedness between the client's questions and attainable records [8].

4.3 Performance Analysis

Since the proposed machine is learner-oriented, they need to get a clear comparison as to how much they have understood in their course. It is very vital to analyze a learner's overall performance as this helps the beginners to tune their overall performance, furnish rightful comments, and take quintessential measures. Important parameters are accessed through online assessments, and the derived parameters are calculated

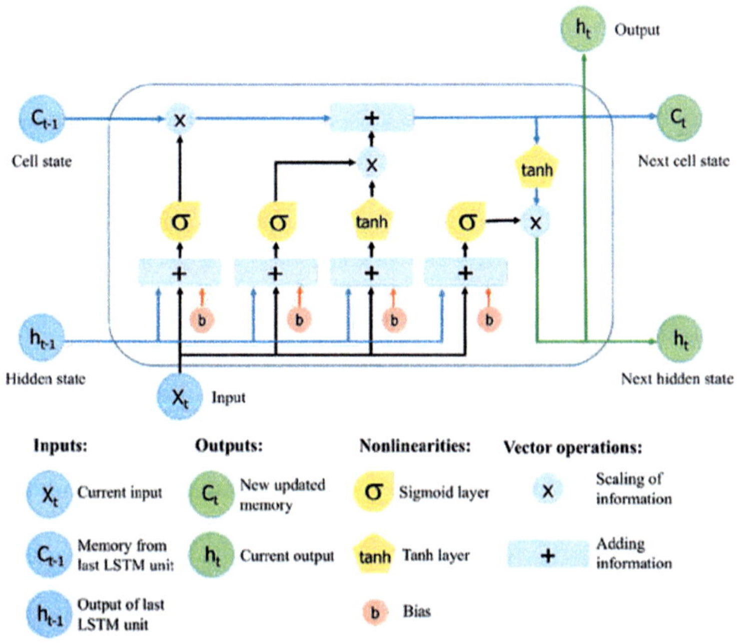

Fig. 8 The structure of LSTM neural network

from the main ones [9]. From these, we have tried to formulate the most environment-friendly parameter to decide the overall performance index of the student. Predictive evaluation is executed over these derived parameters and the overall performance index of the students.

4.3.1 Fundamental Parameters

The necessary and integral parameters regarding our system for overall performance evaluation are as follows:

- Time spent on proper questions—Effective time usage
- Time spent on incorrect questions—Ineffective time usage
- Correct tries—Right attempts
- Wrong tries—Wrong attempts
- Total attempted questions
- The complete active time throughout the examination—fetched by means of cursor motion and AI proctoring
- The total variety of questions
- Total examination time

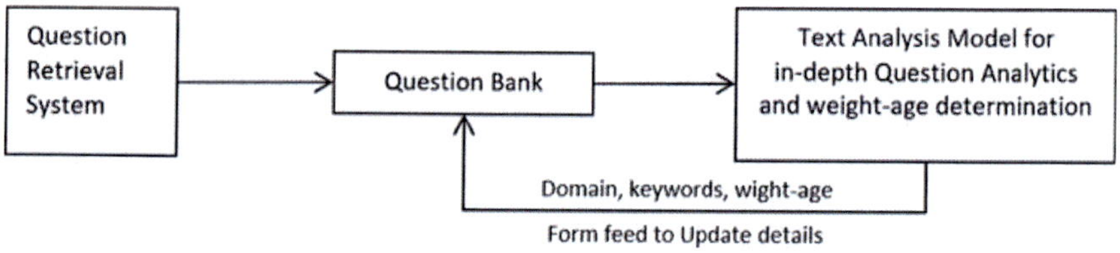

Fig. 9 The initial question retrieval system with question analytics and weightage determination model

Apart from the above parameters, the system is capable of keeping track of search queries, topics, subtopics, subjects, and subject-wise/topic-wise scores and suggests related content available over the World Wide Web. The proposed system also adjusts the questions based on the learner's understanding. For example, let us assume that there are total 5 questions in an assignment. The learner started with ques no.1 which was of easy level. If he/she is able to solve, it then the next question will be of moderate level. And if he is unable to solve the first one, then he/she shall be provided an easy question in ques no.2.

4.3.2 Derived Parameters

- Usage of Time = (Time on right questions) / (Total active time of the candidate)
- Accuracy Index = ((Correct attempt) / (Wrong attempts)) * (Total number of questions attempted)
- Attentive time = (The total active time during the exam) / (Total exam time)
- Effective utility of time = ((Correct attempts) * (Time on right questions)—(Wrong attempts) * (Time on wrong questions))/(Total number of questions) * (Total active time))
- Time of unattempted questions = (Total active time)—((Time on wrong questions) + (Time on right questions))
- Time wasted (ineffective usage) = (Time on wrong questions)^2 + (Time of unattempted questions).

Key Performance Index (KPI) of student = ((Usage of Time) + (Attentive time)) * (Accuracy Index) * (Effective utility of time) / ((Time of unattempt questions) + (Time wasted)) [2] (Figs. 9 and 10).

4.3.3 Facebook Prophet

There are many models currently for the predictive evaluation of time collection like machine getting to know ARIMA (Auto-Regressive Integrated Moving Average model), Auto-Regressive model, Exponential Smoothing, LSTM (Long Short-Term Memory), etc. These models require the facts to be fed, and with positive tweaking

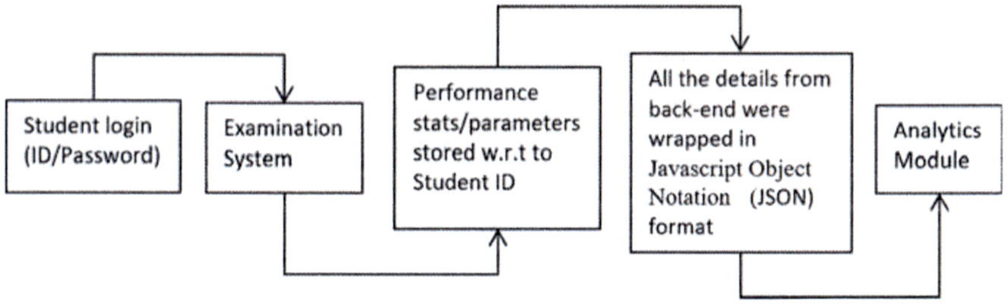

Fig. 10 Examination and information analytics module flow

and fine-tuning, they assist us to make predictions. Thus, our model makes use of Facebook prophet as a third-party library that can function all the fine-tuning components within [3]. The prophet is a system for forecasting time-series statistics which is totally, completely, and simply based on an additive model; the neighborhood nonlinear dispositions are in shape with yearly, weekly, and each and every day seasonality, plus excursion effects. It works great with time collections that have sturdy seasonal results and quite a few seasons of historic data. Prophet is strong at lacking information, shifts in the trend, and usually handles outliers well. The prophet is an open supply software program launched by means of Facebook's Core Data Science team. FBProphet makes use of a decomposable time-series model [10]. It consists of three major components: seasonal, trends, and vacation trips or activities having an effect on and error which are mixed into the below equation:

$$f(x) = s(x) + h(x) + g(x) + e(t)$$

FBProphet utilizes time as a regressor and attempts to coordinate with many straight and nonlinear elements of time as segments.

5 Conclusion

To put it briefly, the proposed system is a full-stack open environment whose objective is to come up with a recommender system for self-directed learning that can be used by learners. This system helps in developing an environment whereby novices can get personalized content material from the Internet, and it is a dynamic environment where the questions are shifted or modified as per the accuracy of the candidate, thus determining the topics and subtopics to be covered again for better performance and also suggesting the learner some new topics and subjects related to the current learning.

6 Future Scope

Our system being no exception to the above rule, there are viable strains of upgrades such as a performance tracker graph on the dashboard, protected exams, implementing association rule mining in order to see what the fellow mates are learning after completing the current topic, helpdesk to answer the learner's queries by the instructors and the administrators, etc.

References

1. Lamberti F (2009) A relation-based page rank algorithm for semantic web search engines. IEEE Trans Knowl Data Eng 21(1):123–136
2. Sengupta S, Banerjee A, Chakrabarti S (2020) Educational data mining and students' academic performance prediction. In: IEM-ICDC 2020: international conference on computational intelligence, data science and cloud computing 2020, Sep 2020, Kolkata
3. Riad AM et al (2010) PSSE: an architecture for a personalized semantic search engine. Int J Adv Inf Sci Service Sci 2(1):102–112
4. Sengupta S, Banerjee A, Chakrabarti S (2020) Prediction of future career path using different machine learning models. In: 2nd International conference on emerging technologies in data mining and information security, July 2020, Kolkata
5. Lee M et al (2010) Semantic association-based search and visualization method on the semantic web portal. Int J Comput Netw Commun (IJCNC) 2(1):140–152
6. Mondal, A, Mukherjee J (2018) An approach to predict a student's academic performance using recurrent neural network (RNN). Int J Comput Appl 181(6): 0975–8887
7. Dang NT, Tuyen DTT (2009) Natural language question answering model applied to document retrieval system, vol 51. World Academy of Science, Engineering and Technology, pp 36 – 39
8. Yang H, Qiu R, Chen W (2020) LSTM-based neural network model for semantic search, smart service systems, operations management, and analytics, pp 17–25
9. Francis BK, Babu SS (2019) Predicting academic performance of students using a hybrid data mining approach. J Med Syst Article number 162
10. Roessler I, Catacutan K (2020) Diversification around Europe—Performance measuring with regard to different missions. Tert Educ Manage 26:265–279

Chapter 37
Facial Recognition-Based Automated Classroom Attendance System

Nirmalya Misra, Sreejit Ray, Subhajit Pal, and Ruchira Dey

1 Introduction

A beholder can extract a lot of information just from a human face. It can tell a lot about the intentions, mood and attentiveness of a person. We identify other humans by their face. Identifying a person can be done in a lot of ways, of which recognizing the face is just one. Other methods such as testing DNA sequences, biometric methods such as fingerprint identification can tell an even greater deal about a person's identity. These methods can uniquely identify individuals and thus find usage in criminal investigation. With increase in compute, machine learning has been observed to be quite accurate in vision-based tasks.

Identifying faces in an automated fashion is the goal of facial recognition. This technology is applied to recognize faces in all types of media such as photographs and videos, and even in real-time footage. Recognizing faces is a task, if accomplished, finds huge application in identifying and confirming an individual's presence.

Facial recognition finds huge applications in the field of biometric security. This technology is mostly used in law enforcement and the maintenance of security. One of its common usage scenarios in our daily-to-day life lies in our smartphones, which use this technology to unlock our phones. Facial recognition can be utilized to make a certain place secure by checking the faces of people walking past cameras, to the images of people on a watch list. These watch lists can contain pictures of anyone who the organization wants to keep an eye on.

The field of pattern recognition is able to provide such abilities to a basic computer is computer vision [1]. The purpose of this field is literally "to make a computer see". It helps computers identify and understand meanings in images and hence in videos.

N. Misra (✉) · S. Ray · S. Pal · R. Dey
Institute of Engineering and Management, Salt Lake Electronics Complex, Y-12, Gurukul, Street Number 18, EP Block, Sector V, Bidhannagar, Kolkata, West Bengal 700091, India
e-mail: nirmalya14misra@gmail.com

L. Mandal et al. (eds.), *Proceedings of International Conference on Computational Intelligence, Data Science and Cloud Computing*, Algorithms for Intelligent Systems,
https://doi.org/10.1007/978-981-19-1657-1_38

It is an old area of research, up until recently dominated by specialized hand-crafted methods designed by digital signal processing experts and various available statistical tools. Recently, there has been progress in this field especially with commodity systems for optical character recognition and face detection in cameras and smartphones. Another factor that has driven computer vision is the increase in the computational ability of computers and the coming up of novel architectures which use convolutions, an operation previously used in signal analysis, now on images.

Computer vision has a wide variety of applications, both old (e.g., mobile robot navigation, industrial inspection and military intelligence) and modern (e.g., human–computer interaction, medical image analysis, etc.) tasks.

Our paper aims at using the technology of recognizing faces to build attendance systems out of this technology [2]. The pandemic has shown us a lot of ways in which time could have been saved earlier, and we feel attendance is one of those tasks which can be automated and thus made better. Manually taking the attendance of students in a class is not very ideal for both students and teachers. What does not help is the creeping in of human errors in this process. With technological advancements, an automated system can be the way forward.

An automatic attendance system is an educational ERP system that records the students' attendance. Unlike any typical system for attendance, this would enable the faculty to record, store and view students' attendance history.

The following sections of the paper throw light on the previous works done in the field of facial recognition, before moving into the methodology used to build the attendance system. Finally, we state the future scopes and further improvements which could be done to take this idea further forward (Fig. 1).

2 Related Works

Capturing the attendance of students in a class has traditionally been a cumbersome task, with teachers marking the attendances on sheets [3]. Moreover, it is tough to authenticate the presence of all the students on a one-to-one basis in a large classroom. The computation of attendance percentage becomes a huge task, leaving room for error as the calculations and recordings have to be done manually. An automated attendance management system using biometrics or maybe using the face of the person would provide a better and more accurate solution. The results showed greater accuracy over the historically used systems for attendance. Biometric-based methods have come up as the go-to option for recognizing individuals in recent years. Instead of using passwords and security keys to identify people and grant them access to numerous virtual and physical domains, this method assesses an individual's physiological and behavioral traits to determine and confirm a person's identity. Identification based on physiological parameters and behavioral attributes is one example of biometric-based technology. Identification using faces has shown to outperform biometric methods in a few ways. Face recognition systems outperform biometric technologies in a number of ways. To begin with, the cost of installation

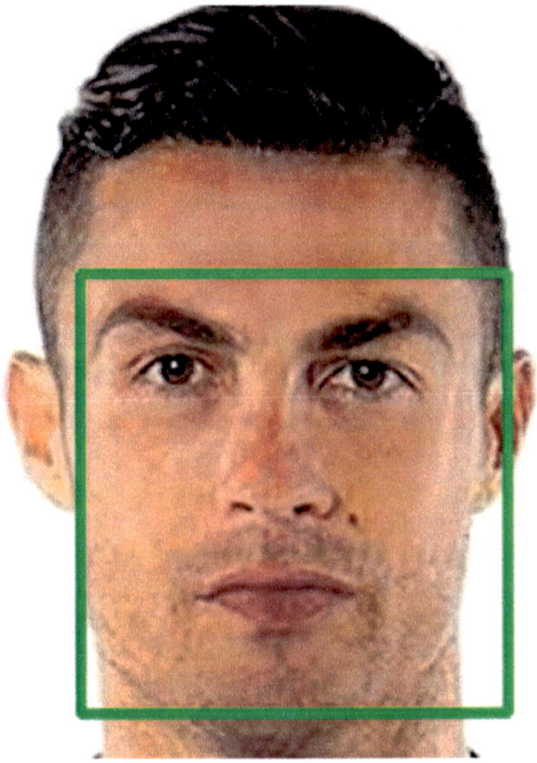

Fig. 1 Facial recognition. The green box encloses the face in the image

would be minimal because all that is required to capture a person's image is a camera. Second, except for a picture of their face, persons are not required to provide any biometric data to their business. This has particularly proven to be beneficial for surveillance purposes.

The initial attempts to use face recognition for various tasks were in the 1960s with a system that was semi-automated [4]. Major facial features were located using marks made on photographs. It used features such as nose, ears, eyes and mouth among others. Then, the distances and ratios were calculated from these marks to a common reference point and compared to reference data. Goldstein, Lesk and Harmon devised a system of 21 subjective indicators in the early 1970s, which included characteristics such as the color of the hair and the thickness of the lip. This proved to be even tougher to automate due to the subjective nature of many of the measurements, which were still made completely by hand. Fisher and Elschlagerb's approach consisted of measuring different pieces of the face and then mapping them all onto a global template. It was found that these features do not comprise of enough unique data to represent an adult face. The earliest system which used automation used some very minimal pattern recognition. It compared faces to a face model of certain features and created a series of patterns, which an image containing faces could have relative to this model. This approach was mainly statistical and relied on techniques like histograms and the value in grayscale.

The paper by Xin Geng states that most face recognition systems need the faces to be fed into it in a certain manner so that there are no inherent biases that might force the system to predict incorrectly [5]. Because these conditions cannot be satisfied, face recognition cannot be used in many real-time applications. Real-time applications need techniques that do not need any strict control over human beings for recognizing the face. These types of systems need these inherent biases, which might be learned from training data to be removed and hence can be called as "uncontrolled conditions".

The work by Turk et al. [6] treats facial recognition as a two-dimensional problem, instead of a three-dimensional one, considering the fact that faces are normally upright. The main features of the face are referred to as eigenfaces, because they are actually the eigen vectors of the face. These eigenfaces are then used for facial recognition. Building on this work, a study by Lata et al. [7] used principal component analysis (PCA) for recognizing faces, using the technique of eigenfaces in their methodology.

An article by Andrejevic et al. [8] explored into the ethical side of using facial recognition-based systems in schools. Using this technology can be harmful to privacy at any level, and hence, discussions regarding how institutions use such data have to be checked, with proper norms in place for the same.

A study by Diaa et al. [9] explored the usage of CNNs or convolutional neural networks, which were deep enough to extract features and recognize faces in an end-to-end manner. The system can be trained for every new person it wants to recognize. It provided better results than traditional algorithms such as decision trees and SVMs.

3 Methodology

Our work can be divided into a few steps. The first would be to identify faces in a picture. The second step would involve encoding these faces into some sort of numbers. Finally, given a test image, we can compare the encodings to see which person the face belongs to.

3.1 Data Collection

The beginning of our methodology starts with collecting data for each student of the class. First, we start by collecting the details of all students and a picture of them each. These images along with their details are stored in the database of the class. Another thing we will store, something we will look at in further sections, is the facial encodings. These facial encodings will come up handy in the process of facial

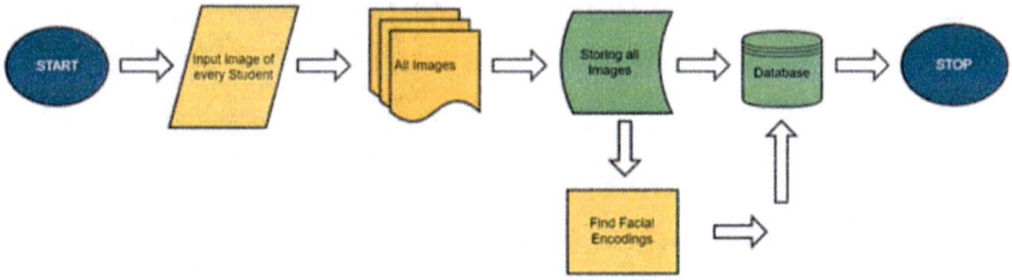

Fig. 2 Data collection and storage pipeline

recognition. The flowchart of the data collection and data storage pipeline is given below (Fig. 2).

3.2 Facial Recognition

The first and very obvious step is to identify faces. If we are able to locate where the faces lie, then we can check only those parts of the image. This would help in denoising the image and would also help our system to understand faces better, without developing biases about the background. Off late, a solution has been found for facial recognition called histogram of oriented gradients (HOG) [10, 11]. In this method, we first convert the colored image into grayscale. Then, for each pixel, an arrow is drawn pointing in the orientation in which the pixels become darker. Gradients are what these arrows are referred to as. If we replace all the pixels with arrows, then we will obtain a HOG representation of the image [12] (Fig. 3).

The next step involves coming up with 68 specific points, called facial landmarks [13]. This step is called facial landmark estimation. This is important because faces can be oriented in multiple directions. Facial landmark estimation helps to center the face, thus helping improve the performance of the system. The last step for facial recognition is to bring out a few basic estimations from each face. Then, we could measure a test face the same way and find the known face with the closest measurements. This step is called finding facial encodings, which gives 128 measurements of a face (Fig. 4).

3.3 Methodology Using Facial Recognition

Initially, we have to take in the images of all people in a class or in the organization along with their names. We have labeled the names as the file names of the respective images. The first job is to find the location of the face [14, 15] in all these images.

Fig. 3 Histogram of gradients (HOG) representation of an image

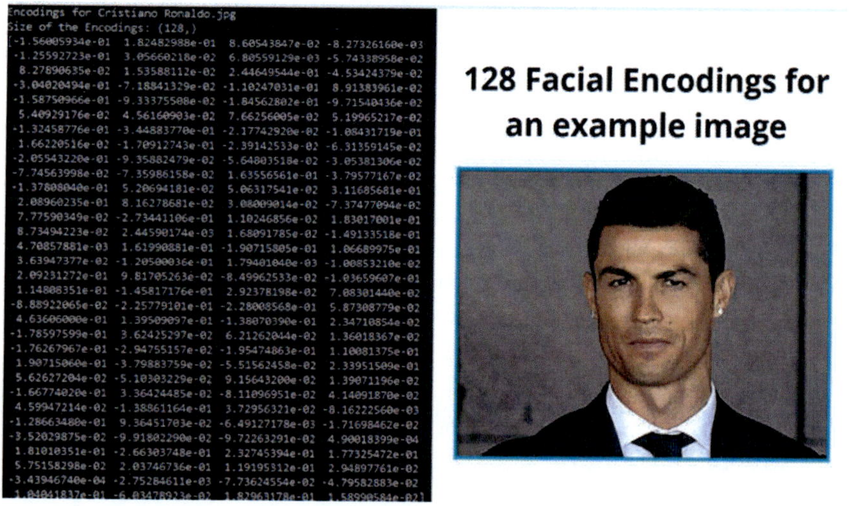

Fig. 4 128-numbered encoding of an example image

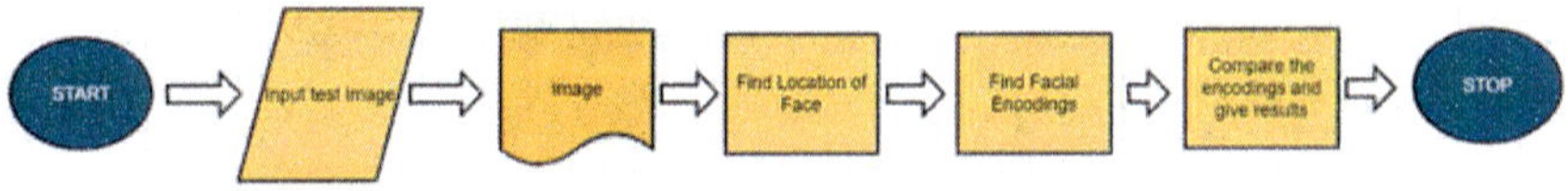

Fig. 5 Pipeline for testing images

After getting the locations of the faces, we can go forward with finding the 128-number encodings for each of the faces. The names of the people involved can be obtained by looping over the image file names and just removing the extension.

When a test image comes in, we would first have to find the location of the face in the image and then find out its encodings. The final step would involve comparing the encodings with the previously recorded encodings of all the people in the database. The person to whom the encodings match will be marked as present.

The full pipeline is given below (Fig. 5).

4 Results and Discussions

The following images show the results of the work. We have taken the picture of four footballers: Cristiano Ronaldo, Lionel Messi, Neymar Jr, and David Villa to test our work (Fig. 6).

Further improvements could be made such as noting down the time at which attendance was recorded. Deployment could be done on a Raspberry Pi, with a camera module attached so that the whole system could be automated. The only problem with this kind of system is the possibility of having identical twins, whose faces are the same. Thus, recognition by facial features is very tough, and even our human eye can fail at it.

The results of our methodology for 15 images each of two faces taken in different orientations are provided below (Table 1).

5 Conclusion

The accuracy of the facial recognition system is high as it uses OpenCV and the face_recognition Python module, which is made out of pre-trained models. This paper just presents the idea of how face recognition can be applied in recording attendance. This idea can be used to build an end-to-end system for recording attendance in an automated manner.

The question of ethics still remain and will mostly depend on how the different institutions intend to use the data.

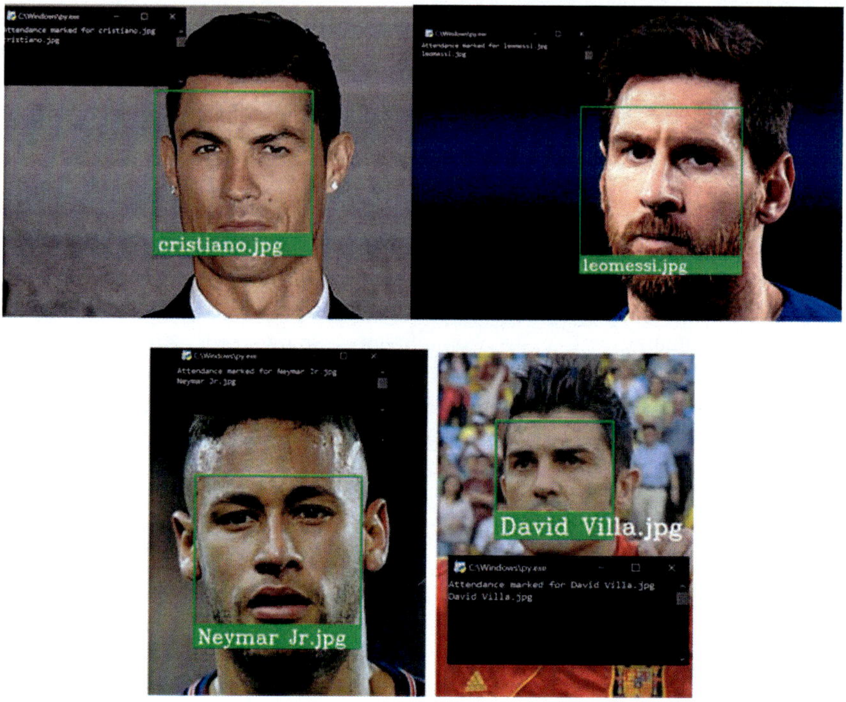

Fig. 6 Outputs of the four test images predicted correctly: Cristiano Ronaldo, Lionel Messi, Neymar Jr, and David Villa

Table 1 Results taking two test images

Image	Upright face correctly detected out of 15	Side face correctly detected out of 15
FACE 1	14	8
FACE 2	13	10

References

1. Brownlee J (2019) Deep learning for computer vision: image classification, object detection, and face recognition in python. Machine Learning Mastery
2. MasterSoft (n.d.). Advantages of student automated attendance management system for schools and colleges. Retrieved 20 Sept 2021, from Iitms.co.in website: https://www.iitms.co.in/blog/automated-student-attendance-management-system-for-schools-and-colleges.html
3. Rekha AL, Chethan HK (n.d.) Automated attendance system using face recognition through video surveillance. Retrieved 20 Sept 2021, from Ijtre.com website: https://www.ijtre.com/images/scripts/2014011113.pdf
4. Vishal K, Keertiraj BR, Dubey NK, Pooja MR (n.d.). Face recognition based attendance system. Retrieved 20 Sept 2021, from Ijert.org website: https://www.ijert.org/research/face-recognition-based-attendance-system-IJERTV9IS060615.pdf
5. Kar N, Debbarma M, Saha A, Pal D (2012) Study of implementing automated attendance

system using face recognition technique. Int J Comput Commun Eng, pp 100–103. https://doi.org/10.7763/IJCCE.2012.V1.28
6. Turk M, Pentland A (1991) Eigenfaces for recognition. J Cogn Neurosci; 3(1):71–86. https://doi.org/10.1162/jocn.1991.3.1.71
7. Lata YV et al (2009) Facial recognition using eigenfaces by PCA. Int J Recent Trends Eng 1(1):587
8. Andrejevic M, Selwyn N (2020) Facial recognition technology in schools: critical questions and concerns. Learn Media Technol 45(2):115–128
9. Salama AbdELminaam D, Almansori AM, Taha M, Badr E (2020) A deep facial recognition system using computational intelligent algorithms. PLoS ONE 15(12): e0242269. https://doi.org/10.1371/journal.pone.0242269
10. Zhou W, Gao S, Zhang L, Lou X (2020) Histogram of oriented gradients feature extraction from raw bayer pattern images. IEEE Trans Circuits Syst II Express Briefs 67(5):946–950. https://doi.org/10.1109/TCSII.2020.2980557
11. Rahmad C et al (2020) Comparison of Viola-Jones Haar Cascade classifier and histogram of oriented gradients (HOG) for face detection. IOP Conf Ser: Mater Sci Eng 732(1)
12. Geitgey A (2016) Machine learning is fun! Part 4: modern face recognition with deep learning. Retrieved 20 Sept 2021, from Medium website: https://medium.com/@ageitgey/machine-learning-is-fun-part-4-modern-face-recognition-with-deep-learning-c3cffc121d78
13. Lee Se A et al (2020) Automatic facial recognition system assisted-facial asymmetry scale using facial landmarks. Otol Neurotol 41(8):1140–1148
14. Hemalatha G, Sumathi CP (2014) A study of techniques for facial detection and expression classification. Int J Comput Sci Eng Surv 5(2):27
15. Sivaram M et al (2019) Detection of accurate facial detection using hybrid deep convolutional recurrent neural network. ICTACT J Soft Comput 9(2)

Chapter 38
Generation of Non-isomorphic Connected Graphs by Successive Edge Removals from a Complete Graph

Sumit Chakraborty, Maumita Chakraborty, and Rajat Kumar Pal

1 Introduction

Two graphs are said to be isomorphic [8] if there exists a one-to-one correspondence between their vertices and between their edges such that the incidence relationship is preserved. Those graphs which do not satisfy the isomorphic property are called non-isomorphic graphs. A set of connected non-isomorphic graphs are very useful in providing a test bed for researchers to determine the performance of the algorithms working with graphs. To determine the performance of approximation algorithms, graph algorithms, modeling and simulation algorithms, database applications-related algorithms, and non-isomorphic connected graphs are very useful.

In this paper, we have devised an algorithm, named *non-isomorphic connected graph generator* (*NICGG*), which generates non-isomorphic connected subgraphs of a complete graph of a given order. Starting from the complete graph (of a given order), the algorithm removes edges successively to get connected subgraphs of the same order but of different sizes. Among these generated subgraphs, there are several subgraphs having the same size but they are also non-isomorphic. The algorithm *NICGG* generates non-isomorphic spanning trees of a complete graph by removing maximum possible edges. The paper is organized as follows:

S. Chakraborty (✉)
Department of Computer Science, Ramsaday College, Amta, Howrah 711401, India
e-mail: sm.chakraborty09@gmail.com

M. Chakraborty
Department of Computer Science and Technology, University of Engineering and Management Kolkata, University Area, Plot No. III, B/5, New Town, Action Area III, Kolkata 700160, India

R. K. Pal
Department of Computer Science and Engineering, University of Calcutta, JD-2, Sector III, Salt Lake, Kolkata 700106, India

L. Mandal et al. (eds.), *Proceedings of International Conference on Computational Intelligence, Data Science and Cloud Computing*, Algorithms for Intelligent Systems,
https://doi.org/10.1007/978-981-19-1657-1_39

- Literature survey in Sect. 2: In this section, we have discussed about some existing methods related to different types of graph generation.
- Proposed algorithm in Sect. 3: In the beginning of this section, we have given the working procedure of our proposed algorithm *NICGG* and later we have given the algorithm.
- Data structures and complexity issues in Sect. 4: It contains the data structures used in the proposed algorithm as well as time and space complexity of the algorithm.
- Illustration with an example in Sect. 5: We have made a dry run of the proposed algorithm by taking a suitable example.
- Experimental results in Sect. 6: In this section, we have discussed about the results of the execution of our proposed algorithm.
- Conclusion in Sect. 7: In this section, we have discussed about the performance, applicability, and future scopes of the proposed algorithm.

2 Literature Survey

In the last few years, many researchers provided different graph generation algorithms. Some of these generate random graphs [16], some generate non-isomorphic graphs, etc. Some of these graph generation algorithms are discussed below:

a. The algorithm proposed by Bayati et al. [2] generates random graphs by considering a given degree sequence. The algorithm can generate random graphs with large girth by generating random bipartite graphs with an optimized degree sequence. These graphs are useful in the design of high-performance low-density parity-check (LDPC) codes.
b. Bhuiyan et al. [3] have proposed a parallel algorithm which generates random graphs with a given degree sequence. They also estimated the number of possible graphs with a given degree sequence. It can be helpful in the study of structural properties and dynamics over a network.
c. The algorithm proposed by Chakraborty et al. [4] takes the order of the graphs as an input and generates the complete graph first and successively removes an edge to produce non-isomorphic connected graphs. It keeps creating such desired graphs until a pendent vertex occurs. By considering the minimum degree of the vertices, it ensures that no disconnected graphs are being produced after removal of edges from the graph. They have not considered the disconnected graphs since they were interested only in finding spanning trees instead of generating spanning forests.
d. Chakraborty et al. [5] proposed the algorithm *random connected graph generator* (*RCGG*) to generate random connected graphs. The generated graphs are also simple and undirected. Initially, it considers a complete graph of order n. It iterates $m - n + 1$ times and each time it generates a random connected subgraph of the graph produced at the previous iteration, by removing an edge chosen randomly, where m is the size of the complete graph of order n. The last graph obtained by this process of *RCGG* is a random spanning tree.

e. A probabilistic approach proposed by Erdös and Rényi [9] computes a random graph of order n and size e. There are ${}^{n}C_{2}$ number of possible edges, and hence, e edges can be chosen in ${}^{n}C_{e}$ different ways. The authors talked about the possible types of random graphs that may be generated, like connected and disconnected graphs, trees, graphs with isolated vertices, and so on. The authors also discussed about the probability of a random graph being a connected graph, probability that the largest connected component of a random graph contains $n - k$ ($k = 0$, 1…) vertices, probability that a random graph should consist of exactly k (1, 2, …) connected components, and many others.
f. The algorithm proposed by Horn et al. [10] generates random graphs by considering the parameters like number of vertices, number of levels, degree of every vertex, and also the height and width of each graph. The graph instances generated by the algorithm are useful in the development of approximation algorithms.
g. A data parallel algorithm, proposed by Nobari et al. [12], generates random graphs. The algorithm focuses on distributing the data across different vertices and operates concurrently on the data, and the generated graphs can fit into memory for further processing. The graph instances generated by the algorithm can be used in database applications, simulations, and sampling.
h. Viger and Latapy [14] have proposed a simple and efficient algorithm to generate large graphs. It is much useful in simulations for the problems that have graph theoretic formulation as the algorithm generates large graphs.
i. Wang et al. [15] have proposed an algorithm to generate large-scale clustered random graphs. The algorithm has the advantage in random evaluation and computational efficiency. These large-scale random graphs generated by the algorithm are suitable for real networks.

3 The Proposed Algorithm

The proposed algorithm, *non-isomorphic connected graph generator* (*NICGG*), begins its execution by taking the following inputs:

- Number of vertices (n): This is the order of every generated graph instance.
- The number of levels (p): This is the number of removable edges, which is less than $(m - n + 2)$ for $m = n \times (n - 1)/2$, where m is the number of edges in the complete graph of order n.

At first (i.e., in the 0th level), the algorithm starts with the complete graph K_n of order n. In the next level, it generates the first subgraph (G_1) by removing any one edge from K_n, since initially every vertex is identical (all the vertices have the same and maximum possible degree $n - 1$), and there is only one non-isomorphic subgraph of size, lesser by one, of the complete graph. Now all the vertices in the first subgraph (G_1) are not identical since the degrees of all the vertices are not alike. Then *NICGG* removes an edge in two ways: (a) from intra-degree vertices and (b) from inter-degree vertices. After removing an edge following the above method, the derived subgraph is tested for connectivity and to ensure the non-isomorphic property, we have measured only unique degree sequences for the subgraphs having same size (i.e., only one subgraph of identical degree sequence is to be considered). Since a connected graph of order n has at least $n - 1$ edges, it is not possible to remove more than $m - n + 1$ number of edges from K_n to get connected subgraphs, i.e., the maximum possible value of p is $m - n + 1$. All p levels of generation of non-isomorphic connected graphs by *NICGG* are described below:

Level 0: *NICGG* considers the complete graph, K_n (G_0), of order n.

Level 1: Input to this level is the complete graph, K_n. The degree of each of the vertices of K_n is $n - 1$. Removing an edge $\{v_1, v_2\}$ from K_n, we get a subgraph $G_{1,1}$. In $G_{1,1}$, the two vertices, v_1 and v_2, are adjacent to all the remaining vertices except each other. The other $n - 2$ vertices are adjacent to each other. If we remove an edge other than $\{v_1, v_2\}$ from K_n, the same structure as $G_{1,1}$ will be generated. Hence, $G_{1,1}$ is the only non-isomorphic graph in this level.

Level 2: Input to this level is the graph G1,1. $G_{1,1}$ has $n - 2$ vertices of degree $n - 1$ each and other two vertices of degree $n - 2$ each. *NICGG* removes an edge $\{v_i, v_j\}$ to get a subgraph $G_{2,1}$, where $deg(v_i) = n - 2$ and $deg(v_j) = n - 1$. *NICGG* removes an edge $\{v_x, v_y\}$ to get a subgraph $G_{2,2}$, where $deg(v_x) = n - 1$ and $deg(v_y) = n - 1$ too. Since, there is no edge between the vertices v_1 and v_2 of degree $n - 2$, it is not possible to generate any more subgraphs in this level.

Level k ($\leq p$): Inputs to this level are the graphs, $G_{k-1,1}$, $G_{k-1,2}$,...,$G_{k-1,q}$ that have been generated in the $(k - 1)$-th level of generating non-isomorphic connected graphs of size $m - k - 1$ each. *NICGG* computes the minimum ($d_{\min}$) and maximum ($d_{\max}$) degree of the vertices of $G_{k-1,1}$. For each degree, d, $d_{\min} \leq d \leq d_{\max}$, *NICGG* removes an edge (if exists) from $G_{k-1,1}$ incident to a vertex of degree d and a vertex of degree $d + e$, for $e = 0, 1, 2, \ldots, d_{\max} - d_{\min}$, to obtain the subgraphs of $G_{k-1,1}$. The disconnected subgraphs of $G_{k-1,1}$ are discarded. The connected subgraphs of $G_{k-1,1}$, with distinct degree sequences are preserved (as well as printed). *NICGG* applies the above procedure on the graphs, $G_{k-1,2}$, $G_{k-1,3}$, ..., $G_{k-1,q}$ to generate all non-isomorphic connected subgraphs in this level. We may note that two or more subgraphs of level $k - 1$ may produce the same subgraph at the k-th level; such subgraphs are eventually deleted to remove redundancy. At each level of this generation, we eliminate all such generated duplicate subgraphs, and then move to the next level of computation.

3.1 The Algorithm, NICGG()

// It uses a queue S to store the graphs temporarily; *insert*() and *delete*() are two operations to insert a new graph on and delete a graph from the queue, S, respectively. Initially, S is empty. //

Step 1: *Begin*

Step 2: Read n, p

Step 3: Compute the complete graph, G_0, of order n

Step 4: Print G_0

Step 5: $m \leftarrow n\times(n-1)/2$, $G\leftarrow G_0$, $c\leftarrow 1$

Step 6: *While* the number of edges in G is greater than $m-p$ *do* // i.e. level is $m-|E_G|+1$ //

Step 7: Compute the minimum (d_{min}) and maximum (d_{max}) degrees of the vertices in G

Step 8: *If* $d_{min} = 1$, *then* take the second least degree as d_{min}

Step 9: *For* $d = d_{min}$ to d_{max} *do*

Step 10: *For* $e = 0$ to $(d_{max}-d)$ *do*

Step 11: Take an edge $x = (u, v) \in G$ such that $deg(u) = d$ and $deg(v) = d + e$

Step 12: $c \leftarrow c+1$

Step 13: Compute the subgraph, G_c, by removing the edge x from G

Step 14: *If* G_c is not a connected graph, *then* go to Step 18

Step 15: *If* G_c has a unique degree sequence, *then*

Step 16: Print G_c, *insert*(S, G_c) and go to Step 23

Step 17: *End If*

Step 18: Reject G_c and $c \leftarrow c-1$

Step 19: *If* edge $(p, q) \in G$ (next to x), such that $deg(p) = d$ and $deg(q) = d + e$ exist, *then*

Step 20: $x \leftarrow (p, q)$ //assume the edge, (p, q) as x //

Step 21: Go to Step 12

Step 22: *End If*

Step 23: *End For*

Step 24: *End For*

Step 25: $G \leftarrow$ *delete*(S) //places the front element of S in G for generating its children //

Step 26: *End While*

Step 27: *End*

4 Data Structures and Complexity Issues

NICGG uses adjacency matrices [8] to store the generated graph instances and these adjacency matrices are kept in a queue [11] *S*. After every completed level, *S* holds only the graphs generated in the last level, so as to generate their children. To perform the connectivity test, we have taken the help of *breath first search* (*BFS*) algorithm [1] and it takes $O(n^2)$ time.

4.1 Time Complexity

Here the inputs to *NICGG* are the order (n) of the graphs to be generated and the number of required levels (p). Thus, we determine the time complexity in terms of n and p.

Step 9 takes at most $n - 2$ iterations since the maximum value of $d_{\max}$ is $n - 1$ and minimum value of $d_{\min}$ is 2. Steps 10 and 11 jointly take at most $(m - k + 1)$ iterations since there are $(m - k + 1)$ number of edges available in the k-th level.

The connectivity testing takes $O(n^2)$ time in Step 14 and the searching time in Step 15 is at most $(m - k + 1)$ times since at that level only $(m - k + 1)$number of edges are available. Then, jointly these two steps take $O(n^2)$ time.

In the k-th level, a subgraph that was generated at the $(k - 1)$-th level will generate its subgraphs by removing its edges by taking at most $((n - 2) \times (m - k + 1) \times n^2)$ time.

Let $f(k - 1)$ be the upper bound to the number of subgraphs generated at the$(k - 1)$-th level. Then *NICGG* will take at most $(f(k - 1) \times (n - 2) \times (m - k + 1) \times n^2)$ time to generate all subgraphs at level k. Now in the $(k - 1)$-th level, the minimum and maximum degree of a vertex will be $n - k$ and $n - 1$, respectively, i.e., there are k different degrees in the $(k - 1)$-th level. Then $f(k - 1) = k \times (k + 1)/2$.Hence, the upper bound to the time complexity, $T(n, p)$, of *NICGG* up to level p is given by:

$$T(n, p) = \sum_{k=1}^{p} \left\{ k \times \frac{(k+1)}{2} \times (n-2) \times (m-k+1) \times n^2 \right\}$$

$$T(n, p) = \frac{n^2 \times (n-2)}{2} \sum_{k=1}^{p} \{k \times (\mathrm{k}+1) \times (m-k+1)\}$$

$$T(n, p) \cong \frac{n^2 \times (n-2)}{2} \sum_{k=1}^{p} k^3$$

$$T(n, p) \cong \frac{n^2 \times (n-2)}{2} \times \left\{ \frac{p \times (p+1)}{2} \right\}^2$$

Hence, *NICGG* will take at most $O(n^3 p^4)$ time to complete its execution.

4.2 Space Complexity

The algorithm *NICGG* stores the graphs in queue S generated in a level and when the instances of the next level are computed, it deletes the earlier instances. Finally, it contains only the graph instances generated in the last level, p. Then S can contain at most $f(k)$ number of graphs computed after level k $(1 \le k \le p)$. Since $f(k-1) = (k+1) \times (k+2)/2$, then for any k $(1 \le k \le p), f(k) = (p+1) \times (p+2)/2$. Hence, the space needed to store the graphs is $O(p^2 \times n^2)$.

5 Illustration of *NICGG* with an Example

In this section, we have tried to explain our proposed algorithm *NICGG* with the help of a suitable example. Let us consider the order (n) of the graphs is 5 and the number of levels (p) is 6. Here we are assuming the maximum possible levels to generate the minimum sized connected subgraphs, i.e., the non-isomorphic spanning trees.

At first, *NICGG* starts with the complete graph K5 (of order 5), which is renamed as G_0, shown in Fig. 1a.The set of vertices of G_0 is $\{v_1, v_2, v_3, v_4, v_5\}$ and the corresponding set of degrees of the vertices is $\{4, 4, 4, 4, 4\}$.

Level 1(i.e., $k = 1$)**:** Here G_0 is the only graph (as input), and hence, the subgraph at this level will be generated from G_0. Now, $d_{\min} = 4$ and $d_{\max} = 4$, i.e., $4 \le d \le 4$. Then edge $\{v_1, v_2\}$ is removed from G_0 to obtain a subgraph $G_{1,1}$, since $deg(v_1) = 4$ and $deg(v_2) = 4$. $G_{1,1}$ is the only subgraph, shown in Fig. 1b, generated by *NICGG*, since $d_{\max} - d_{\min} = 0$.

Level 2 (i.e., $k = 2$)**:** Here $G_{1,1}$ is the only graph generated by *NICGG* in the previous level, and hence, the subgraphs at this level will be generated from $G_{1,1}$. The degree sequence of the vertices of $G_{1,1}$ is $\{3, 3, 4, 4, 4\}$. Now $d_{\min} = 3$ and $d_{\max} = 4$, i.e., $3 \le d \le 4$. So, there are three types of removable edges: (a) edges connecting two vertices of degree 3 each, (b) edges connecting a vertex of degree 3 to a vertex of degree 4, and (c) edges connecting two vertices of degree 4 each. Now,

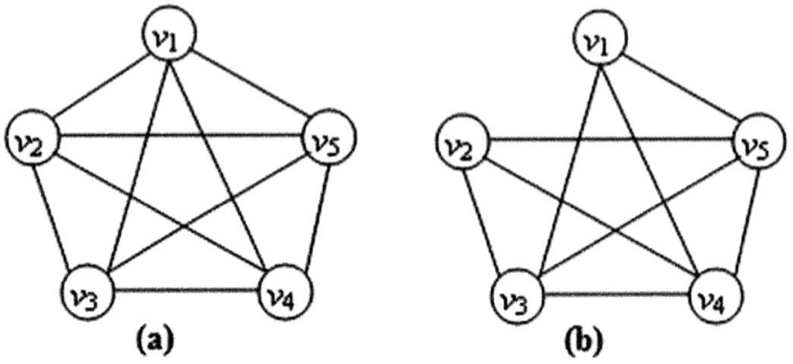

Fig. 1 **a** Complete graph K5 of order 5, renamed as G_0, and **b** Subgraph $G_{1,1}$ is produced at the first level

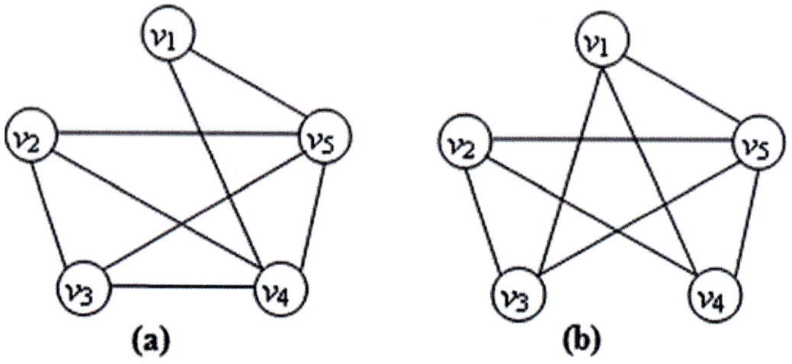

Fig. 2 **a** Subgraph $G_{2,1}$ from $G_{1,1}$ and **b** Subgraph $G_{2,2}$ from $G_{1,1}$

since there is no edge of type (a) in $G_{1,1}$, no subgraph will be generated by *NICGG* by removing an edge of type (a).$\{v_1, v_3\}$ is an edge of type (b) and $\{v_3, v_4\}$ is an edge of type (c). *NICGG* removes the above edges from $G_{1,1}$ to generate two non-isomorphic connected subgraphs $G_{2,1}$ and $G_{2,2}$, as shown in Fig. 2a, b, respectively.

Level 3 (i.e., $k = 3$)**:** Here $G_{2,1}$ and $G_{2,2}$ are the two graphs generated by *NICGG* in the previous level, and hence, the subgraphs at this level will be generated from these two graphs.

The degrees of the vertices of $G_{2,1}$ are $\{2, 3, 3, 4, 4\}$. Now $d_{\min} = 2$ and $d_{\max} = 4$, i.e., $2 \leq d \leq 4$. So, there are six types of edges possible that might be eliminated: (a) edges connecting two vertices each of degree 2, (b) edges connecting a vertex of degree 2 to a vertex of degree 3, (c) edges connecting a vertex of degree 2 to a vertex of degree 4, (d) edges connecting two vertices of degree 3 each, (e) edges connecting a vertex of degree 3 to a vertex of degree 4, and (f) edges connecting two vertices each of degree 4. Since there is no edge of types (a) and (b) in $G_{2,1}$, no subgraph will be obtained by the algorithm after removing an edge of these two types. On the other hand, $\{v_1, v_4\}$, $\{v_2, v_3\}$, $\{v_2, v_4\}$, and $\{v_4, v_5\}$ are the edges of types (c), (d), (e), and (f), respectively. Thus, the algorithm creates the subgraphs $G_{3,1}$, $G_{3,2}$, $G_{3,3}$, and $G_{3,4}$ by removing these edges from $G_{2,1}$. Since the graphs G3,1, G3,2, G3,3, and G3,4 have distinct degree sequences, they are thus printed by *NICGG*; these have been shown in Fig. 3a–d.

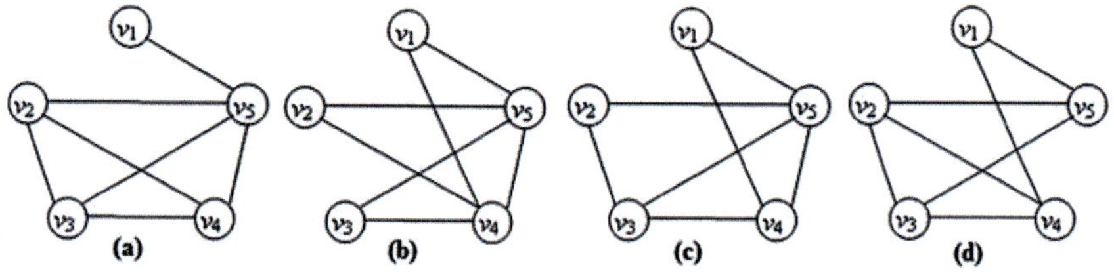

Fig. 3 **a** Subgraph $G_{3,1}$ from $G_{2,1}$, **b** Subgraph $G_{3,2}$ from $G_{2,1}$, **c** Subgraph $G_{3,3}$ from $G_{2,1}$, and **d** Subgraph $G_{3,4}$ from $G_{2,1}$

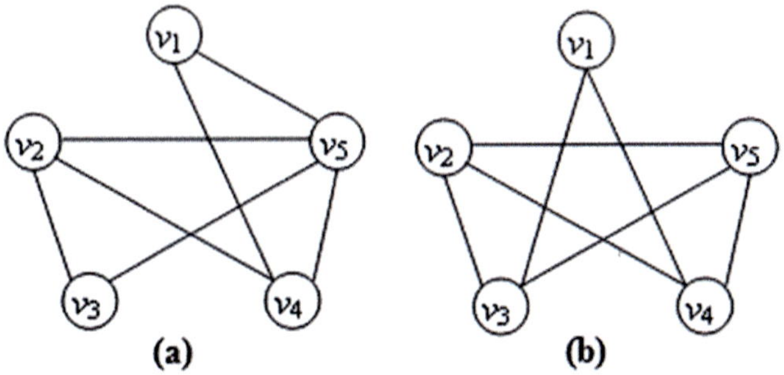

Fig. 4 **a** Subgraph $H_{3,1}$ of $G_{2,2}$ and **b** Subgraph $H_{3,2}$ of $G_{2,2}$ are discarded by *NICGG*

The degrees of the vertices of $G_{2,2}$ are $\{3, 3, 3, 3, 4\}$. Now $d_{\min} = 3$ and $d_{\max} = 4$, i.e., $3 \leq d \leq 4$. So, there are three types of edges possible that might be eliminated: (a) edges joining two vertices of degree 3 each, (b) edges connecting a vertex of degree 3 to a vertex of degree 4, and (c) edges joining two vertices each of degree 4. Since there is no edge of type (c) in $G_{2,2}$, no subgraph will be generated by *NICGG* by removing an edge of type (c) from $G_{2,2}$. However, $\{v_1, v_3\}$ and $\{v_1, v_5\}$ are edges of types (a) and (b), respectively and we eventually create connected subgraphs $H_{3,1}$ and $H_{3,2}$ by removing the said edges from $G_{2,2}$.Yet the graphs H3,1 and H3,2 have the same degree sequences as the graphs G3,3 and G3,4, respectively. Thus, the subgraphs $H_{3,1}$ and $H_{3,2}$ (shown in Fig. 4) are discarded by *NICGG*.

Level 4 (i.e., $k = 4$)**:** The graphs G3,1, G3,2, G3,3, and G3,4 are generated by *NICGG* in the previous level, and hence, the subgraphs at this level will be generated from these four graphs.

The degrees of the vertices of $G_{3,1}$ are $\{1, 3, 3, 3, 4\}$. Now $d_{\min} = 3$ (if an edge leading to a pendant vertex is deleted, the vertex becomes isolated; that is why $d_{\min}$ must be greater than 1 after Step 8 of the proposed algorithm) and $d_{\max} = 4$, i.e., $3 \leq d \leq 4$. So, there are three types of possible edges: (a) edges joining two vertices each of degree 3, (b) edges connecting a vertex of degree 3 to a vertex of degree 4, and (c) edges joining two vertices of degree 4 each. Now, since there is no edge of type (c) in $G_{3,1}$, no subgraph will be generated by the algorithm by removing an edge of type (c) from this graph. However, $\{v_2, v_3\}$ and $\{v_2, v_5\}$ are edges of types (a) and (b), respectively, and we get two connected subgraphs $G_{4,1}$ and $G_{4,2}$ (with distinct degree sequences) by removing these two edges from $G_{3,1}$.

The degrees of the vertices of $G_{3,2}$ are $\{2, 2, 2, 4, 4\}$. Now $d_{\min} = 2$ and $d_{\max} = 4$, i.e., $2 \leq d \leq 4$. Since there is no vertex of degree 3, there are three types of edges possible: (a) edges joining two vertices each of degree 2, (b) edges connecting a vertex of degree 2 to a vertex of degree 4, and (c) edges connecting two vertices of degree 4 each. Since there is no edge of type (a) in $G_{3,2}$, no subgraph will be generated by *NICGG* by removing an edge of type (a) from $G_{3,2}$. $\{v_1, v_4\}$ and $\{v_4, v_5\}$ are edges of types (b) and (c), respectively, and we get two connected subgraphs $H_{4,1}$ and $G_{4,3}$ by removing these two edges from $G_{3,2}$. However, the subgraph $H_{4,1}$ is isomorphic to the graph G4,1, which is already generated by *NICGG*. Hence, $H_{4,1}$

is discarded by *NICGG* and $G_{4,3}$ is the only non-isomorphic connected subgraph of $G_{3,2}$ generated by *NICGG*.

The degrees of the vertices of $G_{3,3}$ are {2, 2, 3, 3, 4}. Now $d_{\min} = 2$ and $d_{\max} = 4$, i.e., $2 \le d \le 4$. So, there are six types of edges possible: (a) edges connecting two vertices each of degree 2, (b) edges joining a vertex of degree 2 to a vertex of degree 3, (c) edges connecting a vertex of degree 2 to a vertex of degree 4, (d) edges joining two vertices of degree 3 each, (e) edges connecting a vertex of degree 3 to a vertex of degree 4, and (f) edges joining two vertices each of degree 4. Now, since there is no edge of types (a) and (f) in $G_{3,3}$, no subgraph will be generated by *NICGG* by removing an edge of types (a) and (f) from $G_{3,3}$.$\{v_1, v_4\}$, $\{v_1, v_5\}$, $\{v_3, v_4\}$, and $\{v_3, v_5\}$ are edges of types (b), (c), (d) and (e), respectively. We get four connected subgraphs $H_{4,2}$, $H_{4,3}$, $G_{4,4}$, and $H_{4,4}$ by removing these edges from $G_{3,3}$. Among them, $H_{4,2}$, $H_{4,3}$, and $H_{4,4}$ have same degree sequences as the graphs G4,1, G4,2, and G4,3, respectively. Hence, the subgraphs $H_{4,2}$, $H_{4,3}$, and $H_{4,4}$ are discarded by *NICGG* and $G_{4,4}$ is the only non-isomorphic connected subgraph of $G_{3,3}$ generated by *NICGG*.

Now, the degrees of the vertices of $G_{3,4}$ are {2, 3, 3, 3, 3}. Here, $d_{\min} = 2$ and $d_{\max} = 3$, i.e., $2 \le d \le 3$. So, there are only two types of edges possible: (a) edges connecting two vertices each of degree 2, (b) edges joining a vertex of degree 2 to a vertex of degree 3, and (c) edges connecting two vertices of degree 3 each. Now, since there is only one vertex of degree 2, then no edge of type (a) exists and hence no subgraph will be generated by *NICGG* by removing an edge of type (a) from $G_{3,4}$.The edges $\{v_1, v_4\}$and $\{v_2, v_3\}$ are of types (b) and (c), respectively. So, we get two connected subgraphs $H_{4,5}$ and $H_{4,6}$ by removing these edges from $G_{3,4}$. Incidentally, the subgraphs $H_{4,5}$ and $H_{4,6}$ have same degree sequences as the graphs G4,2 and G4,3, respectively. Hence, the subgraphs $H_{4,5}$ and $H_{4,6}$ are discarded by *NICGG*.

The connected non-isomorphic subgraphs $G_{4,1}$, $G_{4,2}$, $G_{4,3}$, and $G_{4,4}$ generated by *NICGG* in the fourth level are shown in Fig. 5a–d, and the subgraphs $H_{4,1}$, $H_{4,2}$, $H_{4,3}$, $H_{4,5}$, $H_{4,6}$, and $H_{4,4}$ discarded by *NICGG* in this level are shown in Fig. 6.

Level 5 (i.e., $k = 5$)**:** The graphs G4,1, G4,2, G4,3, and G4,4 are generated by *NICGG* in the previous level, and hence, the subgraphs at this level will be generated from these four graphs.

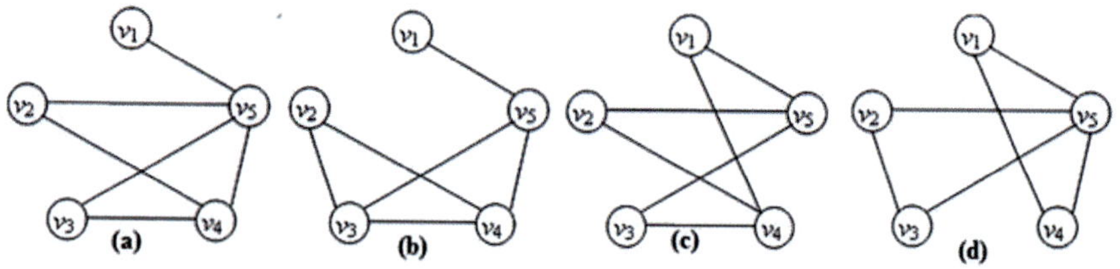

Fig. 5 **a** Subgraph $G_{4,1}$ of $G_{3,1}$, **b** Subgraph $G_{4,2}$ of $G_{3,1}$, **c** Subgraph $G_{4,3}$ of $G_{3,2}$, and **d** Subgraph $G_{4,4}$ of $G_{3,3}$ are generated in the fourth level

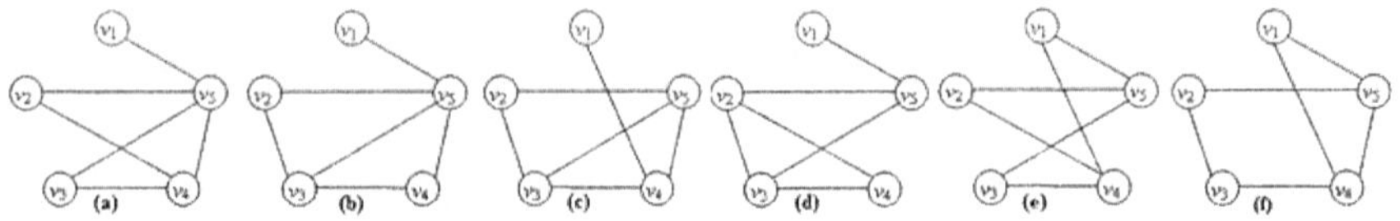

Fig. 6 **a** Subgraphs $H_{4,1}$ of $G_{3,2}$, **b** $H_{4,2}$of $G_{3,3}$, **c** $H_{4,3}$ of $G_{3,3}$, **d** $H_{4,5}$ of $G_{3,4}$, **e** $H_{4,6}$ of $G_{3,4}$ and **f** $H_{4,4}$ of $G_{3,3}$ are generated in the fourth level and discarded accordingly by the algorithm

The degrees of the vertices of $G_{4,1}$ are $\{1, 2, 2, 3, 4\}$. Now $d_{\min} = 2$ and $d_{\max} = 4$, i.e., $2 \leq d \leq 4$. Since the number of vertices of degree 3 and degree 4 are only one, there are only four types of edges possible that might be excluded: (a) edges joining two vertices each of degree 2, (b) edges connecting a vertex of degree 2 to a vertex of degree 3, (c) edges connecting a vertex of degree 2 to a vertex of degree 4, and (d) edges joining a vertex of degree 3 to a vertex of degree 4. Now, since there is no edge of type (a) in $G_{4,1}$, no subgraph will be generated by the algorithm by removing an edge of type (a) from $G_{4,1}$. Edges $\{v_2, v_4\}$, $\{v_2, v_5\}$, and $\{v_4, v_5\}$ are edges of types (b), (c), and (d), respectively, and we get three connected non-isomorphic subgraphs $G_{5,1}$, $G_{5,2}$, and $G_{5,3}$ by removing these three edges from $G_{4,1}$.

Now the degrees of the vertices of $G_{4,2}$ are $\{1, 2, 3, 3, 3\}$. Here, $d_{\min} = 2$ and $d_{\max} = 3$, i.e., $2 \leq d \leq 3$. Since there is only one vertex of degree 2, there are two types of edges possible: (a) edges connecting a vertex of degree 2 to a vertex of degree 3 and (b) edges joining two vertices of degree 3 each. The edges $\{v_2, v_3\}$ and $\{v_3, v_4\}$ are of types (a) and (b), respectively. We get two connected subgraphs $H_{5,1}$ and $H_{5,2}$ by removing these two edges from $G_{4,2}$. The subgraphs $H_{5,1}$ and $H_{5,2}$ have same degree sequences as the graphs G5,2 and G5,3, respectively. Hence, the subgraphs $H_{5,1}$ and $H_{5,2}$ are discarded by *NICGG*.

The degrees of the vertices of $G_{4,3}$ are $\{2, 2, 2, 3, 3\}$. Now $d_{\min} = 2$ and $d_{\max} = 3$, i.e., $2 \leq d \leq 3$. There are three types of edges possible that might be eliminated: (a) edges connecting two vertices each of degree 2, (b) edges connecting a vertex of degree 2 to a vertex of degree 3, and (c) edges joining two vertices of degree 3 each. Now, since there is no edge of types (a) and (c) in $G_{4,3}$, no subgraph will be generated by the algorithm by removing edges of types (a) and (c) from $G_{4,3}$. However, $\{v_1, v_4\}$ is an edge of type (b), and subsequently, we get one connected subgraph $H_{5,3}$ by removing this edge from $G_{4,3}$. By the way, the subgraph $H_{5,3}$ has same degree sequence as the graph G5,3. Hence, the subgraph $H_{5,3}$ is discarded by *NICGG*.

The degrees of the vertices of $G_{4,4}$ are $\{2, 2, 2, 2, 4\}$. Thus, $d_{\min} = 2$ and $d_{\max} = 4$, i.e., $2 \leq d \leq 4$. Since, the number of vertices of degree 3 and 4 are zero and one, respectively, there are two types of edges possible only: (a) edges connecting two vertices of degree 2 each and (b) edges connecting a vertex of degree 2 to a vertex of degree 4. The edges $\{v_1, v_4\}$ and $\{v_1,v_5\}$ are of types (a) and (b), respectively, and we get two connected subgraphs $H_{5,4}$ and $H_{5,5}$ by removing these two edges from $G_{4,4}$. However, the subgraphs $H_{5,4}$ and $H_{5,5}$ have same degree sequences as the graphs G5,1 and G5,3, respectively. Hence, the subgraphs $H_{5,4}$ and $H_{5,5}$ are discarded by the algorithm.

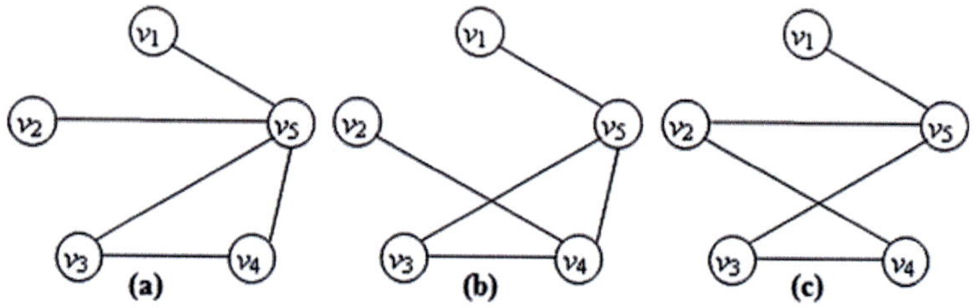

Fig. 7 **a** Valid subgraphs $G_{5,1}$ of $G_{4,1}$, **b** $G_{5,2}$ of $G_{4,1}$, and **c** $G_{5,3}$ of $G_{4,1}$ are produced in the fifth level

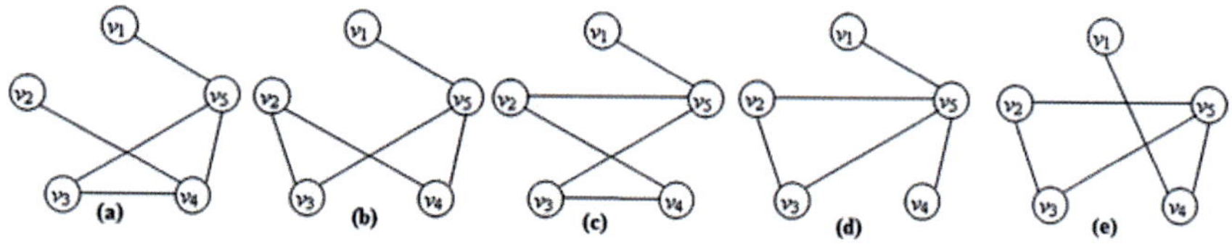

Fig. 8 **a** Subgraphs $H_{5,1}$ of $G_{4,2}$, **b** $H_{5,2}$ of $G_{4,2}$, **c** $H_{5,3}$ of $G_{4,3}$, **d** $H_{5,4}$ of $G_{4,4}$, and **e** $H_{5,5}$ of $G_{4,4}$ are generated in the fifth level and discarded by the algorithm

The connected non-isomorphic subgraphs $G_{5,1}$, $G_{5,2}$, and $G_{5,3}$ generated by *NICGG* in the fifth level are shown in Fig. 7 and the subgraphs $H_{5,1}$, $H_{5,2}$, $H_{5,3}$, $H_{5,4}$, and $H_{5,5}$ discarded by *NICGG* in this level are shown in Fig. 8.

Level 6 (i.e., $k = 6$)**:** Now the connected non-isomorphic subgraphs $G_{5,1}$, $G_{5,2}$, and $G_{5,3}$ are created by the algorithm in the previous level, and hence, the subgraphs at this level will be generated from these three graphs.

The degrees of the vertices of $G_{5,1}$ are $\{1, 1, 2, 2, 4\}$. Now $d_{\min} = 2$ and $d_{\max} = 4$, i.e., $2 \leq d \leq 4$. Since, the number of vertices of degree 3 and 4 are zero and one, respectively, there are two types of edges possible: (a) edges connecting two vertices each of degree 2 and (b) edges connecting a vertex of degree 2 to a vertex of degree 4. Now, the edges $\{v_3, v_4\}$ and $\{v_3, v_5\}$ are two edges of types (a) and (b), respectively. Hence, algorithm *NICGG* produces two connected non-isomorphic subgraphs $G_{6,1}$ and $G_{6,2}$ by removing these two edges from $G_{5,1}$.

Next, the degrees of the vertices of $G_{5,2}$ are $\{1, 1, 2, 3, 3\}$. Now $d_{\min} = 2$ and $d_{\max} = 3$, i.e., $2 \leq d \leq 3$. Since the number of vertices of degree 2 is one, there are two types of edges possible that might be omitted: (a) edges connecting a vertex of degree 2 to a vertex of degree 3 and (b) edges joining two vertices of degree 3 each. Now, $\{v_3, v_4\}$ and $\{v_4, v_5\}$ are two such edges of types (a) and (b), respectively. Thus, we get two connected subgraphs $H_{6,1}$ and $G_{6,3}$ by removing these two edges from $G_{5,2}$. The subgraph $H_{6,1}$ has same degree sequence as the graph G6,2. Hence, the subgraph $H_{6,1}$ is discarded by *NICGG* and $G_{6,3}$ is the only non-isomorphic connected subgraph of $G_{5,2}$ generated by the algorithm.

To finish, the degrees of the vertices of $G_{5,3}$ are $\{1, 2, 2, 2, 3\}$. Now $d_{\min} = 2$ and $d_{\max} = 3$, i.e., $2 \leq d \leq 3$. Since the number of vertices of degree 3 is one, there are

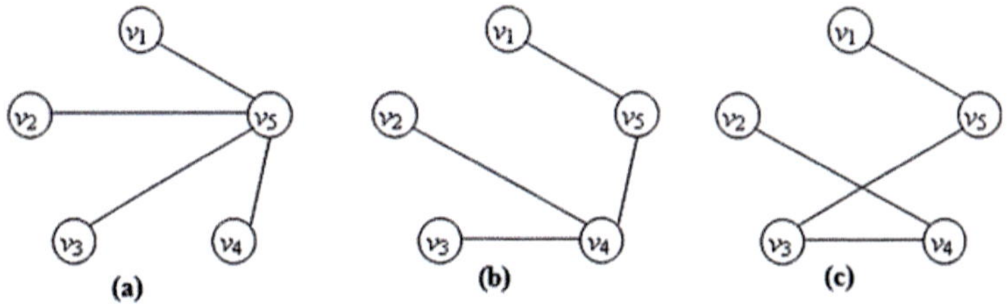

Fig. 9 **a** Permissible subgraphs $G_{6,1}$ of $G_{5,1}$, **b** $G_{6,2}$ of $G_{5,1}$, and **c** $G_{6,3}$ of $G_{5,2}$ are produced in the sixth level

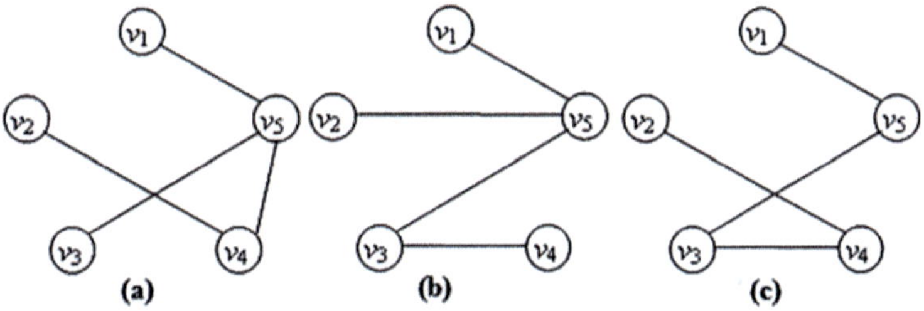

Fig. 10 **a** Subgraphs $H_{6,1}$ of $G_{5,2}$, **b** $H_{6,2}$ of $G_{5,3}$, and **c** $H_{6,3}$ of $G_{5,3}$ are generated in the sixth level by the algorithm and discarded accordingly

two types of edges possible: (a) edges connecting two vertices each of degree 2 and (b) edges connecting a vertex of degree 2 to a vertex of degree 3. Now, the edges $\{v_2, v_4\}$ and $\{v_2, v_5\}$ are two such edges of types (a) and (b), respectively. We get two connected subgraphs $H_{6,2}$ and $H_{6,3}$ by removing these two edges from $G_{5,3}$. The subgraphs $H_{6,2}$ and $H_{6,3}$ have same degree sequences as the graphs G6,2 and G6,3, respectively. Hence, the subgraphs $H_{6,2}$ and $H_{6,3}$ are discarded by *NICGG*.

The connected non-isomorphic subgraphs $G_{6,1}$, $G_{6,2}$, and $G_{6,3}$ generated by *NICGG* in the sixth level are shown in Fig. 9a–c, and the subgraphs $H_{6,1}$, $H_{6,2}$, and $H_{6,3}$ discarded by the algorithm in this level are shown in Fig. 10.

6 Experimental Results

The proposed algorithm, *NICGG* has been implemented on an Intel Dual Core processor running at 3.00 GHz, with 4 GB RAM capacity. The non-isomorphic connected graph instances have been generated for graphs of a given order (i.e., n) ranging from 4 to 11. In the execution, we have also taken the maximum possible value for another input (p), the number of levels, i.e., $p = m - n + 1$, where $m = n \times (n - 1)/2$ is the number of edges in a complete graph of order n. The CPU time, taken by the algorithm to execute, for each such generation is given in minutes and seconds in Table 1.The total number of graphs and the number of trees among them

Table 1 Results after the execution of the proposed algorithm, *NICGG*

Order (n)	# of levels (p)	# of subgraphs generated	# of trees	CPU time (min: sec)
4	3	5	2	0: 0
5	6	17	3	0: 0
6	10	64	5	0: 0
7	15	226	6	0: 0
8	21	844	11	0: 0
9	28	3096	14	0: 1
10	36	11,548	21	0: 5
11	45	43,120	28	1:13

are also included in table. The observations from the results after the execution of *NICGG* are as follows:

- The generated graphs are the connected non-isomorphic subgraphs of the complete graph of the given order (n).
- It produces subgraphs of different sizes like $m - 1, m - 2, \ldots, n - 1$.
- There is at least one non-isomorphic subgraph of size $m - 1, m - 2, \ldots, n - 1$.
- The subgraphs generated at the last level ($p = m - n + 1$) are non-isomorphic trees.

7 Conclusion

Generation of connected graphs is a topic of interest for researchers for different purposes like developing spanning tree generation algorithms [6, 13], graph algorithms, approximation algorithms, simulation [7] and sampling, scheduling algorithms, algorithms for database applications, etc. Non-isomorphic graphs are often required to determine the performance of various other algorithms. In this paper, we have developed a non-isomorphic connected graph generation algorithm, *non-isomorphic connected graph generator* (*NICGG*), to provide a test bed for the above algorithms. *NICGG* is able to generate a set of simple, undirected, connected, and non-isomorphic subgraphs of different sizes of a complete graph of a given order. In the result section of this paper, we have generated such graphs with order of the graphs ranging from 4 to 11. In terms of computation time and a large class of desired non-isomorphic graph generation, the efficiency of the devised algorithm has also been established.

In this paper, we have claimed to generate only a set of connected graphs with n vertices and e edges, where e varies from $n - 1$ to m and $m = n(n - 1)/2$. For a particular size e, though *NICGG* generates many non-isomorphic connected graphs, there may be many more non-isomorphic connected graphs with size e, which have not been generated since *NICGG* generates only one graph of a particular degree sequence. On the other hand, the algorithm *RCGG* [5] generates only one connected graph with n vertices and e edges. There is a future scope of comparing these two algorithms in terms of complexity and throughput. There are few more future research possibilities to generate all non-isomorphic connected graphs with a fixed order as well as a fixed size. So, there are scopes for doing a lot toward generating non-isomorphic connected graph instances. In that direction, this paper has attempted to frame a guideline for the future researchers in this domain of work.

References

1. Aho AV, Hopcroft JE, Ulman JD (1983) Data structures and algorithms, 1st edn. Pearson
2. Bayati M, Kim JH, Saberi A (2010) A sequential algorithm for generating random graphs. Algorithmica 58(4):860–910
3. Bhuiyan H, Khan M, Marathe M (2017) A parallel algorithm for generating a random graph with a prescribed degree sequence. In: Proceedings of the IEEE international conference on Big Data, pp 3312–3321
4. Chakraborty M, Chowdhury S, Pal RK (2019) Generation of simple, connected, non-isomorphic random graphs. In: Proceedings of the sixth international doctoral symposium on applied computation and security systems (ACSS), Proceedings published as Book Chapter of the series Advances in Intelligent Systems and Computing, Springer, vol 995, pp 69–77
5. Chakraborty S, Chakraborty M, Pal RK (2020) Generation of simple undirected connected random graphs. In: Proceedings of international conference on computational intelligence, data science and cloud computing, pp 197–206
6. Char JP (1968) Generation of trees, two-trees, and storage of master forests. IEEE Trans Circ Theory 15(3):228–238
7. Cordeiro D, Mounié G, Perarnau S, Trystram D, Vincent J-M, Wagner F (2010) Random graph generation for scheduling simulations. In: Proceedings of the third international conference on simulation tools and techniques (SIMUTools), pp 1–10, Article No. 60
8. Deo N (2005) Graph theory with applications to engineering and computer science. Prentice Hall of India Pvt. Ltd., New Delhi
9. Erdös P, Rényi A (1960) On the evolution of random graphs. In Publication of the Institute of Mathematical of the Hungarian Academy of Sciences, vol 5, pp 343–347
10. Horn MV, Richter A, Lopez D (2003) A random graph generator. In: Proceedings of 36th annual midwest instruction and computing symposium, Six pages
11. Horowitz E, Sahni S, Anderson S (2008) Fundamentals of data structures in C, 2nd edn. Universities Press Pvt. Ltd., Hyderabad, India
12. Nobari S, Lu X, Karras P, Bressan S (2011) Fast random graph generation. In: Proceedings of the 14th international conference on extending database technology, pp 330–342
13. Rakshit A, SenSarma S, Sen RK, Choudhury AK (1981) An efficient tree generation algorithm. IETE J Res, Inst Electron Telecommun Eng (IETE) 27(3):105–109
14. Viger F, Latapy M (2005) Efficient and simple generation of random simple connected graphs with prescribed degree sequence. In: Proceedings of the 11th annual international conference on computing and combinatorics (COCOON), Springer, pp 440–449
15. Wang C, Lizardo O, Hachen D (2014) Algorithms for generating large-scale clustered random graphs. Netw Sci 2(3):403–415
16. Weisstein EW (2010) Random graph. Wolfram MathWorld Built with Mathematica Technology

Chapter 39
monitorYou: COVID-19 Screening IoT-Based Android Application

Puja Sarkar, Noni Deori, Mojasir Hussain, and Amitava Nag

1 Introduction

COVID-19 pandemic has led to huge loss of lives. Severe acute respiratory syndrome coronavirus-2 (SARS-CoV-2) is a contagious disease. First primary human case of COVID-19 was reported in Wuhan, China which then spread like wildfire after the end of 2019. COVID-19 being highly contagious led to the soaring in infection and mortality rates. Common symptoms include fever, sore throat, nasal congestion [1], tiredness and loss of taste [2]. The disease was than declared as pandemic by World Health Organization (WHO) in January 2020, calling for a global public health emergency [3]. COVID-19 has spread globally which has affected countries and territories around the world socially and economically. Variety of strict measures were imposed such as self-quarantining and isolation, social distancing and restricted mobility by respective nations to confront the spread of the virus which resulted in significant adverse impact on global economics, resources and population [4]. As mentioned earlier, COVID-19 has caused an adverse impact on social and economical sectors. Billions of individuals worldwide across different domains are suffering due to this pandemic which has resulted in embracing severe social and economic reforms. Lockdown imposed by countries is now being gradually relaxed, and organizations are slowly starting to operate. In this post-lockdown period, ensuring safe workspace for the workers/employees is mandatory. Many industrial sectors and

P. Sarkar (✉) · N. Deori · M. Hussain · A. Nag
Central Institute of Technology, Kokrajhar, India
e-mail: p.sarkar@cit.ac.in

A. Nag
e-mail: amitava.nag@cit.ac.in

L. Mandal et al. (eds.), *Proceedings of International Conference on Computational Intelligence, Data Science and Cloud Computing*, Algorithms for Intelligent Systems,
https://doi.org/10.1007/978-981-19-1657-1_40

organizations have reopened during post-lockdown. These organizations and companies may become the hot-spot of COVID-19 due to the shortage of facilities to maintain social distancing during the working hours. This is a huge concern as these industries are known for its crowded nature.

Internet of Things or IoT-enabled appliances have made remote monitoring possible within the health sector and provide the potential to keep in check of patients health and safety. This empowers the physicians to deliver superlative care and has also increased patient engagement and satisfaction as interactions with doctors became easier and more efficient. Remote monitoring of patients' health helps in reducing the tenure of hospital stay and prevents readmission of the patients. IoT is cost efficient and improves treatment outcomes [5–9].

To provide proper health and work safety to the employees while maintaining effective COVID-19 control measures, we are developing an Android health monitoring system with an IoT-based diagnosis for an organization to monitor their employees' health status.

The objective of monitorYou is to ensure that the employees of an organization are safe from COVID-19 and ensure a healthy workspace. By using monitorYou, organizations will be able to achieve the following benefits:

- Safety and security of the members and employees of the organization
- Proper health monitoring of the employees during pandemic
- Maintenance of data recorded through the app
- Cost efficient and time saving.

We propose to implement self-assessment of health through the app against COVID-19 symptoms by the guidelines provided by ICMR and measure the oxygen level, body temperature and heart rate through our IoT device. The health data collected from the employee will be saved in the employee database which will help organizations to keep track of their employees' health status to maintain a healthy and safe workspace.

Key contributions of our proposed system

1. Develop an IoT-based health monitoring Android app
2. Regular check of employee health status
3. Providing proper medical assistance to the employee.

This paper is systematized as follows. The related applications are available against COVID-19, and the technology used and the facilities provided are elaborated in Sect. 2. In Sect. 3, the advantages, system workflow, system architecture, workflow of the IoT device, implementation and decision matrix are followed and the comparative analysis of the proposed prototype is discussed. Future work and conclusions are studied in Sect. 4.

2 Related Works

To keep track of COVID-19, several mobile applications have been launched. They contain mainly features on providing COVID-19 related information, self-screening and contact tracing to monitor and minimize the spread of the virus. One such app is Aarogya Setu [10]. Aarogya Setu alerts users when they come within six feet of COVID-19 patients by using a smartphone's Bluetooth and GPS systems. These alerts are generated by scanning through government-owned, location specific databases which are amid instructions from the Ministry of Health on the way to self-isolate, and therefore, the course of action to take when citizens develop symptoms of coronavirus. HealthCheck [11], launched by Stratum Technology, is a secured cloud-based application. Employees need to take a self-health-assessment where they answer a health-related questionnaire and accordingly receives real-time result on whether they are fit for work. COVID19Tracker [12], launched by KOKOMO24/7, allows employees to partake regular self-screening and accordingly provides badges (Red and Green) that indicates employee health status. The app have other features like case management where it includes contact listing, contact identification, contact follow ups and notify, collect and manage pre-emptively mitigate the risk of an outbreak. ProtectWell [13], launched by Microsoft Corporation and UnitedHealth Group, includes a health screening tool which is AI-powered for screening of COVID-19 symptoms in employees. Employers are directed to COVID-19 testing process and they are notified, if any complications are identified.

There are many other related systems [14–17] released mainly concentrating on industrial and co-worker safety where self-screening is provided and employee need to input their health-related data. But only self-screening is not enough for an organization to ensure worker safety and company well-being, as it may mislead data and which in can turn great risk for the organization. Keeping this in mind, *monitorYou* focuses on providing self-screening as well as diagnosing with sensors to collect vital signs of employees ensuring data integrity.

3 Proposed System

In order to achieve a safe workplace, our aim is to provide an Android health monitoring system with an IoT-based oximeter for an organization to monitor their employees' health status during their work hours. The proposed solution based on smartphone app uses a smart remote health for industries which identifies users of possible COVID-19 infection through self-assessment and health diagnosis, and alerts in case of emergency. The health assessment data is recorded and analysed by our app which helps in identification of the workers who are possibly exposed to COVID-19. The health record will be saved in the cloud server of the respective organization. The employee health record can be shared only by the approval of the respective employee.

Advantages as an employee in an organization

- Self-health-assessment—the employee can self-assess himself.
- Alerting proper health status.
- Emergency helpline and nearby health centres.

Advantages as an administrator in an organization

- Safety and security of the members and employees of the organization
- Proper health monitoring of the employees during pandemic
- Maintenance of data recorded through the app
- Cost efficient and time saving.

3.1 System Workflow and Architecture

As shown in Fig. 1, the system workflow and architecture is designed into three parts:

(A) The app user registers to the system with the office ID, after the registration, the system will receive an encrypted temporary ID. The user is redirected to self-assessment activity. Here he/she will be self-assessed.

(B) After self-assessment, the employee will be redirected to body oxygen and temp check activity. Here employee's body oxygen and temp will be measured with our IoT device.

(C) After the self-assessment analysis is completed, the risk factor is determined. If no symptom is found in the user, a "fit to work" pass is generated for the user. If symptoms are found in the user, the respective authority of the organization is informed. Further he/she is directed for the COVID test. If the user tests positive, he/she will be hospitalized. If the user is found negative, then the user will be directed to 14 days home quarantine and keep monitoring him/herself through the app. In case the user finds no symptoms during the quarantine period, the user is set to "fit for work". If the user finds symptoms during the quarantine, the user is directed to the COVID test. In the test, if the user is found COVID-19 positive he/she is hospitalized and if the user is found COVID-19 negative he/she is recorded as "fit for work". The entire process is updated in the organization database.

3.2 Workflow of the IoT Device

In Fig. 2, system workflow and architecture of the IoT device is as follows:

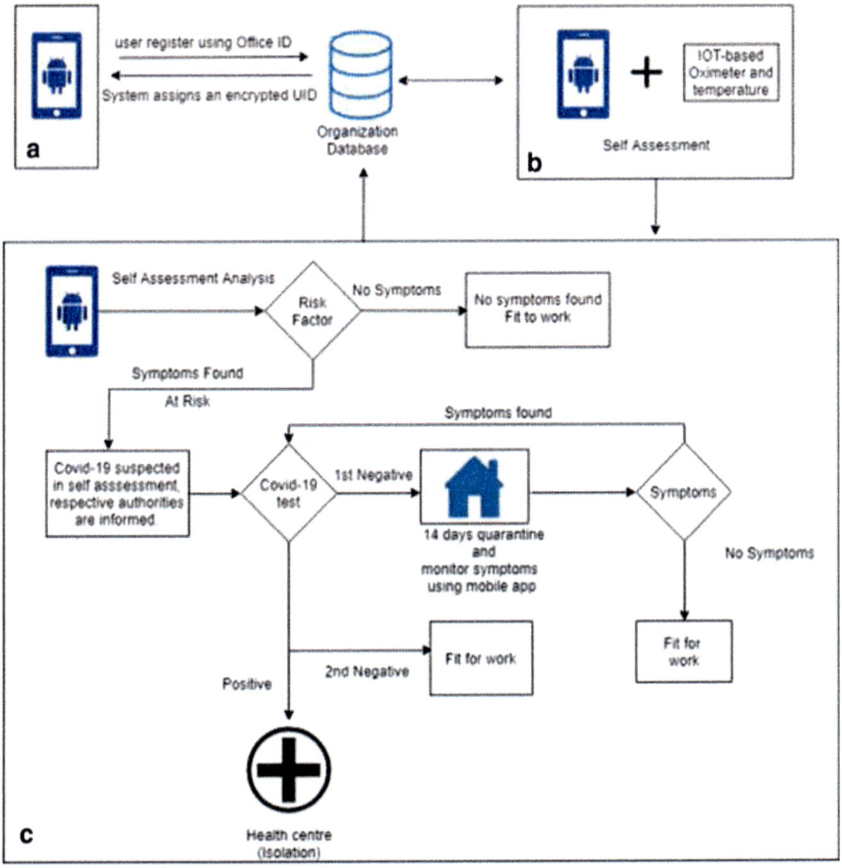

Fig. 1 System workflow and architecture

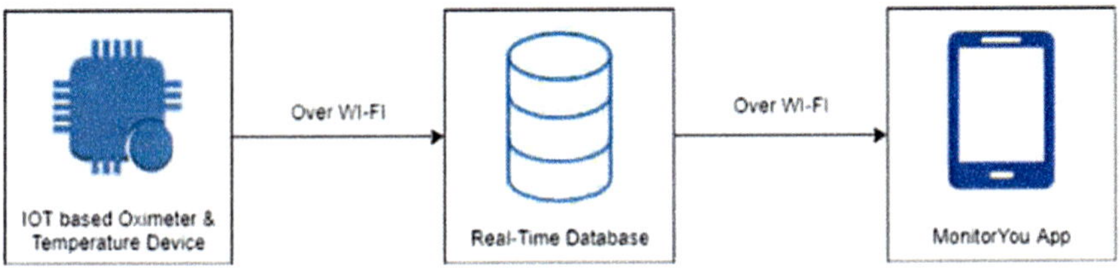

Fig. 2 System workflow and architecture of the IoT device

1. The IoT device collects the necessary sensor data and sends it to the firebase real-time database over Wi-Fi.
2. The data sent by the IoT device is then stored in the database for the current user. It receives the heart rate, blood oxygen and temperature of the user accessed through the device.

3. The stored data in the real-time database is called by our app to display in the health assessment layout. This helps in determining the fit/unfit status of the user.

3.3 Implementation

Figure 3 shows the graphical user interface of our system. In Fig. 3a, the employee after installing our app needs to register him/herself in the register layout. This will create a user account for the employee. The employee is then required to login to the app with the registered email ID and password. These are the necessary credentials required to login (Fig. 3b). After successful login, the user lands on the homepage of our app, where signed in employee details are displayed. To generate a work pass, employee needs to click the "Work Pass" button where he will be redirected to the self-assessment layout. In case of any emergency, below helpline numbers are provided (see Fig. 3c). In Fig. 3d, here the employee is needed to fill the quick questionnaire and proceed further by pressing the submit button. After that, the employee is needed to diagnose him/herself through our IoT device connected over Wi-Fi, which sends the recorded data to our firebase database. When clicking the "*Diagnose*" button the sensor data is retrieved from the database and set to the respective text fields in the layout. After successful diagnosis, the employee can generate a work pass by clicking the "GENERATE PASS" button (see Fig. 3e). Based on the decision matrix followed by the app in (Fig. 3f), "fit for work" work pass for employees is displayed. Similarly in (Fig. 3g), "unfit for work" work pass for employees is displayed based on the decision matrix followed by the app.

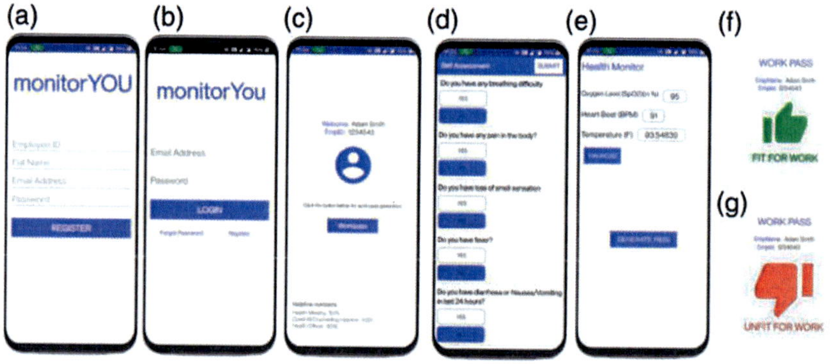

Fig. 3 **a** Employee register screen, **b** login screen, **c** monitorYou homepage, **d** self-assessment layout, **e** diagnostics layout, **f** fit for work work pass layout, **g** unfit for work work pass layout

3.4 Decision Matrix

Our app monitorYou follows the following decision matrix. Every option of every question is assigned a weightage and the final score is calculated as an aggregate sum of the weights of the chosen options. The final score is then compared with the threshold values to determine the suggestion. The questions and weights are assigned to each individual option of each question and also the thresholds of the final score are depicted below.

The questionnaire shown in Fig. 4 helps in determining the health condition of the employees, and the questions asked are about the symptoms shown when affected by COVID-19. The response of the employee can be either in yes or no. And each response has its credit point as shown in Fig. 5. This helps in calculation of the credit score obtained by the employees, after successful completion of self-assessment through the app. If the credit score of the employees is below 5, then he is fit to work, else above or equal to 5 then he might be suffering from COVID-19. Now to clarify this, our IoT device comes into play. After successful diagnosis, the received data from the employee has to go through the IoT decision matrix as shown in Fig. 6.

Questions	Response from User	
Do you have fever?	Yes	No
Do you have any breathing difficulty	Yes	No
Do you have any pain in the body?	Yes	No
Are you having any fatigue/weakness?	Yes	No
Do you have a sore throat?	Yes	No
Do you have a dry cough?	Yes	No
Do you have diarrhoea or Nausea/Vomiting in last 24 hours?	Yes	No
Do you have loss of taste sensation?	Yes	No
Do you have loss of smell sensation?	Yes	No

total credit score	9
credit score obtained	<5
self assessment output	Fit For Work

total credit score	9
credit score obtained	>5
self assessment output	Unfit For Work

Fig. 4 Questionnaire decision matrix

Fig. 5 Credit point

	Credit point
Yes	1
No	0

Fig. 6 IoT decision matrix

Normal body temp	less than 100 F
Normal HeartRate	(60-100)BPM
Normal Oxygenlevel	(93-100)%

Identifier	Providers	Created ↓	Signed In	User UID
random@abc.com		Aug 15, 2021	Aug 15, 2021	kZFqwjp6LgbSr9lM6OataUM4xYC3
dsa@as.con		Aug 15, 2021	Aug 15, 2021	DGnbJLqMaxalogQkb8aMPeVWk
jkl@kl.com		Aug 15, 2021	Aug 15, 2021	jMn1dg0TwlXcuTPGY9xQm6INJb
asd@ds.com		Aug 14, 2021	Aug 15, 2021	aKgh5zLSLOgj3on39M6dvYINC2G3
qwe@qw.com		Aug 13, 2021	Aug 14, 2021	lMqpDyb1S2PWU3yqIKV5faGkOlv1

Fig. 7 Firebase authentication database

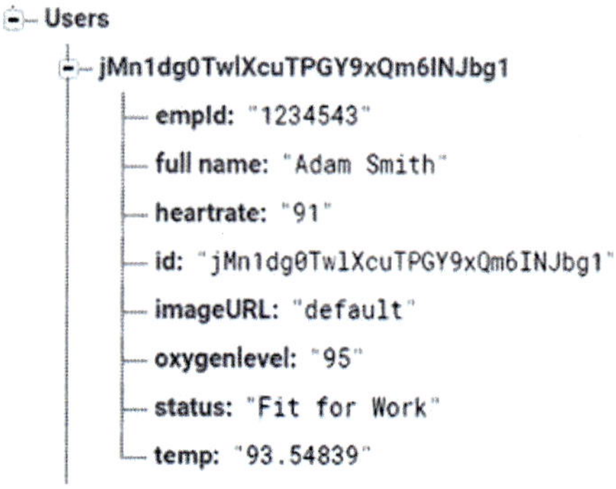

Fig. 8 Questionnaire decision matrix

If the employee data does not match our decision matrix, he/she is suffering from COVID-19 symptoms and this determines if they are fit or unfit to continue to work.

3.5 *Result and Analysis*

After successful register and login, the employee data is recorded in our Firebase database as shown in Fig. 7. Every user is assigned with unique user ID, and the employee ID is provided by the organization itself. Then the user self-assesses him/herself through the self-assessment layout of the app and is diagnosed with our IoT device. The decision matrix of our application determines his/her fit status for work. The collected data is stored in the database with the necessary records of the employee as shown in Fig. 8.

3.6 *Comparative Analysis*

Similar applications launched to fight COVID-19 with the use of digital technology is given in Table 1. Our app monitorYou provides a more functional and reliable

Table 1 Comparative analysis of different apps and monitorYou

App name	Self-screening	Hardware offered	Update real time	Remote monitoring
HealthCheck	X		X	
ProtetWell	X		X	
COVID-19 Tracker	X		X	X
Aarogya Setu	X		X	X
monitorYou	X	X	X	X

IoT framework than others and is beneficial for use to control the infection after the pandemic. Many contact tracing apps have been introduced by different countries to fight against the pandemic and relay information to the general public. However, the mentioned applications merely trace a patient's location history, and alerts users if anyone is diagnosed with COVID-19 in the areas the users have recently visited. On the other hand, monitorYou offers the following benefits as given in Table 1.

4 Conclusion and Future Work

All the individual modules like heartbeat detection module, temperature module gave out the intended results. Our application collects the data from the self-assessment and diagnosis done by the IoT device which is required to analyse the risk estimation of the user. The risk estimation process involves the following steps which are (1) analysing the data received from sensors (2) estimating the risk score through the decision matrix which generates the work pass of the employee. The employee's fitness to work (i.e. fit for work) is denoted by a low value of score; whereas, high value of score indicates the necessity of medical attention. This will help industries/organizations in maintaining safe workspace for the workers, predict possible infection and health status of employees and maintain business continuity. In future, it is planned to implement contact tracing via Bluetooth low energy which will help in determining the number of people the user came in contact. Further it is planned to apply machine learning algorithms on sensors like voice-coughing detector, to constantly detect and monitor the user's voice and record the severity of coughing. This is will make the system more robust and useful for organizations.

References

1. Zhai P et al (2020) The epidemiology, diagnosis and treatment of COVID19. Int J Antimicrob Agents 55(5):1–13, 105955. https://doi.org/10.1016/j.ijantimicag.2020.105955
2. Dawson P et al (2020) Loss of taste and smell as distinguishing symptoms of COVID-19. Clin Inf Dis, pp 1–4. https://doi.org/10.1093/cid/ciaa799
3. https://www.who.int/emergencies/diseases/novel-coronavirus-2019

4. Jesmin SK, Shamim M, Mahmud M (2020) Artificial and Internet of Healthcare Things based Alzheimer care during COVID 19. https://doi.org/10.1007/978-3-030-59277-6-24
5. Jasti Madhu NR, KANSAS STATE UNIVERSITY Manhattan, Kansas (2017) IoT based remote patient health monitoring system
6. Al Bassam N, Hussain SA, Al Qaraghuli A, Jibreal Khan, Sumesh EP, Lavanya V (2021) IoT based wearable device to monitor the signs of quarantined remote patients of COVID-19. Inf Med Unlocked 24:100588
7. Petrović N, Kocić Đ (2020), IoT-based System for COVID-19 indoor safety monitoring
8. Hassanalieragh M et al (2015) Health monitoring and management using Internet-of-Things (IoT) sensing with cloud-based processing: opportunities and challenges
9. Venusamyand K, Basha RFK (2020) An enhanced, contactless, IOT based operation of power appliances for Covid-19 isolation zone. J Green Eng (JGE) 10(9)
10. https://www.aarogyasetu.gov.in/
11. Stratumhealth, How does HealthCheck work? 2020, library Catalog: www.stratumhealth.io. [Online]. Available: https://www.stratumhealth.io/info, Access date: Dec 25, 2020
12. Lee D, COVID19Tracker (contact tracing) 2020, library Catalog: www.kokomo247.com. [Online]. Available: https://www.kokomo247.com/covid19tracker, Access date: Dec 25, 2020
13. United Health Group, The ProtectWell Platform 2020, library Catalog: weprotectwell.com. [Online]. Available: https://www.unitedhealthgroup.com/protectwell/platform.html
14. DROR (2018) DROR—your personal safety social distancing—Apps, library Catalog: play.google.com. [Online]. Available: https://www.drorapp.com/
15. Alight (2020) New Alight app helps companies bring their employees back to their workspaces, library Catalog: alight.com. [Online]. Available: shorturl.at/nKRWZ
16. Hall V (2020) SoCo COVID-19 employee check—Mobile App. 2020, library catalog: socoemer-gency.org. [Online]. Available: shorturl.at/vDHN5
17. SafetyTek (2020) SafetyTek COVID-19 workforce health analysis solution [Online]. Available: https://safetytek.io/covid-19

Author Index

L. Mandal et al. (eds.), *Proceedings of International Conference on Computational Intelligence, Data Science and Cloud Computing*, Algorithms for Intelligent Systems,
https://doi.org/10.1007/978-981-19-1657-1